ENVIRONMENTAL PROBLEMS IN MARINE BIOLOGY

Methodological Aspects and Applications

ENVIRONMENTAL PROBLEMS IN MARINE BIOLOGY

Methodological Aspects and Applications

Editors

Tamara García Barrera

Department of Chemistry
Faculty of Experimental Sciences
University of Huelva
Huelva
Spain

José Luis Gómez Ariza

Department of Chemistry
Faculty of Experimental Sciences
University of Huelva
Huelva
Spain

CRC Press
Taylor & Francis Group
Boca Raton London New York

CRC Press is an imprint of the
Taylor & Francis Group, an **informa** business
A SCIENCE PUBLISHERS BOOK

Cover illustrations reproduced by permission of Isidro Castaño Rojas and José Manuel Sayago Robles.

CRC Press
Taylor & Francis Group
6000 Broken Sound Parkway NW, Suite 300
Boca Raton, FL 33487-2742

Printed on acid-free paper
Version Date: 20170119

International Standard Book Number-13: 978-1-4822-6450-0 (Hardback)

Library of Congress Cataloging-in-Publication Data

Names: García Barrera, Tamara, editor. | Gómez Ariza, José Luis, editor.
Title: Environmental problems in marine biology : methodological aspects and applications / editors, Tamara García Barrera, Department of Chemistry, Faculty of Experimental Sciences, University of Huelva, Huelva, Spain, José Luis Gómez Ariza, Department of Chemistry, Faculty of Experimental Sciences, University of Huelva, Huelva, Spain.
Description: Boca Raton, FL : CRC Press, 2017. | "A Science Publishers book." | Includes bibliographical references and index.
Identifiers: LCCN 2016046133| ISBN 9781482264500 (hardback : alk. paper) | ISBN 9781482264517 (e-book : alk. paper)
Subjects: LCSH: Aquatic organisms--Effect of water pollution on. | Marine biology. | Marine pollution. | Chemical oceanography.
Classification: LCC QH545.W3 E68 2017 | DDC 578.76--dc23
LC record available at https://lccn.loc.gov/2016046133

Visit the Taylor & Francis Web site at
http://www.taylorandfrancis.com

and the CRC Press Web site at
http://www.crcpress.com

Preface

Marine environment has enormous intrinsic value regarding traditional uses (e.g., navigation, fishing, transport of goods, resources source and others) together with more recent activities (e.g., mineral and oil extraction, wind farms) as well as its pivotal role assuring climate stability. However, the unsustainable use of our seas and oceans compromises the fragile balance of marine ecosystems which is actually affected by the presence of chemical metal species (e.g., tributyltin, methylmercury, inorganic selenium and arsenic), pharmaceutically active compounds (e.g., antibiotics, psychiatric drugs, analgesics/anti-inflammatories, tranquilizers, hormones, β-blockers, diuretics), nanoparticles (e.g., colloids, volcanic eruptions, forest fires, engineered nanoparticles) and others, which have important ecological and human risks. In addition, a special mention should be made in relation to the effects of climate change, namely sea warming and ocean acidification, which cause changes in the physiology, phenology and biogeographical distribution of organisms.

Methodological analytical approaches to deal with this issue range from chemical speciation methods for trace element in sea water (Ch. 1) or marine biota (Ch. 2) based on hyphenated instrumental techniques combining HPLC, FFF, and electrophoresis with ICP-MS to massive information techniques such as metallomics and metabolomics also used for this purpose, especially when toxic and protector elements are involved, this is the case of Hg and Se (Ch. 3 and Ch. 4) that remarks the importance to elucidate metals interactions and bioaccessibility in marine organisms and the potential transference to human through marine origin foods. To deep insight on these topics powerful analytical methodologies (the omics) are required in order to complement the information provided by conventional speciation methods. Other toxic metal or metalloid species are also relevant in marine environment such as As (Ch. 5) and Sn (Ch. 6) and suitable speciation methods are necessary to establish their toxicity and risk, assessing to policy makers in the proposal of new regulations that assure the health of population. Nevertheless, the need of methods for massive information is becoming more and more important in marine environmental assessment and related issues, as aquaculture in which functional genomics provides information about biomolecules involved in biological defense mechanisms and restoration of processes altered by contamination (Ch. 7). Moreover, proteomics provides information about the response of bioindicators to contamination and the modulation of this response via epigenetic changes (Ch. 8), the changes of protein expression caused by engineered nanoparticles (Ch. 9 and 12) and those promoted by climate change (Ch. 10).

On the other hand, environmental metabolomics provides global information about free-living organisms since it reflects the last changes in phenotype and therefore accounts for the influence of external factors. This methodology has especial interest in the marine environment where many variables affect the organisms. One important point is related to the overriding role of metals in the metabolism of living organisms whose fluxes and homeostasis have to be maintaining, which generates a host of metal-containing molecules whose characterization is fundamental to understanding the changes and evolution of metabolic cycles. Therefore, metabolomic techniques for metal-containing metabolites (metallomics) are also included in most of the recent approaches in this field, incorporating atomic mass spectrometry to the used workflow (Ch. 11). Finally, the biological effects or emergent contaminants such as pharmaceuticals in marine environment is reviewed, considering the suitable instrumentation to assess the lower concentrations in this medium and different aspects related to the distribution of these pollutants in the environment, geographical variations, and environmental risk assessment (Ch. 13 and 14).

In this volume, we draw together experts in the fields of analytical chemistry, environmental sciences, biochemistry, biology, genomic and toxicology. This book includes together the methodological aspects and applications related to chemical speciation and –omics strategies applied to the dynamic issue of marine environment.

Tamara García Barrera
José Luis Gómez Ariza

Contents

Abbreviations

1DE	:	one-dimensional gel electrophoresis
^{1}HNMR	:	Proton nuclear magnetic resonance
2DE	:	Two-dimensional gel electrophoresis
2D-PAGE	:	Two dimensional polyacrylamide gel electrophoresis
AAS	:	Atomic absorption spectrometry
AB	:	Arsenobetaine
AC	:	Arsenocholine
α–XHXA	:	α–χψανο 4–ηψδροξψ χινναμιχ αχιδ
AChE	:	acethylcholinesterase
ACN	:	Acetonitrile
AdCSV	:	Adsortive cathodic stripping voltammetry
ADP	:	Adenosine diphosphate
AECh	:	Adenylate energy charge
AEC	:	Anion exchange chromatography
AES	:	Atomic emmision spectrometry
AFM	:	Atomic force microscopy
AFS	:	Atomic fluorescence spectroscopy
Ag NPs	:	Silver nanoparticles
AJO	:	El Ajolí, DNP surroundings, a polluted site
ALAD	:	δ-aminolevulinic acid dehydratase
AOP	:	Adverse outcome pathway
APAP	:	Acetaminophen
APCI	:	Atmospheric pressure chemical ionization
APDC	:	Ammonium pyrrolidine dithiocarbamate
API	:	Atmospheric Pressure Ionization
APPI	:	Atmospheric pressure photoionization
As^{3+}	:	Arsenite
As^{5+}	:	Arsenate
AsFlFFF	:	Asymmetrical flow field flow fractionation
AT	:	Atenolol
ATP	:	Adenosine triphosphate
ATS	:	Activated thiol-sepharose
ATSDR	:	Agency for toxic substance and disease registry
B[a]P	:	Benzo[a]Pyrene
BAF	:	Bioaccumulation factors
$BaTiO_3$	:	Barium titanate
BCA	:	Bicinchinonic acid assay
BCAA	:	Branched chain amino acids

BE	:	Back Extraction
BER	:	Bernabé, DNP surroundings, a polluted site
Bi_2O_3	:	Bismuth trioxide
BP	:	Biological process
BQL	:	Below quantification limit
BSA	:	Bovine serum albumin
C60	:	Fullerens with 60 carbon atoms
CAF	:	caffeine
CAT	:	Catalase
CbE	:	carboxyl esterase
CBZ	:	carbamazepine
CC	:	Cellular compartment
cDNA	:	Complementary DNA
CDOM	:	Chromophoric Dissolved Organic Matter
CE	:	Capillary electrophoresis
CEC	:	Cation Exchange Chromatography
CE-MS	:	Capillary electrophoresis mass spectrometry
CeO2 NPs	:	Cerium oxide nanoparticles
CEWAF	:	Chemically-dispersed crude oil
CGC	:	Capillary Gas Chromatography
CHAPS	:	3-[(3-cholamidopropyl)dimethylammonio]-1-propanesulfonate
ChAT	:	Choline acetyltransferase
Chlp	:	Chlorpyrifos
CLE	:	Competitive Ligand Exchange
CNTs	:	Carbon nanotubes
CPMP	:	Committee for property medicinal products
CRM	:	Certified reference material
CrO_2	:	Chromium dioxide nanoparticles
CSV	:	Cathodic Stripping voltammetry
CuO Nps	:	Copper oxide nanoparticles
CVAAS	:	Cold vapor atomic absorption spectroscopy
CV-rf-GD-OES	:	Cold vapor radiofrequency glow-discharge optical-emission spectrometry with on-line microwave oxidation
Cys	:	Cysteine
CZE	:	Capillary zone electrophoresis
DAD	:	Diode array detection
DBF	:	Dibenzylfluorescein
DBR	:	Doñana biological reserve
DBT	:	Dibutyltin
DCF	:	Diclofenac
DCS	:	Differential centrifugal sedimentation
DDT	:	1,1,1-Trichloro-2,2-bis(4-chlorophenyl)ethane
DDTC	:	Diethyldithiocarbamate
DEMOCOPHES	:	DEMOnstration of a study to COordinate and Perform Human biomonitoring on a European Scale
DGT	:	Diffusive gradients in thin films
DHA	:	Docosahexaenoic acid
DHB	:	2,5-dihydroxybenzoic acid
DI-ESI-QqQ-TOF	:	Direct infusion triple quadrupole time-of-flight mass spectrometer with electrospray ionization source

DIGE	:	Difference in-gel electrophoresis
DLLME	:	Dispersive Liquid-Liquid Microextraction
DMDSe	:	Dimethyldiselenide
DMSe:		Dimethylselenide
DIMS	:	Direct infusion mass spectrometry
DIN	:	Direct Injection Nebulizer
DI-QqQ-TOF	:	direct infusion triple quadrupole mass spectrometer
DLLME	:	Dispersive liquid-liquid microextraction
DLS	:	Dynamic light scattering
DMA	:	Dimethylarsinate
DMAS	:	Dimethylarsinothioic acid
DMMA	:	Dimethylmethoxyarsonium
DNA	:	Deoxyribonucleic acid
DNP	:	Doñana National Park
DNPH	:	2,4-dinitrophenylhydrazyne
DOC	:	Na-deoxycholate
DOM	:	Dissolved organic matter
DPASV	:	Differential pulse anodic stripping voltammetry
DRC	:	Dynamic reaction cell
DTT	:	Dithiothreitol
E2	:	Estradiol
EC10	:	Effective concentration 10%
EC50	:	Effective concentration 50%
ECFA	:	Joint Expert Committee on Food Additives
EDS	:	Energy dispersive spectroscopy
EE2	:	17α-ethinylestradiol
EFSA	:	European Food Safety Authority
eGPx	:	Extracellular glutathione peroxidase
ENM	:	Engineered nanomaterial
ENP	:	Engineered nanoparticle
ENPs	:	Engineer nanoparticles
EP	:	Environmental proteomics
EPA	:	Environmental Protection Agency
EPAc	:	Eicosapentaenoic acid
ERA	:	Environmental risk assessment
EROD	:	Ethoxyresorufin O-deethylase
ESI	:	Electrospray ionization
EST	:	Expressed sequence tag
ETAAS	:	Electrothermal atomic absorption spectroscopy
ETV	:	Electrothermal vaporisation
EXAFS	:	Extended X-Ray Absorption Fine Structure
FAO	:	Food and Agriculture Organization of the United Nations
FFF	:	Field flow fractionation
FIA	:	Flow injection analysis
FI-CL	:	Flow Injection with chemiluminescence detection
Fluorescence-EEMs	:	Fluorescence excitation-emission matrix
FTC	:	Fluorescein-5-thisemicarbazide
FTICR	:	Fourier transform ion cyclotron resonance
FX	:	Fluoxetine
G6PDH	:	Glucose-6-P dehydrogenase

GBD	:	Gas bubble disease
GBR	:	Great barrier reef
GC	:	Gas chromatography
GC-MS	:	Gas chromatography mass spectrometry
GFAAS	:	Graphite furnace atomic absorption spectroscopy
Gly-sug	:	Glycerol sugar
GO	:	Gene ontology
GPAsC	:	Glyceryl phosphoryl arsenocholine
GPx	:	Glutathione peroxidase
GR	:	Glutathione reductase
GSH	:	Reduced glutathione
GSHPx	:	Glutathione peroxidase
GSSG	:	Oxidized glutathione
GST	:	Glutathione S-transferase
HDC	:	Hydrodynamic chromatography
HF-LPME	:	Hollow-fiber liquid phase microextraction
HG	:	Hydride generation
HGAAS	:	Hydride generation atomic absorption spectroscopy
HILIC	:	Hydrophilic interaction liquid chromatography
HMW	:	High molecular weight
HPA axis	:	Hypothalamic pituitary adrenal axis
HPLC	:	High performance liquid chromatography
HQ	:	Hazard or risk quotient
HS	:	Humic substances
Hsc	:	Heat cognate protein
Hsp	:	Heat shock protein
IAF	:	5-Iodoacetamido fluorescein
IAM	:	Iodoacetamide
IARC	:	International Agency for Research on Cancer
iAs	:	Inorganic arsenic
IBF	:	Ibuprofen
IC	:	Ion chromatography
IC50	:	Inhibition Concentration 50%
ICP-AES	:	Inductively coupled plasma atomic emission spectroscopy
ICP-MS	:	Inductively coupled plasma mass spectrometry
ICP-OES	:	Inductively coupled plasma optical emission spectrometry
IDA	:	Isotope dilution analysis
IEF	:	isoelectric focusing
IM	:	Isla Mayor
IMAC	:	Immobilized metal affinity chromatography
INAA	:	Instrumental neutron activation analysis
InP	:	Indium phosphide
InSnO	:	Indium tin oxide
IPG	:	Immobilized pH gradient
IRMM	:	Institute for Reference Materials and Measurements
iTRAQ	:	Isobaric tags for relative and absolute quantification
KEGG	:	Kyoto encyclopedia of genes and genomes
LA	:	Laser ablation
LC/ESI-MS	:	Liquid chromatography electrospray mass spectrometry
LC-MS-MS	:	Liquid chromatography tandem mass spectrometry

LC/LC–MS/MS	:	Multidimensional chromatography coupled with tandem mass spectrometry
LC	:	Liquid Chromatography
LC50	:	Lethal concentration 50%
LC-MS/MS	:	Liquid chromatography tandem mass spectrometry
LC-MS	:	Liquid chromatography mass spectrometry
LCPUFAs	:	n-3 Long-chain polyunsaturated fatty acids
LC-Q-TOF	:	Liquid Chromatography quadrupole time of flight
LD	:	Laguna Dulce
LD50%	:	Lethal doses 50%
LDP	:	Lucio del Palacio sampling site
$LiCoO_2$	:	Lithium cobalt dioxide
LMS	:	Lysosome membrane stability
LMW	:	Low molecular weight
LOD	:	Limit of detection
LOEC	:	Lowest effective concentration
LOQ	:	Limit of quantification
LPO	:	Lipid peroxidation
LPS	:	Lipopolysaccharide
MA	:	Methylarsonate
MAE	:	Microwave-assisted extraction
MALDI-TOF	:	Matrix-assisted laser desorption-ionization time-of flight
MALDI-TOF-PMF	:	Matrix-assisted laser desorption ionization-time of flight-mass spectrometry-Peptide mass fingerprinting
MALDI-TOF-TOF	:	Matrix-assisted laser desorption ionization tandem time of flight
MALS	:	Multi-angle light scattering
MAT	:	El Matochal, DNP surroundings, a polluted site
MBD	:	Methylated DNA-binding protein
MBT	:	Monobutyltin
MDA	:	Malondialdehyde
$MeHg^+$	:	Methylmercury
MF	:	Molecular function
MOA	:	Mode of action
MoO_3	:	Molybdenum trioxide
mRNA	:	Messenger RNA
MS	:	Mass spectrometry
MS/MS	:	Tandem mass spectrometry
MSFD	:	Marine Strategy Framework Directive
MT	:	Metallothionein
MTX	:	Methotrexate
MWCNT	:	Multiwall carbon nanotubes
MWCO	:	Molecular-weight cutoff
MWs	:	Molecular weights
NAD(H)	:	Nicotinamide adenine dinucleotide (reduced form)
NADP(H)	:	Nicotinamide adenine dinucleotide phosphate (reduced form)
NEM	:	*N*-ethylmaleimide
nESI	:	nano-Electrospray ionization
nESI-MS/MS	:	Negative ion nano-electrospray ionisation tandem mass spectrometry with ion mobility spectrometry coupled with tandem mass spectrometry
NGS	:	New-generation sequencing

NM	:	Nanomaterial
NMR	:	Nuclear magnetic resonance
Q-TOF-MS	:	Quadrupole time of flight mass spectrometry
NMS	:	Nanomaterials
NOEC	:	Not observed effective concentration
NOM	:	Natural organic matter
NP	:	Nanoparticle
NPS	:	Nanoparticles
NRR	:	Neutral retention rate
NSAID	:	Non steroidal anti-inflammatory drug
NTA	:	Nanoparticle Tracking Analysis
OA	:	Ocean acidification
OECD	:	Organisation for Economic Co-operation and Development
O-PLS-DA	:	Orthogonal partial least square discriminant analysis
OTC	:	Oxytetracycline
OTs	:	Organotin compounds
PAGE	:	Polyacrylamide gel electrophoresis
PAH	:	polycyclic aromatic hydrocarbon
PAHs	:	Polyaromatic hydrocarbons
PAR	:	El Partido, DNP surroundings, a polluted site
PARAFAC	:	Parallel factor analysis
PBB, PCB	:	Polychlorinated (or poly-Br) biphenyl
PBS	:	Phosphate buffered saline
PCA	:	Principal component analysis
PCR	:	Polymerase chain reaction
PDMS	:	Polydimethylsiloxane
PEGs	:	Polyethylen glycols
PEOs	:	Polyethylene oxides
PES	:	Protein expression signatures
PGE	:	Platinum Group Elements
PhACs	:	Pharmaceutically Active Compounds
pI	:	Isoelectric point
pIs	:	Isoelectric points
PLE	:	Pressurize liquid extraction
PMF	:	Peptide mass fingerprint
PO_4-sug	:	Phosphate sugar
POC	:	Particulate organic carbon
pp'-DDD	:	1,1-Dichloro-2,2-bis (p-chlorophenyl)ethane, a breakdown product of DDT
pp'-DDE	:	1,1-Dichloro-2,2-bis (4-chlorophenyl)ethene, a breakdown product of DDT
PRDX	:	Peroxiredoxin
PRO	:	Propanolol
PSU	:	Practical salinity units
PT	:	Proficiency tests
PTM	:	post-translational modification
PTMs	:	Post translational modifications
PVC	:	Polyvinyl chloride
QD	:	Quantum dots
RBC	:	Red blood cells

RfD	:	Reference Dose established by EPA
RM	:	Reference material
RNAseq	:	RNA sequencing
RNS	:	Reactive nitrogen species
ROS	:	Reactive oxygen
ROC	:	La Rocina, DNP surroundings, a polluted site
RPC	:	Reverse phase chromatography
SCP	:	Stripping chronopotentiometry
SCX	:	Strong cation exchange
SDS	:	Sodium dodecyl sulphate
SDS-PAGE	:	Sodium dodecyl sulfate polyacrylamide gel electrophoresis
SeAlb	:	Selenoalbumin
SEC	:	Size exclusion chromatography
SeCys	:	Selenocysteine
SedFFF	:	Sedimentation field flow fractionation
SF	:	Sector Field
SELDI	:	Surface-excision laser-desorption ionisation
Se(VI)	:	Selenate
Se(IV)	:	Selenite
SeCys	:	Selenocysteine
SeMet	:	Selenomethionine
SeMetO	:	Selenomethionine Se-oxide
SEM	:	Scanning electron microscopy
SeP	:	Selenoprotein P
SPE	:	Solid phase extraction
SO_3-sug	:	Sulfonate sugar
SO_4-sug	:	Sulfate sugar
SOD	:	Superoxide dismutase
SPE	:	Solid phase extraction
SPM	:	Suspended particulate matter
SPME	:	Solid phase microextraction
SSH	:	Suppression subtractive hybridization
SUID	:	Species-unspecific isotope dilution
SWCNT	:	Single wall carbon nanotubes
TBARS	:	Thiobarbituric acid reactive substance
TBT	:	Tributyltin
TCAr	:	Tricarboxylic acid
TCA	:	Trichloroacetic acid
TCEP	:	tris(2-carboxyethyl) phosphine
TEL	:	Triethyl lead
TEM	:	Transmission electron microscopy
TETRA	:	Tetramethylarsonium Ion
TFA	:	Trifluoroacetic acid
TH	:	Tyrosine hydroxylase
ThxR	:	Thioredoxin reductase
TMAH	:	Tetramethylammonium hydroxide
TMAO	:	Trimethylarsine oxide
TMF	:	Tamixofen
TML	:	Trimethyl lead
TOF	:	Time of flight

TPT	:	Triphenyltin
TRV	:	Toxicity reference value
TUF	:	Tangential flow ultrafiltration
TWI	:	Tolerable weekly intake
UF	:	Ultrafiltration
UHPLC-QTOF-MS	:	Ultra high performance liquid chromatography quadrupole time-of-flight mass spectrometry
US EPA	:	United States Environmental Protection Agency
UV	:	Ultraviolet
VG	:	Vapour generation
WAF	:	Water-accommodated fraction
WFD	:	Water Framework Directive
WHO	:	World Health Organization
WWTP	:	Waste Water Treatment Plant
XANES	:	X-ray absorption near edge structure
XAS	:	X-Ray absorption spectroscopy
XPS	:	X-Ray photoelectron spectrometry
XRD	:	X-Ray diffraction
ZnO NPs	:	Zinc oxide nanoparticles

1

Element Speciation in Seawater

From Free Metal Determination to Proteomic Analysis

María Carmen Barciela-Alonso,[a] *Pilar Bermejo-Barrera,*[b,*]
Antonio Moreda-Piñeiro[c] and *Elena Peña-Vázquez*[d]

ABSTRACT

Various methods for trace element speciation in seawater samples are summarized in this chapter, ranging from free ions to metalloprotein analysis. These methods include separation and preconcentration techniques, such as SPE or HPLC, to avoid the problems due to the low concentration of the analytes and the interferences caused by the salinity of the samples. The combined use of both separation (HPLC, FFF, SDS-PAGE, or OFFGEL electrophoresis) and detection techniques (electrochemical methods, AAS, ICP-MS, or ESI-MS) provides a wide number of applications to quantify and identify the different species in the marine environment.

Introduction

Trace elements are present in seawater originating from either natural or anthropogenic sources. The latter include chemicals used in the manufacture of pigments, paper, wood preservatives, pharmaceutical compounds, pesticides, etc. These trace elements enter the food chain, thus posing serious health hazards in some cases.

Trace elements can be classified as essential (Cu, Co, Fe, Mn, Mo, Zn, Cr, Ni, and Se) or toxic (Cd, Hg, Pb, Sb, Bi, Sn, and Tl) elements. The toxicity, mobility, and bioavailability of these elements depend on their concentration as well as their chemical form which determine their physical and chemical behavior

Department of Analytical Chemistry, Nutrition and Bromatology, Faculty of Chemistry, University of Santiago de Compostela 15782 Santiago de Compostela, Spain.
[a] Email: mcarmen.barciela@usc.es
[b] Email: pilar.bermejo@usc.es
[c] Email: antonio.moreda@usc.es
[d] Email: elenamaria.pena@usc.es
* Corresponding author

in natural systems (Ure and Davison 2001). Trace elements can be present in natural aquatic systems in the form of free ions with different oxidation states and be associated to colloids and particles.

Speciation of Free Ions

Trace elements occur in seawater samples as free ions in different stages of oxidation. Some examples of species reported in the literature for elements such as antimony, arsenic, chromium, lead, mercury, selenium, tellurium, tin, and iodine are shown in Table 1. These species are present in seawater at very low concentrations usually at levels of few ng L^{-1} or $\mu g\ L^{-1}$.

Table 1: Element species in seawater samples.

Element	Species
Antimony	Inorganic antimony Sb(III) and Sb(V)
Arsenic	arsenite (As(III)), arsenate (As(V)), monomethylarsonic (MMA), dimethylarsinic (DMA)
Chromium	Cr(III), Cr(VI)
Lead	Pb^{2+}, triethyllead (TEL), trimethyllead (TML)
Mercury	$MeHg^{+}$, EtHg, and inorganic Hg
Selenium	Se(IV), Se(VI)
Tellurium	Te(IV) and Te(VI)
Tin	MBT, DBT, TBT
Iodine	Iodide, iodate

The determination of these species in seawater samples presents some difficulties associated with their low concentrations and the salinity of the sample, which can produce interferences in analytical techniques. In order to solve these problems, highly sensitive analytical techniques are needed as well as preconcentration and separation steps to remove the matrix interferences.

Sample pretreatment

Several separation and preconcentration methods have been reported in the literature for element speciation in seawater. These methods include liquid-liquid extraction, solid phase extraction (SPE), solid phase microextraction (SPME), dispersive liquid-liquid microextraction (DLLME), preconcentration using knotted reactors, headspace-stir bar sorptive extraction, etc.

For SPE, a mini column filled with Amberlite XAD-8 resin has been used for Sb(III) preconcentration (Ozdemir et al. 2004). In this case, antimony was complexed with ammonium pyrrolidine dithiocarbamate (APDC) and quantitatively adsorbed in the column. The elution was performed using 10 mL of acetone. In a study by Calvo et al. (2011) [1,5-bis(2-pyrydil)-3-sulfophenylmethylene] thiocarbonohydrazyde immobilized on aminopropyl-controlled pore glass was used for Sb(III) preconcentration and Amberlite IRA-910 resin for Sb(V) preconcentration (Calvo et al. 2011). On-line preconcentration on an microcolumn packed with C_{18} bound silica gel has been reported in the literature for selective preconcentration of Cr(VI) as diethyldithiocarbamate complex (Prasada-Rao et al. 1998). An iminodiacetate resin (Muromac A-1) has also been used for Cr(III) separation and preconcentration (Hirati et al. 2000).

Solid phase extraction using ion-imprinted polymethacrylic microbeads has been used for mercury speciation before its determination by cold vapor atomic absorption spectrometry (Dakova et al. 2009). Metal ion-imprinted polymer particles were prepared by copolymerization of methacrylic acid as monomer, trimethylopropane trimethacrylate as cross-linking agent, and 2,2-azobisisobutyronitrile as initiator, in the presence of Hg(II)-1-(2-thizaolylazo)-2-naphthol complex. The adsorbed inorganic mercury was desorbed using 4 M HNO_3. SPE using *S. aureus* loaded Dowex optipore V-493 columns has also been used for Hg speciation (Tuzen et al. 2009). Sequential elution with 0.1 M HCl and 2M HCl was performed after sample loading for $MeHg^+$ and Hg^{2+}, respectively.

Preconcentration of Hg(II) using poly(acrylamide) grafted onto cross-linked poly(4-vinyl pyridine) (P4-VP-*g*-PAm), in presence of $MeHg^+$(I), has been investigated. The sorbent showed excellent selectivity for Hg(II) in presence of other ions such as, Pb(II), Zn(II), Cu(II), Cd(II) and Fe(III) (Yayayürük et al. 2011).

SPE has been used for Se(IV) separation and preconcentration (Saygi et al. 2007) in the form of Se(IV)-ammonium pyrrolidine dithiocarbamate chelate on Diaion HP-2MG. Total Se was determined after reduction of Se(VI) by heating the samples in a microwave oven with 4 M HCl.

Solid phase micro extraction (SPME) with fused silica fiber coated with polydimethylsiloxane (PDMS) has been used after derivatization of mercury species using sodium tetraphenylborate ($Na(BPh_4)$) (Carro et al. 2002; Bravo Sánchez et al. 2004; Mishra et al. 2005) allowing the determination of these species at very low concentrations (ng L^{-1}). The only problem with this methodology is the blank contamination which can affect inorganic mercury evaluation. This problem was overcome by setting up cleaning procedures using minicolumns packed with 8-hydroxyquinoline to trap and remove mercury traces present in all the analytical reagents used (Bravo Sánchez et al. 2004). Head space SPME was further used for simultaneous determination of organometallic compounds of mercury, lead, and tin in seawater using a Divinyl benzene/carboxen/polydimethylsiloxane fiber (Beceiro et al. 2009). Selective magnetic SPE (MSPE) separation was used for speciation of inorganic tellurium from seawater (Huang and Hu 2008). In these case, Te(IV) was quantitatively adsorbed on γ-mercaptopropyltrimethoxysilane modified silica coated magnetic nanoparticles, within the pH range of 2–9, while Te(VI) was not retained thus remaining in the solution. The magnetic nanoparticles were then separated from the solution by applying an external magnetic field, and Te(IV), was recovered using a solution containing 2M HCl and $K_2Cr_2O_7$.

Herbello et al. have used on-line sorption preconcentration in a knotted reactor for As(III) and Cr(VI) determination (Herbello et al. 2005, 2011). In this case, As(III) and Cr(VI) was complexed with ammonium pyrrolidine dithiocarbamate and adsorbed into the inner wall of the knotted reactor. The complexes were then eluted with ethanol (40 μL). Enrichment factors of 44 and 31 were obtained using this method for As(III) and Cr(VI) determination, respectively.

Organotin species have been extracted into toluene after addition of sodium ethyldithiocarbamate (DDTC) and NaCl to seawater samples (Tsunoi et al. 2000). Dispersive liquid-liquid microextraction (DLLME) and back extraction has been used for Hg^{2+} preconcentration (Li et al. 2011). Mercury was extracted with 1-(2-pyridylazo)-2-naphthol (PAN) in ethanol and back extraction was performed with L-cysteine. This method resulted in an enrichment factor of 625.

Analytical techniques

Several analytical techniques have been used for the determination of trace element species in seawater samples, including electrochemical methods, catalytic spectrophotometry, atomic absorption spectrometry (AAS), atomic fluorescence spectroscopy (AFS), inductively coupled plasma atomic emission spectroscopy (ICP-AES), inductively coupled plasma mass spectrometry and high performance liquid chromatography (HPLC), capillary electrophoresis (CE), and gas chromatography (GC) coupled with various detectors.

Electrochemical methods present certain advantages over other methods for element speciation in seawater, including high sensitivity and selectivity associated with low detection thresholds. The seawater matrix is an "ideal" electrolyte to perform the measurements with minimal sample preparation, low-cost of analysis, and requiring minimal analysis times (Riso et al. 2004). Electrochemical methods used for trace speciation analysis are based on differential pulse anodic/cathodic stripping voltammetry (Quentel and Filella 2002; Papoff et al. 1998; Quentel and Elleouet 1999; Nascimiento et al. 2009), anodic stripping voltammetry with a gold microelectrode (Salaün et al. 2007), adsorptive cathodic stripping voltammetry (AdCSV) (Carballo et al. 2008), stripping chronopotentiometry (SPC) (Riso et al. 2004), and stripping chronopotentiometry with a mercury film electrode (Tanguy et al. 2010). Limits of detection (LOD) in within the range of a few ng L^{-1} were obtained using these techniques. As an example, LOD of 8 ng L^{-1} was obtained for Sb(III) determination using stripping chronopotentiometry with a mercury film electrode (Tanguy et al. 2010), and an LOD of 0.16 ng L^{-1} for Se(IV) using differential pulse cathodic stripping voltammetry (Papoff et al. 1998).

Catalytic spectrophotometry is an analytical technique used for iodine speciation (Truesdale et al. 2003a; Truesdale et al. 2003b). Total iodine is determined by catalytic spectrophotometry (arsenious acid plus Ce(IV) sulphate reaction catalyzed by iodide at 34°C). Iodate is determined by spectrophotometry (conversion of iodate to I_3^- ion with potassium iodide in sulphamic acid).

Atomic absorption spectroscopy techniques including flame atomic absorption spectroscopy (FAAS), electrothermal atomic absorption spectroscopy (ETAAS) and cold vapor atomic absorption spectroscopy (CVAAS), Hydride Generation Atomic Absorption Spectroscopy (HGAAS), have also been used in speciation studies. In these cases, preconcentration and separation steps are needed before the determination. This can be performed off-line or on-line using solid phase extraction procedures, with different sorbents (ion-imprinted polymethacrylic microbeads, C18 bounded silica gel, etc.), coprecipitation, or using knotted reactors. Some methods reported in the literature for the determination of As, Cr, Hg, Se, Sb, and Sn using AAS, are summarized in Table 2. This table also shows the sample preparation step and the LODs obtained with these methods.

Atomic fluorescence spectroscopy (AFS) is a highly sensitive technique also used in speciation studies. Limits of detection of 13 and 15 ng L^{-1} were obtained for As(III) and As(V), respectively, using this technique. The method was based on the generation of arsine (AsH_3) from the reaction between the arsenic species in the injected solution and tetrahydroborate immobilized on a strong anion-exchange resin (Amberlite IRA-400). Speciation was based on two different measurement conditions: (i) acidification to 0.7 M with HCl, (ii) acidification to 0.1 M with HCl in the presence of 0.5% L-cysteine, resulting in two calibration equations with different sensitivities for each species (Wang and Tyson 2014).

Cold vapor atomic absorption spectroscopy has been used for Hg^{2+} and $MeHg^+$. In this case, mercury species were preconcentrated using poly(acrilamide) grafted onto cross-linked poly(4-vinyl pyridine) and eluted using 2 mL of HNO_3 (14.3 M). An LOD of 2 ng L^{-1} was obtained (Yayayürük et al. 2011).

Chromatographic techniques have been widely used in speciation studies coupled with different detectors, such as AFS, ICP-AES, ICP-MS, MS, MS/MS. HPLC coupled to ICP-MS offered certain advantages in these studies due to their possibility of multielemental determination, as well as high sensitivity. Some applications using HPLC coupled with different detectors are summarized in Table 3. Gas chromatography has a large number of applications for lead, mercury, and tin speciation in seawater samples. A derivatization step, using sodium tetra phenyl borate or sodium tetra ethyl borate, is usually included before the chromatographic separation. Some applications reported in the literature using gas chromatography for trace element speciation in seawater samples are summarized in Table 4. As can be observed in Tables 3 and 4, LODs of a few ng L^{-1} are reported using these techniques.

Capillary Zone Electrophoresis (CZE) with diode array detection (DAD) has been used for Hg^{2+} determination in seawater samples. Hg^{2+} was preconcentrated by dispersive liquid-liquid microextraction-back extraction (DLLME-BE) using 1-(2-Pyridylazo)-2-naphthol (PAN) and L-cysteine as chelating reagents for DLLME and BE, respectively. An LOD of 0.62 µg L^{-1} has been obtained (Li et al. 2011). CE-UV has been used for direct analysis of iodine species in seawater (Huang et al. 2004a,b), obtaining LODs of 0.23 and 10 µg L^{-1} for iodide and iodate, respectively.

Complexes of Trace Metals with Organic Ligands

Dissolved organic matter (DOM) plays a crucial role in the transport and fate of nutrients and trace micropollutants (trace metals included) in surface waters and also in seawater (Ogawa and Tanoue 2003; Park 2009). The presence of several oxygen, nitrogen, and sulfur-containing functional groups such as carboxylic, phenolic, alcohol, amino acid, and thiol, is responsible for the DOM complexation properties (Wu et al. 2004).

Trace elements can exist as different species in natural waters: free aquatic ionic species, dissolved inorganic or organic complexes, complexes with colloidal or particulate matter (inorganic or organic),

Table 2: Speciation studies using Atomic Absorption Spectrometry.

Species	Sample preparation	Detector	Detection limit	Ref.
As(III) and As(V) inorganic species; MMA and DMA	Flow injection hydride generation: As(III): seawater at pH 7 using Tris/HCl buffer. Hydride reactive arsenic species (As(III), As(V), MMA, DMA): Seawater mixed with persulphate or alkaline persulphate solution and hydride generation with $NaBH_4$ and HCl	HG-GFAAS	1.5 ng L^{-1}	Cabon and Cabon 2000
As(III), As_T	On-line sorption preconcentration in a knotted reactor. On-line formation of As(III)-pyrrolidine dithiocarbamate complex in an acid medium. Elution with ethanol	ETAAS	0.005 μg L^{-1}	Herbello et al. 2005
Cr(III) and Cr(VI)	On-line preconcentration on a microcolumn packed with C_{18} bonded silica gel. Selective formation of diethyl ditiocarbamate complexes with Cr(VI). Elution with methanol	FAAS	0.02 μg L^{-1}	Prasada-Rao et al. 1998
Cr(VI)	On-line sorption preconcentration in a knotted reactor: On-line formation of Cr(VI)-ammonium pirrolidine dithiocarbamate complex in an acid medium. Elution with ethanol	ETAAS	0.007 μg L^{-1}	Herbello et al. 2011
$MeHg^+$, Hg^{2+}	SPE using ion-imprinted polymethacrylic microbeads	CVAAS	0.006 μg L^{-1} for Hg^{2+}	Dakova et al. 2009
$MeHg^+$, Hg^{2+}	Solid Phase Extraction using *S. aureus* loaded Dowex optipore V-493 column. Sequential elution with 0.1 M HCl for $MeHg^+$ and 2M HCl for Hg^{2+}	CVAAS	1.7 and 2.5 ng L^{-1}	Tuzen et al. 2009
Se(IV)	Hydride generation. Determination of total Se: thermal treatment (100ºC) in 5M HCl or UV irradiation in basic medium. Se(-II) and selenomethioneine conversion to Se(IV) in seawater 0.01 M HCl and 30 min irrradiation time	HG- ETAAS	1.5 ng L^{-1}	Cabon and Erler 1998
Se(IV) and Se_T	Hydride generation. Total Se: 1) 20 mL of sample and UV irradiation for 45 min. 2) Addition of 0.1 mL 1% (m/v) $NaNH_4$ 3) H_2Se is trapped in a 3 μL drop containing 20 ng L^{-1} $Pd(NO_3)_2$ in 1.5% m/v HNO_3. 4) Injection into the graphite tube	HG- ETAAS	0.15 ng L^{-1}	Fragueiro et al. 2006
Se(IV) and Se(VI)	Solid phase extraction of Se(IV)-ammonium pyrrolidine dithiocarbamate chelate on Diaion HP-2MG. Reduction of Se(IV) heating the sample in microwave oven with 4M HCl. Elution with 1M HNO_3 in acetone	GFAAS	0.010 μg L^{-1}	Saygi et al. 2007
Sb(III) and Sb_T	Hydride generation. Sb(III): Buffering the seawater sample with 0.2 M Tris-HCl solution. Sb_T: Acidifying the sample with 2 M HCl and irradiated on-line	GFAAS	5 and 10 ng L^{-1}	Cabon and Madec 2004
Sb(III) and Sb_T	Sb_T: Adsorption of inorganic species using tianium dioxide (TiO_2). Sb(III): Coprecipitation with APDC (10 mg L^{-1}) in acid media and in presence of $Pb(NO_3)_2$. Redisolution with 1.0 mL HNO_3 (1:1)	GFAAS	0.14 μg L^{-1}	Zhang et al. 2007
Total organotin compounds	Preconcentration using a Molecular Imprinted Polymer (MIP). Elution with 0.1 M HCl in methanol.	GFAAS	30 ng L^{-1}	Krishan et al. 2004

Table 3: Analytical methods for trace element speciation using HPLC.

Species	Sample preparation and Chromatographic conditions	Detector	Detection limit	Ref.
As(III), As(V), MMA, DMA	Chromatographic column: PRP X-100 (250 x 4.1 mm, 10 μm). Mobile phase: 25 mM potassium hydrogenphosphate (pH 5.8, flow rate 1.1 ml min^{-1}) Hydride generation: 1.5 mM HCl and $NaBH_4$ (1.5% (w/v) stabilized in 1% (w/v) NaOH)	HPLC-(UV)-HG; HPLC-HG-UV AFS ICP-MS	0.3 μg L^{-1}	Gómez-Ariza et al. 2000
As(III), As(V), MMA, DMA	Chromatographic column: PRP X-100 (250 x 4.1 mm, 10 μm). Mobile phase: 20 mM potassium hydrogenphosphate (pH 5.8, flow rate 1.1 ml min^{-1}) Hydride generation: 1.5 mM HCl and $NaBH_4$ (1.5% (w/v) stabilized in 1% (w/v) NaOH)	HPLC-HG-ICP-MS	0.1 μg L^{-1} for arsenite, DMA and MMA; 0.3 μg l^{-1} for arsenate	Sánchez-Rodas et al. 2005
Cr(III) and Cr(VI)	On-line preconcentration of Cr(VI) using tetrabutyl ammonium bromide (TBABr) as ion-pair forming agent Chromatographic column: Lichrospher 100-RP 18, 5 μm particle size, 12.5 cm length, 4.6 mm i.d. Mobile phase: TBABr (3 x 10^{-4} M), NH_4 acetate (1 x 10^{-4} M) and Methanol (40% v/v)	HPLC-ICP-AES HPLC-ICP-MS	0.20 ng mL^{-1} (ICP-AES); 0.12 ng mL^{-1} (ICP_MS)	Posta et al. 1996
Hg(II), $MeHg^+$, $EtHg^+$ and $PhHg^+$	Column: Interchrom ODB (100 mm x 1 mm, 3 μm) Flow rate: 70 μL min^{-1} Mobile Phase A: 1% MeOH, 0.01 M NH_4Ac, 0.05% L-Cysteine, pH 3.0 Mobile phase B: 10% MeOH, 0.1 M NH_4Ac, 0.5% L-Cysteine pH 3.0	microHPLC-microneb-ICP-MS	11, 23, 8, 32 ng L^{-1}	Castillo et al. 2006
Hg(II), $MeHg^+$	On-line pre-concentration micro-column and separation by HPLC Micro-column: Opti-LynxTM trap cartridge with a bed volume of 100 μL packed with a C-18 stationary phase. Elution with the mobile phase. Mobile phase: 0.5% L-cysteine (m/v) and 0.5% 2-mercaptoethanol (v/v) dissolved in ultrapure water. HPLC column: 100 x 2.1 mm All time HP C-18 3 μm particle size	HPLC-ICP-MS m/z: 184	0.07, 0.02 ng L^{-1}	Cairns et al. 2008
Sb(III), Sb(V), and $(CH_3)SbCl_2$	Chromatographic column: PRP X-100 Gradient mode Mobile Phase A: 20 mM EDTA + 2 mM KHP (potassium hydrogen phthalate) Mobile phase B: 50 mM $(NH_4)_2HPO_4$	HPLC-HG-AFS	0.13, 0.07, 0.13 μg L^{-1}	Gregori et al. 2005
MMT, DBT, TBT	Column: Agilent zorbax Eclipse XDB-C18 5 μm, 4.6 x 150 mm Mobile phase: 70% (v/v) methanol, 3% (v/v) HAc, 27% (v/v) water, 0.1% (m/v) tropolone	HPLC-HG-Quarzt surface induced luminiscence-flame photometric detection (QSIL-FPD)	1.69, 0.51, 0.36 ng mL^{-1}	Zhai et al. 2007
Iodide and iodate	Column: Agilent G3154A/101 (porous polymethacrylate resin) 10 μm, 4.6 x 150 mm Mobile phase: 20 mM NH_4NO_3 at pH 5.6	HPLC-ICP-MS	2 and 1.5 μg L^{-1}	Chen et al. 2007

Table 4: Analytical methods for trace element speciation using Gas Chromatography.

Species	Sample preparation	Detector	Detection limit	Ref.
$MeHg^+$, EtHg, inorganic Hg	Derivatization-Solid-Phase-Microextraction: Derivatization with sodium tetraphenylborate ($NaBPh_4$). SPME with fused silica fibre coated with 100-µm film of polydimethylsiloxane (PDMS)	GC-Atomic emission detection	0.1, 0.1 and 0.3 ng mL^{-1}	Carro et al. 2002
$MeHg^+$, Hg^{2+}	Derivatization-SPME: Derivatization with sodium tetraphenylborate ($NaBPh_4$). SPME with fused silica fibre coated with 100-µm film of polydimethylsiloxane (PDMS)	GC-ICP-MS	0.17, 0.35 ng L^{-1}	Bravo-Sánchez et al. 2004
$MeHg^+$, Hg^{2+}	Derivatization with sodium tetraphenylborate ($NaBPh_4$). SPME with fused silica fibre coated with 100-µm film of polydimethylsiloxane (PDMS)	GC-MS	0.02 and 0.05 ng	Mishra et al. 2005
$MeHg^+$, Hg^{2+}	Derivatization with sodium tetraphenylborate ($NaBPh_4$) or sodium tetra ethylborate ($NaBEt_4$)	GC-ICP-MS	0.021 and 0.126 ng L^{-1}	Monperrus et al. 2005
$MeHg^+$	Headspace-stir bar sorptive extraction-thermal desorption gas chromatography Derivatization using $NaBEt_4$	GC-MS	5 ng L^{-1}	Prieto et al. 2008
$MeHg^+$, Hg^{2+}, TML, TEL, MBT, TBT	Derivatization-Headspace-Solid-Phase Microextraction: Derivatization with sodium tetra ethylborate ($NaBEt_4$) SPME with fused silica fiber coated with 100-µm film of polydimethylsiloxane (PDMS)	GC-MS	0.2–16.8 ng L^{-1}	Centineo et al. 2004
Tin species				
Organotin compounds	Ethylation using $NaEt_4B$ and extraction with n-pentane	CGC and pulsed flame photometric detection (PFPD)	0.2–0.4 pg	Jacobsen et al. 1997
Organotin compounds	Ethylation using $NaEt_4B$ and extraction into hexane	GC-ICP-MS	3.8–50 fg	Tao et al. 1999
MBT, DBT, and TBT	Derivatization using sodium tetrahydridoborate followed by head space SPME with a polydimethylsiloxane (PDMS) coated silica fiber	GC-flame photometric detection	19.4, 1.5 and 0.5 ng L^{-1}	Jian et al. 2000
Organotin compounds	Derivatization using $NaEt_4B$ in methanol and extraction into hexane	GC-HR-MS	7–29 ng L^{-1}	Ikonomou et al. 2002
MBT, DBT, TBT, MPT DPT TPT	Addition of sodium ethyldihiocarbamate (DDTC) and NaCl to the samples. Extraction of organotin compounds into Toluene. Derivatization with $C_5H_{11}MgBr$	GD-ion trap-tandem mass spectrometry	0.29, 0.29, 0.35, 0.35 0.20, 0.21 pg	Tsunoi et al. 2002
MBT, DBT, TBT	Addition of ^{119}Sn enriched butyltin mixture. Derivatization with tetraethylborate solution. Extraction into hexane	Isotope dilution GC-MS	0.18–0.25 ng L^{-1}	Centineo et al. 2006
MMT, DBT, TBT	SPME Divinyl benzene/carboxen/polydimethylsiloxane fiber	GC-MS/MS	28, 17, 27 ng L^{-1}	Beceiro et al. 2009

and complexes associated with the biota. Labile metal fractions are usually the most toxic, with some exceptions, such as Hg or Sn. The labile fractions of trace metals determined by ion exchange methods are generally greater than those obtained by voltammetric techniques, owing to the longer contact times in ion exchange preconcentration (Jiann and Presley 2002).

Speciation by electrochemical methods

The formation of complexes of trace metals with organic ligands has been studied mainly using electrochemical methods (Sohrin and Bruland 2011), e.g., Differential Pulse Anodic Stripping Voltammetry (Coale and Bruland 1988; Bruland 1989, 1992; Capodaglio 1990; Jakuba et al. 2008; Baars and Croot 2011), or Competitive Ligand Exchange (CLE) with an added ligand and Cathodic Stripping Voltammetry (CSV) detection of the complex adsorbed to a hanging mercury-drop electrode (Rue and Bruland 1995; Saito and Moffett 2001; Buck and Bruland 2005, 2007; Van den Berg 2006; Wu and Jin 2009; Velasquez et al. 2011; Hassler et al. 2013; Abualhaija and Van der Berg 2014; Monticelli et al. 2014; Thompson et al. 2014). The concentrations of the free metal ions and of the metal organic complexes can be calculated from CSV-labile and total dissolved metal concentrations. The labile fraction consists of the inorganically complexed metal ions and a portion of the natural organic-metal complexes, depending on their relative stability in the presence of the added chelator. Total dissolved metal concentrations can be obtained after UV-photolysis of acidified samples (Nimmo et al. 1989). Organic complexation is significant for elements such as Cd, Co, Cu, Fe, Ni, Pb, and Zn, and the ligands show an important specificity and intensity of complexation (Byrne 2002).

Vertical profiles of Cd show strong covariance with P and N, suggesting that the distribution of Cd is controlled by incorporation in living tissues and biogenic debris, probably due to its similarity to Zn (Byrne 2002). The inorganic chemistry of Cd^{2+} is dominated by chloride complexation in intermediate and deep waters; however, approximately 70% of Cd in North Pacific surface waters is complexed by strong Cd-specific organic ligands (Bruland 1992). Mercury profiles, in contrast to cadmium, exhibit important differences at different locations. Hg^{2+} is mainly present as $HgCl_4^{2-}$, but Hg^{2+} has a strong tendency towards covalent bonding and complexation in general, with affinity for organic ligands. Hg^{2+} is more intensely complexed than Cd^{2+} (Byrne 2002).

Pb exists in seawater as Pb^{2+} forming chloride complexes at low pHs or complexed with carbonate at high pHs in surface waters. Capodaglio (1990) has also reported the association of 30–50% of the element with specific organic ligands in surface waters from the eastern North Pacific.

Van den Berg et al. (1994) studied the chemical speciation of Al, Cr, and Ti in samples from the NW Mediterranean Sea by adsorptive CSV. On average, 72% of Al and 67% of Ti occurred in a non-reactive form, either organically complexed or as inorganic colloidal matter. The organically bound fraction was released after an UV-irradiation of the acidified sample. In another study, in the western Mediterranean, it was observed that 10–20% of the total dissolved Ni appeared to be strongly organically complexed. Both labile and total dissolved Ni concentrations were lower in surface Atlantic waters than in Mediterranean deeper waters, and the influence of seasonal changes on the speciation of Ni was small (Achterberg and Van den Berg 1997).

Some of the elements of the period 4, especially Fe and Zn, are vital in oceanic biogeochemical cycles, and affect plankton production (Byrne 2002). A very strong complexation has been reported in open ocean for Fe (99.97% of the Fe(III) was reported to be organically complexed in North Pacific surface waters by natural organic ligands) (Rue and Bruland 1995). Gledhill et al. (1998) have used a combination of techniques to study the speciation of iron in the northern North Sea in an area covering at least two different water masses and an algal bloom. Catalytic CSV was used to measure the concentration of reactive Fe that complexes 1-nitroso-2-napthol, and total Fe was determined after an UV digestion at pH 2.4 in unfiltered samples. Dissolved Fe was analyzed by GFAAS after extraction of filtered seawater. The concentration of natural organic ligands was determined in unfiltered samples by titration. Fe was fully (99.9%) complexed by the organic ligands at a pH of 6.9 and largely complexed (82–96%) at pH 8. The concentration of Fe strongly bound to organic ligands was also greater than 99% in most of the samples of seawater from the South and Equatorial Atlantic (Powell and Donat 2001).

The speciation of dissolved Fe in the oceanic and shelf domains of the southeastern Bering Sea was evaluated in surface and subsurface samples using CLE adsorptive CSV with the added ligand salicylaldoxime (Buck and Bruland 2007). Two ligand classes were measured in all samples, a stronger L_1 ligand class and a weaker L_2 ligand class. The concentrations of dissolved Fe were strongly correlated with ambient stronger L_1 ligand concentrations for all samples with dissolved Fe concentrations > 0.2 nmol L^{-1}. In samples with dissolved Fe concentrations < 0.2 nmol L^{-1}, large and variable excesses of L_1 ligand were measured, suggesting that the phytoplankton community is able to access dissolved Fe from the FeL_1 complex, resulting in excess L_1 in these waters. These ligands could be polysaccharides, organic colloids, or a fraction of the uncharacterized component of dissolved organic carbon (Buck and Bruland 2007).

Van der Berg (2006) performed the chemical speciation of Fe in seawater by CLE-CSV using 2,3-dihidroxynaphtalene as an adsorptive and competing ligand. The main advantage of this method was the better sensitivity for Fe and the complexing ligands, and the reduction of analysis time to one tenth. Analysis of the results obtained in a two-ligand system with the Scatchard linearization method showed artifacts that can be eliminated using a non-linear regression approach with Turoczy and Sherwood's iteration (Wu and Jin 2009). Speciation is also dependent on the temperature (Hassler et al. 2013), and several of the CLE-CSV procedures suffer from interferences of humic compounds ubiquitous in coastal and ocean waters. CSV methods of analysis should therefore be reoptimized (Abualhaija and Van den Berg 2014).

Cu and Zn were also reported to be strongly complexed by organic, reducing the inorganic fraction to approximately 0.2% of the total Cu in waters shallower than 200 m of the central Northeast Pacific (Coale and Bruland 1988), or to 2% of the total Zn in the upper 600 m of the central North Pacific (Bruland 1989). The CLE adsorptive CSV method, with salicydaldoxime as the added competitive ligand, was used to determine dissolved Cu speciation and total dissolved Cu concentrations at six San Francisco Bay sites (Buck and Bruland 2005). Total dissolved Cu was determined using 100 mL of each sample, UV-oxidized at ambient pH in a Jelight Model 342 UVO-cleaner for 2 h at 10 mV/cm^2. The strong type L_1 ligand complexed 99.9% of all the dissolved Cu. A similar result (99.0% complexation) was obtained by Thompson et al. (2014) and Thompson and Elwood (2014) in most of the samples from the Tasman Sea.

An ASV method was adapted for Zn speciation using a fresh Hg film plated with each sample aliquot; the results were comparable to CLE-CSV when studying a profile from the North Atlantic Ocean, though the ligand concentrations obtained by fresh film ASV were slightly higher (Jakuba et al. 2008). Baars and Croot (2011) studied the speciation of dissolved Zn in the Atlantic sector of the Southern Ocean along two transects across the major frontal systems: along the Zero Meridian and across the Drake Passage. The authors found detectable concentrations of labile inorganic Zn throughout the surface waters in contrast to studies performed at lower latitudes, and electrochemically inert Zn ligands throughout the Southern Ocean. The concentration of Zn complexing ligands exceeded the dissolved Zn concentration only in surface waters, suggesting the presence of a biological source.

Recently Monticelli et al. (2014) developed an apparatus to perform CLE-CSV that shows several advantages: 20-fold reduction in sample volume, decrease in analysis time, and 20-fold drop in reagent consumption. The analytical capabilities were not affected, and the method was validated for the speciation of Cu concentration in seawater samples from a surface-bottom vertical profile from Mahon bay (Spain), showing no differences resulting from reduction in sample volume.

Use of solid phase extraction

Solid phase extraction has also been used for speciation because it offers the possibility to perform preconcentration increasing the sensitivity of the techniques. Iminodiacetate resins have been widely used (Canizzaro et al. 2000; Abollino et al. 2000; Point et al. 2007; Hurst and Bruland 2008; Milne et al. 2010; Lagerström et al. 2013), and it has to be taken into account that without an UV pretreatment only the dissolved labile element can be measured. Abollino et al. (2000) used three columns packed with an anion exchange resin (AG MP-1), a reversed phase sorbent (RP-C18) and a chelating ion exchange resin (Chelex-100) to study the speciation of Cd, Cu, and Pb onto the different substrates. The authors performed experiments with the free metal ions, their anionic complexes with EDTA, 8-hydroxyquinoline, and 8-hydroxyquinoline-5-sulphonic acid. They applied the technique to seawater spiked with known amounts of the three elements and found that cadmium was almost completely retained by AG MP-1 owing to the formation of chloro-complexes. Sixty five percent of Pb was in a cationic form and was retained by Chelex-100, and the remaining fraction interacted with AG MP-1. In the case of Cu, half of the metal was collected by AG MP-1, and the remaining fraction was distributed between C18 and Chelex-100.

Point et al. (2007) have developed an integrated approach for the determination of total, labile, and organically bound dissolved trace metal concentration in the field to avoid sample storage prior to ICP-MS analysis of Cd, Cu, Mn, Ni, Pb, U, and Zn. The authors developed a UV on-line unit including a 254 nm low pressure lamp (6 W, 5400 μW cm^{-2}) surrounded by a 0.8 mm i.d. quartz coil (25 cm, 0,10 mL), and a chelation/preconcentration/matrix elimination module using two Metpac cc1 chelating columns. They used a 5 mL-sample and achieved LODs ranging between 0.6 ng L^{-1} for Cd and 33 ng L^{-1} for Ni. Natural water samples spiked with humic standards were used to study the influence of UV photolysis on organic matter using fluorescence spectroscopy. The speciation scheme was applied to two natural freshwater and seawater samples from the Adour Estuary (southwestern France) and the results indicated that the organic complexation levels were high and similar (82% for the freshwater sample vs. 89% for the seawater sample) for Cu in both samples, whereas different patterns were observed for Cd, Mn, Ni, Pb, U, and Zn, probably due the existence of different organic ligands and transformations along the estuary.

Hurst and Bruland (2008) studied the distribution of particulate elements, dissolved trace metals, and dissolved nutrients in the Gulf of the Farallones. The authors performed the analysis of the total amount of the element in the separated fractions combining the on-line flow injection of UV oxidized samples using iminodiacetate Toyopearl AF-Chelate resin with analysis by ICP-MS.

Milne et al. (2010) performed the off-line extraction of Mn, Fe, Co, Ni, Cu, Zn, Cd, and Pb from seawater using a iminodiacetate Toyopearl AF-Chelate-650M resin followed by analysis using isotope dilution and high resolution magnetic sector ICP-MS. The acidified seawater samples (sample volume = 12 mL, pH 1.7) were spiked with the multi-element standard of trace metal isotopes enriched over natural abundance and UV irradiated using a low power source (119 mV cm^{-2}, 254 nm) for at least one hour. The samples were left overnight for the isotopes to equilibrate and were buffered to pH 6.4 prior to loading onto the resin and elution with 1 M ultrapure nitric acid. Since trace metals such as Fe, Zn, Co, Cu, and Cd can be present as chelates in seawater, the authors tested UV irradiation (30 W; 0.5, 1, 1.5, and 3 h) and microwave treatment at 60°C in addition to the acidification of the samples. UV irradiation produced an increase in Co (50%) and Cu (10%) concentrations while the concentrations of the other metals remained unaffected. These changes were observed after the treatment of a SAFe (sampling and analysis of Fe) deep-water sample while no variation was observed in surface waters probably as a consequence of the lower concentration of the elements near the LOD of the method.

It seems that Cu and specially Co are strongly bound to organic matter, and the complexes can only be broken after UV irradiation, as previously proposed by several authors that observed the decrease in the recovery of Co (Vega and Van der Berg 1997) and Cu (Achterberg et al. 2001) in non UV-irradiated seawater samples. Batch and on-line digestion using a 400 W medium pressure Hg lamp and 15 mM H_2O_2 was effective to break Cu complexing organic ligands previously to preconcentration in a 8-hydroxyquinoline microcolumn and determination of the element by flow injection with chemiluminescence detection (FI-CL); the release of Cu from complexing organic matter was independent of the irradiation time (on line method: 5.6 min; batch method: 8 hours) (Achterberg et al. 2001). Another FI-CL method modified after Canizzaro et al. (2000) used an iminodiacetate Toyopearl AF chelate resin for the determination of dissolved Co in open ocean samples and was suitable for shipboard use (Shelley et al. 2010). The authors used the resin for on line preconcentration, ammonium acetate (pH 4.0) to condition the column, and UV irradiation of acidified water seawater samples to determine total dissolved Co. The accuracy of the method was evaluated by analysis of acidified North Pacific deep seawater samples from the SAFe program and the certified reference material NASS-5. The dissolved Co concentration measured in those samples was 38% lower without UV irradiation.

Biller and Bruland (2012) carried out an off-line concentration using the Nobias PA-1 EDTri-A-type chelating resin of Mn, Fe, Co, Ni, Cu, Zn, Cd, and Pb with subsequent analysis by magnetic sector ICP-MS. The authors based their method on Sohrin et al. (2008) from the international GEOTRACES research project.

The samples (40 mL) were UV oxidized for 1.5 hours (18 mW/cm^2 using a mercury lamp), and buffered to pH 6.2 with ammonium acetate. Afterwards, the preconcentration step was carried out. A manifold with eight columns was used to process eight samples simultaneously. These resin columns were rinsed with ammonium acetate buffer and eluted with 3–4 mL of 1 N nitric acid for a preconcentration of factor approximately 13. Co and Cu showed a significant increase in concentration after UV-oxidation, and the complete recovery of the elements and organics destruction took place after at least one hour of irradiation.

Identification of organic ligands

Hirose (2006) summarized some of the characteristics of the organic ligands binding bioactive metals, such as Fe or Cu. The strong organic ligands dissolved in seawater can be classified into three groups depending on the conditional stability constant of metal complexes and the concentration of the organic ligand. The L_1-type ligand is present in both particulate and DOM and originates from marine microorganisms. These organic ligands have several functional groups, including carboxylates. A review (Vraspir and Butler 2009) summarized the knowledge about the chemistry of marine ligands and siderophores. Marine microorganisms produce organic ligands to facilitate uptake of some metals or to mitigate their potential toxic effect. Few advances have been produced in the identification of ligands binding trace metals, with the exception of Fe. In this case, siderophores are low-molecular weight Fe-binding ligands produced by marine bacteria. Knowledge regarding the interaction of the biological processes with Fe chemistry is expected to increase due to the improvements of the methodologies in this field (shipboard incubation systems, highly sensitive electrochemical methods, x-ray spectroscopy with TEM microscopy, molecular techniques, genomics and proteomics, etc.) (Breitbarth et al. 2010). The use of radioisotopes is also useful to examine the exchange between chemical species at environmental relevant concentrations (Croot et al. 2011).

In addition to ferrioxamines and amphibactins, siderophore type chelates (Mawji et al. 2011) were detected in incubated seawater from the Atlantic Ocean enriched with glucose and ammonium, glycine (as a source of carbon and nitrogen) or chitin and ammonium at different concentrations. The detection was performed using high performance liquid chromatography coupled to inductively coupled plasma mass spectrometry (HPLC–ICP–MS) after complexation with Ga. Samples were subsequently analyzed by HPLC–electrospray ionisation mass spectrometry (HPLC–ESI–MS) to confirm the identity of species. A total of twenty two different siderophore type chelates were resolved in the HPLC–ICP–MS chromatograms, and ten different siderophore type chelates were identified by HPLC–ESI–MS. The concentration and diversity of siderophore type chelates was highest in seawater with added glucose, and also varied with the biogeographical area in the Atlantic Ocean. Detection of hidroxamate siderophores was reported in coastal and Sub-Antarctic waters (Velasquez et al. 2011). The siderophore activity was detected using the chrome azurol S assay from organic compounds extracted from 200 to 480 L of seawater by preconcentration on a XAD-resin. After isolation, the compounds were identified by HPLC-MS/MS using the natural Fe-isotope pattern.

Thuróczy et al. (2010) observed that the concentration of Fe in the dissolved fraction (< 0.2 μm) and in the fraction smaller than 1000 kDa followed a nutrient-type profile with depth: depleted at the surface by phytoplankton uptake, increased at 500–1000 m, and approximately constant below. Fe concentrations were measured by flow injection analysis and the organic Fe complexation by voltammetry. The high concentration of empty binding sites for Fe in the smallest fractions was explained due to Fe uptake but also by ligand production due to phytoplankton and bacteria.

Boiteau et al. (2013) described a method to detect low concentrations of Fe-binding ligands using HPLC-tandem multicollector ICP-MS. The authors achieved sensitive detection after combining a Fe free-HPLC system, a hexapole collision cell and introducing oxygen into the sample carrier gas. Fe ligand complexes were detected from the organic extract of surface South Pacific seawater and from culture media of the siderophore producing bacteria *Synechococcus* sp. PCC 7002. The separated components should be later isolated and characterized with complementary techniques such as ESI-MS.

Cottrell et al. (2014) recently identified Cu-binding ligands from seawater that may be potential photosensitizers. Cu immobilized metal affinity chromatography (IMAC) was used to fraction and enrich the seawater DOM, allowing identification by ^{1}H NMR of ninety-seven compounds as possible sources of the excited state species in the IMAC fractions. Sample volume is not limited in this type of chromatography and the ligands are isolated without the need for additional potentially interfering buffers.

Thompson et al. (2014) proposed that the source of organic Cu-binding ligands on the Tasman Sea were the *Synechococcus* spp., cyanobacteria, and eukaryotes (coccolithophores). The authors used on a combination of voltammetric speciation with phototrophic cell counts and biomarker pigment concentrations. Thompson and Ellwood (2014) also used solvent extraction coupled with anion-exchange and multi-collector ICP-MS to understand the dissolved Cu isotope biogeochemistry. They hypothesized that the breakdown of dissolved organic Cu complexes may provide a source of isotopically heavy Cu that could be removed from the oceanic system in certain environments such as anoxic basins.

Trends in Cd and Zn complexing ligands were studied in a near-pristine Irish estuary to identify their ligands in natural waters (Murray et al. 2014). The evidence suggested that the naturally occurring Cd ligands include fulvic acids, whereas the Zn ligands are likely to be exuded from seaweeds.

Metal-Protein Complexes

Most of the published studies deal with metal complexation by natural organic matter (NOM) in surface waters, mainly humic substances (HS) (Park 2009; Wu et al. 2004; Rottmann and Heumann 1994; Cabaniss 2011; Vogl and Heumann 1998; Schmitt et al. 2000). Results suggest that organic compounds of high molecular weight tend to be associated to metals with high binding strength; whereas, organic compounds of low molecular weight are mainly bound to metals exhibiting low strength (Wu et al. 2004).

Little is known about DOM and metal-DOM complexes in the marine environment. This is because the various types of organic substances exhibit very different molecular weights (MWs) in the marine DOM fraction (Hansell and Carlson 2001). As an example, amino acids and large biochemical substances such as sugars, lipids, and proteins from marine algae metabolism, microbial exudation, and cellular lysis, have been identified in surface seawater (Hansell and Carlson 2001). This class of substances has shown to be reactive, influencing toxicity and bioavailability of nutrients and toxic contaminants (Ogawa and Tanoue 2003; Mopper et al. 2007). Regarding dissolved proteins, discrete units have been found in the marine environment (Powell et al. 2005), and are probably the most characterizable component of marine DOM at the molecular level (Powell et al. 2005; Nunn and Timperman 2007).

Some developments have therefore been addressed to demonstrate the presence of discrete dissolved proteins in oceanic waters. Characterization studies have revealed the presence of outer membrane proteins (Tanoue et al. 1995; Yamada and Tanoue 2003) but, more recently, the presence of intracellular enzymes from marine plankton has also been demonstrated (Powell et al. 2005; Yamada and Tanoue 2006).

Sample pretreatment

The assessment of presence of marine DOM and, especially marine proteins, is difficult mainly due to their low concentration and the presence of large amount of inorganic salts. A sample pretreatment implying DOM (protein) pre-concentration as well as salts removal is therefore needed. In addition, the sample pretreatment method must be suitable for coping with large volumes of sample, typically 50 L.

Solid phase extraction

DOM sorption by functionalized solid phases (solid phase extraction, SPE) has been reported as a sample pretreatment (Mopper et al. 2007). However, low DOM recoveries when using certain adsorbents such as XAD resins have been reported. The degree of DOM adsorption is also dependent on the size of DOM, with low molecular weight (LMW) compounds exhibiting the highest analytical recoveries (Sánchez-González et al. 2012). In addition, preparation and purification of functionalized solid phases for DOM extraction are time-consuming, and their use in DOM extraction requires greater control in matrix pH, salinity, and polarity. Other problems arise due to the need for adjusting sample pH at low values (protonation of carboxyl groups on DOM solutes) for enhancing their sorption onto the hydrophobic surfaces, and the subsequent elution by a strong base (0.1 M NaOH, 1 M NH_4OH, or an alkaline methanol mixture). In addition to artifacts resulting from the use of these eluting solutions, the methodology is not adequate for guaranteeing the metal-DOM (metal-protein) complexes' stability (Sánchez-González et al. 2012).

Tangential flow ultrafiltration

Tangential flow ultrafiltration (TUF) is the recommended technique for isolating and fractionating marine DOM by size, yielding LMW and high molecular weight (HMW) fractions (Mopper et al. 2007; Nunn and Timperman 2007; Sánchez-González et al. 2012; García-Otero et al. 2013a). In TUF, the sample is tangentially passed across a filter membrane (typically polyethersulfone, nominal molecular weight cut-off higher than 1 kDa) at positive pressure relative to the permeate side. A proportion of the material which is smaller than the membrane pore size passes through the membrane as permeate or filtrate; whereas, substances that are retained on the feed side of the membrane are pre-concentrated as retentate. Compared to the conventional pretreatment systems, TUF is advantageous because it operates in full-automatic operation mode, and also offers a smaller footprint and less chemical dosage (Xu et al. 2012). However, the most challenging problem affecting with TUF performance is membrane fouling, which implies loss of DOM due to increased sorption onto membrane surfaces. In addition, DOM loss can also be attributed to hydrophobic organic-organic interactions with the TUF system plumbing (Mopper et al. 2007). Losses of marine DOM by adsorption onto the UF membranes were within the 11–16% range (Minor et al. 2002), and as reported by Powell and Timperman, this phenomena is especially important for dissolved proteins because of their highly adsorptive nature (Powell and Timperman 2005). Regarding protein isolation, sodium dodecyl sulphate (SDS) is highly recommended to prevent protein adsorption onto the UF membrane (Powell and Timperman 2005). Typical SDS concentrations of 0.01% (m/v) have been reported (Powell and Timperman 2005; Yamada and Tanoue 2003; Saijo and Tanoue 2004; Yamada and Tanoue 2006; Yamada and Tanoue 2009; Yamashita and Tanoue 2004; García-Otero et al. 2013b).

Retentate (volumes ranging from 400 to 600 mL) after TUF can be directly subjected to desalting and/or protein precipitation. However, a further pre-concentration stage by centrifugal ultrafiltration (ultrafiltration tubes with polyethersulfone membrane of 10 kDa molecular cut-off) can be required in some cases, especially when the concentrations of protein are very low (García-Otero et al. 2013b). After successive centrifugal ultrafiltration (4000 rpm, 10–30 min), the retentate can be reduced up to 20 mL, leading to pre-concentration factors up to 3000 (García-Otero et al. 2013b).

Desalting: Protein isolation

Desalting by preparative size-exclusion chromatography (SEC) has commonly been reported when assessing marine DOM. Substances with the largest sizes migrate through the SEC column at the greatest rate, followed by solutes with progressively decreasing sizes (Mopper et al. 2007). Based on several SEC applications when separating DOM solutes (either whole seawater or HMW isolates) (Minor et al. 2002; Dittmar and Kattner 2003), commercial Hi Trap Desalting minicolumns can be used for retentate desalting purposes (García-Otero et al. 2013a). A fractionation range between 1 and 5 kDa (exclusion limit of approximately 5 kDa) ensures the separation of bio-molecules exhibiting a MW larger than 5 kDa from those molecules with MW less than 1 kDa. Under UV monitoring and using buffered mobile phases (a 25 mM/25 mM ammonium sulphate/diammonium hydrogen phosphate buffer solution at pH 6.5 (García-Otero et al. 2013a)), the excluded fraction can be collected for further studies.

Protein precipitation from the retentate fraction implies a desalting process itself since the protein pellet can be separated from the liquid phase containing salts. However, the need for water soluble interferences (salts) removal when precipitating the isolate protein has been reported when using certain electrophoresis techniques for protein separation. Therefore, a protein precipitation method by using methanol/water/

chloroform (4:3:1) has been proposed for water and methanol soluble interferences removal (chloroform layer containing the isolated protein) before protein precipitation by cold methanol (–20°C) (García-Otero et al. 2013b). This procedure is more efficient than conventional protein precipitation procedures such as those based on ice-cold acetone (protein pellet formation at –20°C for 2 h) (Wang et al. 2006).

Analytical techniques

High performance liquid chromatography

Size exclusion chromatography (SEC) has been widely used for DOM characterization on the basis of the apparent molecular weight (Allpike et al. 2005; Her et al. 2008). Developments using UV, fluorescence, and DOC detection have been typically reported when assessing NOM (Her et al. 2002; Her et al. 2003) and also marine DOM (García-Otero et al. 2013a; García-Otero et al. 2013c). However, SEC presents some drawbacks in determining the molecular size distribution of DOM. First, SEC column calibration is difficult due to the lack of knowledge of the DOM nature which is being separated (globular and non-globular nature) and the use of globular or non-globular proteins for column calibration (Kostanski et al. 2004). In addition, other non-proteic standards would be preferred when separating non-proteic DOM. In these cases, a mixture of polyethylene oxides (PEOs) and polyethylene glycols (PEGs) (Sánchez-González et al. 2012), and also polystyrene sulfonates (PSSs) (Her et al. 2002) standards has been proposed. However, chromatographic conditions for PEOs/PEGs standards elution are not the same as those needed for marine DOM elution (García-Otero et al. 2013a), which requires the use of proteins as standards for SEC column calibration when separating non-proteic HMW compounds. Regarding PSSs, the strong interaction of PSSs with the SEC column in the presence of sodium sulphate (typically used in mobile phases) is the main reason for the different eluting characteristics than those used for globular or non-globular proteins as standards (Her et al. 2002). Other problems arise from non-ideal interactions between DOM and the SEC resin, and as explained by Specht and Frimmel (2000), hydrophobic interactions appear to prevail at high ionic strength; whereas, electrostatic repulsion prevails at low ionic strength values. In addition, artifacts caused by differences between the ionic strength of the samples and the mobile phase (salt boundary peak produced by a gradient in the ionic strength between the sample and the mobile phase) have been reported (Schmitt et al. 2000; Garcia-Otero et al. 2013a). Finally, UV detectors offer poor sensitivity (Allpike et al. 2005; Her et al. 2002; Her et al. 2003), although fluorescence spectrometry has been shown to offer satisfactory results for dissolved humic substances and protein-like compounds (Her et al. 2003; Baker 2001; Akagi et al. 2007; Lu et al. 2009).

On other occasions, strategies based on two-dimensional chromatography have been proposed for characterizing marine DOM (García-Otero et al. 2013a; García-Otero et al. 2013c). SEC and anion exchange chromatography (AEC) have been proposed. The SEC fraction encompassing various MWs was collected and further separated by AEC leading to four different groups of substances (García-Otero et al. 2013a; García-Otero et al. 2013c). Two-dimensional chromatography has also proved adequate for elucidating artifacts attributed to salt boundary peaks by the gradient in the ionic strength between the sample and the mobile phase (García-Otero et al. 2013a). Other two-dimensional HPLC (strong cation exchange, SCX; and reverse phase chromatography, RPC) with tandem mass spectrometry (MS/MS) for detection has also been described for assessing dissolved proteins in seawater (Powell et al. 2005).

Finally, few works have focused on assessing metal-DOM complexes. Most of the developments were performed for metal-NOM complexes in lake water (Schmitt et al. 2000; Wu et al. 2004; Rottmann and Heumann 1994), in bog and river water (Rottmann and Heumann 1994), and in sewage samples, groundwater, brown water, and seepage water samples from soil (Vogl and Heumann 1998) by SEC-ICP-MS. Recently, SEC/AEC-ICP-MS developments have been proposed for assessing metal-marine DOM complexes in estuarine water (García-Otero et al. 2013a; García-Otero et al. 2013c). Table 5 summarizes molecular weights (MWs) of the assessed DOM as well as metals (and concentrations) reported.

Table 5: Molecular size distribution characteristics of the metal–DOM complexes in seawater.

Sample pretreatment	Separation technique	Elemental detection	Organic matter type	Molecular weight/isoelectric point	Metals	Reference
TUF (10 kDa cut-off) followed by freeze drying and Hi Trap desalting. Preconcentration factor of 400	SEC-AEC	ICP-MS (on-line)	Marine DOM	1.7–16 kDa	Co, Mn, Sr and Zn	García-Otero et al. 2013a
TUF (3.0 and 10 kDa cut-off) followed by Hi Trap desalting. Preconcentration factor of 200	SEC-AEC	ICP-MS (on-line)	Marine DOM	1.6–21.6 kDa	Cu, Mn, Mo, Sr and Zn	García-Otero et al. 2013c
TUF (10 kDa cut-off) followed by centrifugal ultrafiltration (10 kDa cut-off) and protein precipitation. Preconcentration factor of 3000	Offgel electrophoresis	ETAAS, ICP-OES (off-line)	Dissolved proteins	Surface seawater: 15–63 kDa (pIs from 4.82 to 8.37) Deep seawater: 21–24 kDa (pIs from 3.30 to 4.22)	Cd, Cu, Fe, Mn, Ni and Zn Cu and Mn	García-Otero et al. 2013b
TUF (10 kDa cut-off) followed by centrifugal ultrafiltration (10 kDa cut-off) and protein precipitation. Preconcentration factor of 3000	2DE	LA-ICP-MS (off-line)	Dissolved proteins	10 kDa (pI 5.8) 13 kDa (pI 5.8) 14 kDa (pI 6.3) 13 kDa (pI 7.0) 12 kDa (pI 7.3)	Cd, Cr, and Zn Zn Cd, Cu, Cr and Zn Cd, Cu, Cr and Zn Cd	García-Otero et al. 2013d
Ultrafiltration (1 kDa cut-off). Preconcentration factor of 26	FFFF	ICP-MS (on-line)	CDOM-colloids protein-like colloids Fe-rich colloids	0.5–4 nm 3–8 nm >40 nm 5–40 nm	Fe, Mn, Cu, Zn, Pb, and U P Fe, P, Mn and Pb P, Mn, Zn, and Pb	Stolpe et al. 2010

AEC–anion exchange chromatography; CDOM–chromophoric dissolved organic matter; DOM–dissolved organic matter; ETAAS–electrothermal atomic absorption spectrometry; FFFF–flow field flow fractionation; ICP-OES–inductively coupled plasma–optical emission spectrometry; ICP-MS–inductively coupled plasma–mass spectrometry; LA–laser ablation; SEC–size exclusion chromatography; TUF–tangential ultrafiltration.

Electrophoresis

One-dimensional sodium dodecyl sulphate polyacrylamide gel electrophoresis (SDS-PAGE) (Yamada and Tanoue 2003; Powell and Timperman 2005; Suzuki et al. 1997), and two-dimensional electrophoresis (2DE) (Yamada and Tanoue 2006; Saijo and Tanoue 2004; Yamada and Tanoue 2009) have mainly been used for characterizing dissolved proteins. In addition, other two-dimensional approaches such as those based on the combination of immobilized pH gradient isoelectric focusing (IPG IEF) for the first dimensional separation according to the proteins' isoelectric point (pIs) and RPC (second dimensional separation according to the proteins' MWs) have also been developed (Cargile et al. 2002; Essader et al. 2005). These procedures were performed under denaturing and reducing conditions for assessing/characterizing marine dissolved proteins. Denaturing or non-denaturing SDS-PAGE conditions, as well as applied current and post-separation gel treatment, have been reported to affect the stability of metal-protein complexes. In addition, gel staining can also alter the stability of the metal–protein complexes and prevent detection of metals bound to proteins (Jiménez et al. 2009; Jiménez et al. 2010a). The use of non-reducing SDS-PAGE (nrSDS-PAGE) is therefore preferred for guaranteeing the stability of the metal-protein complexes. Some recent developments imply the use of thiol-free reducing agents, such as tris(2-carboxyethyl) phosphine (TCEP) for nrSDS-PAGE when separating oxaliplatin metalloproteins (Mena et al. 2013). On other occasions, offgel electrophoresis (preparative IPG IEF where proteins can be recovered from the liquid phase) (Hörth et al. 2006) has been reported to ensure stability of metal-protein complexes (García-Otero et al. 2013b; Moreda-Piñeiro et al. 2014).

SDS-PAGE, IEF, and 2DE as separation techniques and laser ablation (LA)–ICP-MS as detection system, was applied to study the distribution of metal–humic acids (metal–HA) complexes in compost (Jiménez et al. 2010b). This methodology, mainly 2DE, is advantageous when assessing metal-protein complexes because metals can be assessed by direct LA–ICP-MS analysis of the dehydrated gels. 2DE-LA-ICP-MS has therefore been applied for assessing metal bound to dissolved proteins and proteins from marine plankton (García-Otero et al. 2013d). Results showed the presence of Cd, Cr, Cu, and Zn in five protein spots (proteins exhibiting MWs within the 10–14 kDa range and pIs from 5.8 to 7.3) as listed in Table 5. Other approaches have implied the direct analysis of offgel fractions for elucidating metals bound to proteins of similar pIs (García-Otero et al. 2013b). Due to the less drastic conditions required for offgel electrophoresis, the stability of the metal-protein complexes isolated from seawater was greater than when using 2DE (García-Otero et al. 2013d), and the presence of Fe, Mn, and Ni, together with Cu and Zn was verified after electrothermal atomic absorption spectrometry (ETAAS) and inductively coupled plasma—optical emission spectrometry (ICP-OES) in the offgel fraction containing proteins (the presence of proteins, and also the concentration, was verified by lab-on-a-chip electrophoresis) (García-Otero et al. 2013b). Similarly to the use of nrSDS-PAGE for guaranteeing the stability of the metal-protein complexes, the careful selection of offgel solutions can minimize the metal-protein complex breakdown. Recent developments propose the use of denaturing solutions without thiourea and without dithiothreitol (DTT) for guaranteeing the integrity of platinum-binding proteins during offgel-IEF (Mena et al. 2011).

Field flow fractionation

Field flow fractionation (FFF) is a hydrodynamic fractionation-based technique suitable for analyzing complex samples, from particles in the range of micrometers to macromolecules down to a few kDa (Laborda et al. 2011). The high resolution offered by FFF techniques together with the allowed size fractionation ranges and the possibility of hyphenation with ICP-MS, makes FFF a powerful technique for characterizing a large number of materials (Dubascoux et al. 2010).

Regarding environmental samples, recent developments by asymmetrical flow field flow fractionation (AsFlFFF) have focused on assessing NOM from leachates from leaf litter (Cuss and Guéguen 2012; Cuss and Guéguen 2015) and compost (Bolea et al. 2010; Laborda et al. 2011), and also NOM in surface waters (Cuss and Guéguen 2015; Alasonati et al. 2010; Guéguen and Cuss 2011; Pifer et al. 2011; Jirsa et al. 2013; Balch and Guéguen 2013; Guégen et al. 2013; Luan and Vadas 2015; Kuhn et al. 2015). Applications based on flow field flow fractionation (FFFF) have also been performed for NOM in river waters (Stolpe et al. 2014) and peat bog drainage (Neubauer et al. 2011).

UV (Cuss and Guéguen 2012; Cuss and Guéguen 2015; Bolea et al. 2010; Laborda et al. 2011; Alasonati et al. 2010; Guéguen and Cuss 2011; Jirsa et al. 2013; Balch and Guéguen 2013; Neubauer et al. 2011; Luan and Vadas 2015; Kuhn et al. 2015) and fluorescence (Cuss and Guéguen 2012; Cuss and Guéguen 2015; Guéguen and Cuss 2011; Guégen et al. 2013; Jirsa et al. 2013; Pifer et al. 2011; Stolpe et al. 2014) spectrometry have been used as detection systems. On other occasions, chemometric tools such as principal component analysis (PCA) (Guéguen and Cuss 2011) and parallel factor analysis (PARAFAC) (Cuss and Guéguen 2012; Cuss and Guéguen 2015; Guégen et al. 2013; Pifer et al. 2011) have been applied in the fluorescence excitation–emission matrix (fluorescence-EEMs) decomposition and interpretation. AsFlFFF hyphenated with ICP-MS has been used for assessing metal associations to microparticles, nanocolloids, and macromolecules in compost leachates (Bolea et al. 2010; Laborda et al. 2011), and also in peat bog runoff (Jirsa et al. 2013), wastewater treatment plant effluents, stormwater runoff (Luan and Vadas 2015), and river waters (Luan and Vadas 2015; Kuhn et al. 2015). Similarly, FFFF-ICP-MS has also been used for characterizing uranium-humic complexes (standard solutions) (Lesher et al. 2013), trace metals associated to NOM in wetland runoff from peat bog (Neubauer et al. 2013), and in river water (Stolpe et al. 2010; Stolpe et al. 2013a; Stolpe et al. 2013b).

Regarding marine DOM and trace metal-binding marine DOM, developments based on FFFF with UV and fluorescence detection and ICP-MS have been applied for sizing of colloidal organic matter (chromophoric dissolved organic matter (CDOM), humic-type and protein-type fluorescent organic matter) in seawater (Gulf of Mexico) (Stolpe et al. 2010). As listed in Table 5, Fe, Mn, Cu, Zn, Pb, and U are the main trace metals associated with CDOM, while P is mainly associated to protein-like colloids.

Conclusion

Advances in chemical speciation of trace elements in seawater have been reviewed in this chapter. A great advance has been seen in quantifying chemical species and identifying organic ligands binding bioactive metals and protein-metal complexes. These improvements have been possible because of the combination of different techniques such as, sensitive CLE-CSV, incubation methods, HPLC, non reducing SDS-PAGE, OFFGEL electrophoresis, FFF, ICP-MS, and ESI-MS/MS.

Acknowledgements

The authors wish to thank the *Xunta de Galicia* (Grupo de Referencia Competitiva 6RC2014/2016) and *Ministerio de Economía y Competitividad* (Project number CTQ2012-38901-C02-02) for financial support.

Keywords: Trace elements, speciation, metalloproteins, dissolved organic matter, separation techniques, preconcentration methods, inductively coupled plasma-mass spectrometry, electrochemical methods, electrospray ionization-mass spectrometry

References

Abollino, O., M. Aceto, C. Sarzanini and E. Mentasti. 2000. The retention of metal species by different solid sorbents. Mechanisms for heavy metal speciation by sequential three column uptake. Anal. Chim. Acta 411: 223–237.

Abualhaija, M.M. and C.M.G. Van den Berg. 2014. Chemical speciation of iron in seawater using catalytic cathodic stripping voltammetry with ligand competition against salicylaldoxime. Mar. Chem. 164: 60–74.

Achterberg, E. and C.M.G. Van Den Berg. 1997. Chemical speciation of chromium and nickel in the western Mediterranean. Deep-Sea Res. II 44: 693–720.

Achterberg, E.P., C.B. Braungardt, R.C. Sandford and P.J. Worsfold. 2001. UV digestion of seawater samples prior to the determination of copper using flow injection with chemiluminescence detection. Anal. Chim. Acta 440: 27–36.

Akagi, J., A. Zsolnay and F. Bastida. 2007. Quantity and spectroscopic properties of soil dissolved organic matter (DOM) as a function of soil sample treatments: air drying and pre-incubation. Chemosphere 69: 1040–1046.

Alasonati, E., V.I. Slaveykova, H. Gallard, J.P. Croué and M.F. Benedetti. 2010. Characterization of the colloidal organic matter from the Amazonian basin by asymmetrical flow field-flow fractionation and size exclusion chromatography. Water Res. 44: 223–231.

Allpike, B.P., A. Heitz, C.A. Joll, R.I. Kagi, G. Abbt-Braun, F.H. Frimmel, T. Brinkmann, N. Her and G. Amy. 2005. Size exclusion chromatography to characterize DOC removal in drinking water treatment. Environ. Sci. Technol. 39: 2334–2342.

Baars, O. and P.L. Croot. 2011. The speciation of dissolved zinc in the Atlantic sector of the Southern Ocean. Deep-Sea Res. II 58: 2720–2732.

Baker, A. 2001. Fluorescence excitation–emission matrix characterization of some sewage-impacted rivers. Environ. Sci. Technol. 35: 948–953.

Balch, J. and C. Guéguen. 2015. Effects of molecular weight on the diffusion coefficient of aquatic dissolved organic matter and humic substances. Chemosphere 119: 498–503.

Beceiro González, E., A. Guimaraes and M.F. Alpendurad. 2009. Optimization of a headspace-solid phase microextraction method for simultaneous determination of organometallic compounds of mercury, lead and tin in water by gas chromatography-tandem mass spectrometry. Journal of Chromatography A 1216: 5563–69.

Biller, D.V. and K.W. Bruland. 2012. Analysis of Mn, Fe, Co, Ni, Cu, Zn, Cd and Pb in seawater using the Nobias-chelate PA1 resin and magnetic sector inductively coupled plasma mass spectrometry (ICP-MS). Mar. Chem. 130-131: 12–20.

Boiteau, R.M., J.N. Fitzsimmons, D.J. Repeta and E.A. Boyle. 2013. Detection of iron ligands in seawater and marine cyanobacteria cultures by high performance liquid chromatography-inductively coupled plasma-mass spectrometry. Anal. Chem. 85: 4357–4362.

Bolea, E., F. Laborda and J.R. Castillo. 2010. Metal associations to microparticles, nanocolloids and macromolecules in compost leachates: Size characterization by asymmetrical flow field-flow fractionation coupled to ICP-MS. Anal. Chim. Acta 661: 206–214.

Bravo Sánchez, L.R., J. Ruiz Enzinar, J.I. Fidalgo Martínez and A. Sanz Medel. 2004. Mercury speciation analysis in sea water by solid phase microextraction-gas chromatography coupled plasma mass spectrometry using ethyl and propyl derivatization. Matrix effects evaluation. Spectrochim. Acta Part B. Atom. Spectrom. 59: 59–66.

Breitbarth, E., E.P. Achterberg, M.V. Ardelan, A.R. Baker, E. Bucciarelli, F. Chever, P.L. Croot, S. Duggen, M. Gledhill, M. Hassellöv, C. Hassler, L.J. Hoffmann, K.A. Hunter, D.A. Hutchins, J. Ingri, T. Jickells, M.C. Lohan, M.C. Nielsdóttir, G. Sarthou, V. Schoemann, J.M. Trapp, D.R. Turner and Y. Ye. 2010. Iron biogeochemistry across marine systems–progress from the past decade. Biogeosci. 7: 1075–1097.

Bruland, K.W. 1989. Complexation of zinc by natural organic ligands in the Central North Pacific. Limnol. Oceanogr. 34: 269–285.

Bruland, K.W. 1992. Complexation of cadmium by natural organic ligands in the Central North Pacific. Limnol. Oceanogr. 37: 1008–1017.

Buck, K.N. and K.W. Bruland. 2005. Copper speciation in San Francisco Bay: a novel approach using multiple analytical windows. Mar. Chem. 96: 185–198.

Buck, K.N. and K.W. Bruland. 2007. The physicochemical speciation of dissolved iron in the Bering Sea, Alaska. Limnol. Oceanogr. 52: 1800–1808.

Byrne, R.H. 2002. Speciation in seawater. pp. 322–357. *In*: A.M. Ure and C.M. Davidson [eds.]. Chemical Speciation in the Environment. Blackwell Science, London, UK.

Cabaniss, S.E. 2011. Forward modeling of metal complexation by NOM: II. Prediction of binding site properties. Environ. Sci. Technol. 45: 3202–3209.

Cabon, J.Y. and W. Erler. 1998. Determination of selenium species in seawater by flow injection hydride generation *in situ* trapping followed by electrothermal atomic absorption spectrometry. Analyst. 123: 1565–1569.

Cabon, J.Y. and N. Cabon. 2000. Determination of arsenic species in seawater by flow injection hydride generation *in situ* collection followed by graphite furnace atomic absoprtion spectrometry stability of As (III). Anal. Chim. Acta 418: 19–31.

Cabon, J.Y. and C.L. Madec. 2004. Determination of major antimony species in seawater by continuous flow injection hydride generation atomic absorption spectrometry. Anal. Chim. Acta 504: 209–215.

Calvo Fornieles, A., A. García de Torres, E. Vereda Alonso, M.T. Silles Cordero and J.M. Cano Pavón. 2011. Speciation of antimony (III) and antimony (V) in seawater by flow injection and solid phase extraction coupled with online hydride generation inductively coupled plasma mass spectrometry. J. Anal. At. Spectrom. 59: 1619–1626.

Cairns, W.R.L., M. Ranaldo, R. Hennebelle, C. Turretta, G. Capodaglio, C.F. Ferrari, A. Dommergue, P. Cescon and C. Barbante. 2008. Speciation analysis of mercury in seawater from the lagoon of Venice by on-line pre-concentration HPLC-ICP-MS. Anal. Chim. Acta 622: 62–69.

Canizzaro, V., A.R. Bowie, A. Sax, E.P. Achterberg and P.J. Worsfold. 2000. Determination of cobalt and iron in estuarine and coastal waters using flow injection with chemiluminescence detection. Analyst 125: 51–57.

Capodaglio, G. 1990. Lead speciation in surface waters of the Eastern North Pacific. Mar. Chem. 29: 221–233.

Cargile, B.J., D.L. Talley and J.L. Jr. Stephenson. 2004. Immobilized pH gradients as a first dimension in shotgun proteomics and analysis of the accuracy of pI predictability of peptides. Electrophoresis 25: 936–945.

Carro, A.M., I. Neira, R. Rodil and R.A. Lorenzo. 2002. Speciation of Mercury compounds by gas chromatography with atomic emission detection. Simultaneous optimization of a headspace solid-phase microextraction and derivatization procedure by use of chemometric techniques. Chromatographia 56: 733–738.

Carvalho, L.M., P.C. Nascimento, D. Bohrer, R. Stefanello, E.J. Pilau and M.B. Rosa. 2008. Redox speciation of inorganic arsenic in water and saline samples by adsorptive cathodic stripping voltammetry in the presence of sodium diethyl dithiocarbamate. Electroanalysis 20(7): 776–781.

Castillo, A., A.F. Roig-Navarro and O.J. Pozo. 2006. Method optimization for the determination of four mercury species by micro-liquid chromatography-inductively coupled plasma mass spectrometry coupling in environmental samples. Anal. Chim. Acta 577: 18–25.

Centieno, G., E. Blanco-González and A. Sanz-Medel. 2004. Multielemental speciation analysis of organometallic compounds of mercury, lead and tin in natural water samples by headspace–solid phase microextraction followed by gas chromatography-mass spectrometry. J. Chromatography A 1034: 191–197.

Centineo, G., P. Rodríguez-González, E. Blanco Gonzalez, J.I. García Alonso, A. Sanz Medel, N. Font Cardama, J.L. Aranda Mares and S. Ballester Nebot. 2006. Isotope dilution GC-MS routine method for the determination of butyltin compounds in water. Anal. Bioanal. Chem. 384: 906–914.

Chen, Z., M. Megharaj and R. Naidu. 2007. Speciation of iodate and iodide in seawater by non-suppressed ion chromatography with inductively coupled plasma mass spectrometry. Talanta 72: 1842–1846.

Coale, K.J. and K.W. Bruland. 1988. Copper complexation in the Northeast Pacific. Limnol. Oceanogr. 33: 1084–1101.

Cottrell, B.A., M. Gonsior, S.A. Timko, A.J. Simpson, W.J. Cooper and W. Van der Veer. 2014. Photochemistry of marine and fresh waters: A role for copper-dissolved organic matter ligands. Mar. Chem. 162: 77–88.

Croot, P.L., M.I. Heller, C. Schlosser and K. Wuttig. 2011. Utilizing radioisotopes for trace metal speciation measurements in seawater. *In*: N. Singh [ed.]. Radioisotopes—Applications in Physical Sciences. InTech. http://cdn.intechopen.com/pdfs-wm/21654.pdf. Last visited: 30/03/2015.

Cuss, C.W. and C. Guéguen. 2012. Determination of relative molecular weights of fluorescent components in dissolved organic matter using asymmetrical flow field-flow fractionation and parallel factor analysis. Anal. Chim. Acta 733: 98–102.

Cuss, C.W. and C. Guéguen. 2015. Relationships between molecular weight and fluorescence properties for size-fractionated dissolved organic matter from fresh and aged sources. Water Res. 68: 487–497.

Dakova, I., I. Karadjova, V. Georgieva and G. Georgiev. 2009. Ion-imprinted polymethacrylic microbeads as a new sorbent for preconcentration and speciation of mercury. Talanta 78: 523–529.

Dittmar, T. and G. Kattner. 2003. Recalcitrant dissolved organic matter in the ocean: major contribution of small amphiphilics. Mar. Chem. 82: 115–123.

Dubascoux, S., I. Le Hécho, M. Hassellöv, F. Von Der Kammer, M. Potin Gautier and G. Lespes. 2010. Field-flow fractionation and inductively coupled plasma mass spectrometer coupling: History, development and applications. J. Anal. At. Spectrom. 25: 613–623.

Essader, A.S., B.J. Cargile, J.L. Bundy and J.L. Stephenson, Jr. 2005. A comparison of immobilized pH gradient isoelectric focusing and strong-cation-exchange chromatography as a first dimension in shotgun proteomics. Proteomics 5: 24–34.

Fragueiro, S., I. Lavilla and C. Bendicho. 2006. Hydride generation-headspace single-drop microextraction-electrothermal atomic absorption spectrometry method for determination of selenium in waters after photoassisted prereduction. Talanta 68: 1096–1101.

García-Otero, N., P. Bermejo-Barrera and A. Moreda-Piñeiro. 2013a. Size exclusion and anion exchange high performance liquid chromatography for characterizing metals bound to marine dissolved organic matter. Anal. Chim. Acta 760: 83–92.

García-Otero, N., E. Peña-Vázquez, M.C. Barciela-Alonso, P. Bermejo-Barrera and A. Moreda-Piñeiro. 2013b. Two-dimensional isoelectric focusing OFFGEL and microfluidic labon-chip electrophoresis for assessing dissolved proteins in seawater. Anal. Chem. 85: 5909–5916.

García-Otero, N., P. Herbello-Hermelo, R. Domínguez-Gonzalez, M. Aboal-Somoza, A. Moreda Piñeiro and P. Bermejo-Barrera. 2013c. Evaluation of tangential flow ultrafiltration procedures to assess trace metals bound to marine dissolved organic matter. Microchem. J. 110: 501–509.

García-Otero, N., M.C. Barciela-Alonso, P. Bermejo-Barrera, A. Moreda-Piñeiro, M.S. Jiménez, M.T. Gómez and J.R. Castillo. 2013d. Assessment of metals bound to marine plankton proteins and to dissolved proteins in seawater. Anal. Chim. Acta 804: 59–65.

Gómez-Ariza, J.L., D. Sánchez-Rodas, I. Giráldez and E. Morales. 2000. A comparison between ICP-MS and AFS detection for arsenic speciation in environmental samples. Talanta 51: 257–268.

Gregori, I.D., W. Quiroz, H. Pinochet, F. Pannier and M. Potin-Gautier. 2005. Simultaneous speciation of Sb(III), Sb(V) and $(CH_3)_3SbCl_2$ by high performance liquid chromatography-hydride generation-atomic fluorescence spectrometry detection (HPLC-HG-AFS): Application to antimony speciation in seawater. J. Chromatogr. A 1091: 94–101.

Guéguen, C. and C.W. Cuss. 2011. Characterization of aquatic dissolved organic matter by asymmetrical flow field-flow fractionation coupled to UV–Visible diode array and excitation emission matrix fluorescence. J. Chromatogr. A 1218: 4188–4198.

Guéguen, C., C.W. Cuss and W. Chen. 2013. Asymmetrical flow field-flow fractionation and excitation-emission matrix spectroscopy combined with parallel factor analyses of riverine dissolved organic matter isolated by tangential flow ultrafiltration. Int. J. Environ. Anal. Chem. 93: 1428–1440.

Hansell, D.A. and C.A. Carlson. 2011. Marine dissolved organic matter and the carbon cycle. Oceanography 14: 41–49.

Hassler, C.S., F.E. Legiret and E.C.V. Butler. 2013. Measurement of iron chemical speciation in seawater at 4°C: The use of competitive ligand exchange-adsorptive cathodic stripping voltammetry. Mar. Chem. 149: 63–73.

Her, N., G. Amy, D. Foss, J. Cho, Y. Yoon and P. Kosenka. 2002. Optimization of method for detecting and characterizing NOM by HPLC–size exclusion chromatography with UV and on-line DOC detection. Environ. Sci. Technol. 36: 1069–1076.

Her, N., G. Amy, D. McKnight, J. Sohn and Y. Yoon. 2003. Characterization of DOM as a function of MW by fluorescence EEM and HPLC–SEC using UVA, DOC, and fluorescence detection. Water Res. 37: 4295–4303.

Her, N., G. Amy, J. Chung, J. Yoon and Y. Yoon. 2008. Characterizing dissolved organic matter and evaluating associated nanofiltration membrane fouling. Chemosphere 70: 495–502.

Herbello-Hermelo, P., M.C. Barciela Alonso, A. Bermejo Barrera and P. Bermejo Barrera. 2005. Flow on-line sorption in a knotted reactor coupled with electrothermal atomic absorption spectrometry for selective As(III) determination in seawater samples. J. Anal. Atom. Spectrom. 20: 662–664.

Herbello-Hermelo, P., M.C. Barciela Alonso and P. Bermejo Barrera. 2011. Cr(VI) determination in seawater samples using an on-line sorption preconcentration in a knotted reactor coupled with electrothermal atomic absorption spectrometry. Atom. Spectroscopy 32: 27–33.

Hirata, S., K. Honda, O. Shikino, N. Maekawa and M. Aihara. 2000. Determination of chromium (III) and total chromium in seawater by on-line column preconcentration inductively coupled plasma mass spectrometry. Spectrochim. Acta 55: 1089–1099.

Hirose, H. 2006. The chemical speciation of trace metals in seawater: a review. Anal. Sci. 22: 1055–1063.

Hörth, P., C.A. Miller, T. Preckel and C. Wenz. 2006. Efficient fractionation and improved protein identification by peptide OFFGEL electrophoresis. Mol. Cell. Proteomics 5: 1968–1974.

Huang, C. and B. Hu. 2008. Speciation of inorganic tellurium from seawater by ICP-MS following magnetic SPE separation and preconcentration. J. Sep. Sci. 31: 760–767.

Huang, Z., K. Ito, A.R. Timerbaev and T. Hirokawa. 2004a. Speciation studies by capillary electrophoresis-simultaneous determination of iodide and iodate in seawater. Anal. Bioanal. Chem. 378: 1836–1841.

Huang, Z., K. Ito and T. Hirokawa. 2004b. Further research on iodine speciation in seawater by capillary zone electrophoresis with isotachophoresis preconcentration. J. Chromatography A 1055: 229–234.

Hurst, M.P. and K.W. Bruland. 2008. The effects of the San Francisco Bay Plume on trace metal and nutrient distributions in the Gulf of the Farallones. Geochim. Cosmochim. Acta 72: 395–411.

Gledhill, M., C.M.G. Van den Berg, R.F. Nolting and K.R. Timmermans. 1998. Variability in the speciation of iron in the northern North Sea. Mar. Chem. 59: 283–300.

Ikonomou, M.G., M.P. Fernandez, T. He and D. Cullon. 2002. Gas chromatography-high-resolution mass spectrometry based method for simultaneous determination of nine organotin compounds in water, sediment and tissue. J. Chromatogr. A 975: 319–333.

Jacobsen, J.A., F. Stuer-Lauridsen and G. Pritzl. 1997. Organotin speciation in environmental samples by capillary gas chromatography and pulsed flame photometric detection (PFPD). Applied Organometallic Chemistry 11: 737–741.

Jakuba, W., J.W. Moffett and M.A. Saito. 2008. Use of a modified, high-sensitivity, anodic stripping voltammetry method for determination of zinc speciation in the North Atlantic Ocean. Anal. Chim. Acta 614: 143–152.

Jian, G.B., J. Liu and K.W. Yan. 2000. Speciation analysis of butyltin compounds in chinese seawater by capillary gas chromatography with flame photometric detection using *in-situ* hydride derivatization followed by head space solid phase microextraction. Anal. Chim. Acta 421: 67–74.

Jiann, K.T. and B.J. Presley. 2002. Preservation and determination of trace metal partitioning in river water by a two-column ion exchange method. Anal. Chem. 74: 4716–4724.

Jiménez, M.S., M.T. Gómez, L. Rodríguez, L. Martínez and J.R. Castillo. 2009. Some pitfalls in PAGE-LA-ICP-MS for quantitative elemental speciation of dissolved organic matter and metalomics. Anal. Bioanal. Chem. 393: 699–707.

Jiménez, M.S., L. Rodríguez, M.T. Gómez and J.R. Castillo. 2010a. Metal–protein binding losses in proteomic studies by PAGE–LA-ICP-MS. Talanta 81: 241–247.

Jiménez, M.S., M.T. Gómez, L. Rodriguez, R. Velarte and J.R. Castillo. 2010b. Characterization of metal–humic acid complexes by polyacrylamide gel electrophoresis–laser ablation-inductively coupled plasma mass spectrometry. Anal. Chim. Acta 676: 9–14.

Jirsa, F., E. Neubauer, R. Kittinger, T. Hofmann, R. Krachler, F. von der Kammer and B.K. Keppler. 2013. Natural organic matter and iron export from the Tanner Moor, Austria. Limnologica 43: 239–244.

Kostanski, L.K., D.M. Keller and A.E. Hamielec. 2004. Size-exclusion chromatography–a review of calibration methodologies. J. Biochem. Biophys. Methods 58: 159–186.

Krishan Puri, B., R. Muñoz Olivas and C. Camara. 2004. A new polymeric adsorbent for screening and pre-concentration of organotin compounds in sediments and seawater samples. Spectrochim. Acta Part B 59: 209–214.

Kuhn, K.M., E. Neubauer, T. Hofmann, F. von der Kammer, G.R. Aiken and P.A. Maurice. 2015. Concentration and distributions of metals associated with dissolved organic matter from the Suwannee River (GA, USA). Environ. Eng. Sci. 32: 54–65.

Laborda, F., S. Ruiz-Beguería, E. Bolea and J.R. Castillo. 2011. Study of the size-based environmental availability of metals associated to natural organic matter by stable isotope exchange and quadrupole inductively coupled plasma mass spectrometry coupled to asymmetrical flow field flow fractionation. J. Chromatogr. A 1218: 4199–4205.

Lagerström, M.E., M.P. Field, M. Seguret, L. Fischer, S. Hann and R.M. Sherrell. 2013. Automated on-line flow-injection ICP-MS determination of trace metals (Mn, Fe, Co, Ni, Cu and Zn) in open ocean seawater: Application to the GEOTRACES program. Mar. Chem. 155: 71–80.

Lesher, E.K., B.D. Honeyman and J.F. Ranville. 2013. Detection and characterization of uranium–húmica complexes during 1D transport studies. Geochim. Cosmochim. Acta 109: 127–142.

Li, J., W. Lu, J. Ma and L. Chen. 2011. Determination of mercury (II) in water samples using dispersive liquid-liquid microextraction and back extraction along with capillary zone electrophoresis. Microchim. Acta 175: 301–308.

Lu, F., C.-H. Chang, D.-J. Lee, P.-J. He, L.-M. Shao and A. Su. 2009. Dissolved organic matter with multi-peak fluorophores in landfill leachate. Chemosphere 74: 575–582.

Luan, H. and T.M. Vadas. 2015. Size characterization of dissolved metals and organic matter in source waters to streams in developed landscapes. Environ. Pollut. 197: 76–83.

Mawji, E., M. Gledhill, J.A. Milton, M.V. Zubkov, A. Thompson, G.A. Wolff and E.P. Achterberg. 2011. Production of siderophore type chelates in Atlantic Ocean waters enriched with different carbon and nitrogen sources. Mar. Chem. 124: 90–99.

Mena, M.L., E. Moreno-Gordaliza, I. Moraleja, B. Cañas and M.M. Gómez-Gómez. 2011. OFFGEL isoelectric focusing and polyacrylamide gel electrophoresis separation of platinum-binding proteins. J. Chromatogr. A 1218: 1281–1290.

Mena, M.L., E. Moreno-Gordaliza and M.M. Gómez-Gómez. 2013. TCEP-based rSDS–PAGE and nLC–ESI-LTQ-MS/MS for oxaliplatin metalloproteomic analysis. Talanta 116: 581–592.

Minor, E.C., J.-P. Simjouw, J.J. Boon, A.E. Kerkhoff and J. van der Horst. 2002. Estuarine/marine UDOM as characterized by size-exclusion chromatography and organic mass spectrometry. Mar. Chem. 78: 75–102.

Milne, A., W. Landing, M. Bizimis and P. Morton. 2010. Determination of Mn, Fe, Co, Ni, Cu, Zn, Cd and Pb in seawater using high resolution magnetic sector inductively coupled mass spectrometry (HR-ICP-MS). Anal. Chim. Acta 665: 200–207.

Mishra, S., R.M. Triphathi, S. Bhalke, V.K. Shukla and V.D. Puranik. 2005. Determination of methylmercury and mercury(II) in a marine ecosystem using solid-phase microextraction gas chromatography-mass spectrometry. Anal. Chim. Acta 551: 192–198.

Monperrus, M., E. Tessier, S. Veschanbre, D. Amoroux and O. Donard. 2005. Simultaneous speciation of mercury and butyltin compounds in natural waters and snow by propylation and species-specific dilution mass spectrometry. Anal. Bioanal. Chem. 381: 854–862.

Monticelli, D., L.M. Laglera and S. Caprara. 2014. Miniaturization in voltammetry: Ultratrace elements in analysis and speciation with twenty-fold sample size reduction. Talanta 128: 273–277.

Mopper, K., A. Stubbins, J.D. Ritchie, H.M. Bialk and P.G. Hatcher. 2007. Advanced instrumental approaches for characterization of marine dissolved organic matter: extraction techniques, mass spectrometry, and nuclear magnetic resonance spectroscopy. Chem. Rev. 107: 419–442.

Moreda-Piñeiro, A., N. García-Otero and P. Bermejo-Barrera. 2014. A review on preparative and semi-preparative offgel electrophoresis for multidimensional protein/peptide assessment. Anal. Chim. Acta 836: 1–17.

Murray, H., G. Meunier, D.B. Stengel and R. Cave. 2014. Metal complexation by organic ligands (L) in near-pristine estuarine waters: evidence for the identity of L. Environ. Chem. 11: 879–99.

Nascimiento, P.C. do., C.L. Jost, L.M. Carvalho, D. Boher, Koschinsky. 2009. Voltammetric determination of Se(IV) and Se(VI) in saline samples-studies with seawater, Hydrothermal and hemodialysis fluids. Anal. Chim. Acta 648: 162–166.

Neubauer, E., F. van der Kammer and T. Hofmann. 2011. Influence of carrier solution ionic strength and injected sample load on retention and recovery of natural nanoparticles using flow field-flow fractionation. J. Chromatogr. A 1218: 6763–6773.

Neubauer, E., F. van der Kammer and T. Hofmann. 2013. Using FLOWFFF and HPSEC to determine trace metalecolloid associations in wetland runoff. Water Res. 47: 2757–2769.

Nimmo, M., C.M.G. Van den Berg and J. Brown. 1989. The chemical speciation of dissolved nickel, copper, vanadium and iron in Liverpool Bay, Irish Sea. Estuar. Coast. Shelf Sci. 29: 57–74.

Nunn, B.L. and A.T. Timperman. 2007. Marine proteomics. Mar. Ecol. Prog. Ser. 332: 281–289.

Ogawa, H. and E. Tanoue. 2003. Dissolved organic matter in oceanic waters. J. Oceanogr. 59: 129–147.

Ozdemir, N., M. Soylak, L. Elci and M. Dogan. 2004. Speciation analysis of inorganic Sb(III) and Sb(V) ions by using mini column filled with Amberlite XAD-8 resin. Anal. Chim. Acta 505: 37–41.

Park, J.-H. 2009. Spectroscopic characterization of dissolved organic matter and its interactions with metals in surface waters using size exclusion chromatography. Chemosphere 77: 485–494.

Papoff, P., F. Bocci and F. Lanza. 1998. Speciation of selenium in natural waters and snow by DPCSV at hanging mercury drop electrode. Microchemical Journal 59: 50–76.

Pifer, A.D., D.R. Miskin, S.L. Cousins and J.L. Fairey. 2011. Coupling asymmetric flow-field flow fractionation and fluorescence parallel factor analysis reveals stratification of dissolved organic matter in a drinking water reservoir. J. Chromatogr. A 1218: 4167–4178.

Point, D., G. Bareille, H. Pinally, C. Belin and O.F.X. Donard. 2007. Multielemental speciation of trace elements in estuarine waters with automated on-site UV photolysis and resin chelation coupled to inductively coupled plasma mass spectrometry. Talanta 72: 1207–1216.

Posta, J., A. Alimonti, F. Petrucci and S. Caroli. 1996. On-line separation and preconcentration of chromium species in seawater. Anal. Chim. Acta 325: 185–193.

Powell, R.T. and J.R. Donat. 2001. Organic complexation and speciation of iron in the South and Equatorial Atlantic. Deep-Sea Res. II 48: 2877–2893.

Powell, M.J. and A.T. Timperman. 2005. Quantitative analysis of protein recovery from dilute, large volume samples by tangential flow ultrafiltration. J. Memb. Sci. 252: 227–236.

Powell, M.J., J.N. Sutton, C.E. del Castillo and A.T. Timperman. 2005. Marine proteomics: generation of sequence tags for dissolved proteins in seawater using tandem mass spectrometry. Mar. Chem. 95: 183–198.

Prasada-Rao, T., S. Karthikeyan, B. Vijayalekshmy and C.S.P. Iyer. 1998. Speciative determination of chromium (VI) and Chromium (III) using flow-injection on-line preconcentration and flame-atomic-absorption spectrometry. Anal. Chim. Acta 369: 69–77.

Prieto, A., O. Zuloaga, A. Usobiaga, N. Etxebarria, L.A. Fernández, C. Marcic and A. de Diego. 2008. Simultaneous speciation of methylmercury and butyltin species in environmental samples by headspace-stir bar sorptive extraction-thermal desoption-gas chromatography-mass spectrometry. J. Chromatogr. A 1185: 130–138.

Quentel, F. and C. Elleouet. 1999. Speciation analysis of selenium in seawater by cathodic stripping voltammetry. Electroanalysis 11: 47–51.

Quentel, F. and M. Filella. 2002. Determination of inorganic antimony species in seawater by differential pulse anodic stripping voltammetry: stability of the trivalent state. Anal. Chim. Acta 452: 237–244.

Riso, R.D., M. Waeles, S. Garbarino and P. Le Corre. 2004. Measurements of total selenium and selenium (IV) in seawater by stripping chronopotentiometry. Anal. Bioanal. Chem. 379: 1113–1119.

Rottmann, L. and K.G. Heumann. 1994. Determination of heavy metal interactions with dissolved organic materials in natural aquatic systems by coupling a high-performance liquid chromatography system with an inductively coupled plasma mass spectrometer. Anal. Chem. 66: 3709–3715.

Rue, E.L. and K.W. Bruland. 1995. Complexation of Fe(III) by natural organic ligands in the Central North Pacific as determined by a new competitive ligand equilibration/adsorptive cathodic stripping voltammetric method. Mar. Chem. 50: 117–138.

Saijo, S. and E. Tanoue. 2004. Characterization of particulate proteins in Pacific surface waters. Limnol. Oceanogr. 49: 953–963.

Saito, M.A. and J.W. Moffet. 2001. Complexation of cobalt by natural organic ligands in the Sargasso Sea as determined by a new high-sensitivity electrochemical cobalt speciation method suitable for open ocean work. Mar. Chem. 75: 49–68.

Salaün, P., B. Planer-Friedrich and C.M.G. Van den Berg. 2007. Inorganic arsenic speciation in water and seawater by anodic stripping voltammetry with a gold microelectrode. Anal. Chim. Acta 585: 312–322.

Sánchez-González, J., N. García-Otero, A. Moreda-Piñeiro and P. Bermejo-Barrera. 2012. Multi-walled carbon nanotubes-solid phase extraction for isolating marine dissolved organic matter before characterization by size exclusion chromatography. Microchem. J. 102: 75–82.

Sánchez-Rodas, D., J.L. Gómez-Ariza, I. Giráldez, A. Velasco and E. Morales. 2005. Arsenic speciation in river and estuarine waters from southwest Spain. Science of the Total Environment 345: 207–217.

Saygi, K.O., E. Melek, M. Tuzen and M. Soylak. 2007. Speciation of selenium (IV) and selenium (VI) in environmental samples by the combination of graphite furnace atomic absorption spectrometric determination and solid phase extraction on Diaion HP-2MP. Talanta 71: 1375–1381.

Schmitt, D., M.B. Müller and F.H. Frimmel. 2000. Metal distribution in different size fractions of natural organic matter. Acta Hydrochim. Hydrobiol. 28: 400–410.

Shelley, R.U., B. Zachhuber, P.N. Sedwick, P.J. Worsfold and M.C. Lohan. 2010. Determination of total disolved cobalt in UV-irradiated seawater using flow injection with chemiluminescence detection. Limnol. Oceanogr. Meth. 8: 352–362.

Sohrin, Y. and K.W. Bruland. 2011. Global status of trace elements in the ocean. Trend. Anal. Chem. 30: 1291–1307.

Sohrin, Y., S. Urushihara, S. Nakatsuka, T. Kono, E. Higo, T. Minami, K. Norisuye and S. Umetani. 2008. Multielemental determination of GEOTRACES key trace metals in seawater by ICP-MS after preconcentration using an ethylenediaminetriacetic acid chelating resin. Anal. Chem. 80: 6267–6273.

Specht, C.H. and F.H. Frimmel. 2000. Specific interactions of organic substances in size exclusion chromatography. Environ. Sci. Technol. 34: 2361–2366.

Stolpe, B., L. Guo, A.M. Shiller and M. Hassellöv. 2010. Size and composition of colloidal organic matter and trace elements in the Mississippi River, Pearl River and the northern Gulf of Mexico, as characterized by flow field-flow fractionation. Mar. Chem. 118: 119–128.

Stolpe, B., L. Guo and A.M. Shiller. 2013a. Binding and transport of rare earth elements by organic and iron-rich nanocolloids in Alaskan rivers, as revealed by field-flow fractionation and ICP-MS. Geochim. Cosmochim. Acta 106: 446–462.

Stolpe, B., L. Guo, A.M. Shiller and G.R. Aiken. 2013b. Abundance, size distributions and trace-element binding of organic and iron-rich nanocolloids in Alaskan rivers, as revealed by field-flow fractionation and ICP-MS. Geochim. Cosmochim. Acta 105: 221–239.

Stolpe, B., Z. Zhou, L. Guo and A.M. Shiller. 2014. Colloidal size distribution of humic- and protein-like fluorescent organic matter in the northern Gulf of Mexico. Mar. Chem. 164: 25–37.

Suzuki, S., K. Kogure and E. Tanoue. 1997. Immunochemical detection of dissolved proteins and their source bacteria in marine environments. Mar. Ecol. Prog. Ser. 158: 1–9.

Tanguy, V., M. Waeles, J. Vandenhecke and R.D. Riso. 2010. Determination of ultra-trace Sb(III) in seawater by stripping chronopotentiometry (SCP) with a mercury film electrode in presence of copper. Talanta 81: 614–620.

Tao, H., R. Babu Rajendran, C.R. Quetel, T. Nakazato, M. Tominaga and A. Miyazaki. 1999. Tin speciation in the fentogram range in open ocean seawater by gas chromatography/inductively coupled plasma mass spectrometry using a shield torch at normal conditions. Anal. Chem. 71: 4208–4215.

Tanoue, E., S. Nishiyama, M. Kamo and A. Tsugita. 1995. Bacterial membranes: possible source of a major dissolved protein in seawater. Geochim. Cosmochim. Acta 59: 2643–2648.

Thompson, C.M. and M.J. Ellwood. 2014. Dissolved copper isotope biogeochemistry in the Tasman Sea, SW Pacific Ocean. Mar. Chem. 165: 1–9.

Thompson, C.M., M.J. Ellwood and S.G. Sander. 2014. Dissolved copper speciation in the Tasman Sea, SW Pacific ocean. Mar. Chem. 164: 84–94.

Thuróczy, C.E., L.J.A. Gerringa, M.B. Blunder, R. Middag, P. Laan, K.R. Timmermans and H.J.W. de Baar. 2010. Speciation of Fe in the Eastern North Atlantic Ocean. Deep-Sea Res. I 57: 1444–1453.

Truesdale, V.W., D.S. Danielson and T.J. Waite. 2003a. Summer and winter distribution of dissolved iodine in Skagerrak. Est. Coast. Shelf Sci. 57: 701–713.

Truesdale, V.W. and R. Upstill-Goddard. 2003b. Dissolved iodate and total iodine along the British east coast. Est. Coast. Shelf Sci. 56: 261–270.

Tsunoi, S., T. Matoba, H. Shioji, L.T. Huong, H. Harino and M. Tanaki. 2002. Analysis of organotin compounds by Grignard dervatization and gas chromatography-ion trap tandem mass spectrometry. J. Chromatogr. A 962: 197–206.

Tuzen, M., I. Karaman, D. Citak and M. Soylak. 2009. Mercury (II) and methyl mercury determinations in water and fish samples by using solid phase extraction and cold vapour atomic absorption spectrometry combination. Food and Chemical Toxicology 47: 1648–1552.

Ure, A.M. and C.M. Davison. 2001. Chemical speciation in the environment. 2nd edition. 464.

Van den Berg, C.M.G., M. Boussemart, K. Yokoi, T. Prartono, M. Campos and A.M. Lucia. 1994. Speciation of aluminum, chromium and titanium in the NW Mediterranean. Mar. Chem. 45: 267–82.

Van den Berg, C.M.G. 2006. Chemical speciation of iron in seawater by cathodic stripping voltammetry with dihydroxynaphthalene. Anal. Chem. 78: 156–163.

Vega, M. and C.M.G. Van der Berg. 1997. Determination of cobalt in seawater by catalytic adsorptive cathodic stripping voltammetry. Anal. Chem. 69: 874–881.

Velasquez, I., B.L. Nunn, E. Ibisanni, D.R. Goodlett, K.A. Hunter and S.G. Sander. 2011. Detection of hydroxamate siderophores in coastal and Sub-Antartic waters off the South Eastern Coast of New Zealand. Mar. Chem. 126: 97–107.

Vogl, J. and K.G. Heumann. 1998. Development of an ICP–IDMS method for dissolved organic carbon determinations and its application to chromatographic fractions of heavy metal complexes with humic substances. Anal. Chem. 70: 2038–2043.

Vraspir, J.M. and A. Butler. 2009. Chemistry of marine ligands and siderophores. Ann. Rev. Mar. Sci. 1: 43–63.

Wang, N. and J. Tyson. 2014. Non-chromatographic speciation of inorganic arsenic by atomic fluorescence spectrometry with flow injection hydride generation with a tetrahydroborate-form anion-exchanger. J. Anal. At. Spectrom. 29: 665–673.

Wang, W., R. Vignani, M. Scali and M. Cresti. 2006. A universal and rapid protocol for protein extraction from recalcitrant plant tissues for proteomic analysis. Electrophoresis 27: 2782–2786.

Wu, F., D. Evans, P. Dillon and S. Schiff. 2004. Molecular size distribution characteristics of the metal–DOM complexes in stream waters by high-performance size-exclusion chromatography (HPSEC) and high-resolution inductively coupled plasma mass spectrometry (ICP-MS). J. Anal. At. Spectrom. 19: 979–983.

Wu, J. and M. Jin. 2009. Competitive ligand exchange voltammetric determination of iron organic complexation in seawater: examination of accuracy using computer simulation and elimination of artifacts in ideal two-ligand case using iterative non-linear multiple regression. Mar. Chem. 114: 1–10.

Xu, J., L.G. Ruan, X. Wang, Y.Y. Jiang, L.X. Gao and J.C. Gao. 2012. Ultrafiltration as pretreatment of seawater desalination: Critical flux, rejection and resistance analysis. Sep. Purif. Technol. 85: 45–53.

Yamada, N. and E. Tanoue. 2003. Detection and partial characterization of dissolved glycoproteins in oceanic waters. Limnol. Oceanogr. 48: 1037–1048.

Yamada, N. and E. Tanoue. 2006. The inventory and chemical characterization of dissolved proteins in oceanic waters. Prog. Oceanogr. 69: 1–18.

Yamada, N. and E. Tanoue. 2009. Similarity of electrophoretic dissolved protein spectra from coastal to pelagic seawaters. J. Oceanogr. 65: 223–233.

Yamashita, Y. and E. Tanoue. 2004. Chemical characteristics of amino acid-containing dissolved organic matter in seawater. Org. Geochem. 35: 679–692.

Yayayürük, O., E. Henden and N. Bicak. 2011. Determination of mercury (II) in the presence of methylmercury after preconcentration using poly(acrylamide) grafted onto cross-linked poly(4-vynil piridine): Application to mercury speciation. Analytical Sciences 27: 833–838.

Zhai, G., J. Liun, G. Jiang, B. He and Q. Zhou. 2007. On-line coupling HPLC and quartz surface-induced luminiscence FPD with hydride generation and microporous membrane gas-liquid separator as interface for speciation of methyltins. J. Anal. At. Spectrom. 22: 1420–1426.

Zhang, L., Y. Morita, A. Sakuragawa and A. Asozaki. 2007. Inorganic speciation of As(III, V) and Sb(III, V) in natural water with GF-AAS using solid phase technology. Talanta 72: 723–729.

2

The Decisive Role of ICPMS and Related Hyphenated Techniques in Multielement and Speciation Analysis of Marine Biota

George Zachariadis[a,*] and *Evangelos Trikas*[b]

ABSTRACT

Nowadays, mass spectrometry (MS) is considered as one of the most important and widely employed analytical techniques for the analysis of any kind of sample matrices, especially those of environmental origin. Moreover, Inductively Coupled Plasma Mass Spectrometry (ICP-MS) is a powerful conjugation of an effective atomization technique with a highly selective one and therefore, can aid in the measurement of ultra trace concentrations of almost any kind of analyte at trace and ultra trace levels.

In this respect, this article aims to describe the most recent advances and applications of the above technique in the speciation and metallomic analysis of marine organisms and several kinds of biota. Since the species-specific analysis is the target of this chapter, most of the applications make use of the hyphenated technique of LC-ICPMS and to a lesser extend of other similar techniques. As metal species and metallomes such as metalloproteins, organometallics, and metal-bound polymers do not have significant volatility, liquid chromatography instead of gas chromatography is utilized. The twin technique of ICP-AES, although simple, is employed less because of insufficient detectability for biometals in ultratrace analysis.

Introduction to ICP Atomization

Inductively Coupled Plasma is the ultimate atomization solution in elemental analysis both for atomic and mass spectrometers today (Broekaert and Bowmans 1987; Sharp 1987; Nolte 2003; Browner et al. 2007). Apparently, any molecular information of the samples is not available any more due to the hard atomization conditions. However, by conjugating the ICP with a separation technique it becomes possible to retrieve

Lab. of Analytical Chemistry, Department of Chemistry, Aristotle University of Thessaloniki, 54124 Thessaloniki, Greece.
[a] Email: zacharia@chem.auth.gr
[b] Email: trikas.vag@hotmail.com
* Corresponding author

quite a lot of the molecular profile of the sample. In this context, plasma atomization is very well suited as an interface between liquid chromatography and an element- or species- or molecule-selective detector like a mass spectrometric one or an atomic emission one.

In mass spectrometry, the aim is to collect all the analyte particles if possible and there is no concern regarding the photons emitted. Consequently, in ICP-MS hyphenation, a sampling cone with an orifice is employed, emitted photons are stopped in a reflector, and only ions are driven to the mass filter through appropriate ion lenses. In contrast, in atomic spectrometry, the interest is focused on the emitted radiation. Therefore, in ICP-AES hyphenation, only the emmited radiation from plasma is directed to the polychromator through a quartz window and not the ions or any particle, which are removed from the system after suction through the hood (Taylor 2001). The ionization induced by the plasma produces mainly single-ionized ions therefore their successive discrimination in a mass analyzer is facilitated (Zachariadis 2012).

The main parts of an inductively coupled plasma atomizer/ionizer are the sample introduction system, the radiofrequency generator, the nebulization system, and the plasma torch. Despite of the development of several new types of nebulization systems by instrument manufacturers, the basic philosophy of the nebulization process and the plasma torch design remains unaltered. Slight dimensional variations of the torches have been developed by various manufacturers as well as several new nebulization improvements are introduced like ultrasonic nebulization, membrane desolvation (Carr et al. 2006), etc.

A suitable sample propulsion system is employed for liquid samples, compatible with the plasma capabilities. The goal is to deliver the sample into the nebulizer and finally to the plasma with a non-pulsated and uninterrupted precise flow. The sample introduction procedure should keep the solution homogeneity unaffected, and preserve it from any potential contamination. Peristaltic and syringe pumps are required in the ICP-MS/AES mode, however in LC-ICPMS the column eluate is simply propelled by the single- or dual-piston LC pump. Common mobile phase flow rates of liquid chromatography are more or less compatible with the interface requirements of the plasma atomizer. In case of direct interface of LC to MS, only part of the flow may be delivered to the MS ionization system, in order to comply with the ESI or API requirements.

Although injection of liquid samples in ICP-MS is more convenient, in recent decades the development of numerous methods for slurry injection allowed the employment of methods without previous sample decomposition (Santos et al. 2006; Todoli and Mermet 2006). It is clear that besides the apparent advantages there are also important limitations of slurry analysis. A critical step in slurry preparation is the grinding of the material, then the stabilization of slurry solution in order to avoid sedimentation or coagulation, and finally special calibration strategies required for this purpose.

The development and the wide use of laser ablation systems and their combination with the ICP atomizer enabled the analysis of various solid treated or untreated biological matrices with satisfactory reliability. The high cost of the additional equipment required is balanced by the benefit from avoiding tedious sample decomposition procedures and also by the ability to make practically non-destructive analysis in important samples like those of biological specimens which are available in small quantities. A further point to mention is the ability to make spatial resolved analysis in very small sites on the sample surface. The laser ablation procedure with its various types has an increasing impact on solid elemental analysis either in combination with ICP-AES (LA-ICP-AES) but mainly with ICP-MS techniques (LA-ICP-MS) as described thoroughly by Becker (Becker 2007). For this purpose, the commonly used laser sources are cubic crystals of neodymium-doped yttrium aluminium garnet, but several other lasers may also be employed. Nevertheless, there are some disadvantages in the LA-ICP-MS technique as compared to other sample introduction techniques, e.g., unreliable calibration, reproducibility of the effect of the beam on various matrices, stability in mass transportation, etc. But careful control of laser ablation parameters allows for precision depth sampling and even depth profiling under favorable conditions.

The purpose of using a nebulizer is to transform a liquid flow into a fine mist, containing suspended droplets in a carrier gas. In this way, an aerosol with various sized droplets is formed which is compatible with plasma and can be injected into it. For simplicity it is generally accepted that rapidly the shape of all droplets becomes spherical, therefore their surface and volume can be calculated. Normally droplets having diameter of 5–25 μm are of interest for subsequent atomization, although the particle size distribution in an aerosol may vary in wide ranges (Moore 1989). The common nebulizers for analytical use can be classified in three types: (i) pneumatic nebulizers, (ii) ultrasonic nebulizers, and (iii) thermal injection nebulizers. The most popular types of ICP pneumatic nebulizers are concentric, fixed or adjustable cross-flow, and high-solids nebulizers of Babington type. The microconcentric nebulizers (MCN) are constructed by glass or preferably polymeric material and they can be used for low sample flow rates, 10–200 μL/min, as is case of LC to ICP hyphenation. To gain from their efficiency they are commonly attached to a low-volume cyclonic spray chamber but can be coupled to low-volume Scott spray chambers also. Transport efficiencies of 50% are typical for MCNs therefore are quite useful to interface liquid chromatography with the plasma atomizer. The fixed and the adjustable cross-flow nebulizers are used with a Scott spray chamber, and provide moderate to good sensitivity. The argon gas flow is cross-sectioned with the outlet of the sample solution tube and creates a droplets dispersion of the solutions. In general they perform better than the concentric design for solutions containing high dissolved solids concentrations, yet less efficiently than the Babington type nebulizers for this purpose. The range of droplet size in a mist created by a nebulizer is usually from very small ones up to 100–200 μm (Moore 1989). Immediately after, a spray chamber is placed in order to physically discriminate between the large and the fine droplets and discard the first ones. Ideally, only droplets of < 10 μm (or practically < 25 μm) in size should be allowed to proceed to plasma in order to maintain a satisfactory stability. Various designs of spray chambers were reported in literature (Todoli et al. 2000). While the primary function of the spray chamber is to remove large droplets from the aerosol, a secondary function of it is to smooth out any pulsation which may occur in the sample delivery during nebulization.

The plasma torch employed in analytical instruments is constructed as an arrangement of three concentric tubes through which three separate gases flows past. Argon gas is most commonly used although helium plasmas are also emloyed in some ICP-MS applications having the characteristic of an increased ionization energy (Evans et al. 1995). The tubes are manufactured from a heat-resistant material, usually quartz glass for the intermediate and the outer tube and quartz or alumina or other for the inner tube. The torch is surrounded by an induction copper coil, creating a radiofrequency field near to the end of the torch. The RF power induces an oscillating magnetic field whose force lines are axially orientated in the plasma torch and follow elliptical paths outside the induction. The intermediate quartz tube supplies argon flow as make-up gas to maintain the plasma, while the inner tube delivers the aerosol in the central zone of the plasma. The carrier argon gas creates and delivers the aerosol of the sample solution usually in a flow rate range between 0.7–1.3 L/min (Dean 1997; Becker 2007).

Table 1 summarizes comparatively the main ICP-MS operating conditions usually employed. It is worth emphasizing that the slight differences in the plasma atomization settings are ruled not only by the selected analytes to be measured but also from the type and the model of the instrumentation.

The torch can be placed in the ICP compartment in two different positions. When axial viewing of the plasma is applied, the torch is arranged in a horizontal position. Alternatively, when radial viewing is required, the torch is arranged in a vertical position. Both arrangements have advantages and drawbacks. Although in ICP-AES, the viewing of the emitted radiation can be vertical or horizontal, in ICP-MS, the torch is always placed horizontally in line with the sampling cone in the entrance of the mass analyzer. It is worth mentioning here that although the configuration of torches used in ICP-AES and ICP-MS are quite

Table 1: Typical ICP-MS operating conditions applied in marine biota analysis.

Reference	RF power (W)	Plasma gas flow (L min^{-1})	Auxiliary gas flow (L min^{-1})	Nebulizer gas flow (L min^{-1})	Dwell time	Sweeps
Terol et al. 2012	1500	14.5	1.65	1.00	400 ms	20
Contreras-Acuna et al. 2013	1500	15.0	1.00	0.90	–	–
Brito et al. 2012	1300	13.0	0.70	0.87	100 ms	10
Chang et al. 2007	1350	15.0	1.20	0.15	100 ms	10
Chen et al. 2013	1200	13.0	0.75	0.85	100 ms	–
Lee and Jiang 2005	1300	15.0	1.13	1.05	70 ms	5

similar, their dimensions are slightly different. Larger diameter torches are used in ICP-AES to allow for longer stay of the atoms in the plasma, while in MS this is unwanted, since double-charged ions may be formatted which complicate the MS spectrum obtained (Wrobel and Wrobel 2007).

Multielement Analysis of Metals in Marine Biota with ICP-MS

One of the biggest advantages of the ICP-MS system is the potential for simultaneous multielement analysis. This way a sample can be analyzed for various analytes very rapidly, reducing not only the necessary time for a thorough analysis, but also the total cost of the procedure. It is precisely this strategic leverage that is exploited by the researchers in the papers of this subchapter.

Two works that could be considered as characteristic examples of this is the analysis of giant kelp (*Macrocystispyrifera*) sieve tube sap (STS) for the determination of 17 different metals (Al, V, Cr, Mn, Ni, Co, Cu, Zn, As, Se, Rb, Sr, Ag, Cd, Cs, Hg, and Pb) with scope to use these data as an indication of the pollution of the environment where this creature lives (Fink and Manley 2011), as well as, the determination of 12 elements (As, Ba, Cd, Co, Cr, Cu, Li, Mn, Ni, Pb, V, and Zn) in ten microalgae species (Brito et al. 2012). The ICP-MS method proved to be ideal for the fulfillment of both projects, achieving the accurate determination of the desired analytes.

Using data of more than 10 years from the German Environmental Specimen, Rudel et al. (2010) tried to monitor how the heavy metals pollution fluctuating in the Baltic and North seas through the years. The samples they used were bladder wrack (*Fucusvesiculosus*), blue mussel (*Mytilus edulis*), eelpout (*Zoarcesviviparus*), and eggs of herring gulls (*Larusargentatus*). After the data evaluation, it became apparent that there are differences depending on the place of sampling. Furthermore, a decrease in the metals concentration were observed over time in the North Sea, while in the Baltic sea the Cd, Hg, and Pb concentrations were relatively stable.

Except microalgae, clams, and fishes, in the aqua world there is also another category of organisms. These are the mammals that live in oceans, like dolphins and whales. The potential unwanted impact of inorganic content, such as metals, on the health of the Indo-Pacific humpback dolphin (*Sousa chinensis*) and the finless porpoise (*Neophocaenaphocaenoides*) were investigated (Hung et al. 2007). The stomachs of these animals were analyzed by ICP-MS, and the total content in Ag, As, Cd, Co, Cr, Cs, Cu, Hg, Mn, Ni, Se, V, and Zn were measured. In order to correlate these data with potential health risks, two toxicity guideline values were used, namely the Reference Dose and the Toxicity Reference Value. Applying the TRV test, the results were no reason forconcern, but applying the stricter Rfd, the concentrations of Cr, Cu, Ni, Cd, Hg were at the verge of warning limits. Especially for As, the total concentration may be high but the probably dominant As species is not considered toxic. The same two toxicity guideline values (TRV, RfD) were also used during a study that was investigating the possible dangers to the humpback dolphins from the consumption of contaminated fishes (Hung et al. 2006). Thirteen trace elements (Ag, As, Cd, Co, Cr, Cs, Cu, Hg, Mn, Ni, Se, V, and Zn) were analyzed. The highest risk quotient (RQ), as above, was calculated for the As.

The use of zooplankton as an alternative protein source for fishes diet was investigated (Moren et al. 2006). The parameters studied was not only the muscle growth and the wellbeing of the fishes when the whole or part of their protein nutrition had replaced with the zooplankton, but also the way this change affected the metal's concentration in fishes tissues, considering the fact that the final destination of many of these fishes could be someone's plate. Study meals were produced from antarctic krill (*Euphausiasuperba*), arctic krill (*Thysanoessainermis*), and the arctic amphipod *Themistolibellula*. The fishes that tried those diets were Atlantic salmon and Atlantic cod. Results showed no problems regarding the growth and the survival of the fishes after the alternation of their diet, but ICP-MS analyses indicated that the processed Antarctic krill resulted in a dietary Cu level higher than the upper allowable limits of the EU, while the Cd level was 6 times higher in the amphipods processed meals. As it becomes obvious, it is not as simple as it might seem to use zooplankton as an alternative protein source, and there are a lot of parameters that should be taken into consideration before that.

The use of simple organisms as bioindicators is an area that has attracted increased interest the recent years. In order for an organism to be considered as good candidate for a bioindicator, it has to be easily affected by small variation in the conditions of its environment, and these variations have to be reflected on a specific index of this organism. In this case, the use of marine sponge, *Haliclonatenuiramosa*, as a bioindicator of heavy metal pollution in India was investigated (Venkateswara et al. 2009). The heavy metal concentrations in the sea near and far from the coast was analyzed by ICP-MS and the same analyses were done on tissue from the marine sponge population living in the specific areas. It was proven that the specific organism reflects the increased concentration of metals between the two areas and, consequently, it can be used as a bioindicator at the current case study. In another case, eight fish species were analyzed for their Zn, Cu, and Pb content, in an attempt to use fishes as biomonitors for heavy metal pollution (Kamaruzzaman et al. 2010). The results showed a correlation between fish size and heavy metals concentration in the fish organs.

The combination of a typical ICP-MS with a dynamic reaction cell is applied in a study of 16 elements (Al, As, Cd, Co, Cr, Cu, Fe, Hg, Mn, Mo, Ni, Pb, Se, Sn, V, and Zn) in seafood analyses. The DRC is a system that was introduced by Perkin-Elmer and it is connected with the ICP-MS system in order to eliminate some of the possible interferences. It is a chamber with a quadrupole. The chamber is filled with one or more reaction gases (ammonia, methane, oxygen, or hydrogen). These gases reacts with some of the possible interferences (the reason for their name), resulting in a much more accurate and "clearer" analyses without the need for complicated and time wasting sample pretreatment. Furthermore, with the use of the quadrupole, some possible unwanted reaction products could be ejected before they entered the analyses quadrupole. Specifically in marine biota multielement analyses, the interferences could be quite a few, as listed in Table 2 (Cubadda et al. 2006; May and Wiedmeyer 1998). In this specific case, ammonia was used as reactive gas in order to remove polyatomic ions interfering with ^{27}Al, ^{52}Cr, ^{56}Fe, and ^{51}V.

There are a large number of further studies using the ICP-MS to determine the multielement content in various marine organisms. In a research done by Das et al. (2009), the heavy metal levels in marine organisms of the Black Sea were determined. The analysis of rare earth elements (Y, La, Ce, Pr, Nd, Sm, Eu, Gd, Tb, Dy, Er, Ho, Tm, and Lu) in mollusc, barnacle, shrimp, and fish samples was the subject of related paper (Zhang et al. 2009), while Sn and Hg were determined in wild shrimps and crabs in Yantai coastal areas (Shao et al. 2011), with the results showing that the Sn and Hg pollution in the studied area has not reached dangerous levels. The combination of ICP-MS with ICP-AES was used for the analysis of 17 elements in the radular teeth of limpets, chitons, and other marine molluscs (Okoshi and Ishii 1996). The baseline metal concentration in the gammaridean amphipod *Paramoerawalker* has been also investigated and during this procedure the LODs of various elements were estimated (Palmer et al. 2006). The LODs for the As, Cd, Cr, Hg, Cu, Ni, Zn, and Pb using ICP-MS have been calculated also by Eira et al., 2008, in two standard reference materials, the Dogfish (*Squalusacanthias*) liver and the Dogfish muscle (Eira et al. 2008).

To investigate the marine biochemistry of PGE accumulation by the marine microalga *Chlorella stigmatophora*, a research group from Plymouth (Shams et al. 2013) added Rh, Pd, and Pt at ppb concentrations to cultures that remained in sea water at 15°C for two periods of time, one short term (24 h) and one long term (156 h). The accumulation results for the short term period were in the order: Rh > Pd > Pt, while for the long term, Rh > Pd = Pt. These data indicate that Rh participate in biological processes in marine environment to a larger extent as compared to Pd and Pt.

Last but not least, a multielement analysis has been made, not directly in marine organisms, but in ocean water at various sea depths at Lihir Island, Papua New Guinea (Sherwood et al. 2009). The samples were collected using a thin film technique (DGT). Nine metals (Al, Cd, Cr, Cu, Fe, Pb, Mn, Ni, and Zn) were determined at each site. For the sampling procedure two units were used, one containing a 0.4mm and the other a 0.8mm diffusive gel layer. These two units were submerged in three different depths (50–70, 105–130, and 135–155 m). The DGT technique was selected for this particular research not only because it was an ideal way to collect samples from the seawater but also because the studied analytes could be easily concentrated before the analysis and without any additional sample pretreatment.

However as has already been mentioned above in the "Basics of the ICP-MS", there is more than one way for the introduction of the sample in the ICP-MS system. So except the more frequently used nebulizers, which are ideal for liquid samples, there is also the Laser Ablation approach. This uses a pulsed laser beam to hit the surface of the sample and has proved optimum for solid samples analysis. A research group from

Table 2: ICP-MS polyatomic spectral interferences to some characteristic analyte isotopes.

Isotope	Abundance (%)	Interference
^{27}Al	100.0	$^{12}C^{15}N$, $^{13}C^{14}N$, $^{12}C^{14}NH$
^{75}As	100.0	$^{40}Ar^{35}Cl$, $^{38}Ar^{37}Cl$, $^{37}Cl_2H$, $^{40}Ca^{35}Cl$, $^{43}Ca^{16}O_2$, $^{59}Co^{16}O$, $^{36}Ar^{38}ArH$, $^{36}Ar^{39}K$, $^{23}Na^{12}C^{40}Ar$, $^{12}C^{31}P^{16}O_2$, $^{40}Ar^{34}SH$
^{110}Cd	12.5	$^{39}K_2{}^{16}O$
^{111}Cd	12.8	$^{95}Mo^{16}O$, $^{94}Zr^{16}O^1H$, $^{39}K_2{}^{16}O_2{}^1H$
^{112}Cd	24.1	$^{40}Ca_2{}^{16}O_2$, $^{40}Ar_2{}^{16}O_2$, $^{96}Ru^{16}O$
^{113}Cd	24.22	$^{96}Zr^{16}O^1H$.$^{40}Ca_2{}^{16}O_2{}^1H$, $^{40}Ar_2{}^{16}O_2{}^1H$, $^{96}Ru^{17}O$
^{114}Cd	28.7	$^{98}Mo^{16}O$, $^{98}Ru^{16}O$
^{116}Cd	7.49	$^{100}Ru^{16}O$
^{59}Co	100.0	$^{42}Ca^{16}OH$, $^{43}Ca^{16}O$, $^{24}Mg^{35}Cl$, $^{36}Ar^{23}Na$, $^{40}Ar^{18}OH$, $^{40}Ar^{19}F$
^{50}Cr	4.35	$^{34}S^{16}O$, $^{36}Ar^{14}N$, $^{35}C^{115}N$ $^{36}S^{14}N$, $^{32}S^{18}O$, $^{33}S^{17}O$
^{52}Cr	83.76	$^{35}Cl^{16}O^1H$, $^{40}Ar^{12}C$, $^{36}Ar^{16}O$, $^{37}Cl^{15}N$, $^{34}S^{18}O$, $^{36}S^{16}O$, $^{38}Ar^{14}N$, $^{36}Ar^{15}N^1H$, $^{35}Cl^{17}O$
^{53}Cr	9.51	$^{37}Cl^{16}O$, $^{38}Ar^{15}N$, $^{38}Ar^{14}N^1H$, $^{36}Ar^{17}O$, $^{36}Ar^{16}O^1H$, $^{35}Cl^{17}O^1H$, $^{35}Cl^{18}O$, $^{36}S^{17}O$, $^{40}Ar^{13}C$
^{54}Cr	2.39	$^{37}Cl^{16}O^1H$, $^{40}Ar^{14}N$, $^{38}Ar^{15}N^1H$, $^{36}Ar^{18}O$, $^{38}Ar^{16}O$, $^{36}Ar^{17}O^1H+$, $^{37}Cl^{17}O+$, $^{19}F_2{}^{16}O$
^{63}Cu	69.2	$^{31}P^{16}O_2$, $^{40}Ar^{23}Na$, $^{23}Na_2{}^{16}OH$, $^{46}Ca^{16}OH$, $^{46}Ca^{17}O$, $^{36}Ar^{12}C^{14}NH$, $^{14}N^{12}C^{37}Cl$, $^{12}C^{16}O^{35}Cl$
^{65}Cu	30.8	$^{32}S^{16}O_2H$, $^{32}S^{33}S$, $^{32}S^{16}O^{17}O$, $^{33}S^{16}O_2$, $^{48}Ca^{16}OH$, $^{31}P^{16}O^{18}O$, $^{40}Ar^{25}Mg$, $^{36}Ar^{14}N_2H$, $^{12}C^{16}O^{37}Cl$, $^{12}C^{18}O^{35}Cl$
^{54}Fe	5.8	$^{36}Ar^{18}O$, $^{38}Ar^{16}O$, $^{36}Ar^{17}OH$, $^{40}Ar^{14}N$, $^{38}Ar^{15}NH$, $^{36}S^{18}O$, $^{37}Cl^{16}OH$, $^{35}Cl^{18}OH$, $^{37}Cl^{17}O$
^{56}Fe	91.7	$^{40}Ar^{16}O$, $^{38}Ar^{18}O$, $^{38}Ar^{17}OH$, $^{40}Ca^{16}O$, $^{40}Ar^{15}NH$, $^{37}Cl^{18}OH$
^{57}Fe	2.2	$^{40}Ar^{16}OH$, $^{40}Ar^{17}O$, $^{38}Ar^{18}OH$, $^{40}Ca^{16}OH$, $^{38}Ar^{19}F$
^{24}Mg	78.7	$^{12}C_2$
^{25}Mg	10.13	$^{12}C_2{}^1H$
^{26}Mg	11.17	$^{12}C^{14}N$, $^{12}C_2{}^1H_2$, $^{12}C^{13}C^1H$
^{55}Mn	100	$^{39}K^{16}O$, $^{37}Cl^{18}O$, $^{37}Cl^{17}OH$, $^{40}Ar^{14}NH$, $^{40}Ar^{15}N$, $^{38}Ar^{17}O$, $^{36}Ar^{18}OH$, $^{38}Ar^{16}OH$, $^{23}Na^{32}S$, $^{36}Ar^{19}F$
^{58}Ni	67.77	$^{23}Na^{35}Cl$, $^{40}Ca^{18}O$, $^{40}Ca^{17}OH$, $^{42}Ca^{16}O$, $^{40}Ar^{18}O$, $^{40}Ar^{17}OH$, $^{29}Si_2$
^{60}Ni	26.16	$^{44}Ca^{16}O$, $^{43}Ca^{16}OH$, $^{23}Na^{37}Cl$
^{61}Ni	1.25	$^{44}Ca^{16}O^1H$, $^{45}Sc^{16}O$
^{62}Ni	3.66	$^{46}Ti^{16}O$, $^{23}Na^{39}K$, $^{46}Ca^{16}O$
^{64}Ni	1.16	$^{32}S^{16}O_2$, $^{32}S_2$
^{206}Pb	24.1	$^{190}Pt^{16}O$
^{207}Pb	22.1	$^{191}Ir^{16}O$
^{208}Pb	52.4	$^{192}Pt^{16}O$

Table 2 contd....

Table 2 contd...

Isotope	Abundance (%)	Interference
^{76}Se	9.02	$^{40}Ar^{36}Ar$, $^{38}Ar^{38}Ar$
^{77}Se	7.58	$^{40}Ar^{37}Cl$, $^{36}Ar^{40}Ar^{1}H$, $^{38}Ar_2{}^{1}H$, $^{12}C^{19}F^{14}N^{16}O_2$
^{78}Se	23.52	$^{40}Ar^{38}Ar$, $^{38}Ar^{40}Ca$
^{80}Se	49.82	$^{40}Ar_2$, $^{32}S^{16}O_3$
^{82}Se	9.19	$^{12}C^{35}Cl_2$, $^{34}S^{16}O_3$, $^{40}Ar_2{}^{1}H_2$
^{51}V	99.8	$^{35}Cl^{16}O$, $^{37}Cl^{14}N$, $^{34}S^{16}OH$, $^{32}S^{18}OH$, $^{33}S^{18}O$, $^{34}S^{17}O$, $^{36}S^{15}N$, $^{38}Ar^{13}C$, $^{36}Ar^{15}N$, $^{36}Ar^{14}NH$
^{64}Zn	48.89	$^{32}S^{16}O_2$, $^{48}Ti^{16}O+$, $^{31}P^{16}O_2{}^{1}H$, $^{48}Ca^{16}O+$, $^{32}S_2$, $^{31}P^{16}O^{17}O$, $^{34}S^{16}O_2$, $^{36}Ar^{14}N_2$
^{66}Zn	27.81	$^{50}Ti^{16}O+$, $^{34}S^{16}O_2$, $^{33}S^{16}O_2{}^{1}H$, $^{32}S^{16}O^{18}O$, $^{32}S^{17}O_2$, $^{33}S^{16}O^{17}O$, $^{32}S^{34}S$, $^{33}S_2$
^{67}Zn	4.11	$^{35}Cl^{16}O_2$, $^{33}S^{34}S$, $^{34}S^{16}O_2{}^{1}H$, $^{32}S^{16}O^{18}O^{1}H$, $^{33}S^{34}S$, $^{34}S^{16}O^{17}O$, $^{33}S^{16}O^{18}O$, $^{32}S^{17}O^{18}O$, $^{33}S^{17}O_2$, $^{35}Cl^{16}O_2$
^{68}Zn	18.57	$^{36}S^{16}O_2$, $^{34}S^{16}O^{18}O$, $^{40}Ar^{14}N_2$, $^{35}Cl^{16}O^{17}O$, $^{34}S_2$, $^{36}Ar^{32}S$, $^{34}S^{17}O_2$, $^{33}S^{17}O^{18}O$, $^{32}S^{18}O_2$, $^{32}S^{36}S$

Czech has engaged intensively in the marina biota analysis with the use of LA-ICP-MS system and has published two papers regarding the analysis of metals in fish scales with the specific system. In the first paper that was published in 2009 (Hola et al. 2009), analysis of heavy metals (Fe, Mn, Cd, Zn, Cr, V, Co, Pb, Cu, Ni, and As) was done in the scales of recent and subrecent (2-25 years) fish buried in the oxbow lake sediments of the Morava River. The results revealed a large variation in heavy metals concentration between those two sample matrices and also that the collagen concentration was decreasing very rapidly resulting in a significant difference between the determined concentration among fishes with only a few years difference. The second research (Hola et al. 2011) investigated the specific matter more thoroughly, applying two different LA methodologies to recent common carp (*Cyprinuscarpio*) scales. One was a line scan across the whole fish scale, and the other was a depth scan using the stable spot analysis. The combined results from both of these approaches led to plenary set of information about the distribution of the analyzed elements (Zn, Mg, Mn, Sr, Ca, P, Fe, and Pb) in fish scales.

Another very interesting field where this technology could be applied is the dual isotopes marking. This is a technique that, especially the recent years, finds increasing application in fish marking. This is an alternative way for the research groups, institutes, fishermen, and any other that would like to physically mark some fish populations, and could help for the monitoring of those fishes in order to study their reproductive rate or their migration movements. In a relevant project (Huelga-Suarez et al. 2013), the effectiveness of marking with barium the salmon otoliths was investigated for use in transgenerational studies (the analogy between ^{137}Ba and ^{135}Ba was monitored). Apart from the simple LA, the 2D LA was tested, in addition. A scan area of ca. 1.53 x 1.30 mm was selected. The acquired data were evaluated and the final molar fraction data for natural Xe, natural Ba, Ba-135, and Ba-137 were designed as an image. The accuracy of the proposed method indicated that it wouldn't be hard to separate fishes that have been marked with a different isotopic analogy for more advanced and complicated studies.

Single Element Analysis in Marine Biota with ICP-MS

Inspite of the fact that one of the main advantages of the ICP-MS technique is the capability of simultaneous multielement analysis, there are several cases where this is not necessary. In these cases, only one analyte is being determined either as the total concentration or as concentration of its various species (speciation analysis). Therefore, ICP-MS is also important for single-element analysis.

Copper is a trace element that is essential to the human organism. Copper is comprehended in many foods and drinks that humans consume. But while in low quantities it is necessary, if its uptake is increased significantly, it can lead to more or less serious health problems. Long-term exposure to copper can cause headaches, stomachaches and dizziness, liver and kidney failure, DNA damage, and in some extreme cases even death. The effects of copper on the marine bivalve mollusc, *Mytilus edulis* were studied (Al-Subiai et al. 2011) by exposing these organisms for five days to "realistic" copper concentrations. The maximum tolerated concentration (MTC) of copper was found to be 100 μg L^{-1}, above which the death percentage approached 100% for the specific period of exposure.

A paper that grappled with total iodine determination in Japanese coastal seawater in 14 estuaries was published by Zheng et al. in 2012. The achieved detection limit in seawater was 0.23 μg L^{-1}. A different approach for the mass analyzer of the ICP-MS apparatus was followed there in contrast to the more widely used quadrupole analyzer. The mass analyzer used was a sector-field consisting of a magnetic and an electric sector field. The SF-ICP-MS was selected for the iodine determination because it provides two times higher sensitivity compared to the one achieved with quadrupole ICP-MS due to the low ionization efficiency of iodine (33.9%). Generally the pros of these analyzers are their high resolution, high sensitivity, and high dynamic range. On the other hand, the disadvantages are that they are not well-suited for pulsed ionization methods (e.g., MALDI), and usually are larger, less convenient, and more expensive than other mass analyzers.

Uranium was the target analyte of two studies that appeared in 2009 (Charmasson et al. 2009) and 2011 (Takata et al. 2011) respectively. According to these studies, uranium could enter the marine ecosystem through two paths. The first was from deep-sea hydrothermal vents. The magma that emerges from these vents could contain notable amounts of uranium that could be transferred to the ocean when this hot hydrothermal fluid is mixed with the cold seawater. The other (not "natural" way) for uranium to reach marine biota is due to the human activities. Unfortunately in some cases radioactive wastes are disposed directly into the oceans, or are buried in disposal areas which communicate with underground rivers, travelling this way and reaching the sea. In the paper by Charmasson et al. 2009, the samples collected were biological samples from the East Pacific Rise and the Mid-Atlantic Ridge in 1996, 2001, and 2002 and were analyzed for their uranium content (^{238}U, ^{235}U, and ^{234}U). Additionally in two samples collected in 2002, the content in Po and Pb were also determined. Results confirmed the original assumption that organisms from these areas would exhibit higher uranium content compared with organisms outside of these areas. Specifically, the polychaetes from the East Pacific Rise showed high levels of uranium isotopes while the highest levels were found in *Paralvinellagrasslei* mucous tubes. In contrast, the deep-sea amphipod sample (*Orchomenella* sp.) that was collected outside hydrothermal vent areas and was used as a control sample showed the lowest values. In the latter paper (Takata et al. 2011), the samples analyzed were seawater, sediment, and marine organisms from Japanese estuarine areas. However one problem that these researchers faced was that elements such as Na, K, Mg, Ca, and Cl, could act as interference agents. In order to overcome the potential problems that could arise, they used a NOBIAS-CHELATE PA1 resin. This specific resin had been used in the past for the separation of heavy metals from alkali and alkaline earth metal elements in seawater, but it was the first time that this material was adopted in the uranium determination. The results showed that the U extractability on the resin is strongly pH dependent. At low pH (< 2), less than 20% of U was extracted, but at higher pH values more than 95% of U was retained on the resin, so the selected working pH value was at pH = 5.7. At the same time, at this specific pH value, other elements such as Na, Mg, K, and Ca passed through the resin and could not be found in the U fraction.

A study with a subject quite different than the previous ones was performed by Kasemann et al. in 2009. The topic was the analysis of boron isotope in marine carbonates in order to use these data to determine the palaeo-pH and CO_2 atmospheric conditions of the oceans thousands years ago. The potential of boron isotopic analogue use for this purpose is something that has been already demonstrated. In order for the analysis to be as complete as possible, three analytical techniques were combined for the determination of the boron isotope composition of carbonate material. These were the multicollector inductively coupled plasma mass spectrometry (MC-ICP-MS), the thermal ionization mass spectrometry (TIMS), and the secondary ionization mass spectrometry (SIMS). An MC-ICP-MS system is a mass spectrometer that combines the advantages of superior ionization of inductively coupled plasma and the precision of a magnetic sector multi-collector mass spectrometer. The main advantage of the MC-ICP-MS compared with the TIMS, is that it has the potential to analyze elements with even higher ionization potential.

Speciation Analysis by ICP-MS Hyphenated Techniques in Marine Organisms

Up till now, several studies were described in marine biota, which employed ICP-MS to determine various elemental concentrations in a pleiad of matrices in the aquatic environment. Liquid chromatography is a common and quite efficient technique that is combined with the ICP-MS not to determine the total concentration of an element, but to achieve the determination of each of its different species. This procedure is called speciation analysis and can find a wide application in marine biota analysis. The need behind this more sophisticated analysis is the tremendous difference between the toxicities of an element's species. In many cases it is not enough to know an element's total concentration in order to evaluate the danger that arise from its consumption, but rather to determine each different element species.

This is the subject of the book written by Di Carro and Magi (2010) which investigates the organotin compounds in the marine environment. The need for this investigation emerged from the different toxicities of the various organotin compounds. The butyltins have been found to cause various problems in marine organisms (see Chapter 13 in this book). One of the main ways that these compounds entered the underwater ecosystem was the use of TBT as a component in ships antifouling paints. Fortunately in the ecosystem TBT degrades to the more polar compounds, DBT and MBT, that are less toxic compared with the TBT. So with the purpose to determine not only the total tin, but to distinguish and quantify the different tin species, the coupling of chromatography with ICP-MS (among other techniques) was employed in this study.

Selective papers referring to speciation analysis with ICP-MS in marine biota are summarized in Table 3, listing the examined biological matrix and the determined analytes.

Speciation of Arsenic in Marine Organisms and Aquatic Ecosystems

Arsenic is a widespread element, introduced to the environment from natural and anthropogenic sources. Owing to the natural metabolic processes in the biosphere, arsenic exists in a large number of different inorganic and organic species. Today more than 25 different arsenic species have been identified in the environment, including inorganic forms as arsenite [As(III)] and arsenate [As(V)] and organic forms such as methylarsonic acid, arsenobetaine, arsenocholine, and a series of arsenolipids and arsenosugars. Inorganic arsenic exhibits high toxic levels, simple methylated species are less toxic, and arsenobetaine and arsenocholine are not toxic. The need for arsenic speciation analysis has led to an extensive use of chromatographic techniques coupled with ICP-MS in environmental samples analysis.

Arsenosugars and arsenate are dominant arsenic forms in marine algea, while in biomass, it is primarily in an organic form (arsenobetaine is the main arsenic species found in fish and shellfish). However, despite the fact that arsenobetaine is considered as a compound of low toxicity, it can transform into more toxic forms. This can happen for instance after interaction of AB aqueous solutions with natural zeolites. Related studies proved that these interactions could lead to the methylated arsenic species with the following two having the higher concentration ratios: DMMA (37%) and DMA (18%)(Mattusch et al. 2008).

The concentration of arsenic species in marine food is a subject that has been studied extensively due to the danger that could derive from its presence in high concentrations. Three papers that engage with this matter are by Contreras-Acuña et al. 2013; Thomas and Sniatecki 1995; Kohlmeyer et al. 2002. The first one uses *Anemoniasulcate* as a sample. This is an anemonia species that lives in the Mediterranean Sea and is widely consumed in nearby countries. The HPLC-(AEC)–ICP-MS, HPLC-(CEC)–ICP-MS, and HPLC-(AEC)–ICP-MS techniques were combined in this study. The achieved separation of As species after AEC and CEC is illustrated in the chromatograms of Fig. 1. The order of dominant arsenic species that was detected in the raw samples of *anemoniasulcata* was found to be as follows: DMA > AB > DMAS > AC > As^{5+} > TETRA > GPAsC > MA > TMAO.

In the work of Thomas and Sniatecki (1995), the studied samples were fishes and mussels and the determined arsenic species were the arsenious acid, arsenic acid, monomethylarsinic acid, dimethylarsinic acid, arsenocholine, and arsenobetaine, with detection limits ranging from 6 μgL^{-1} As (As^{3+}, AB) to 25 $\mu g\ L^{-1}$ (As^{5+}) in standards aqueous arsenic solutions. The apparatus used was an ion-pair reversed phase chromatography-ICP-MS system.

Table 3: Collective table with biological matrices and analytes in selective references of the last two decades.

Reference	Sample	Analyte
Multielement Analysis		
Fink and Manley 2011	giantkelp	Al, V, Cr, Mn, Ni, Co, Cu, Zn, As, Se, Rb, Sr, Ag, Cd, Cs, Hg, Pb
Brito et al. 2012	10 microalgae species	As, Ba, Cd, Co, Cr, Cu, Li, Mn, Ni, Pb, V, Zn
Rudel et al. 2010	bladder wrack, blue mussel, eelpout, and eggs of herring gulls	Cd, Hg, Pb
Hung et al. 2007	*humpback dolphin* and *finless porpoise*	Ag, As, Cd, Co, Cr, Cs, Cu, Hg, Mn, Ni, Se, V, Zn
Hung et al. 2006	*humpback dolphins*	Ag, As, Cd, Co, Cr, Cs, Cu, Hg, Mn, Ni, Se, V, Zn
Moren et al. 2006	Antarctic krill, Arctic krill, Arctic amphipod *Themistolibellula*	Cu, Zn, As, Cd, Hg, Pb
Venkateswara et al. 2009	marine sponge	As, Cd, Co, Cu, Fe, Mn, Ni
Kamaruzzaman et al. 2010	8 fish species	Zn, Cu, Pb
Cubadda et al. 2006	mussels and various fish species	Al, As, Cd, Co, Cr, Cu, Fe, Hg, Mn,Mo, Ni, Pb, Se, Sn, V, Zn
Das et al. 2009	red mullet, whiting and mussel, turbot, veined rapa whelk, and halibut	Pb, Cd, Hg, As
Zhang et al. 2009	mollusk, barnacle, shrimp, and fish samples	Y, La, Ce, Pr, Nd, Sm, Eu, Gd, Tb, Dy, Er, Ho, Tm, Lu
Shao et al. 2011	wild shrimps and crabs	Sn, Hg
Okoshi and Ishii 1996	radular teeth of limpets, chitons, and other marine mollusks	17 elements (Fe, Si, Al, Ca, K, Mg, Na, P, Ca, Sr, and more)
Palmer et al. 2006	gammaridean amphipod	Mg, Sb, Sn, V, Zn, Cu, Cd, Fe, Mn, Pb, La, Ce, Pr, Nd, Sm, Eu, Gd, Dy, Er, Yb, Lu
Eira et al. 2008	nematoda and cestoda	As, Cd, Cr, Cu, Hg, Ni, Pb, Zn, Pd, Pt
Shams et al. 2013	marine microalga	Rh, Pd, Pt
Sherwood et al. 2009	ocean water in various depths	Al, Cd, Cr, Cu, Fe, Pb, Mn, Ni, Zn
Hola et al. 2009	scales of fish	Fe, Mn, Cd, Zn, Cr, V, Co, Pb, Cu, Ni, and As
Hola et al. 2011	scales of common carp (*Cyprinus carpio*)	Zn, Mg, Mn, Sr, Ca, P, Fe, and Pb
Huelga-Suarez et al. 2013	salmon otoliths	Ba isotopes

Table 3 contd....

Table 3 contd...

Reference	Sample	Analyte
Single Element Analysis		
Al-Subiai et al. 2011	marine bivalve mollusc	Cu
Zheng et al. 2012	coastal seawater	I
Charmasson et al. 2009	*Paralvinella grassleis*, vestimentiferanworms, bivalve molluscs, shrimps, *amphipod Orchomenella* sp.	U
Takata et al. 2011	seawater, sediment, macroalgae, crustaceans, molluscs	U
Kasemann et al. 2009	marine carbonates	B
Metal Speciation Analysis		
Di Carro and Magi 2010	marine environment	Sn speciation
Mattusch et al. 2008	natural zeolites	As speciation
Contreras-Acuna et al. 2013	anemoniasulcate	As speciation
Kohlmeyer et al. 2002	fishes, mussels, oysters, and marine algae	As speciation
Thomas and Sniatecki 1995	fishes and mussels	As speciation
Terol et al. 2012	algae and crustaceans	As speciation
Goessler et al. 1997	seaweed *Hormosirabanksii*, the gastropod *Austrocochleaconstricta*, and gastropod *Morulamarginalba*	As speciation
Takeuchi et al. 2005	coastal marine sediments	As speciation
Koch and Reimer 2012	marine snails	As speciation
Carioni et al. 2014	tuna fish tissue and robalo liver tissue	As speciation

The most thorough study exclusively for arsenic speciation in marine biota was done by Kohlmeyer et al. in 2002. Fish, mussel, oyster, and marine algae samples were analyzed with an HPLC-ICP-MS system and 17 different arsenic species were determined. The limits of detection were calculated for eight arsenic species ranging between 0.008–0.500 $\mu g\ AsL^{-1}$. An interesting observation is that in both works (Kohlmeyer and Jantzen 2002; Thomas and Sniatecki 1995) the lowest LOD was obtained for arsenite, while the higher LOD was obtained for arsenate. The tendency of arsenate to give higher LOD is something that can be observed also in the Price et al. (2013) study where the uptake and bioaccumulation of geothermally derived inorganic As by the soft coral *Clavularia* sp., the calcareous algae *Halimeda* sp., and the sea squirt *Polycarpa* sp. was investigated.

A variation of liquid chromatographic techniques is used for the analysis of arsenosugars in marine biological samples, such as algae and crustaceans. This is the high temperature liquid chromatography that coupled with the ICP-MS system (Terol et al. 2012) allowed researchers to determine four different sugars in this category (glycerol, phosphate, sulfonate, and sulfate). Some of the advantages of this approach for the arsenosugars analysis are the reduction of retention times and the improved sensitivity, while in correlation with the conventional HPLC-ICP-MS method, the main advantage is the avoiding of the saline buffers and high organic solvent content use in the mobile phase. As a result the generated waste is less toxic and the ICP-MS matrix effects are reduced.

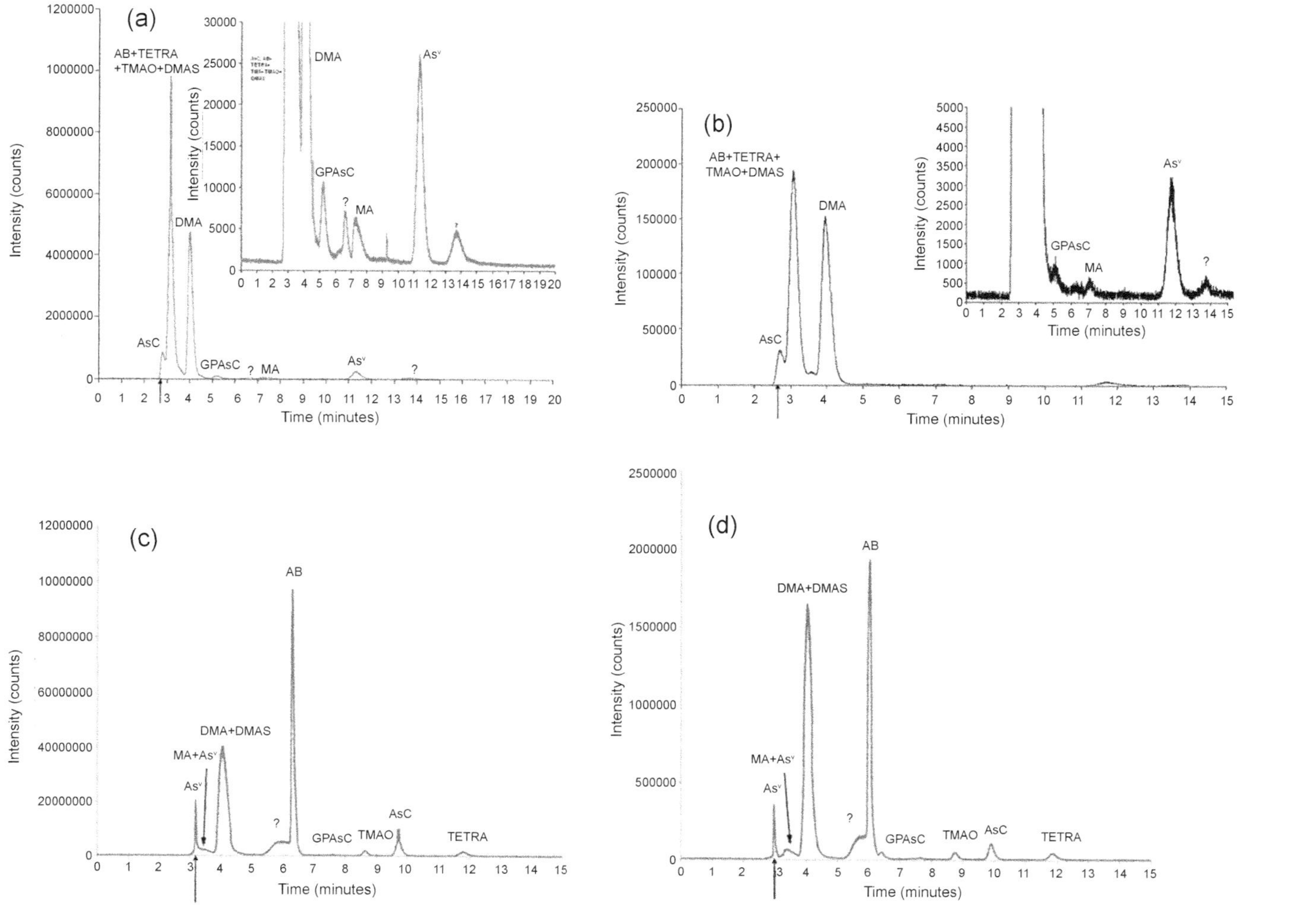

Fig. 1: Chromatograms obtained from a pool of anemones raw (a) and cooked (b) using HPLC-(AEC)–ICP-MS and HPLC-(CEC)–ICP-MS (raw, c and cooked, d). The black arrow in the abscissa indicates the void volume of the columns. Operating conditions of HPLC–ICP-ORC-MS are summarized in Table 1. The insets of (a) and (b), represent the same chromatograms magnified. (Reprinted from Contreras-Acuña et al. 2013, with permission from Elsevier.)

A more spherical approach would be to study not only the arsenic concentration into one or two different marine species, but to follow the course of it inside a food chain. That was exactly what Goessler et al. (1997) had done by studying the seaweed *Hormosirabanksii*, the gastropod *Austrocochleaconstricta* which consumes the seaweed, and the gastropod *Morulamarginalba* which eats *Austrocochleaconstricta*. The results showed an increase of total arsenic concentration as moving to the top of the food chain, in the seaweed *Hormosirabanksii* 27.2 μg g^{-1} As dry mass, in the gastropod *Austrocochleaconstricta* 74.4 μg g^{-1}As dry mass, and in the gastropod *Morulamarginalba* 233 μg g^{-1} As dry mass were determined. These results indicate clearly that biomagnification occurs in the studied samples.

Except in marine biota, the As levels have also been studied in coastal marine sediments. The concentration of As in the living environment is certainly a parameter that should be taken into consideration when someone studies marine biota, not only because it could affect the concentration in the living underwater organisms, but also because these organoarsenicals are mainly formed by marine organisms that are delivered to the sediment and can be degraded within several decades. Takeuchi et al. in 2005 found that the arsenobetaine was the dominant organoarsenical at four of the seven stations. Furthermore, core analysis showed that concentrations of organoarsenicals were inversely proportional with depth, and that they are degraded within 60 years of deposition.

In order to have accurate results, the extraction of the analytes should be quantitative and the procedure repeatable. This is the topic of the paper by Koch and Reimer (2012), which evaluates (with the use of ICP-MS) the effectiveness of three procedures for the As extraction from marine snails.

This increasing demand for arsenic analysis in marine biota has led to the need for the production of new reference materials, such as tuna fish tissue and robalo liver tissue. Testing the RMs was the aim of a recently reported study by Carioni et al. (2014). In order to do so, the researchers' first step was the elemental characterization of the candidate RMs by INAA. The INAA is a nuclear process used for determining the total concentration of various elements. INAA relies on excitation by neutrons so that the treated sample emits gamma-rays. Afterwards, the extraction procedure was optimized, and the arsenic speciation in the extracts was performed with the use of cation exchange liquid chromatography-ICP-MS. The results demonstrated that there was excellent agreement between calculated and reported values of arsenic concentration in the specific CRMs.

Speciation of Mercury and Lead in Marine Organisms and Aquatic Ecosystems

Mercury is an element of high toxicity that is associated with various health problems in humans. Even moderate Hg concentrations could cause problems in the neurological and immune systems, lungs, kidneys, skin, and eyes. It is especially dangerous for fetus development, and studies have shown that it could adversely affect a baby's growing brain and nervous system. There are two main pathways through which people can be exposed to Hg. One is drinking water and the other is the consumption of seafood. The primary Hg species present in these substrates are inorganic Hg^{2+}, and organic mercury in the form of MeHg. The second one, which is formed by methylation of inorganic Hg ions, is much more toxic than the Hg^{2+} and this is the main reason why Hg speciation is of such great importance.

The separation and determination of at least the two of these mercury species was investigated by numerous researchers groups such as Balarama Krishna et al. (2010) and Santoyo et al. (2009). In the first report, the studied samples included lake water, seawater, and tuna fish tissue. A polyaniline microcolumn was used for the separation and preconcentration of mercury species. This column was synthesized by chemical oxidation aniline using as medium ammonium peroxydisulphate in HCl. This column exhibited different behavior on different pH levels, and this was the characteristic that allowed the successful separation. At pH < 3 the MeHg was passing through the column while the inorganic Hg was absorbed and could be eluted afterwards with a proper elution solvent. But when the pH value was approaching 7, both of the species were sorbed by the column. The subsequent determination of the separated Hg species was made by a chemical vapour generation ICP-MS. The researchers of the second group did the separation and analysis of the L-cysteine mercury complexes formed in an HPLC-ICP-MS system but they also used Bi^{3+} as an internal standard in order to improve the relatively poor reliability that had been observed in some previous studies. The reason behind bismuth selection was that it was considered that it would resemble the behavior of inorganic Hg during extraction and HPLC separation procedures. The studied samples were two kinds of fish tissues, one was red snapper liver, and the other was king mackerel muscle.

The speciation of three Hg species was the goal of another two papers published in 2007 and 2012 respectively (Chang et al. 2007; Zhao et al. 2012). In both cases, the investigated species were inorganic Hg, MeHg, and EtHg, and both papers did not follow the usual approach in an aspect of their research. The first group chose a less common nebulizer, while the second group selected the capillary electrophoresis as the separation technique applied. In the study of 2007, the analyzed samples were reference materials of dogfish and swordfish muscles, the separation was done by RP liquid chromatography with a C18 column, and with the exception of Hg and Pb, speciation was performed with the detected lead species such as inorganic lead ions, trimethyl lead, and triethyl lead. The alteration was in the nebulizer type. Instead of the more common pneumatic nebulizer, the DIN nebulizer was chosen. That decision was based on the main advantages of this kind of nebulizer that are the almost quantitatively nebulizing ability, the tolerance for high concentrations of organic solvents, and the decreased impact of the memory effect. The second study chose to experiment with the separation step. They examined capillary electrophoresis rather than gas or liquid chromatography. Gas chromatography has the disadvantage that it needs a prior derivatization step before the injection of the sample, while liquid chromatography required formation of complexes, usually with L-cysteine, that in some cases are not stable enough to survive until the determination step. Furthermore CE outclasses the other chromatographic techniques at operating cost, sample and reagent consumption, and the absence of synergic effects between sample and stationary phase. The analyzed matrices were dried fish muscle (*T. anchovy*) and river water.

A similar study considered the use of CE for Pb speciation (Lee and Jiang 2005). Lead species included Pb^{2+}, tri-MePb, and tri-EtPb. Two reference materials, oyster tissue and dogfish liver, and two real samples, i.e., fish muscle and oysters were analysed. The capillary material used was of fused silica and the extraction efficiency was calculated between 97% and 102%.

The last study reported on the speciation of four Hg species (Chen et al. 2013). These are Hg^{2+}, MeHg, EtHg, and PhHg, and the substrate analyzed was seawater and fish. Except L-cysteine, thiourea was also tested as possible complexing reagent and eluent. The results showed that both of these reagents could successfully be used in the above research. Selective limits of detection for Hg species are summarized in Table 4.

Metalloproteins Analysis in Marine Organisms and Aquatic Ecosystems

Metalloprotein is a generic term for a protein that contains a metal ion cofactor. The presence of this metal is essential for protein functions. Metalloproteins are responsible for many of the biochemical procedures in a cell, such as photosynthesis, respiration, and storage and transport of proteins. Nearly 50% of the known proteins have been found to belong to that category. Some of the most common metals found in metalloproteins complexes are Co, Mn, Zn, and Cu.

The procedure of photosynthesis has already been mentioned as a procedure in which metalloproteins are involved. Specifically the chlorophyll that can be found in chloroplasts of algae and plants contains a porphyrin ring, which is a group of magnesium complexes, and this is the reason why a study have been made trying to identify the magnesium isotope measurement in the different chlorophyll forms (a, b, c1 + c2, and c3) of phytoplankton (Ra and Kitagawa 2007). Magnesium has three naturally occurring isotopes, ^{24}Mg, ^{25}Mg, and ^{26}Mg, in a 78.99%, 10.00%, and 11.01% analogy, respectively. An HPLC-ICP-MS method was applied constituted from two separation steps. The first was the separation of chlorophyll forms, and the second was the separation of Mg from sea salts (Na, K, and Ca) with the use of a cation exchange column. The results suggested biological processing mass fractionation of Mg.

The metalloproteins profile of the marine cyanobacterium *Synechococcus* sp. WH8102 was investigated by Barnett et al. 2012. The analytical procedure included three parts. Firstly, multidimensional liquid chromatography was applied with scope to fractionate the proteome of the cyanobacterium. Secondly, the fractions were analyzed in the ICP-MS system so that those containing specific metals (Fe, Ni, Mn, and

Table 4: LODs of mercury species determined in the papers refered to in the section "Speciation of mercury and lead in marine organisms and aquatic ecosystems".

Reference	Sample matrix	Analyte/LOD
Balarama Krishna et al. 2010	tuna fish tissue	Hg^{2+} = 0.252 μg L^{-1} MeHg = 0.324 μg L^{-1}
Santoyo et al. 2009	king mackerel muscle and red snapper liver	Hg^{2+} = 0.8 μg L^{-1} MeHg = 0.7 μg L^{-1}
Chang et al. 2007	dogfish muscle and a swordfish	Hg^{2+} = 0.2 μg L^{-1} MeHg = 0.2 μg L^{-1} EtHg = 0.3 μg L^{-1} Pb^{2+} = 0.1 μg L^{-1} tri-MePb = 0.1 μg L^{-1} tri-EtPb = 0.3 μg L^{-1}
Zhao et al. 2012	dried fish (*T. anchovy*) muscle and river water	Hg^{2+} = 0.027 μg L^{-1} MeHg = 0.021 μg L^{-1} EtHg = 0.032 μg L^{-1}
Lee and Jiang 2005	oyster tissue and dogfish liver	Pb^{2+} = 0.2 μg L^{-1} tri-MePb = 0.5 μg L^{-1} tri-EtPb = 0.6 μg L^{-1}
Chen et al. 2013	pomfret (*pampusargenteus*), hairtail (*trichiurusjaponicus*), small yellow croaker (*larimichthyspolyactis*), brown croaker (*miichthysmiiuy*) and japanese seabass (*lateolabraxjaponicus*)	Hg^{2+} = 0.019 μg L^{-1} MeHg = 0.027 μg L^{-1} EtHg = 0.031 μg L^{-1} PhHg = 0.022 μg L^{-1}

Co) could be identified. Thirdly, the protein fractions that were found to be bound with metals were isolated from the soluble fraction of cyanobacterium cells by binding them on IMAC columns. This combination of 2D-LC-ICP-MS with IMAC led to the identification of various metal binding proteins in the cell lysates of cyanobacterium *Synechococcus* sp. WH8102.

The same approach was also reported in another paper dealing with the distribution of selected trace elements, i.e., essential (Co, Cu, Fe, Mn, Mo, Se, and Zn) and nonessential (Cd, Pb), among protein fractions from hepatic cytosol of European chub (*Squaliuscephalus* L.) published by Krasnici et al. (2013). The whole procedure is constituted from the fractionation step and the subsequent analysis step. At first, size exclusion liquid chromatography is used to fractionate chub's hepatic cytosols (Fig. 2). Next, the fractions were analyzed by high resolution ICP-MS. The obtained results confirmed the presence of several of the studied elements in the hepatic cytosol in various concentrations (maximum to minimum ratio, 7–20). Interestingly, Pb was identified in a significant number of proteins varying from 30 to 100 kDa.

A comparative study between two HPLC-ICP-MS methods for the separation and determination of transferrin glycoforms in blood samples of North Sea harbour seals have been described (Grebe et al. 2012). The quantification of the identified glycoforms was based on their iron content. In the first approach, the quantification of transferrin glycoforms was achieved using a FIA system, a one point calibration, and a certified Fe standard. The second tested method was based on the determination of different stable isotopes of Fe by applying isotope dilution analysis (IDA) (Fig. 3). In total 15 organisms were tested and the results of transferrin concentration for the 11 of them were practically the same with both methods. Furthermore, the two methods identified the S_4 as the most abundant glycoform. In conclusion, both methods exhibited satisfactory agreement, proving that both could be applied for the transferrin glycoforms determination. The FIA method was less complicated, while the IDA method provides additional information rather than the different glycoform concentrations.

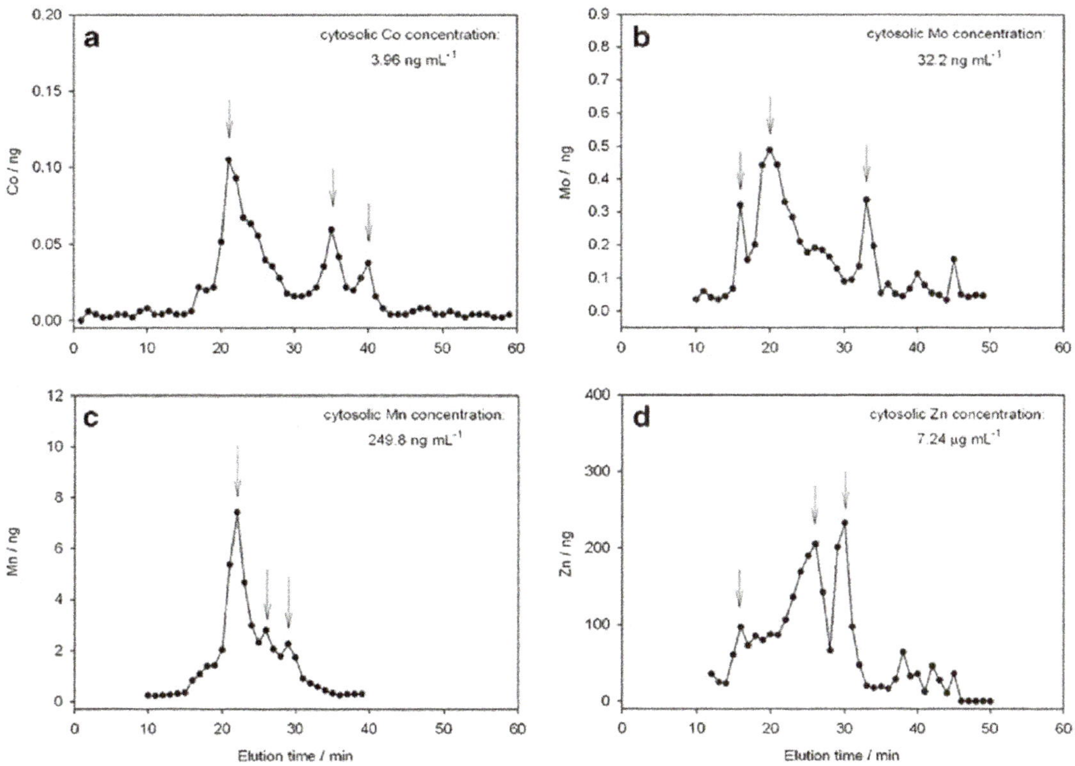

Fig. 2: Distribution profiles of essential trace elements: (a) Co; (b) Mo; (c) Mn; and (d) Zn, among cytosolic fractions of chub liver containing proteins of different molecular masses, separated by SE-HPLC with Superdex™ 200 10/300 GL column; the results are presented as nanograms of trace element eluted at specific elution times, after passing 200 µL of hepatic cytosol through the chromatographic column. (Reprinted from Krasnici et al. 2013, with permission from Springer.)

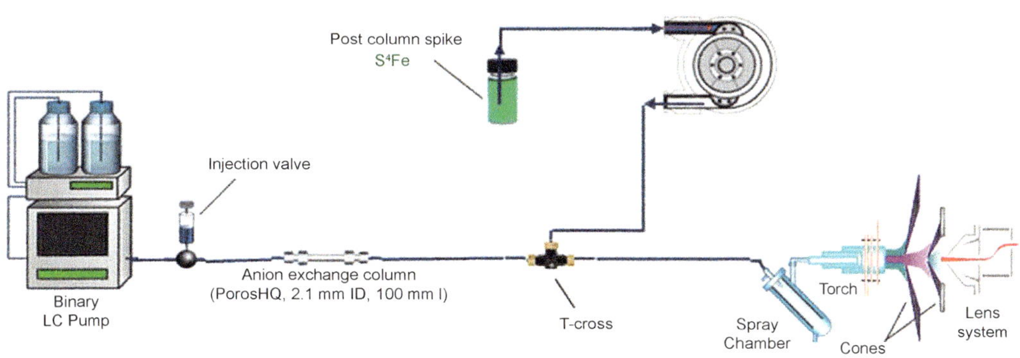

Fig. 3: Instrumental setup for the quantification of transferrin for both post-column IDA and double spiking IDA. (Reprinted from Grebe et al. 2012, with permission from the Royal Society of Chemistry.)

Mercury is known to bind with many cysteine containing proteins in cells. The speciation of Hg in salmon cell egg cytoplasm was investigated with an HPLC-ICP-MS system (Hasegawa et al. 2005). The total Hg concentration in cytoplasm was initially determined in FIA mode. For increased precision, the standard addition method was employed for four concentration levels and each spiked sample was measured three times (Fig. 4). The total Hg determination was followed by the fractionation which was done using a CHAPS-coated ODS column and the subsequent Hg determination using an HPLC system with dual detection systems, UV-absorption, and an ICP-MS.

The obtained data indicated that the Hg in salmon egg cell cytoplasm tended to bind with proteins containing cysteine or/and selenocysteine, probably due to the high affinity of mercury to sulphur traces. In another study about mercury, the methylmercury tendency towards proteins was explored (Kutscher et al. 2012). The studied matrices were tuna fish and muscle tissue, while various techniques were combined in this research to obtain the most thorough results possible. These included size-exclusion chromatography, liquid chromatography, ICP-MS as well as other complementary mass spectrometric techniques, ESI-MS/MS, and alternative separation techniques like SDS-PAGE. It was demonstrated that the major binding occurs to a protein above 200 kDa, leaving as potential candidate a skeletal muscle myosin chain.

There is another study that deals with mercury, along with Cd, Cu, and Zn, in the identification of MT complexes in a white-sided dolphin (*Lagenorhynchusacutus*) liver homogenate (Pedrero et al. 2012). What makes this study different from the previous one (Kutscher et al. 2012) is the fact that they decided to not apply the more common reversed phase HPLC but to use the hydrophilic interaction liquid chromatography (HILIC). The reason behind this choice was the intention to keep stable the fragile and rare metallothionein zinc–mercury mixed complexes (MT-Zn_6Hg). Metallothionein is known to exhibit high affinity for metals like Cd, Zn, and Cu, and this was also verified in the results which showed mixed complexes of Cd, Zn, Cu and Hg. Specifically the presence of Hg had never been detected before.

The Cd complexes with various cytosolic fractions, like MT, were also studied by Rodriguez-Cea et al. in 2006. More specifically, the way that Cd is accumulated into a fish kidney and liver was investigated. For that purpose, the stable ^{111}Cd isotope was used during this project. The selected organism was the European eel (*Anguilla anguilla*) and the analytical technique chosen was ICP-MS for determination of total Cd concentration, and anion-exchange ICP-MS for the speciation studies. In this way, researchers were able for the first time to define the redistribution mechanism of Cd between liver and kidney.

The combination of UV and MS detectors was used again by Vitoulova et al. in 2011. The aim of this project was the study of clam *Chameleagallina* for its total metal concentration (due to reason that its living

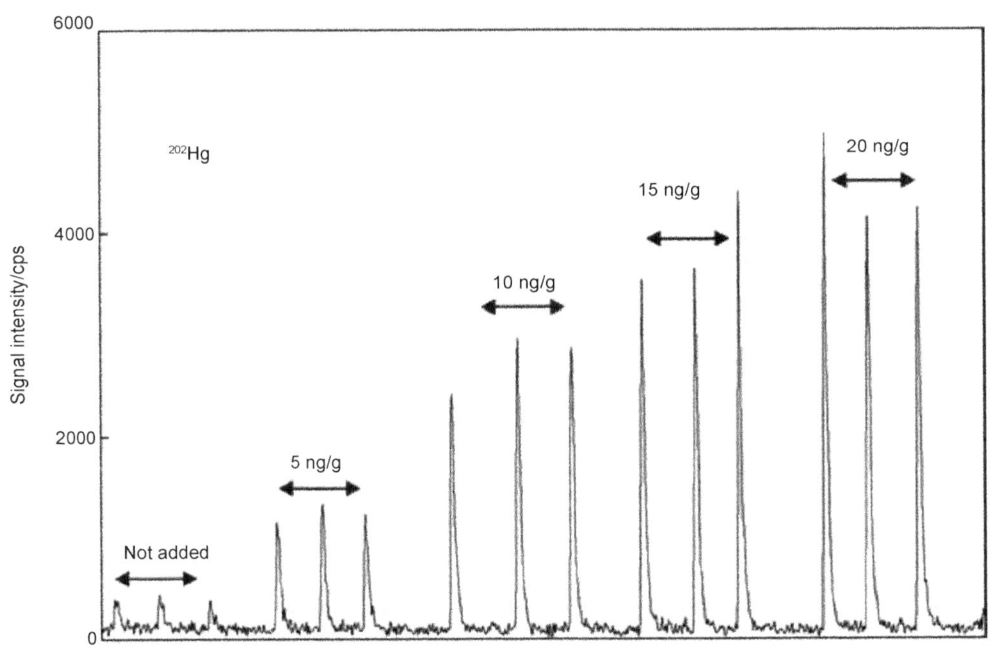

Fig. 4: Flow injection signals for ^{202}Hg in salmon egg cell cytoplasm, observed by the standard addition method. Samples: salmon egg cell cytoplasm diluted five-fold with 0.1 MTris–HNO_3 buffer solution (pH 7.4) with and without addition of Hg. Added amount of Hg: 5, 10, 15 and 20 ng g^{-1}. (Reprinted from Hasegawa et al. 2005, with permission from Elsevier.)

environment was heavily polluted due to mining activities that were taking place in the specific territory, the Atlantic southwest coast of Spain) and also for its metallomic content. After extensive analysis, two metal-binding protein fractions were isolated. Three metals were found to participate in these metalloproteins molecules, and these were As, Cu, and Zn.

Three very interesting papers were published from the research group of Gomez-Jacinto (Gomez-Jacinto et al. 2010, 2012a,b) and all three of them are dealing with a field that is expected to attract a lot of interest in the future, the biotechnological production of enriched foods. It is worth mentioning that many people believe that these types of foods will be the dominant type in the diet of people in the future because of the additional nutritional value they could possess. In order to investigate this specific field, the bioaccumulation of Se and I in algae species was studied. In the earlier research of 2010 by Gomez-Jacinto et al. the microalgae *Chlorella vulgaris* was used as a sample, while in the other two the microalgae *Chlorella sorokiniana* was investigated. The first study deals with I accumulation and applies SEC-ICP-MS and AEC to fractionate, analyze, and isolate peaks from the fractions. The final results confirmed the effectiveness of the specific separation procedure, detecting in the isolated fractions about 80% of the total iodine amount. In the second study (Gomez-Jacinto et al. 2012a), an HPLC-ICP-MS system with two chromatographic columns, one C18 type and one chiral type, was used for Se metallomics determination in the algae and culture medium. Two main conclusions emerged from the obtained data. The first was that except the remains of the selenium source used (selenate), three selenium binding proteins were detected, namely selenocystine (SeCys), selenomethyl-selenocysteine (SeMeSeCys), and primarily selenomethionine (SeMet). The second was that in the absence of sulfur, the selenium, due to their chemical resemblance, had the tendency to take its place, increasing this way the SeMet and Se(VI) concentration in the algae. In the third research (Gomez-Jacinto et al. 2012b), both the enrichments with Se and I (KI) were investigated. The well being of the colony were checked daily and the results showed that the I accumulation was much more explosive, accumulating 1.2 mg g^{-1} within 24 h in comparison to Se which needed 100 h to reach just up to 3 $\mu g\ g^{-1}$.

In the research reported by Infante et al. (2003) another mass spectrometry method, the time of flight mass spectrometry method (TOFMS) was used. The TOF mass analyzer measures the times it takes to specific ions to travel the distance between the ion source and the detector. Some of its advantages are its high ion transmission and the facts that it's the fastest mass analyzer, has the highest practical mass range, and it's ideal for pulsed ionization methods. The last one is at the same time one of its drawbacks since it needs the pulsed ionization method or ion beam switching in order to work. The SEC-HPLC-TOF-ICP-MS system was used for the determination of Cu, Zn, and Cd in carp and eel cytosols of fish tissue.

Lastly, an all-element analysis that was done on salmon egg cells should be referred to (Haraguchi et al. 2008). Three techniques were combined, ICP-AES, ICP-MS, and CHN elemental analysis resulting in the identification of 74 elements. This analysis was followed by protein-binding metallic elements determination by an HPLC-ICP-MS system using an octadecylsilyl silica (ODS) column and size exclusion chromatography (SEC).

Conclusions

The hyphenation of inductive coupled plasma with mass spectrometry is a very useful and powerful analytical tool that nowadays holds one of the dominant positions in element and multielement analysis of marine organisms as briefly presented in this review article. Of course there are a lot of reasons for something like this happening. ICP is a powerful atomization technique that can be applied to almost any sample matrice after proper treatment. The MS on the other hand is an ideal detector to be conjugated to ICP, collecting the analyte particles and separating and identifying their ions.

Furthermore, the various alternative modes that can be employed to the ICP-MS apparatus makes it possible for the marine bioanalysts to customize it according to their particular needs. Two of the main parts that can be selected according to the type of analyses are the sample introduction system and the mass analyzer. The selection of the appropriate sample introduction system mainly depends on the sample state. If it is a liquid sample the nebulizer is the system of choice while if it is solid, the LA system could be selected. Regarding the mass analyzer selection, there are few alternatives with different characteristics, such as quadrupole, TOF, SF and some others. However, apart from the already mentioned reasons, may be the main advantage of the ICP-MS system is its easy hyphenation with chromatographic techniques, allowing selective speciation and metallomic analysis. It appears that the current and future trend is focused on the development of ICP-MS based methodologies to investigate unknown biochemical paths in marine biota.

Clearly it is becoming apparent why this technique has found such wide use in marine biota analysis. The samples that have been analyzed with ICP-MS cover almost every aspect of the aquatic ecosystem. This can be applied across a wide range of marine biota, from the simplest organisms, like microalgae, bivalves, marine sponges, amphipods and marine snails, to the most advanced such as crabs, eelpouts, mullets, and even dolphins. The wide application spectrum is not restrained only to the sample type but is also extended to the determination of analytes.

The total concentration of heavy metals and many other elements (Al, V, Cr, Mn, Ni, Co, Cu, Zn, As, Se, Rb, Sr, Ag, Cd, Cs, Hg, Pb, Ba, Li, V, Y, La, Ce, Pr, Nd, Sm, Eu, Gd, Tb, Dy, Er, Ho, Tm, Lu, Sn, Rh, Pd, and Pt) have been determined with the ICP-MS technique. Furthermore, it has been applied to many speciation studies of metals such as As, Hg, Sn, and Pb. The speciation analysis becomes feasible after hyphenation with a separation technique, usually a type of liquid chromatography (HPLC, SEC, AEC, and CEC). This field (speciation) is of great importance in, environmental sample analysis, and mainly in edible organisms, because one factor that depends greatly on the specific species of the element is its possible toxicity and its impact on human health. A characteristic example is that in the two main forms of Hg in marine organisms, MeHg is more toxic while the inorganic Hg^{2+} is far less dangerous.

Last but not least, ICP-MS has also been applied to the study of metalloproteins. Metalloprotein is a generic term for a protein that contains a metal ion cofactor. The presence of such metal is essential for protein functions. The application of ICP-MS to that kind of analyses has helped to shed light on many interesting aspects of element bioaccumulation and how various metals are binding to the specific proteins. ICP-MS have earned its place as one of the most frequently used tools of marine biota analyses and as is becoming obvious, it has offered valuable data and information in more than one or two areas of this field.

Keywords: Marine organisms, metal-biomolecules, metalloenzymes, metallomics, metalloproteins, toxic metals, ICPMS, ICPAES, LC-ICPMS, hyphenated techniques, speciation analysis, bioanalysis, multielemental, bioaccumulation, biomonitoring

References

Al-Subiai, S.N., A.J. Moody, S.A. Mustafa and A.N. Jha. 2011. Ecotoxicology and environmental safety A multiple biomarker approach to investigate the effects of copper on the marine bivalve mollusc, *Mytilus edulis*. Ecotoxicol. Environ. Saf. 74: 1913–1920.

Balarama Krishna, M.V., K. Chandrasekaran and D. Karunasagar. 2010. On-line speciation of inorganic and methyl mercury in waters and fish tissues using polyaniline micro-column and flow injection-chemical vapour generation-inductively coupled plasma mass spectrometry (FI-CVG-ICPMS). Talanta 81: 462–472.

Barnett, J.P., D.J. Scanlan and C.A. Blindauer. 2012. Fractionation and identification of metalloproteins from a marine cyanobacterium. Anal. Bioanal. Chem. 402: 3371–3377.

Becker, J.S. 2007. Inorganic Mass Spectrometry. John Wiley and Sons, Chichester, England.

Brito, G.B., T.L. De Souza, F.C. Bressy, C.W.N. Moura and M.G.A. Korn. 2012. Levels and spatial distribution of trace elements in macroalgae species from the Todos os Santos Bay, Bahia, Brazil. Mar. Pollut. Bull. 64: 2238–2244.

Broekaert, J.A.C. and P.W.J.M. Bowmans. 1987. Sample introduction techniques in ICP-AES. pp. 296–356. *In*: P.W.J. Bomans [ed.]. Inductively Coupled Plasma Emission Spectroscopy, Part I: Methodology, Instrumentation and Performance. Wiley Interscience, Canada.

Broekaert, J.A.C. 2002. Analytical Atomic Spectrometry with Flame and Plasmas. Wiley–VCH Verlag GmbH, Weinheim, Germany.

Browner, R.F., S. Hill and J. Ed. 2007. Inductively Coupled Plasma Spectrometry and its Applications. Blackwell Publishing, Oxford, England.

Carioni, V.M.O., C.S. Nomura, L.L. Yu and R. Zeisler. 2014. Use of neutron activation analysis and LC–ICP-MS in the development of candidate reference materials for as species determination. J. Radioanal. Nucl. Chem. 299: 241–248.

Carr, J.E., K. Kwok, G.K. Webster and J.W. Carnahan. 2006. Effects of liquid chromatography mobile phases and buffer salts on phosphorus inductively coupled plasma atomic emission and mass spectrometries utilizing ultrasonic nebulization and membrane desolvation. J. Pharm. Biomed. Anal. 40: 42–50.

Chang, L.F., S.J. Jiang and A.C. Sahayam. 2007. Speciation analysis of mercury and lead in fish samples using liquid chromatography-inductively coupled plasma mass spectrometry. J. Chromatogr. A 1176: 143–148.

Charmasson, S., P.M. Sarradin, A. Le Faouder, M. Agarande, J. Loyen and D. Desbruyeres. 2009. High levels of natural radioactivity in biota from deep-sea hydrothermal vents: a preliminary communication. J. Environ. Radioact. 100: 522–526.

Chen, X., C. Han, H. Cheng, Y. Wang, J. Liu, Z. Xu and L. Hu. 2013. Rapid speciation analysis of mercury in seawater and marine fish by cation exchange chromatography hyphenated with inductively coupled plasma mass spectrometry. J. Chromatogr. A 1314: 86–93.

Contreras-Acuna, M., T. Garcia-Barrera, M.A. Garcia-Sevillano and J.L. Gomez-Ariza. 2013. Speciation of arsenic in marine food (Anemonia sulcata) by liquid chromatography coupled to inductively coupled plasma mass spectrometry and organic mass spectrometry. J. Chromatogr. A 1282: 133–141.

Cubadda, F., A. Raggi and E. Coni. 2006. Element fingerprinting of marine organisms by dynamic reaction cell inductively coupled plasma mass spectrometry. Anal. Bioanal. Chem. 384: 887–896.

Currell, G. 2000. Analytical Instrumentation, Performance Characteristics and Quality. John Wiley & Sons, Chichester, England.

Das, Y.K., A. Aksoy, R. Baskaya, H.A. Duyar, D. Guvenc and V. Boz. 2009. Heavy metal levels of some marine organisms collected in Samsun and Sinop coasts of Black Sea, in Turkey. J. Anim. Vet. Adv. 8: 496–499.

Dean, J.R. 1997. Atomic Absorption and Plasma Spectroscopy, in Analytical Chemistry by Open Learning. John Wiley and Sons, Chichester, England.

Di Carro, M. and E. Magi. 2010. Chromatography: types, techniques and methods. pp. 369–388. *In*: T.J. Quintin [ed.]. Coupling Chromatography and Mass Spectrometry for the Study of Organotin Compounds in the Marine Environment. Nova Science Publishers, New York, USA.

Eira, C., J. Torres, J. Miquel, J. Vaqueiro, A.M.V.M. Soares and J. Vingada. 2008. Trace element concentrations in Proteocephalus macrocephalus (Cestoda) and Anguillicola crassus (Nematoda) in comparison to their fish host, Anguilla anguilla in Ria de Aveiro, Portugal. Sci. Total Environ. 407: 991–998.

Evans, E.H., J.J. Giglio, T.M. Castillano and J.A. Caruso. 1995. Inductively Coupled and Microwave Induced Plasma Sources for Mass Spectrometry. The Royal Society of Chemistry, Cambridge, England.

Fink, L.A. and S.L. Manley. 2011. The use of kelp sieve tube sap metal composition to characterize urban runoff in southern California coastal waters. Mar. Pollut. Bull. 62: 2619–2632.

Goessler, W., W. Maher, K.J. Irgolic, D. Kuehnelt, C. Schlagenhaufen and T. Kaise. 1997. Arsenic compounds in a marine food chain. Fresenius J. Anal. Chem. 359: 434–437.

Gomez-Jacinto, V., A. Arias-Borrego, T. Garcia-Barrera, I. Garbayo, C. Vilchez and J.L. Gomez-Ariza. 2010. Iodine speciation in iodine-enriched microalgae *Chlorella vulgaris*. Pure Appl. Chem. 82: 473–481.

Gomez-Jacinto, V., T. Garcia-Barrera, I. Garbayo, C. Vilche and J.L. Gomez-Ariza. 2012a. Metallomic study of selenium biomolecules metabolized by the microalgae Chlorella sorkiniana in the biotechnological production of functional foods enriched in selenium. Pure Appl. Chem. 84: 269–280.

Gomez-Jacinto, V., T. Garcia-Barrera, I. Garbayo, C. Vilche and J.L. Gomez-Ariza. 2012b. Metal-metabolomics of microalga Chlorella sorokiniana growing in selenium- and iodine-enriched media. Chem. Pap. 66: 821–828.

Grebe, M., D. Profrock, A. Kakuschke, M.E. del Castillo Busto, M. Montes-Bayon, A. Sanz-Medel, J.A.C. Broekaert and A. Prange. 2012. Comparison of different methods for the absolute quantification of harbor. J. Anal. At. Spectrom. 27: 440–448.

Haraguchi, H., A. Ishii, T. Hasegawa, H. Matsuura and T. Umemura. 2008. Metallomics study on all-elements analysis of salmon egg cells and fractionation analysis of metals in cell cytoplasm. Pure Appl. Chem. 80: 2595–2608.

Hasegawa, T., M. Asano, K. Takatani, H. Matsuura and T. Umemura. 2005. Speciation of mercury in salmon egg cell cytoplasm in relation with metallomics research. Talanta 68: 465–469.

Hola, M., J. Kalvoda, O. Babek, R. Brzobohaty, I. Holoubek, V. Kanicky and R. Skoda. 2009. LA-ICP-MS heavy metal analyses of fish scales from sediments of the Oxbow Lake Certak of the Morava River (Czech Republic). Environ. Geol. 58: 141–151.

Hola, M., J. Kalvoda, H. Novakova, R. Skoda and V. Kanicky. 2011. Possibilities of LA-ICP-MS technique for the spatial elemental analysis of the recent fish scales: Line scan vs. depth profiling. Appl. Surf. Sci. 257: 1932–1940.

Huelga-Suarez, G., B. Fernandez, M. Moldovan and J.I.G. Alonso. 2013. Detection of transgenerational barium dual-isotope marks in salmon otoliths by means of LA-ICP-MS. Anal. Bioanal. Chem. 405: 2901–2909.

Hung, C.L.H., R.K.F. Lau, J.C.W. Lam, T.A. Jefferson, S.K. Hung, M.H.W. Lam and P.K.S. Lam. 2007. Risk assessment of trace elements in the stomach contents of Indo-Pacific Humpback Dolphins and Finless Porpoises in Hong Kong waters. Chemosphere 66: 1175–1182.

Infante, H.G., K.V. Campenhout, D. Schaumloffel, R. Blust and F.C. Adams. 2003. Multi-element speciation of metalloproteins in fish tissue using size-exclusion chromatography coupled "on-line" with ICP-isotope dilution-time-of-flight-mass spectrometry. Analyst 128: 651–657.

Kamaruzzaman, B.Y., M.C. Ong and S.Z. Rina. 2010. Concentration of Zn, Cu and Pb in some selected marine fishes of the Pahang coastal waters, Malaysia. American Journal of Applied Sciences 7: 309–314.

Kasemann, S.A., D.N. Schmidt, J. Bijma and G.L. Foster. 2009. *In situ* boron isotope analysis in marine carbonates and its application for foraminifera and palaeo-pH. Chem. Geol. 260: 138–147.

Koch, I. and K.J. Reimer. 2012. Talanta Arsenic species extraction of biological marine samples (Periwinkles, Littorina littorea) from a highly contaminated site. Talanta 88: 187–192.

Kohlmeyer, U., J. Kuballa and E. Jantzen. 2002. Simultaneous separation of 17 inorganic and organic arsenic compounds in marine biota by means of high-performance liquid chromatography/inductively coupled plasma mass spectrometry. Rapid Commun. Mass Spectrom. 16: 965–974.

Krasnici, N., Z. Dragun, M. Erk and B. Raspor. 2013. Distribution of selected essential (Co, Cu, Fe, Mn, Mo, Se, and Zn) and nonessential (Cd, Pb) trace elements among protein fractions from hepatic cytosol of European chub (Squalius cephalus L.). Environ. Sci. Pollut. Res. 20: 2340–2351.

Kutscher, D.J., A. Sanz-Medel and J. Bettmer. 2012. Metallomics investigations on potential binding partners of methylmercury in tuna fish muscle tissue using complementary mass spectrometric techniques. Metallomics 4: 807–813.

Lee, T. and S.J. Jiang. 2005. Speciation of lead compounds in fish by capillary electrophoresis-inductively coupled plasma mass spectrometry. J. Anal. At. Spectrom. 20: 1270–1274.

Mattusch, J., D. Moller, M.P.E. Gonzalez and R. Wennrich. 2008. Investigation of the degradation of arsenobetaine during its contact with natural zeolites and the identification of metabolites using HPLC coupled with ICP-MS and ESI-MS. Anal. Bioanal. Chem. 390: 1707–1715.

May, T.W. and R.H. Wiedmeyer. 1998. A table of polyatomic interferences in ICP-MS. At. Spectrosc. 19: 150–155.

Moore, G.L. 1989. Introduction to Inductively Coupled Plasma Atomic Emission Spectrometry. Elsevier, Amsterdam, Netherlands.

Moren, M., J. Suontama, G.I. Hemre Karlsen, R.E. Olsen, H. Mundheim and K. Julshamn. 2006. Element concentrations in meals from krill and amphipods. Possible alternative protein sources in complete diets for farmed fish. Aquaculture 261: 174–181.

Nolte, J. 2003. ICP Emission Spectrometry. A Practical Guide. Wiley–VCH Verlag GmbH & Co., Weinheim, Germany.

Okoshi, K. and T. Ishii. 1996. Concentrations of elements in the radular teeth of limpets, chitons, and other marine mollusks. J. Mar. Biotechnol. 3: 252–257.

Palmer, A.S., I. Snape, J.S. Stark, G.J. Johnstone and A.T. Townsend. 2006. Baseline metal concentrations in Paramoera walkeri from East Antarctica. Mar. Pollut. Bull. 52: 1441–1449.

Pedrero, Z., L. Ouerdane, S. Mounicou, R. Lobinski, M. Monperrusa and D. Amourouxa. 2012. Identification of mercury and other metals complexes with metallothioneins in dolphin liver by hydrophilic interaction liquid chromatography with the parallel detection by ICP MS and electrospray hybrid linear/orbital trap MS/MS. Metallomics 4: 473–479.

Price, R.E., J. London, D. Wallschlager, M.J. Ruiz-Chancho and T. Pichler. 2013. Enhanced bioaccumulation and biotransformation of as in coral reef organisms surrounding a marine shallow-water hydrothermal vent system. Chem. Geol. 348: 48–55.

Ra, K. and H. Kitagawa. 2007. Magnesium isotope analysis of different chlorophyll forms in marine phytoplankton using multi-collector ICP-MS. J. Anal. At. Spectrom. 22: 817–821.

Rodriguez-Cea, A., M. Rosario Fernandez, J.I. Garcia Alonso and A. Sanz-Medel. 2006. The use of enriched 111Cd as tracer to study *de novo* cadmium accumulation and quantitative speciation in Anguilla anguilla tissues. J. Anal. At. Spectrom. 21: 270–278.

Rudel, H., A. Fliedner, J. Kosters and C. Schroter-Kermani. 2010. Twenty years of elemental analysis of marine biota within the German environmental specimen bank - a thorough look at the data. Environ. Sci. Pollut. Res. 17: 1025–1034.

Santos, M.C. and J.A. Nobrega. 2006. Slurry nebulization in plasmas for analysis of inorganic materials. Appl. Spectrosc. Rev. 41: 427–448.

Santoyo, M.M., J.A.L. Figueroa, K. Wrobel and K. Wrobel. 2009. Analytical speciation of mercury in fish tissues by reversed phase liquid chromatography-inductively coupled plasma mass spectrometry with Bi3+ as internal standard. Talanta 79: 706–711.

Shams, L., T. Andrew, G.E. Millward and M.T. Brown. 2013. Science direct extra- and intra-cellular accumulation of platinum group elements by the marine microalga, Chlorella stigmatophora. Water Res. 50: 432–440.

Shao, L.N., Z.M. Ren, G.S. Zhang, L.L. Chen, D.Y. Liu, Z.J. Wang and J.P. Zhao. 2011. Study on concentrations of Sn, Hg in some edible marine organisms in Yantai coastal areas. Environ. Sci. 32: 1696–1702.

Sharp, B.L. 1987. Applications: agriculture and food. pp.65–99. *In*: P.W.J. Boumans [ed.]. Inductively Coupled Plasma Emission Spectroscopy, Part II: Applications and Fundamentals. Boumans P.W.J., Ed., Wiley Interscience, Canada Toronto.

Sherwood, J.E., D. Barnett, N.W. Barnett, K. Dover, J. Howitt, H. Ii, P. Kew and J. Mondon. 2009. Deployment of DGT units in marine waters to assess the environmental risk from a deep sea tailings outfall. Anal. Chim. Acta 652: 215–223.

Takata, H., T. Aono, K. Tagami and S. Uchida. 2011. Determination of naturally occurring uranium concentrations in seawater, sediment, and marine organisms in Japanese estuarine areas. J. Radioanal. Nucl. Chem. 287: 795–799.

Takeuchi, M., A. Terada, K. Nanba, Y. Kanai, M. Owaki, T. Yoshida, T. Kuroiwa, H. Nirei and T. Komai. 2005. Distribution and fate of biologically formed organoarsenicals in coastal marine sediment. Appl. Organometal. Chem. 19: 945–951.

Taylor, H.E. 2001. Inductively Coupled Plasma-Mass Spectrometry: Practices and Techniques. Academic Press. Netherlands.

Terol, A., F. Ardini, M. Grotti and J.L. Todoli. 2012. High temperature liquid chromatography – inductively coupled plasma mass spectrometry for the determination of arsenosugars in biological samples. J. Chromatogr. A 1262: 70–76.

Thomas, P. and K. Sniatecki. 1995. Inductively coupled plasma mass spectrometry: Application to the determination of arsenic species. Fresenius J. Anal. Chem. 351: 410–414.

Todoli, J.L., S. Maestre, J. Mora, A. Canals and V. Hernandis. 2000. Comparison of several spray chambers operating at very low liquid flow rates in inductively coupled plasma atomic emission, Spectrometry, Fresen. J. Anal. Chem. 368: 773–779.

Todoli, J.L. and J.M. Mermet. 2006. Sample introduction systems for the analysis of liquid microsamples by ICP-AES and ICP-MS. Spectrochim. Acta B 61: 239–283.

Venkateswara, R.J., K. Srikanth, R. Pallela and T.R. Gnaneshwar. 2009. The use of marine sponge, Haliclona tenuiramosa as bioindicator to monitor heavy metal pollution in the coasts of Gulf of Mannar, India. Environ. Monit. Assess. 156: 451–459.

Vitoulova, E., T. Garcia-Barrera, J.L. Gomez-Ariza and M. Fisera. 2011. Characterisation of metal-binding biomolecules in the clam chamelea gallina by bidimensional liquid chromatography with in series UV and ICP-MS detection, International Journal of Environmental Anal. Chemistry 91: 1282–1295.

Wrobel, K., K. Wrobel and J.A. Caruso. 2007. Metal analysis. pp. 205–226. *In*: Scott, R.A. and C.M. Lukehart [eds.]. Applications of Physical Methods to Inorganic and Bioinorganic Chemistry. Wiley Interscience, Canada, Toronto.

Zachariadis, G. 2012. Inductively Coupled Plasma Atomic Emission Spectrometry. Nova Science Publishers, New York, USA.

Zhang, H., L. Deng, J. Yang, J. Jiang, Z. Shen and J. Xie. 2009. Rare earth elements in marine organisms from Shenzhen coastal region. J. Hyg. Res. 38: 543–545.

Zhao, Y., J. Zheng, L. Fang, Q. Lin, Y. Wu, Z. Xue and F. Fu. 2012. Speciation analysis of mercury in natural water and fish samples by using capillary electrophoresis-inductively coupled plasma mass spectrometry. Talanta 89: 280–285.

Zheng, J., H. Takata, K. Tagami, T. Aono, K. Fujita and S. Uchida. 2012. Rapid determination of total iodine in Japanese coastal seawater using SF-ICP-MS. Microchem. J. 100: 42–47.

3

Mercury Speciation in Marine Biota and its Influence on Human Health

Metallomic and Metabolomic Approaches

Tamara García-Barrera,[1,a,*] *José Luis Gómez Ariza,*[1,b] *Gema Rodríguez Moro,*[1,c] *Cristina Santos Rosa*[2,d] and *Inés Velasco López*[2,e]

ABSTRACT

Mercury is a highly toxic element to marine organisms, especially those situated at the top level of the aquatic trophic web, which developed apparently different detoxification mechanisms as they are known to accumulate it without evidence of toxicity symptoms. However, this fact is of great importance for human health since seafood can contain high levels of mercury, especially methylmercury, which mainly affects the central nervous system and in severe cases, specific areas of the brain causing irreversible damage. The consumption of mercury during pregnancy is especially dangerous since it can cross the placenta barrier and can affect the foetus's health. In this chapter, we discuss recent advances in the research of mercury occurrence in seafood and biofluids, as well as speciation, bioaccessibility, and metabolomic studies in connection to mercury mode of action (MOA). In addition, metallomic studies allows deep insight into the interactions of this element with selenium biomolecules, formation of metallothionein complexes, and elucidation of mercury binding sites with proteins.

Introduction

Eating contaminated fish is the major source of human exposure to methylmercury, foetuses, infants, and young children being the most sensitive populations. The toxicity of mercury depends not only on its

[1] Department of Chemistry, Faculty of Experimental Sciences, Campus El Carmen, University of Huelva, Fuerzas Armadas Ave., 21007 Huelva, Spain.
[2] Hospital de Riotinto La Esquila, Ave., 21660 Minas de Riotinto (Huelva) Spain.
[a] Email: tamara@dqcm.uhu.es
[b] Email: ariza@uhu.es
[c] Email: gema.moro@dqcm.uhu.es
[d] Email: crinsantos@hotmail.com
[e] Email: inesvelas@msn.com
* Corresponding author

concentration, but also on its chemical form. Alkylated species like methylmercury ($MeHg^+$) are more toxic than inorganic forms, due to their easy penetration through biological membranes, high stability, and half-life in tissues (Horvat et al. 2003; Palenzuela et al. 2004; Tao and Willie 1998). While, $MeHg^+$ mainly affects the central nervous system (Leermakers et al. 2005), inorganic mercury is mainly accumulated in the liver and kidney through mechanisms involving its high affinity to sulphur, which leads to its linkage to endogenous thiol-containing molecules, as proteins or metabolites, which is the main mechanism regulating uptake, accumulation, and toxicity of inorganic Hg in the liver and kidney (Gado and Aldahmash 2013). In addition, mercury can interact with other elements which are also present in seafood or in the human body, affecting human health. One of the most known mechanisms of Hg detoxification by marine organisms involves the presence of selenium, probably by the formation of tiemanite (HgSe), a non-bioavailable insoluble compound (Ontario Ministry of Environment 2011).

Although inorganic mercury is the main form of mercury in food, in seafood the most abundant specie is methylmercury ($MeHg^+$), which is a chemical specie highly bioaccumulative and represents more than 60% of the total mercury commonly found in fish (Li et al. 2005). The amount of mercury in fish is related to the age of the fish and its position within the food chain; predatory and older fish have higher concentrations than others (EFSA Scientific Committee 2015). However, the consumption of fish, especially during pregnancy, also confers health benefits such as improved vision and cognitive development in offspring (Koletzko et al. 2008; Oken et al. 2005) and in adults, lower risk of sudden cardiac death (Kris-Etherton 2002) at least in part due to the content of omega 3-fatty acids in fish (Adkins and Kelley 2010). In addition, one key factor affecting human risk of seafood consumption is the bioaccesibility of mercury (the fraction of a contaminant that is soluble in the intestine, and is therefore available for the subsequent processes of absorption (García-Barrera et al. 2015; Thiry et al. 2012) which has been estimated to be < 50% for Hg for most fish species, while bioaccessible $MeHg^+$ can be only quantified in several seafood samples (Afonso et al. 2015; Cano-Sancho et al. 2015; Costa et al. 2015; Matos et al. 2015). Therefore, optimal health requires a balance between risk and benefits, which makes it difficult for public organizations to establish safety thresholds.

Analytical methods for total mercury determination in seafood and biosamples usually involve the use of an atomic spectrometric detector like atomic absorption spectrometry (AAS) (Cano-Sancho et al. 2015; Costa et al. 2016; Mieiro et al. 2014; Nair et al. 2014; Okyere et al. 2015; Spanopoulos-Zarco et al. 2014), atomic fluorescence spectroscopy (AFS) (Gómez-Ariza et al. 2004) or the multielemental inductively coupled plasma mass spectrometry (ICP-MS) (Karimi et al. 2014a,b; Moreno et al. 2010, 2013) coupled to GC or HPLC for speciation. Analytical approaches based in metabolomics have been applied to mammals exposed to mercury to decipher the mode of action of this contaminant (Agrawal et al. 2014; García-Sevillano et al. 2014, 2015a), while metallomic approaches make possible deeper insight into the interactions of this element with biomolecules (Blanusa et al. 1994; García-Sevillano et al. 2014, 2015a; Zalups 1995; Zalups and Koropatnick 2000), formation of metallothionein complexes (Pedrero et al. 2012), and elucidation of mercury binding sites with proteins (Zayas et al. 2014).

In the present chapter, we report the recommended safety mercury thresholds established by public organisms in seafood and biosamples, the occurrence of mercury species and its bioaccessibility as well as metabolomic studies of mercury mode of action (MOA), and metallomic studies to provide deep insight into the interactions of this element with selenium biomolecules, formation of metallothionein complexes, and elucidation of mercury binding sites with proteins.

Toxicity of Mercury Species in Mammals

According to the Agency for Toxic Substance and Disease Registry (ATSDR), mercury is the third most dangerous heavy metal after arsenic and lead (Emsley 2001); the brain, liver, and kidneys being the organs three most affected by mercury exposure (Flora et al. 2008; Magos and Clarkson 2006; Othman et al. 2014).

Alkylated species like methylmercury ($MeHg^+$) are more toxic than inorganic forms, due to their easy penetration through biological membranes, high stability, and half-life in tissues (Horvat et al. 2003; Palenzuela et al. 2004; Tao and Willie 1998). $MeHg^+$ mainly affects the central nervous system and in severe cases, specific areas of the brain causing irreversible damage (Leermakers et al. 2005).

Inorganic Hg is not able to cross the blood brain barrier, but it can be present in the brain due to dealkylation of organic Hg or oxidation of elemental Hg (Lohren et al. 2015). Mercury exposure increases the production of free radicals and oxidative stress that are related to the pathogenesis of acute hepatic and renal disorders (Gado and Aldahmash 2013; Othman et al. 2014). Due to the high affinity of mercury to sulphur, it can be easily bonded to endogenous thiol-containing molecules, as proteins or metabolites, which is the main mechanism regulating uptake, accumulation, and toxicity of inorganic Hg in the liver and kidney (Gado and Aldahmash 2013; Othman et al. 2014). Thus, it is invariably found in cell, tissues, and biological fluids bound to thiol-containing molecules, such as reduced glutathione (GSH) or cysteine (Cys) (Percy et al. 2007). Although organic and inorganic mercury forms have been shown to accumulate in the kidney, this organ appears to be a preferential site for Hg^{2+} accumulation (Bridges 2014).

Due to the hazardous character of mercury species, the FAO/WHO Joint Expert Committee on Food Additives (JECFA) established a Provisional Tolerable Weekly Intake (PTWI) of 1.6 and 4.0 μg/kg body weight per week for methylmercury and inorganic mercury, respectively (EFSA Scientific Committee 2015). However, for methylmercury, new studies indicate that beneficial effects related to long chain omega 3 fatty acids present in fish may have previously led to an underestimation of the potential adverse effects of methylmercury in fish. For this reason, the EFSA Panel has therefore proposed a TWI for methylmercury of 1.3 μg/kg bw/week as mercury, based on prenatal neurodevelopmental toxicity.

Interactions of Mercury with Other Elements

Selenium against mercury toxicity

In general, the simultaneous administration of selenite counteracts the negative impacts of exposure to inorganic mercury, particularly in regard to neurotoxicity, fetotoxicity, and cardiovascular diseases. In humans, a selenite and SeMet-dependent protection against mercury induced apoptosis and growth inhibition in human cells has been previously described. However, other studies reveals that inorganic selenium is ineffective in preventing most of the $MeHg^+$ induced brain biochemical alterations and toxic effects caused by selenite has also been demonstrated without the combination with methylmercury. On the other hand, the selenium to mercury molar ratio is very important since Hg in molar excess over Se is a stronger inducer of metallothionein levels in some animals (García-Barrera et al. 2012, 2015).

As stated previously, selenium can also counteracts the toxicity of methylmercury as demonstrated in an utero study on mice exposed to $MeHg^+$ and Se, in which the group that was given the lowest amount of Se and the highest dose of $MeHg^+$ was mostly adversely affected in neurobehavioural outcome. In rodents, antioxidant nutrients such as Se and vitamin E in a diet may alter $MeHg^+$ reproductive and developmental toxicity (García-Barrera et al. 2012, therein).

A great number of studies have been carried out related to the protective influence of the selenocompounds against $MeHg^+$ toxicity, especially selenomethionine (SeMet). Also, exposition of $MeHg^+$ in rats resulted in a significant increase of urinary porphyrins and a decrease in motor activity that was counteracted by SeMet (García-Barrera et al. 2012, herein).

Mechanisms of Se-Hg interaction

Until now, several mechanisms have been proposed to explain the interaction between these elements, namely:

(i) Redistribution of mercury to less sensible organs by the action of selenium
(ii) Competition for the same cleavages
(iii) Formation of Hg-Se complexes
(iv) Conversion of highly toxic Hg species in less toxic forms induced by selenium
(v) Selenium prevents the oxidative stress caused by Hg

It is believed that a 1:1 Hg-Se compound of low biological availability and activity is formed inside the cells, and cell damage is quite low even in the presence of very high Hg concentrations if both elements are mostly in equimolecular ratio. This has been stated in studies where marine mammals and humans were exposed to high levels of inorganic mercury. In 1978, experiments with marine organisms suggested a direct Hg-Se linkage and the 1:1 molar ratio of mercury and selenium increment holds true in several species including humans (García-Barrera et al. 2012, therein). A tissue with Se:Hg molar ratio higher to 1 is suggested as a threshold for the protective action against Hg toxicity.

The possibility that HgSe formation is responsible for the 1:1 molar ratio is supported by experiments which showed that (i) enzimatically digested liver and plasma fractions with a 1:1 molar ratio release mercury and selenium in insoluble forms and (ii) binding to the same plasma protein is preceded with the conversion of selenite to H_2Se in red blood cells (García-Barrera et al. 2012, therein).

Selenium can also affect the activities of enzymes cleaving the carbon-mercury bond in organic mercury compounds. In this way, experiments with rats show an enhancement of PMA cleavage enzymes in liver when sodium selenite is supplemented in drinking water. It has also been observed that $MeHg^+$ exposure exerts an inhibitory effect on paronaxe 1 activity that can be counteracted by selenium in humans. Other hypotheses are that selenium can promote a redistribution of Hg from more sensitive organs (kidney, central nervous system) to less sensitive ones (muscle), that there is competition of Se for the same receptors, that complexes such as tiemannite or Se-Hg-S are formed, and that $MeHg^+$ conversion into less toxic forms is promoted and oxidative damage prevented. Yang et al. 2002 proposed the involvement of Se in the demethylation of $MeHg^+$ in the liver to form inorganic and less toxic Hg compounds (García-Barrera et al. 2012, therein).

Biochemical interactions between mercury and selenium

The abilities of different selenium compounds and selenium incorporated *in vivo* into liver tissue (biological selenium) to form a Hg-Se compound are different and increase in the following order: biological Se < selenomethionine < selenite; thus the protective effect of the selenium compounds against mercury toxicity might follows the same order. Mercury ions can react with thiols (-SH) and selenols (-SeH) that constitute a part of cysteine and selenocysteine and as a consequence they can be incorporated to proteins, prosthetic groups of enzymes, and peptides. Mercury ions can also react with selenides (Se^{2-}), and with hydrogen selenide they can form complexes together with glutathione which can finally bond to selenoprotein P (García-Barrera et al. 2012, therein).

Similar complexes can be formed in other cells with active selenium metabolism or during degradation of metal bond proteins and metallo(selenoproteins) in lysosomes (biomineralization processes), representing the last step of detoxification. A direct interaction between $MeHg^+$ and the selenol group of GPx (Glutathione peroxidase) has also been reported, but to explain the reduced activity of the enzyme after $MeHg^+$ exposure another molecular mechanism has been proposed, based in the fact that cultured cells showed that $MeHg^+$ induced a "selenium-deficient-like" condition, which affects GPx1 synthesis through a posttranscriptional effect (García-Barrera et al. 2012, therein).

An experiment with mice exposed to Hg^{2+} and selenite provided evidence that low molecular mass selenospecies (selenite) are consumed to form SeP which correlatively increases in plasma and hepatic cytosol as a detoxification mechanism against mercury based in its redistribution and excretion in the form of Hg-SeP (García-Sevillano et al. 2014, 2015a,b). In addition, the redox ThxR and eGPx enzymes are up-regulated in liver cytosolic extracts and serum after mercury exposure, in parallel with the selenium supplementation dose in the former, indicating a mechanism in which the antioxidative properties of selenoproteins help to compromise the reactive oxygen species induced by mercury (García-Sevillano et al. 2015b). Finally, selenium supplementation promote Hg binding to high molecular mass molecules (SeP and SeAlb) (Chen et al. 2006; García-Sevillano et al. 2015b). These findings seem to support the hypothesis that the formation of Hg-SeP and Hg-SeAlb complexes was accompanied by a redistribution of Hg in the organism (García-Sevillano et al. 2015b).

Mercury vapour shows a similar behaviour to $MeHg^+$ in relation with its facility to penetrate cell membranes where it is oxidized to the biological active form (Hg^{2+}) by catalase. Such *in situ* generated ions can react with endogenously generated highly reactive Se metabolites, like HSe^-, and consequently a part of the selenium is unavailable for selenoprotein synthesis. Mercury can also provoke the increase of free radicals that induce lipid, protein, and DNA oxidation (García-Barrera et al. 2012, therein).

Interaction of mercury with cadmium and zinc

Although the antagonistic interaction of mercury with selenium is the most known biological process, other elements act in synergistic or antagonistic fashion with this element. This is the case of cadmium, which induces higher Hg levels in rat blood and lower levels in the heart, muscle, and skeleton (Chaoui et al. 1997) zinc, which protects against low-dose Hg-induced testicular damage in mice (Orisakwe et al. 2001), causes teratogenic and embryopathic effects in hamsters (Gale 1973) and is displaced by Hg in metallothionein in rats (Day et al. 1984).

Finally, a synergistic interaction has also been described between mercury and tellurium in mice since Hg retention is increased by pre-administration of Telike in the case of Se (Khayat and Dencker 1984).

Mercury Occurrence in Seafood and its Impact on Human Health after Consumption

Alkyl mercury is not well metabolized, and due to its high solubility in lipids it is bioaccumulated and biomagnified through the food chain (Río Segade and Tyson 2003). Mercury is usually methylated by benthic microorganisms (bacteria and sulphate reducers) in water (Watras et al. 1995) and sediments (Gómez-Ariza et al. 2005a), increasing its toxicity as commented before. Methylmercury ($MeHg^+$) has a high bioaccumulation factor (106–107) (Li et al. 2005), and represents more than 60% of total mercury commonly found in fish (Ebdon et al. 2002). More precisely, the contribution of methylmercury to total mercury is typically 80–100% in fish and 50–80% in seafood other than fish. In other foods, mercury is presumed to be present mainly as inorganic mercury.

Total mercury concentrations in foods, except for fish and other seafood, generally does not exceed 50 μg/kg. Higher concentrations are observed in fish and other seafood, which therefore provide the major source of dietary exposure to methylmercury in consumers. The amount of mercury in fish is related to the age of the fish and its position within the food chain; predatory and older fish having higher concentrations than others (EFSA Scientific Committee 2015).

Studies performed by the EFSA Scientific Committee, requested by the European Commission, established that fish meat was the dominating contributor to methylmercury dietary exposure for all age classes, followed by fish products. In particular tuna, swordfish, cod, whiting, and pike were major contributors to methylmercury dietary exposure in the adult age groups, while the same species, with the addition of hake, were the most important contributors in the child age groups. On the other hand, it is important to consider the beneficial effects related to long chain omega 3 fatty acids present in fish. Figure 1 shows methylmercury concentration and n-3 LCPUFA content levels in fish and seafood published by EFSA (EFSA Scientific Committee 2015).

An interesting study compares the content of Hg, Se, and omega 3-fatty acid among common fish and shellfish, and concludes that fish of higher trophic levels have higher Hg, but similar Se and EPAc + DHA concentrations when compared to lower trophic levels (Karimi et al. 2014a).

On the other hand, the European Commission Regulation 466/2001/EC (amended by Regulation 221/2002/EC) came into effect on April 2002 and was later modified by the European Commission Regulation 1881/2006/EC of 19 December 2006. This regulation sets maximum levels for mercury in fishery products of 0.5 mg kg^{-1} of their wet weight, excluding some species in which the limited content of mercury allowed is 1 mg kg^{-1} of their wet weight (Commission Regulation (EC) No. 1881/2006), but because the types of fish that generally contain higher levels of mercury are wild, not farmed, it is not possible to reduce the levels by controls on feed (EFSA Scientific Committee 2015).

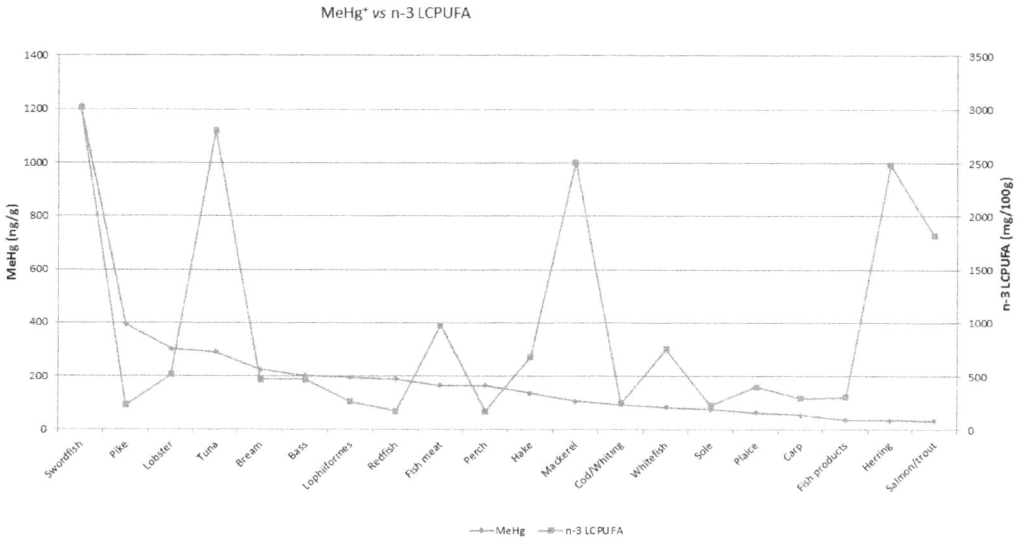

Fig. 1: Methylmercury concentration and n-3 LCPUFA content levels in fish and seafood, published by EFSA.

Mercury bioaccessibility in seafood

It has been established that Hg concentration is generally higher in cooked seafood samples than in raw ones, which can be related to the effect of Hg pre-concentration, formation of complexes involving mercury species and sulfhydryl groups present in tissues, and/or loss of water and fat (Costa et al. 2016; Matos et al. 2015). Other studies concluded that mercury levels are high in raw and cooked (boiled and grilled) tuna (Afonso et al. 2015).

However, before assuming potential toxic effects caused by mercury levels present in cooked food, the pivotal factor is its bioaccessibility, which can be defined as the fraction of a contaminant that is soluble in the intestine, and is therefore available for the subsequent processes of absorption, while the bioavailable and bioactive fractions are defined as the portion that is absorbed by the intestine and reaches the systemic circulation or that is transformed in bioactive metabolites, respectively.

Several studies have been performed related with the bioaccessibility of mercury species in seafood. A study carried out in commercial fish and seafood species widely consumed in Spain incubated samples in a 3-compartmental (mouth, stomach, and small intestine) *in vitro* gastrointestinal model, and showed that the bioaccessibility of Hg was < 50% for most species, while $MeHg^+$ could be only quantified in swordfish and tuna (Cano-Sancho et al. 2015). Other authors, studied mercury bioaccessibility in sharks and also concluded that it is about 50% for both mercuric species in grilled samples (Matos et al. 2015), while Hg bioaccessibility percentages are lower (39–48%) in the cooked tuna, even lower (< 20%) in canned tuna (Afonso et al. 2015), and about 30% in gilled salmon (Costa et al. 2015). So, as Hg and $MeHg^+$ show a low and variable bioaccessibility in marine species, it may suggest an overestimation of health risks for the adult population. However, further studies should assess the bioaccessibility of mercurial species for children. Due to the above commented interaction between mercury and selenium and the beneficial character of the later against the toxicity of the former, several authors study also selenium bioaccessibility and conclude that it was high in tuna (lower in canned than in raw products) and that for the bioaccessible fraction, all molar Se:$MeHg^+$ ratios were higher than one (Afonso et al. 2015). Selenium is naturally present in all foods, but is especially abundant in ocean fish and greatly exceeds $MeHg^+$ levels in all specimens, but a few in predatory species such as the mako shark (Kaneko and Ralston 2007). Moreover, the bioaccessibility of selenium and mercury was found to be related with the type of fish, probably due to the different degradation effectiveness of the food matrix, which affect the release of these elements.

In this sense, several authors found Se in excess together with a low $MeHg^+$ bioaccessibility in all the analysed samples, which can supports a significant antagonistic effect of Se on Hg toxicity (Cabañero et al. 2007). These authors also found that the Se/Hg, [Se/Hg] bioaccessible and [SeMet/Hg] bioaccessible molar ratios decrease in the same order: sardine > tuna > swordfish.

In conclusion, the potential toxicity of fish cannot be evaluated independently by analysing total Hg or $MeHg^+$, since Se, and probably other elements, can significantly influence the Hg bioaccessible fraction.

Mercury levels in biological samples and safety thresholds for mercury exposure

The average daily intake of mercury from food is in the range 2–20 µg, but may be much higher in regions where ambient waters have become contaminated with mercury and where fish constitutes a high proportion of the diet (Galal-Gorchev 1991).

The EFSA Scientific Committee established that the mean middle bound methylmercury dietary exposure across surveys and countries varied from the lowest minimum of 0.06 µg/kg bw/week seen in the elderly to the highest maximum of 1.57 µg/kg bw/week in toddlers. The 95th percentile middle bound dietary exposure across surveys and countries ranged from the lowest minimum of 0.14 µg/kg bw/week in the very elderly to the highest maximum of 5.05 µg/kg bw/week in adolescents (EFSA Scientific Committee 2015). On the other hand, the US EPA reference dose (RfD) for methylmercury, defined as "an estimate of a daily exposure to the human population (including sensitive subgroups) that is likely to be without appreciable risk of deleterious effects during a lifetime" is 0.1 µg/kg per day which is the same as the RfD derived by EPA in 1995 based on an earlier study of a poisoning episode in Iraq, in which data on adverse neurological effects in infants was used as the point of departure for derivation of the RfD (Environmental Protection Agency 10/30/2000).

On the other hand, the US federal biological exposure index (BEI) is currently set at 50 µg/L in urine, but there are obvious problems associated with basing a monitoring index on a measurement which only reflects current or recent exposure, and not overall body burden. In this sense, several clinical studies show objective symptoms below 50 µg/L, with many probed values extending down into the low end of the reference range for urinary mercury excretion (Bernhoft 2012; Echeverria et al. 1995, 1998; Haut et al. 1999; Lucchini et al. 2002; Meyer-Baron et al. 2002; Roels et al. 1999). Similar criticisms have been made about the EPA Reference Dose for methylmercury (Rice 2004) and as summarized by Kazantzis, "it has not been possible to set a level for mercury in blood or urine below which mercury related symptoms will not occur" (Kazantzis 2002). The RfD is equivalent to a blood mercury concentration of 5.8 µg/L in adults and 1 µg/g in hair, but effects below this threshold on the developing brain are under research (Rose and Fernandes 2013).

Seafood consumption and mercury levels in maternal biofluids

Although undoubtedly seafood is an important source of nutrients for foetal development, the neurotoxic effects of this element and the levels found led to the recommendation of the different government agencies to reduce or avoid the consumption of some species during pregnancy and childhood. Epidemiological studies in the Faroe Islands and New Zealand (Debes et al. 2006; Strain et al. 2008) report adverse effects associated with increasing maternal seafood consumption, but other studies demonstrate the beneficial, rather than adverse consequences in pregnancy (Seychelle Islands and Bristol) (Hibbeln et al. 2007; van Wijngaarden et al. 2013). Studies of very high $MeHg^+$ exposures demonstrated that maternal $MeHg^+$ exposure during pregnancy correspond with neurodevelopmental defects (0.1 IQ points) observed in children of exposed mothers (Crump et al. 1998; Grandjean et al. 1997, 1998; Marsh 1987; Marsh et al. 1980; Takeuchi et al. 1989, 1996), but other studies considering low levels of Hg exposure demonstrated beneficial effects on child development associated with higher levels of maternal seafood consumption (Daniels et al. 2004; Davidson et al. 1998; Myers and Davidson 1998; Myers et al. 2000).

Several authors found that fish consumption resulted in differences in mercury levels in placenta and cord blood (Gilman et al. 2015; Soon et al. 2014), but there are also numerous individuals who either had higher mercury with no fish consumption or lower mercury with high fish consumption, indicating a lack of correlation or the occurrence of another source of mercury other than fish (Gilman et al. 2015). An interesting study correlates several genes of children with mercury accumulation in the foetus since the presence of them led to a stronger correlation between maternal seafood consumption and cord blood mercury concentration (Llop et al. 2014). The correlation between the seafood consumption and Hg hair levels has also been demonstrated by the European Project DEMOCOPHES (Castaño et al. 2015) as well as by other studies (Nair et al. 2014). The DEMOCOPHES study demonstrates that 95% of mothers who consume once per week fish only, and no other marine products, have mercury levels of 0.55 μg/g which is around half the US EPA recommended threshold of 1 μg/g mercury in hair (Castaño et al. 2015).

Other studies showed relationships between different bio-samples, as the level of total Hg in maternal blood and breast milk, reaching an average breast milk level of 27% of the blood levels (Oskarsson et al. 1996) or cord blood and maternal venous blood (Kim et al. 2012; Ramirez et al. 2000; Sakamoto et al. 2004, 2007; Stern and Smith 2003). However, maybe due the complicated mechanism of Hg absorption and excretion by kidney, the correlation between the mercury level in urine and other biosamples were rather low (Li et al. 2014; Pesch et al. 2002).

On the other hand, $MeHg^+$ inhibits Se-dependent enzymes causing adverse child effects especially in mothers that eat fish with disproportionally high Hg:Se molar ratios (Gilman et al. 2015). However, a study demonstrated that while the absence of seafood consumption correlates with low placental and blood Hg levels, no strong correlations have been seen between seafood consumption or its absence and the levels of selenoproteins or sepenoenzymes activity (Gilman et al. 2015).

Analytical Techniques for Mercury Speciation in Seafood and Biosamples

Speciation methods in seafood and biofluids

Due to the low levels of mercury that usually occur in seafood and biosamples, highly sensitive analytical methods are required for mercury determination and speciation. To this end, total mercury in biological samples is usually determined after acid digestion by selective techniques such as cold-vaporatomic absorption spectrometry (CV-AAS) for umbilical cord blood (Gilman et al. 2015; Soon et al. 2014), placental mercury (Gilman et al. 2015) and umbilical cord serum, maternal venous blood, breast milk, maternal hair and maternal urine (Li et al. 2014), combustion-gold amalgamation atomic absorption spectroscopy (GA-AAS) has been used for hair mercury determination (Nair et al. 2014), and total mercury in human urine has been determined by HPLC cold vapor radiofrequency glow-discharge optical-emission spectrometry with on-line microwave oxidation (HPLC-CV-rf-GD-OES) (Martínez et al. 2001). Speciation of mercury ($MeHg^+$) has also been performed in cord blood by cold-vaporatomic fluorescence spectrometry (CV-AFS) after purge and trap separation (Llop et al. 2014).

Undoubtedly, ICP-MS is a valuable technique in this field since it allows: (i) multiisotopic analysis (including non-metals such as S, P, Se), (ii) detection capability, (iii) high sensitivity, (iv) tolerance to matrix, and (v) large linearity range. In this sense, ICP-MS has been applied for total mercury determination in blood (Karimi et al. 2014b) and using a column switching HPLC method for mercury and selenium simultaneous speciation in serum (Moreno et al. 2010, 2013) and human urine (Moreno et al. 2010) and total selenium and mercury determination in blood (Karimi et al. 2014a).

Only few methods have been reported for mercury speciation in biological samples since total levels are established by public organisms as safety thresholds which are determined after acid digestion. However, a number of methods for the extraction of mercury have been described. In this sense, miniaturized membrane based extraction techniques have several important advantages such as reproducibility, absence of sample carryover (due to the disposable character of membranes, high analyte enrichments, high throughput, low cost, applicability to many different types of analytes, wide pH tolerable range, and facility for automation and conversion into a greener analytical samples (Pena-Pereira et al. 2010)). In addition, hollow-fiber liquid phase microextraction (HF-LPME) has been applied for mercury and selenium simultaneous speciation in urine before HPLC-ICP-MS determination (Moreno et al. 2013).

In seafood, mercury species, mainly inorganic and methylmercury are usually determined. It is remarkable the use of GC-AFS via pyrolysis after derivatization (ethylation) or formation of chlorides (Gómez-Ariza et al. 2004) or using AAS after extraction with cysteine (Cano-Sancho et al. 2015). Total mercury has also been determined in seafood by CV-AAS (Spanopoulos-Zarco et al. 2014) and AAS with thermal decomposition and gold amalgamation (Costa et al. 2016; Mieiro et al. 2014). Finally, mercury, lead, and cadmium have also been determined in seafood by CV-AAS (Okyere et al. 2015).

Metallomic and metabolomics approaches

In the post genomic era, the use of methods of massive information, the -omics is very useful to study the biological response of an organism against toxic effects and to fully understand the action mechanisms. Likewise, during the last decade, the -omics technologies provide massive information generating methods that allow comprehensive description of nearly all components within the cell. To this end, genomics reveals the characteristics of the information contained in the cellular core that determines the cell function and behaviour, transcriptomics examines the gene expression, and proteomics involves the analysis of protein synthesis and cell signalling. In addition, in 1999, J. Nicholson defines metabonomics as "the quantitative measurement of the dynamic multiparametric metabolic response of living systems to pathophysiological stimuli or genetic modification" (Nicholson et al. 1999), while metabolomics may be defined as the measurement of all the metabolites in a specified biological sample (Nicholson et al. 1999). Whereas metabonomics provides a means for understanding the variation in low molecular mass metabolites in complex multicellular organisms and their response to change, the additional mentioned "-omics" sciences are concerned with cellular macromolecules. Then, to understand a cell, tissue, or living organism behaviour it is also necessary to consider low molecular mass molecules since they represent the last action mechanism of the organisms.

In the case of proteomics, it allows the identifying of important roles of proteins in cell homeostasis, quantitative analysis, and it has been shown to generate protein biomarkers for *in vivo* toxicity, but there are a large number of proteins and post-translational modifications, and not all proteins in a sample can be identified. In relation to transcriptomics and genomics, they allow efficient sequencing of complete genomes and the study of polymorphisms can give insight into the role of genetics in toxicology and it can explain differences in susceptibility, but alterations in gene expression do not always lead to adverse health effects and it is often difficult to translate genomics results to *in vivo* or human toxicity or disease. Moreover, post-translational modifications, phosphorylation and glycosylation of proteins determine their function and it is well-known that a lot of environmental factors or multigenic processes (e.g., aging and disease) cannot be explained only with a genomics basis. Finally, metabolomics is the omic science considered to be closer to the phenotype, allows the simultaneous measurement of hundreds of metabolites *in vitro* cell cultures, *in vivo* tissue and even in non-invasive blood and urine applications, and it has been shown to predict *in vivo* liver and kidney toxicity, but quenching and metabolite extraction procedures as well as the complexity of data analysis and interpretation (e.g., metabolic pathways) limit the detection of metabolites. Moreover, *in vivo* approaches are influenced by variability factors (e.g., age, gender, diet, stress, housing conditions, health status) as well as *in vitro* ones (e.g., cell culture conditions, metabolic competence, media formulations, serum additions, treatment vehicle). In this scenario, the combination of omics seems to be the most suitable approach, but the combination of proteomics with metabolomics has some drawbacks as the temporal space is different (i.e., metabolomics gives information about what happens right now, but it can be related with numerous post-translational modifications that happened previously). In this sense, it seems that the combination of genomics with metabolomics is easier. Thus, when metabolomics data are interpreted in combination with genomic, transcriptomic, and proteomic results, in the so-called systems biology approach, a holistic knowledge of the organism/process under investigation can be achieved.

Although metabolomics and metabonomics, are the general headings for metabolic analysis, they are applied following different approaches or analytical strategies, namely (Dunn and Ellis 2005; Kaderbhai et al. 2003; Ogra and Anan 2012) (i) Metabolite profiling, the identification and quantification of a selected number of pre-defined metabolites, generally related to a specific metabolic pathway, (ii) Metabolic fingerprinting, high throughput, rapid, global analysis of samples to provide their classification, usually without quantification and metabolic identification, (iii) Metabolite target analysis, qualitative, and quantitative analysis of one or few metabolites related to a specific metabolic reaction, (iv) Metabolite footprinting, the study of metabolites in extracellular fluids, (v) Metal-metabolomics, the study of metal or metalloid containing metabolites (i.e., selenometabolomics) (Ge et al. 2011), and (vi) Toxicometabolomics, metabolomic analysis applied to toxicology (i.e., metabolomics analysis of living organisms exposed to xenobiotics) (Bouhifd et al. 2013). In addition, generally, there are two major approaches used in metabolomics studies: targeted and untargeted (global). A targeted approach is used to determine the relative abundances and concentrations of specific sets of metabolites, thus information about the exact structure of the metabolite must be available in advance and involves the comparison of the analytes to the corresponding standards. One major disadvantage of the targeted approach is that this approach is dependent on the availability of a purified form of the known metabolite and is not applicable for the identification of new metabolites. In contrast, untargeted metabolomics is the comprehensive study of all metabolites in a biological sample, which is sometimes called 'global metabolome analysis'. This approach aims to find differing metabolites based on their relative quantitation and annotation of as many chromatographic/spectroscopic peaks ('features') as possible. The disadvantage of untargeted metabolomics is that this approach is a relative quantification, not an absolute one. Furthermore, some of the significant features/peaks are not identifiable (Bouhifd et al. 2013). Figure 2 shows a scheme about the different types of metabolomics studies.

On the other hand, it is necessary to consider that approximately one third of proteins need the presence of metals as cofactors to develop their function (metalloproteins) (Dunn and Ellis 2005; Kaderbhai et al. 2003; Ogra and Anan 2012) and metals influence more than 50% of the proteins (Mounicou et al. 2009). These metals are responsible for catalytic properties or structure of proteins and the presence in molecules is determined in many cases by the genome (Tainer et al. 1991). The metallome was defined by Williams as the distribution of elements, concentration at equilibrium of free metallic ions or free elements in a cellular compartment, cell or organism (Williams 2001), and refers to the identity and/or quantity of metals/metalloids and their species (Koppenaal and Hieftje 2007; Lobinski et al. 2010; Mounicou et al. 2009; Szpunar 2004). Then, metallomics consider that the identification of a metal cofactor in a

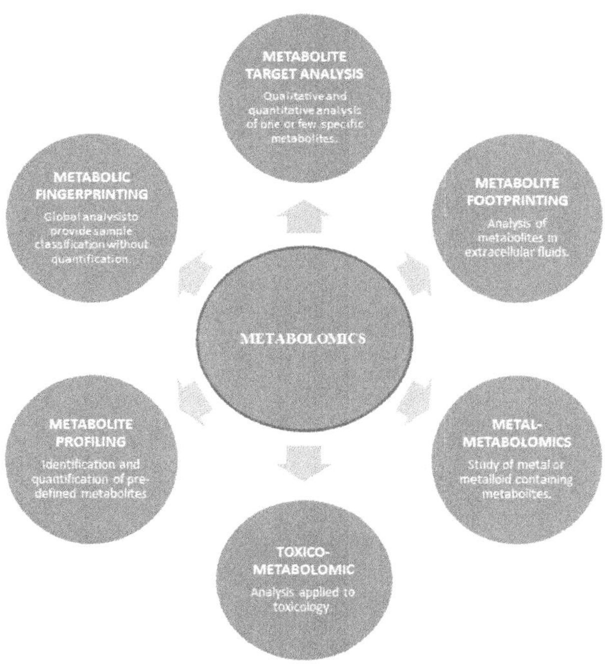

Fig. 2: Types of metabolomic approaches.

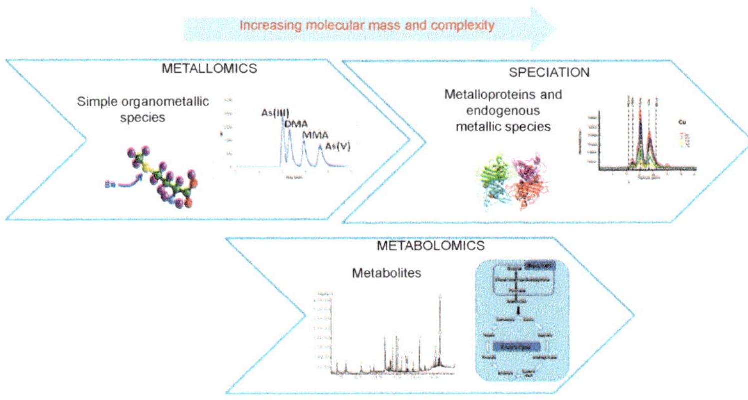

Fig. 3: Scheme of omics the approaches.

protein can greatly assists its functional assignment and help to place it in the context of known cellular pathways (González-Fernández et al. 2008; Gómez-Ariza et al. 2005b) and uses metal or metalloids as heteroatomic markers or tags to track these molecules on complex matrices (González-Fernandez et al. 2008, 2013; Sanz-Medel 2008). Metabolomics provides a good alternative to get a deeper insight into the fate of elements in exposed organisms, and gives information about metal trafficking, interactions, and homeostasis. Since, chemical species are the specific forms of an element defined to isotopic composition, electronic or oxidation state, and/or complex or molecular structure (Templeton et al. 2000), the line between metallomics, metal-metabolomics, and chemical speciation is absolutely thin. Figure 3 shows a scheme of the omics approaches.

Thus, metabolomics, as technology catching the phenotypic change at molecular level, allows the characterization of the impact of an agent in living organisms. Toxicometabolomics, in particular, in combination with proteomics, metallomics, and genomics/transcriptomics is a promising tool in the search for the action mechanisms of chemical species in living organisms.

Metabolic signatures of mercury exposure in mammals

Figure 4 shows the main metabolic cycles affected by mercury exposure. It has been issued that inorganic mercury exposure in rats causes inhibition of blood δ-aminolevulinic acid dehydratase (ALAD) activity and glutathione levels, which has associated with an increase of thiobarbituric acid reactive substance (TBARS). Mercury also increases reactive oxygen species (ROS), TBARS, and glutathione peroxidase (GPx) activity accompanied by a decreased superoxide dismutase (SOD), catalase, and reduced and oxidized glutathione (GSH and GSSG) levels in blood and tissues. Mercury alone produced a significant induction of hepatic and renal metallothionein (MT) concentrations. Serum transaminases, lactate dehydrogenase, and alkaline phosphatase activities increased significantly suggesting liver injury (Agrawal et al. 2014).

Non-targeted metabolomic approaches based on the combined use of direct infusion into a triple quadrupole mass spectrometer (DI-QqQ-TOF) and GC-MS have been applied in mice organs (liver and kidneys) and plasma after inorganic mercury exposure combining organic and inorganic mass spectrometry (García-Sevillano et al. 2014, 2015a). The results reveals mercury caused alterations in energy and amino acid metabolism, phospholipids breakdown (also observed by histopathology) and changes in some oxidative stress-related metabolites in serum. The quantitative analysis of reversible oxidized thiols in serum proteins reinforced the hypothesis of an intense induction of oxidative stress in Hg-treated mice. On the other hand, disturbances of glycolysis, membrane turnover, glutathione, and ascorbate metabolism in RBC are biochemical responses to Hg toxicity, that lead to inflammation and vacuolization of the liver tissue, accumulation of lipids (steatosis) and the dissolution of the chromatinin which suggest hepatic cells apoptosis (García-Sevillano et al. 2015a). Table 1 shows the main metabolites altered in mammals after mercury exposure.

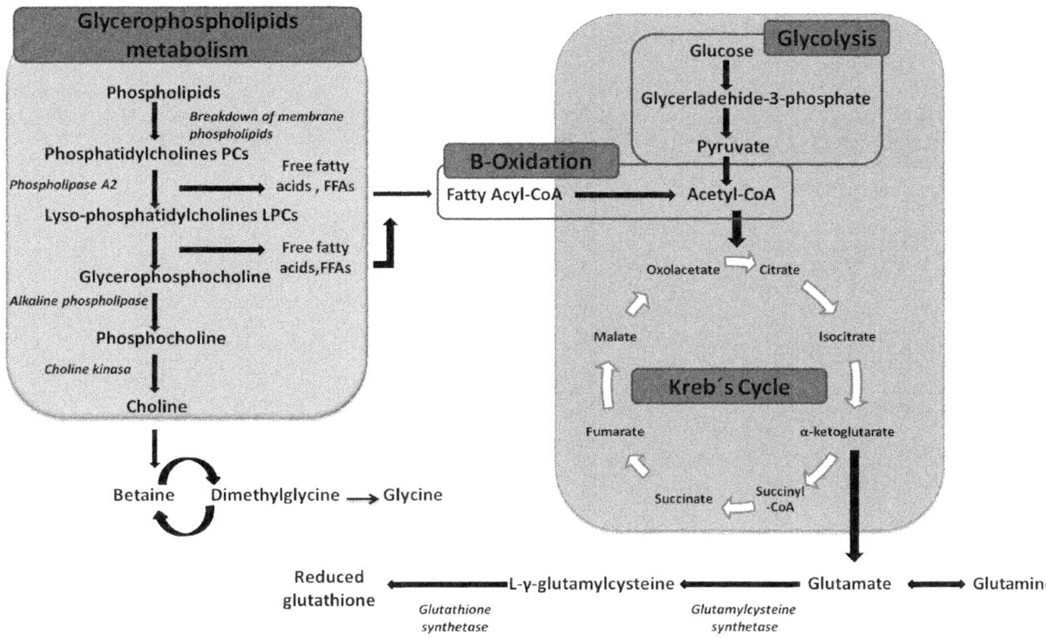

Fig. 4: Main metabolic cycles altered by mercury exposure.

Metal containing biomolecules and metals interactions related to mercury exposure in mammals: Metallomics

Mice exposed to inorganic mercury during 14 days (García-Sevillano et al. 2015a) displayed that liver levels of Cu, Zn-SOD (32 kDa) (tentatively identified by size exclusion chromatography coupled to inductively coupled plasma mass spectrometer-SEC-ICP-MS) decreased with the exposure while those of Hg-SOD increases. This enzyme was later absolutely quantified in the cytosol and mitochondria of mice hepatic cells exposed to mercury by species-unspecific isotope dilution analysis (SUID) using a column switching method based in size exclusion chromatography (SEC), anion exchange chromatography (AEC) coupled to ICP-MS. In this experiment, Cu, Zn, Cd-MTs were detected in kidneys cytosolic extract, increasing the levels along with the mercury exposure, like in the case of Hg-MT. The effect of inorganic mercury in copper accumulation in kidneys has been previously documented (Blanusa et al. 1994) and seems to be regulated by a competition of mercury for Cu containing proteins like SOD (Zalups 1995), mercury being finally eliminated by urine in the form of Hg-MS (Zalups and Koropatnick 2000).

On the other hand, the antagonistic character of selenium against the toxicity of mercury was first described in 1980 (Magos and Webb 1980) in an experiment with rats. In this sense, mice exposed to inorganic mercury during 14 days (García-Sevillano et al. 2015a) revealed that selenium levels are reduced in plasma, especially during the first two days of the exposure experiment suggesting that selenium is initially consumed during the detoxification of mercury. As previously described, inorganic mercury present high affinity for thiol groups present in proteins like Selenoprotein P (SeP) and cysteine (Cys), being the first implicated in the transport of selenium to the brain (Burk and Hill 2009). Then, at the same time that selenium levels decreased in plasma, SeP increased in liver to transport this element, which is consumed in the detoxification process (García-Sevillano et al. 2015a). In general, these results are in good agreement with the formation of a Hg-Se complex that can be latterly linked to SeP (Gailer et al. 2000).

The interactions of mercury with biomolecules and the characterization of its specific binding have been issued in an interesting and recent publication (Zayas et al. 2014). This interaction plays a central role in metabolic pathways controlling its uptake, transformation, and toxicity. In the analyzed white-sided dolphin (*Lagenorhynchusacutus*) liver homogenate sample, approximately 60% of the $MeHg^+$ was

Table 1: Metabolites altered in mammals after mercury exposure.

Metabolite	Effect	Pathway	Organism	Target organ	Mercury species	Reference
2-oxoglutarate	↓	Kreb's cycle	Sprague-Dawley rats	Urine	$HgCl_2$	(Kim et al. 2010)
	↓	Kreb's cycle	Wistar rats	Urine	HgS	(Wei et al. 2008)
3-hydroxybutyrate	↑	Fatty acids beta-oxidation	*R. philippinarum* clams (White clams)	Digestive glands	$HgCl_2$	(Liu et al. 2011a)
Acetate	↑	Glucose metabolism	Sprague-Dawley rats	Urine	$HgCl_2$	(Nicholson et al. 1985) (Kim et al. 2010)
	↑	Glucose metabolism	*Mytilus Galloprovincialis* mussels	Gills	-	(Cappello et al. 2013)
	↑	Glucose metabolism	Wistarrats	Urine	HgS	(Wei et al. 2008)
Acetoacetate	↓	Fatty acids beta-oxidation	*R. philippinarum* clams (White and zebra clams)	Muscle	$HgCl_2$	(Liu et al. 2011b)
	↓	Fatty acids beta-oxidation	*R. philippinarum* clams (White clams)	Digestive glands	$HgCl_2$	(Liu et al. 2011a)
	↑	Fatty acids beta-oxidation	*Mytilus Galloprovincialis* mussels	Gills	-	(Cappello et al. 2013)
	↑	Fatty acids beta-oxidation	Wistar rats	Urine	HgS	(Wei et al. 2008)
Acetylcholine, ACo	↑	Glucose metabolism	*R. philippinarum* clams (White, red and zebra clams)	Muscle	$HgCl_2$	(Liu et al. 2011b)
	↓	Glucose metabolism	*R. philippinarum* clams (Zebraclams)	Digestive glands	$HgCl_2$	(Liu et al. 2011a)
	↑	Glucose metabolism	*Mytilus Galloprovincialis* mussels	Gills	-	(Cappello et al. 2013)
Alanine	↑	Alanine Cycle	Sprague-Dawley rats	Urine	$HgCl_2$	(Nicholson et al. 1985) (Kim et al. 2010)
	↓	Alanine Cycle	*R. philippinarum* clams (White and zebra clams)	Muscle	$HgCl_2$	(Liu et al. 2011b)
	↑	Alanine Cycle	*R. philippinarum* clams (White and red clams)	Digestive glands	$HgCl_2$	(Liu et al. 2011a)
	↓	Alanine Cycle	*R. philippinarum* clams (Zebraclams)	Digestive glands	$HgCl_2$	(Liu et al. 2011a)
	↑	Alanine Cycle	*Mytilus Galloprovincialis* mussels	Gills	-	(Cappello et al. 2013)
	↑	Alanine Cycle	S. salsahalophytes	Above ground tissue	$HgCl_2$	(Wu et al. 2012)
Allantoin	↓	Energy metabolism	Sprague-Dawley rats	Urine	$HgCl_2$	(Kim et al. 2010)
Arginine	↓	Arginine and proline metabolism (amino acid metabolism)	*R. philippinarum* clams (White clams)	Muscle	$HgCl_2$	(Liu et al. 2011b)

Table 1 contd....

Table 1 contd....

Metabolite	Effect	Pathway	Organism	Target organ	Mercury species	Reference
Arginine	↓	Arginine and proline metabolism (amino acid metabolism)	*R. philippinarum* clams (Red and zebra clams)	Digestive glands	$HgCl_2$	(Liu et al. 2011a)
	↑	Arginine and proline metabolism (amino acid metabolism)	*Mytilus Galloprovincialis* mussels	Gills	-	(Cappello et al. 2013)
Aspartate	↓	Alanine, aspartate, and glutamate metabolism	*R. philippinarum* clams (Red clams)	Muscle	$HgCl_2$	(Liu et al. 2011b)
	↑	Alanine, aspartate, and glutamate metabolism	*R. philippinarum* clams (Red clams)	Digestive glands	$HgCl_2$	(Liu et al. 2011a)
	↑	Alanine, aspartate, and glutamate metabolism	*Mytilus Galloprovincialis* mussels	Gills	-	(Cappello et al. 2013)
ATP/ADP	↓	Energy metabolism	*R. philippinarum* clams (White clams)	Muscle	$HgCl_2$	(Liu et al. 2011b)
	↓	Energy metabolism	*R. philippinarum* clams (Red and zebra clams)	Digestive glands	$HgCl_2$	(Liu et al. 2011a)
	↑	Energy metabolism	*Mytilus Galloprovincialis* mussels	Gills	-	(Cappello et al. 2013)
Betaine	↑	Choline and methionine metabolism	*R. philippinarum* clams (White clams)	Muscle	$HgCl_2$	(Liu et al. 2011b)
	↑	Choline and methionine metabolism	*R. philippinarum* clams (Red clams)	Digestive glands	$HgCl_2$	(Liu et al. 2011a)
	↑	Choline and methionine metabolism	*Mytilus Galloprovincialis* mussels	Gills	-	(Cappello et al. 2013)
Choline	↑	Phospholipid metabolism	Wistarrats	Serum	HgS	(Wei et al. 2008)
	↑	Phospholipid metabolism	*M. Musculus* mice	Plasma	$HgCl_2$	(García-Sevillano et al. 2014)
Citrate	↓	Glycolysis and Kreb'scycle	Sprague-Dawley rats	Urine	$HgCl_2$	(Nicholson et al. 1985) (Kim et al. 2010)
	↑	Glycolysis and Kreb's cycle	*R. philippinarum* clams	Digestive glands	$HgCl_2$	(Liu et al. 2011a)
	↓	Glycolysis and Kreb's cycle	Wistar rats	Urine	HgS	(Wei et al. 2008)
	↓	Glycolysis and Kreb's cycle	*M. Musculus* mice	Plasma	$HgCl_2$	(García-Sevillano et al. 2014)
Creatine	↑	Arginine and proline metabolism (amino acid metabolism)	Wistar rats	Serum	HgS	(Wei et al. 2008)
	↑	Arginine and proline metabolism (amino acid metabolism)	*M. Musculus* mice	Plasma	$HgCl_2$	(García-Sevillano et al. 2014)

Table 1 contd....

Table 1 contd....

Metabolite	Effect	Pathway	Organism	Target organ	Mercury species	Reference
Creatinine	↓	Arginine and proline metabolism (amino acid metabolism)	Sprague-Dawley rats	Urine	$HgCl_2$	(Nicholson et al. 1985)
Creatinine	↑	Arginine and proline metabolism (amino acid metabolism)	Sprague-Dawley rats	Plasma	$HgCl_2$	(Nicholson et al. 1985)
Creatinine	↑	Arginine and proline metabolism (amino acid metabolism)	Wistar rats	Urine	HgS	(Wei et al. 2008)
Creatinine	↓	Arginine and proline metabolism (amino acid metabolism)	*M. Musculus* mice	Plasma	$HgCl_2$	(García-Sevillano et al. 2014)
Dimethylglycine	↓	Fatty acids beta-oxidation	Wistar rats	Urine	HgS	(Wei et al. 2008)
FFAs	↑	B-oxidation of fatty acids	*M. Musculus* mice	Plasma	$HgCl_2$	(García-Sevillano et al. 2014)
Formate	↑	Glucose metabolism	Sprague-Dawley rats	Urine	$HgCl_2$	(Nicholson et al. 1985)
Formate	↓	Glucose metabolism	Sprague-Dawley rats	Urine	$HgCl_2$	(Kim et al. 2010)
Fructose	↓	Nucleotide metabolism	S. salsahalophytes	Above ground tissue	$HgCl_2$	(Wu et al. 2012)
Fumarate	↓	Kreb's cycle	S. salsahalophytes	Above ground tissue	$HgCl_2$	(Wu et al. 2012)
Glucose	↑	Glucose metabolism	Sprague-Dawley rats	Urine	$HgCl_2$	(Nicholson et al. 1985) (Kim et al. 2010)
Glucose	↓	Glucose metabolism	*R. philippinarum* clams	Digestive glands	$HgCl_2$	(Liu et al. 2011a)
Glucose	↑	Glucose metabolism	*Mytilus Galloprovincialis* mussels	Gills	-	(Cappello et al. 2013)
Glucose	↓	Glucose metabolism	Wistar rats	Serum	HgS	(Wei et al. 2008)
Glucose	↓	Glucose metabolism	S. salsahalophytes	Aboveground tissue	$HgCl_2$	(Wu et al. 2012)
Glucose	↓	Glucose metabolism	*M. Musculus* mice	Plasma	$HgCl_2$	(García-Sevillano et al. 2014)
Glutamate	↓	Glutathione metabolism (amino acid metabolism)	*R. philippinarum* clams (White clams)	Muscle	$HgCl_2$	(Liu et al. 2011b)
Glutamate	↑	Glutathione metabolism (amino acid metabolism)	*R. philippinarum* clams (White and red clams)	Digestive glands	$HgCl_2$	(Liu et al. 2011a)
Glutamate	↓	Glutathione metabolism (amino acid metabolism)	*R. philippinarum* clams (Zebra clams)	Digestive glands	$HgCl_2$	(Liu et al. 2011a)
Glutamate	↑	Glutathione metabolism (amino acid metabolism)	*Mytilus Galloprovincialis* mussels	Gills	-	(Cappello et al. 2013)

Table 1 contd....

Table 1 contd....

Metabolite	Effect	Pathway	Organism	Target organ	Mercury species	Reference
Glutamine	↓	Glutathione metabolism (amino acid metabolism)	*R. philippinarum* clams (White and red clams)	Muscle	$HgCl_2$	(Liu et al. 2011b)
	↑	Glutathione metabolism (amino acid metabolism)	*Mytilus Galloprovincialis* mussels	Gills	-	(Cappello et al. 2013)
	↑	Glutathione metabolism (amino acid metabolism)	S. salsahalophytes	Above ground tissue	$HgCl_2$	(Wu et al. 2012)
Glycine	↑	Glutathione metabolism (amino acid metabolism)	Sprague-Dawley rats	Urine	$HgCl_2$	(Nicholson et al. 1985)
	↓	Glutathione metabolism (amino acid metabolism)	*R. philippinarum* clams (White, red, and zebra clams)	Muscle	$HgCl_2$	(Liu et al. 2011b)
	↓	Glutathione metabolism (amino acid metabolism)	*R. philippinarum* clams (White and Zebra clams)	Digestive glands	$HgCl_2$	(Liu et al. 2011a)
	↑	Glutathione metabolism (amino acid metabolism)	*Mytilus Galloprovincialis* mussels	Gills	-	(Cappello et al. 2013)
	↑	Glutathione metabolism (amino acid metabolism)	S. salsahalophytes	Roots	$HgCl_2$	(Wu et al. 2012)
Glycogen	↑	Glycolysis	*Mytilus Galloprovincialis* mussels	Gills	-	(Cappello et al. 2013)
Hippurate	↓	Glycine metabolism	Sprague-Dawley rats	Urine	$HgCl_2$	(Kim et al. 2010)
	↑	Glycine metabolism	Wistar rats	Urine	HgS	(Wei et al. 2008)
	↑		*Mytilus Galloprovincialis* mussels	Gills	-	(Cappello et al. 2013)
Hypotaurine	↓	Taurine and hypotaurine metabolism	*R. philippinarum* clams (White clams)	Digestive glands	$HgCl_2$	(Liu et al. 2011a)
	↑	Taurine and hypotaurine metabolism	*Mytilus Galloprovincialis* mussels	Gills	-	(Cappello et al. 2013)
Isoleucine	↑	Valine, leucine, and isoleucine degradation/ propanoate metabolism	*R. philippinarum* clams (Zebra clams)	Muscle	$HgCl_2$	(Liu et al. 2011b)
	↑	Valine, leucine, and isoleucine degradation/ propanoate metabolism	*R. philippinarum* clams (Red clams)	Digestive glands	$HgCl_2$	(Liu et al. 2011a)
	↑	Valine, leucine, and isoleucine degradation/ propanoate metabolism	*Mytilus Galloprovincialis* mussels	Gills	-	(Cappello et al. 2013)
	↑	Valine, leucine, and isoleucine degradation/ propanoate metabolism	Wistar rats	Serum	HgS	(Wei et al. 2008)
	↑	Valine, leucine, and isoleucine degradation/ propanoate metabolism	S. salsahalophytes	Above ground tissue	$HgCl_2$	(Wu et al. 2012)

Table 1 contd....

Table 1 contd....

Metabolite	Effect	Pathway	Organism	Target organ	Mercury species	Reference
Ketonebodies	↑	Fatty acids beta-oxidation	Sprague-Dawleyrats	Urine	$HgCl_2$	(Nicholson et al. 1985)
	↑	Fatty acids beta-oxidation	Wistar rats	Serum	HgS	(Wei et al. 2008)
Lactate	↑	Glucose metabolism	Sprague-Dawley rats	Urine	$HgCl_2$	(Nicholson et al. 1985) (Kim et al. 2010)
	↑	Glucose metabolism	Sprague-Dawley rats	Plasma	$HgCl_2$	(Nicholson et al. 1985) (Kim et al. 2010)
	↑	Glucose metabolism	*R. philippinarum* clams (White and zebra clams)	Muscle	$HgCl_2$	(Liu et al. 2011b)
Leucine	↑	Valine, leucine, and isoleucine degradation/ propanoate metabolism	*R. philippinarum* clams (Zebra clams)	Muscle	$HgCl_2$	(Liu et al. 2011b)
	↑	Valine, leucine, and isoleucine degradation/ propanoate metabolism	*R. philippinarum* clams (Red clams)	Digestive glands	$HgCl_2$	(Liu et al. 2011a)
	↑	Valine, leucine, and isoleucine degradation/ propanoate metabolism	*Mytilus Galloprovincialis* mussels	Gills	-	(Cappello et al. 2013)
	↑	Valine, leucine, and isoleucine degradation/ propanoate metabolism	Wistar rats	Serum	HgS	(Wei et al. 2008)
	↑	Valine, leucine, and isoleucine degradation/ propanoate metabolism	S. salsahalophytes	Above ground tissue	$HgCl_2$	(Wu et al. 2012)
	↑	Valine, leucine, and isoleucine degradation/ propanoate metabolism	S. salsahalophytes	Roots	$HgCl_2$	(Wu et al. 2012)
Lyso-PCs	↑	Phospholipid metabolism	*M. Musculus* mice	Plasma	$HgCl_2$	(García-Sevillano et al. 2014)
Malonate	↑	Glucose metabolism	*Mytilus Galloprovincialis* mussels	Gills	-	(Cappello et al. 2013)
PCs	↓	Phospholipid metabolism	*M. Musculus* mice	Plasma	$HgCl_2$	(García-Sevillano et al. 2014)
Phenylacetylglycine	↑	Fatty acids beta-oxidation	Wistarrats	Urine	HgS	(Wei et al. 2008)
Phenylalanine	↑	Phenylalanine and tyrosine metabolism/ catecholamine biosynthesis	S. salsahalophytes	Above ground tissue	$HgCl_2$	(Wu et al. 2012)
Phospholcoline	↑	Phospholipid metabolism	*M. Musculus* mice	Plasma	$HgCl_2$	(García-Sevillano et al. 2014)
Pyruvate	↓	Glycolysis and Kreb's cycle	*M. Musculus* mice	Plasma	$HgCl_2$	(García-Sevillano etAl. 2014)
Succinate	↑	Kreb's cycle	Sprague-Dawley rats	Urine	$HgCl_2$	(Nicholson et al. 1985) (Kim et al. 2010)

Table 1 contd....

Table 1 contd....

Metabolite	Effect	Pathway	Organism	Target organ	Mercury species	Reference
Succinate	↑	Kreb's cycle	*R. philippinarum* clams (White, red, and zebra clams)	Muscle	$HgCl_2$	(Liu et al. 2011b)
	↑	Kreb's cycle	*R. philippinarum* clams (White and red clams)	Digestive glands	$HgCl_2$	(Liu et al. 2011a)
	↑	Kreb's cycle	*Mytilus Galloprovincialis* mussels	Gills	-	(Cappello et al. 2013)
	↓	Kreb's cycle	Wistar rats	Urine	HgS	(Wei et al. 2008)
	↓	Kreb's cycle	S. salsahalophytes	Above ground tissue	$HgCl_2$	(Wu et al. 2012)
Taurine	↓	Taurine and hypotaurine metabolism	Sprague-Dawley rats	Urine	$HgCl_2$	(Kim et al. 2010)
	↑	Taurine and hypotaurine metabolism	*R. philippinarum* clams (White and red clams)	Muscle	$HgCl_2$	(Liu et al. 2011b)
	↓	Taurine and hypotaurine metabolism	*R. philippinarum* clams (Zebra clams)	Muscle	$HgCl_2$	(Liu et al. 2011b)
	↑	Taurine and hypotaurine metabolism	*R. philippinarum* clams (White and zebra clams)	Digestive glands	$HgCl_2$	(Liu et al. 2011a)
	↑	Taurine and hypotaurine metabolism	*Mytilus Galloprovincialis* mussels	Gills	-	(Cappello et al. 2013)
	↑	Taurine and hypotaurine metabolism	Wistar rats	Urine	HgS	(Wei et al. 2008)
TGs	↑	Lipolysis	*M. Musculus* mice	Plasma	$HgCl_2$	(García-Sevillano et al. 2014)
Threonine	↑	Glycine, serine, and threonine metabolism	S. salsahalophytes	Above ground tissue	$HgCl_2$	(Wu et al. 2012)
Tyrosine	↓	Phenylalanine and tyrosine metabolism/ catecholamine biosynthesis	*Mytilus Galloprovincialis* mussels	Gills	-	(Cappello et al. 2013)
Valine	↑	Valine, leucine, and isoleucine degradation/ propanoate metabolism	*R. philippinarum* clams (Zebra clams)	Muscle	$HgCl_2$	(Liu et al. 2011b)
	↑	Valine, leucine, and isoleucine degradation/ propanoate metabolism	*R. philippinarum* clams (Red clams)	Digestive glands	$HgCl_2$	(Liu et al. 2011a)
	↑	Valine, leucine, and isoleucine degradation/ propanoate metabolism	*Mytilus Galloprovincialis* mussels	Gills	-	(Cappello et al. 2013)
	↑	Valine, leucine, and isoleucine degradation/ propanoate metabolism	Wistar rats	Serum	HgS	(Wei et al. 2008)
	↑	Valine, leucine, and isoleucine degradation/ propanoate metabolism	S. salsahalophytes	Above ground tissue	$HgCl_2$	(Wu et al. 2012)
α-ketoglutarate	↑	Kreb's cycle	Sprague-Dawley rats	Urine	$HgCl_2$	(Nicholson et al. 1985)

Variations compared to control groups: ↑ increasing signal intesity; ↓ decreasing signal intensity.

found in the water soluble fraction, specifically associated with high molecular weight biomolecules. The identity of the involved proteins was investigated after tryptic digestion by reversed phase μHPLC with parallel detection by ICP-MS and ESI-MS/MS. Using this metallomic approach, cysteine residue on the dolphin hemoglobin β chain was found to be the main $MeHg^+$ binding site, suggesting that hemoglobin is a major $MeHg^+$ binding protein in this marine mammal and could be a potential carrier of this $MeHg^+$ from blood to liver prior to its degradation in this organ. In parallel, a significant proportion of selenium was found to be present as selenoneine and a potential role of this compound in Hg detoxification has been reported (Zayas et al. 2014).

Finally, the identification of metal (Hg, Zn, Cd, Cu) complexes with individual metallothionein (MT) isoforms in white-sided dolphin (*Lagenorhynchusacutus*) liver homogenate sample has been performed by ESI-MS/MS (Pedrero et al. 2012). The sample preparation was very simple and consisted of a two-fold dilution of the sample cytosol with acetonitrile, centrifugation, and concentration under a gentle nitrogen stream. The MT complexes were separated by hydrophilic interaction liquid chromatography (HILIC) coupled to ICP-MS that allows the preservation of unstable and low abundant metallothionein zinc-mercury ($MT-Zn_6Hg$) complexes. The identification of MT complexes was completed by on line demetallation and the determination of the molecular mass of the apoform, followed by amino acid sequencing in the top-down mode (Pedrero et al. 2012).

Conclusions

Environmental pollution in waterways has resulted in the contamination of fish with methylmercury that can affect population, especially pregnant women and children. For this reason, in the last years great efforts have been made in relation with the establishment of safety thresholds for mercury in seafood or human biofluids and to establish the risk of fish consumption by bioaccessibility measurements, which on the other hand, is a very healthy food highly recommended for its content in omega-3-fatty acids. Very interesting analytical methods have been recently developed to study the mercury mode of action by using metabolomics approaches and to provide deep insight into the interactions of this element with selenium biomolecules, formation of metallothionein complexes, and elucidation of mercury binding sites with proteins. However, the exact mechanism of mercury toxicity in the body, its interactions with other elements and biomolecules, and the adequate safety thresholds in seafood and human biosamples are still under research.

Acknowledgements

This work has been supported by the projects CTM2012-38720-C03-01 and CTM2015-67902-C2-1-P from the Spanish Ministry of Economy and Competitiveness and P12-FQM-0442 from the Regional Ministry of Economy, Innovation, Science and Employment (Andalusian Government, Spain). Finally, authors are grateful to FEDER (European Community) for financial support.

Keywords: Mercury, seafood, biosamples, interactions, speciation, metallomics, metabolomics, ICP-MS

References

Adkins, Y. and D.S. Kelley. 2010. Mechanisms underlying the cardioprotective effects of omega-3 polyunsaturated fatty acids. J. Nutr. Biochem. 21: 781–92.

Afonso, C., S. Costa, C. Cardoso, R. Oliveira, H.M. Lourenço, A. Viula, I. Batista, I. Coelho and M.L. Nunes. 2015. Benefits and risks associated with consumption of raw, cooked, and canned tuna (Thunnus spp.) based on the bioaccessibility of selenium and methylmercury. Environ. Res. 143: 130–7.

Agrawal, S., G. Flora, P. Bhatnagar and S.J.S. Flora. 2014. Comparative oxidative stress, metallothionein induction and organ toxicity following chronic exposure to arsenic, lead and mercury in rats. Cell. Mol. Biol. (Noisy-le-grand) 60: 13–21.

Bernhoft, R.A. 2012. Mercury toxicity and treatment: a review of the literature. J. Environ. Public Health 2012: 460508.

Blanusa, M., L. Prester, S. Radic and B. Kargacin. 1994. Inorganic mercury exposure, mercury-copper interaction, and DMPS treatment in rats. Environmental Health Perspectives 2012: 305–307.

Bouhifd, M., T. Hartung, H.T. Hogberg, A. Kleensang and L. Zhao. 2013. Review: toxicometabolomics. J. Appl. Toxicol. 33: 1365–83.

Bridges, C.C., L. Joshee and R.K. Zalups. 2014. Experimental Gerontology 53: 31–39.

Burk, R.F. and K.E. Hill. 2009. Selenoprotein P-expression, functions, and roles in mammals. Biochim. Biophys. Acta 1790: 1441–7.

Cabañero, A.I., Y. Madrid and C. Cámara. 2007. Mercury–selenium species ratio in representative fish samples and their bioaccessibility by an *in vitro* digestion method. Biol. Trace Elem. Res. 119: 195–211.

Cano-Sancho, G., G. Perelló, A.L. Maulvault, A. Marques, M. Nadal and J.L. Domingo. 2015. Oral bioaccessibility of arsenic, mercury and methylmercury in marine species commercialized in Catalonia (Spain) and health risks for the consumers. Food Chem. Toxicol. 86: 34–40.

Cappello, T., A. Mauceri, C. Corsaro, M. Maisano, V. Parrino, G. Lo Paro, G. Messina and F. Fasulo. 2013. Impact of environmental pollution on caged mussels Mytilus galloprovincialis using NMR-based metabolomics. Mar. Pollut. Bull. 77: 132–139.

Castaño, A., F. Cutanda, M. Esteban, P. Pärt, C. Navarro, S. Gómez, M. Rosado, A. López, E. López, K. Exley, B.K. Schindler, E. Govarts, L. Casteleyn, M. Kolossa-Gehring, U. Fiddicke, H. Koch, J. Angerer, E. Den Hond, G. Schoeters, O. Sepai, M. Horvat, L.E. Knudsen, D. Aerts, A. Joas, P. Biot, R. Joas, J.A. Jiménez-Guerrero, G. Diaz, C. Pirard, A. Katsonouri, M. Cerna, A.C. Gutleb, D. Ligocka, F.M. Reis, M. Berglund, I.R. Lupsa, K. Halzlová, C. Charlier, E. Cullen, A. Hadjipanayis, A. Krsková, J.F. Jensen, J.K. Nielsen, G. Schwedler, M. Wilhelm, P. Rudnai, S. Középesy, F. Davidson, M.E. Fischer, B. Janasik, S. Namorado, A.E. Gurzau, M. Jajcaj, D. Mazej, J.S. Tratnik, K. Larsson, A. Lehmann, P. Crettaz, G. Lavranos and M. Posada. 2015. Fish consumption patterns and hair mercury levels in children and their mothers in 17 EU countries. Environ. Res. 141: 58–68.

Chaoui, A., M. Habib Ghorbal and E. El Ferjani. 1997. Effects of cadmium-zinc interactions on hydroponically grown bean (Phaseolus vulgaris L.). Plant Sci. 126: 21–28.

Chen, C., H. Yu, J. Zhao, B. Li, L. Qu, S. Liu, P. Zhang and Z. Chai. 2006. The roles of serum selenium and selenoproteins on mercury toxicity in environmental and occupational exposure. Environ. Health Perspect. 114: 297–301.

Commission Regulation (EC) No. 466/2001 of 8 March 2001 setting maximum levels for certain contaminants in foodstuffs.

Commission Regulation (EC) No. 1881/2006 of 19 December 2006 setting maximum levels for certain contaminants in foodstuffs.

Costa, S., C. Afonso, C. Cardoso, I. Batista, N. Chaveiro, M.L. Nunes and N.M. Bandarra. 2015. Fatty acids, mercury, and methylmercury bioaccessibility in salmon (Salmo salar) using an *in vitro* model: Effect of culinary treatment. Food Chem. 185: 268–76.

Costa, F. do N., M.G.A. Korn, G.B. Brito, S. Ferlin and A.H. Fostier. 2016. Preliminary results of mercury levels in raw and cooked seafood and their public health impact. Food Chem. 192: 837–41.

Crump, K.S., T. Kjellström, A.M. Shipp, A. Silvers and A. Stewart. 1998. Influence of prenatal mercury exposure upon scholastic and psychological test performance: benchmark analysis of a New Zealand cohort. Risk Anal. 18: 701–713.

Daniels, J.L., M.P. Longnecker, A.S. Rowland and J. Golding. 2004. Fish intake during pregnancy and early cognitive development of offspring. Epidemiology 15: 394–402.

Davidson, P.W., G.J. Myers, C. Cox, C. Axtell, C. Shamlaye, J. Sloane-Reeves, E. Cernichiari, L. Needham, A. Choi, Y. Wang, M. Berlin and T.W. Clarkson. 1998. Effects of prenatal and postnatal methylmercury exposure from fish consumption on neurodevelopment: Outcomes at 66 months of age in the Seychelles child development study. J. Am. Med. Assoc. 280: 701–707.

Day, F.A., A.E. Funk and F.O. Brady. 1984. *In vivo* and *ex vivo* displacement of zinc from metallothionein by cadmium and by mercury. Chem. Biol. Interact. 50: 159–174.

Debes, F., E. Budtz-Jørgensen, P. Weihe, R.F. White and P. Grandjean. 2006. Impact of prenatal methylmercury exposure on neurobehavioral function at age 14 years. Neurotoxicol. Teratol. 28: 536–47.

Dunn, W.B. and D.I. Ellis. 2005. Metabolomics: current analytical platforms and methodologies. TrAC Trends Anal. Chem. 24: 285–294.

Ebdon, L., M.E. Foulkes, S. Le Roux and R. Muñoz-Olivas. 2002. Cold vapour atomic fluorescence spectrometry and gas chromatography-pyrolysis-atomic fluorescence spectrometry for routine determination of total and organometallic mercury in food samples. Analyst 127: 1108–1114.

Echeverria, D., N.J. Heyer, M.D. Martin, C.A. Naleway, J.S. Woods and A.C. Bittner. 1995. Behavioral effects of low-level exposure to Hg^{o} among dentists. Neurotoxicol. Teratol. 17: 161–168.

Echeverria, D., H. Vasken Aposhian, J.S. Woods, N.J. Heyer, M.M. Aposhian, A.C. Bittner, R.K. Mahurin and M. Cianciola. 1998. Neurobehavioral effects from exposure to dental amalgam Hg^{o}: New distinctions between recent exposure and Hg body burden. FASEB J. 12: 971–980.

EFSA Scientific Committee. 2015. Statement on the benefits of fish/seafood consumption compared to the risks of methylmercury in fish/seafood. EFSA J. 13: 3982.

Emsley, J. 2001. Nature's Building Blocks: An A-Z Guide to the Elements. Oxford University Press, Oxford.

Environmental Protection Agency on 10/30/2000, a notice. URL: https://www.federalregister.gov/articles/2000/10/30/00-27781/reference-dose-for-methylmercury.

Flora, S.J.S., M. Mittal and A. Mehta. 2008. Heavy metal induced oxidative stress & its possible reversal by chelation therapy. Indian J. Med. Res. 4: 501–523.

Gado, A.M. and B.A. Aldahmash. 2013. Antioxidant effect of Arabic gum against mercuric chloride-induced nephrotoxicity. Drug Des. Devel. Ther. 7: 1245–52.

Gailer, J., G.N. George, I.J. Pickering, S. Madden, R.C. Prince, E.Y. Yu, M.B. Denton, H.S. Younis and H.V. Aposhian. 2000. Structural basis of the antagonism between inorganic mercury and selenium in mammals. Chem. Res. Toxicol. 13: 1135–1142.

Galal-Gorchev, H. 1991. Dietary intake of pesticide residues: cadmium, mercury, and lead. Food Addit. Contam. 8: 793–806.

Gale, T.F. 1973. The interaction of mercury with cadmium and zinc in mammalian embryonic development. Environ. Res. 6: 95–105.

García-Barrera, T., J.L. Gómez-Ariza, M. González-Fernández, F. Moreno, M.A. García-Sevillano and V. Gómez-Jacinto. 2012. Biological responses related to agonistic, antagonistic and synergistic interactions of chemical species. Anal. Bioanal. Chem. 403: 2237–53.

García-Barrera, T., J.L. Gómez-Ariza, V. Gómez Jacinto, I. Garbayo Nores, C. Vílchez Lobato and Z. Gojkovica. 2015. Selenium, Food and Nutritional Components in Focus, Food and Nutritional Components in Focus. Royal Society of Chemistry, Cambridge.

García-Sevillano, M.A., T. García-Barrera, F. Navarro, J. Gailer and J.L. Gómez-Ariza. 2014. Use of elemental and molecular-mass spectrometry to assess the toxicological effects of inorganic mercury in the mouse Mus musculus. Anal. Bioanal. Chem. 406: 5853–65.

García-Sevillano, M.A., T. García-Barrera, F. Navarro, N. Abril, C. Pueyo, J. López-Barea and J.L. Gómez-Ariza. 2015a. Combination of direct infusion mass spectrometry and gas chromatography mass spectrometry for toxicometabolomic study of red blood cells and serum of mice Mus musculus after mercury exposure. J. Chromatogr. B. Analyt. Technol. Biomed. Life Sci. 985: 75–84.

García-Sevillano, M.A., G. Rodríguez-Moro, T. García-Barrera, F. Navarro and J.L. Gómez-Ariza. 2015b. Biological interactions between mercury and selenium in distribution and detoxification processes in mice under controlled exposure. Effects on selenoprotein. Chem. Biol. Interact. 229: 82–90.

Ge, R., X. Sun and Q.-Y. He. 2011. Overview of the metallometabolomic methodology for metal-based drug metabolism. Curr. Drug Metab. 12: 287–299.

Gilman, C.L., R. Soon, L. Sauvage, N.V.C. Ralston and M.J. Berry. 2015. Umbilical cord blood and placental mercury, selenium and selenoprotein expression in relation to maternal fish consumption. J. Trace Elem. Med. Biol. 30: 17–24.

Gómez-Ariza, J.L., F. Lorenzo, T. García-Barrera and D. Sánchez-Rodas. 2004. Analytical approach for routine methylmercury determination in seafood using gas chromatography-atomic fluorescence spectrometry. Anal. Chim. Acta 511: 165–173.

Gómez-Ariza, J.L., F. Lorenzo and T. García-Barrera. 2005a. Guidelines for routine mercury speciation analysis in seafood by gas chromatography coupled to a home-modified AFS detector. Application to the Andalusian coast (south Spain). Chemosphere 61: 1401–9.

Gómez-Ariza, J.L., T. García-Barrera, F. Lorenzo and A. Arias. 2005b. Analytical characterization of bioactive metal species in the cellular domain (metallomics) to simplify environmental and biological proteomics. Int. J. Environ. Anal. Chem. 85: 255–266.

González-Fernández, M., T. García-Barrera, J. Jurado, M.J. Prieto-Álamo, C. Pueyo, J. López-Barea and J.L. Gómez-Ariza. 2008. Integrated application of transcriptomics, proteomics, and metallomics in environmental studies. Pure Appl. Chem. 80: 2609–2626.

González-Fernandez, M., M.A. García-Sevillano, R. Jara-Biedma, F. Navarro-Roldán, T. García-Barrera, J. López-Barea, C. Pueyo and J.L. Gomez-Ariza. 2013. Use of metallomics in environmental pollution assessment using mice mus musculus/mus spretus as bioindicators. Curr. Anal. Chem. 9: 229–243.

Grandjean, P., P. Weihe, R.F. White, F. Debes, S. Araki, K. Yokoyama, K. Murata, N. Sorensen, R. Dahl and P.J. Jorgensen. 1997. Cognitive deficit in 7-year-old children with prenatal exposure to methylmercury. Neurotoxicol. Teratol. 19: 417–428.

Grandjean, P., P. Weihe, R.F. White and F. Debes. 1998. Cognitive performance of children prenatally exposed to "safe" levels of methylmercury. Environ. Res. 77: 165–72.

Haut, M.W., L.A. Morrow, D. Pool, T.S. Callahan, J.S. Haut and M.D. Franzen. 1999. Neurobehavioral effects of acute exposure to inorganic mercury vapor. Appl. Neuropsychol. 6: 193–200.

Hibbeln, J.R., J.M. Davis, C. Steer, P. Emmett, I. Rogers, C. Williams and J. Golding. 2007. Maternal seafood consumption in pregnancy and neurodevelopmental outcomes in childhood (ALSPAC Study): an observational cohort study. Obstet. Gynecol. Surv. 62: 437–439.

Horvat, M., N. Nolde, V. Fajon, V. Jereb, M. Logar, S. Lojen, R. Jacimovic, I. Falnoga, Q. Liya, J. Faganeli and D. Drobne. 2003. Total mercury, methylmercury and selenium in mercury polluted areas in the province Guizhou, China. Sci. Total Environ. 304: 231–56.

Kaderbhai, N.N., D.I. Broadhurst, D.I. Ellis, R. Goodacre and D.B. Kell. 2003. Functional genomics via metabolic footprinting: monitoring metabolite secretion by *Escherichia coli* tryptophan metabolism mutants using FT-IR and direct injection electrospray mass spectrometry. Comp. Funct. Genomics 4: 376–91.

Kaneko, J.J. and N.V.C. Ralston. 2007. Selenium and mercury in pelagic fish in the central north pacific near Hawaii. Biol. Trace Elem. Res. 119: 242–54.

Karimi, R., N.S. Fisher and J.R. Meliker. 2014a. Mercury-nutrient signatures in seafood and in the blood of avid seafood consumers. Sci. Total Environ. 496: 636–43.

Karimi, R., S. Silbernagel, N.S. Fisher and J.R. Meliker. 2014b. Elevated blood Hg at recommended seafood consumption rates in adult seafood consumers. Int. J. Hyg. Environ. Health 217: 758–64.

Kazantzis, G. 2002. Mercury exposure and early effects: an overview. Medicina Del Lavoro. pp. 139–147.

Khayat, A. and L. Dencker. 1984. Interactions between tellurium and mercury in murine lung and other organs after metallic mercury inhalation: A comparison with selenium. Chem. Biol. Interact. 50: 123–133.

Kim, D.S., G.B. Kim, T. Kang and S. Ahn. 2012. Total and methyl mercury in maternal and cord blood of pregnant women in Korea. Toxicol. Environ. Health Sci. 3: 254–257.

Kim, K.B., S.Y. Um, M.W. Chung, S.C. Jung, J.S. Oh, S.H. Kim, H.S. Na, B.M. Lee and K.H. Choi. 2010. Toxicometabolomics approach to urinary biomarkers for mercuric chloride (HgCl2)-induced nephrotoxicity using proton nuclear magnetic resonance (1H NMR) in rats. Toxicol. Appl. Pharmacol. 249: 114–126.

Koletzko, B., E. Lien, C. Agostoni, H. Böhles, C. Campoy, I. Cetin, T. Decsi, J.W. Dudenhausen, C. Dupont, S. Forsyth, I. Hoesli, W. Holzgreve, A. Lapillonne, G. Putet, N.J. Secher, M. Symonds, H. Szajewska, P. Willatts and R. Uauy. 2008. The roles of long-chain polyunsaturated fatty acids in pregnancy, lactation and infancy: review of current knowledge and consensus recommendations. J. Perinat. Med. 36: 5–14.

Koppenaal, D.W. and G.M. Hieftje. 2007. Metallomics: an interdisciplinary and evolving field. J. Anal. At. Spectrom. 22: 855.

Kris-Etherton, P.M. 2002. Fish consumption, fish oil, omega-3 fatty acids, and cardiovascular disease. Circulation 106: 2747–2757.

Leermakers, M., W. Baeyens, P. Quevauviller and M. Horvat. 2005. Mercury in environmental samples: Speciation, artifacts and validation. TrAC Trends Anal. Chem. 24: 383–393.

Li, M.-M., M.-Q. Wu, J. Xu, J. Du and C.-H. Yan. 2014. Body burden of Hg in different bio-samples of mothers in Shenyang city, China. PLoS One 9: e98121.

Li, Y., Y. Jiang and X.-P. Yan. 2005. On-line hyphenation of capillary electrophoresis with flame-heated furnace atomic absorption spectrometry for trace mercury speciation. Electrophoresis 26: 661–7.

Liu, X., L. Zhang, L. You, J. Yu, M. Cong, Q. Wang, F. Li, L. Li, J. Zhao, C. Li and H. Wu. 2011a. Assessment of clam ruditapes philippinarum as heavy metal bioindicators using NMR-based metabolomics. CLEAN - Soil, Air, Water 39: 759–766.

Liu, X., L. Zhang, L. You, M. Cong, J. Zhao, H. Wu, C. Li, D. Liu and J. Yu. 2011b. Toxicological responses to acute mercury exposure for three species of Manila clam ruditapes philippinarum by NMR-based metabolomics. Environ. Toxicol. Pharmacol. 31: 323–332.

Llop, S., K. Engström, F. Ballester, E. Franforte, A. Alhamdow, F. Pisa, J.S. Tratnik, D. Mazej, M. Murcia, M. Rebagliato, M. Bustamante, J. Sunyer, A. Sofianou-Katsoulis, A. Prasouli, E. Antonopoulou, I. Antoniadou, S. Nakou, F. Barbone, M. Horvat and K. Broberg. 2014. Polymorphisms in ABC transporter genes and concentrations of mercury in newborns – evidence from two Mediterranean birth cohorts. PLoS One 9: e97172.

Lobinski, R., J.S. Becker, H. Haraguchi and B. Sarkar. 2010. Metallomics: Guidelines for terminology and critical evaluation of analytical chemistry approaches (IUPAC Technical Report). Pure Appl. Chem. 82: 493–504.

Lohren, H., L. Blagojevic, R. Fitkau, F. Ebert, S. Schildknecht, M. Leist and T. Schwerdtle. 2015. Toxicity of organic and inorganic mercury species in differentiated human neurons and human astrocytes. J. Trace Elem. Med. Biol. 32: 200–8.

Lucchini, R., I. Cortesi, P. Facco, L. Benedetti, D. Camerino, P. Carta, M.L. Urbano, A. Zaccheo and L. Alessio. 2002. Effetti neurotossici da esposizione a basse dosi di mercurio, in: Medicina Del Lavoro. pp. 202–214.

Magos, L. and M. Webb. 1980. The interactions of selenium with cadmium and mercury. Crit. Rev. Toxicol. 8: 1–42.

Magos, L. and T.W. Clarkson. 2006. Overview of the clinical toxicity of mercury. Ann. Clin. Biochem. 43: 257–68.

Marsh, D.O., G.J. Myers, T.W. Clarkson, L. Amin-Zaki, S. Tikriti and M.A. Majeed. 1980. Fetal methylmercury poisoning: clinical and toxicological data on 29 cases. Ann. Neurol. 7: 348–353.

Marsh, D.O. 1987. Fetal methylmercury poisoning. Arch. Neurol. 44: 1017.

Martínez, R., C. Pérez, N. Bordel, R. Pereiro, J.L.F. Martín, J.B. Cannata-Andía and A. Sanz-Medel. 2001. Exploratory investigations on the potential of radiofrequency glow discharge-optical emission spectrometry for the direct elemental analysis of bone. J. Anal. At. Spectrom. 16: 250–255.

Matos, J., H.M. Lourenço, P. Brito, A.L. Maulvault, L.L. Martins and C. Afonso. 2015. Influence of bioaccessibility of total mercury, methyl-mercury and selenium on the risk/benefit associated to the consumption of raw and cooked blue shark (Prionace glauca). Environ. Res. 143: 123–9.

Meyer-Baron, M., M. Schaeper and A. Seeber. 2002. A meta-analysis for neurobehavioural results due to occupational mercury exposure. Arch. Toxicol. 76: 127–36.

Mieiro, C.L., M. Dolbeth, T.A. Marques, A.C. Duarte, M.E. Pereira and M. Pacheco. 2014. Mercury accumulation and tissue-specific antioxidant efficiency in the wild European sea bass (Dicentrarchus labrax) with emphasis on seasonality. Environ. Sci. Pollut. Res. Int. 21: 10638–51.

Moreno, F., T. García-Barrera and J.L. Gómez-Ariza. 2010. Simultaneous analysis of mercury and selenium species including chiral forms of selenomethionine in human urine and serum by HPLC column-switching coupled to ICP-MS. Analyst 135: 2700–5.

Moreno, F., T. García-Barrera and J.L. Gómez-Ariza. 2013. Simultaneous speciation and preconcentration of ultra trace concentrations of mercury and selenium species in environmental and biological samples by hollow fiber liquid phase microextraction prior to high performance liquid chromatography coupled to indu. J. Chromatogr. A 1300: 43–50.

Mounicou, S., J. Szpunar and R. Lobinski. 2009. Metallomics: the concept and methodology. Chem. Soc. Rev. 38: 1119–38.

Myers, G.J. and P.W. Davidson. 1998. Prenatal methylmercury exposure and children: Neurologic, developmental, and behavioral research. Environ. Health Perspect. 106: 841–847.

Myers, G.J., P.W. Davidson, D. Palumbo, C. Shamlaye, C. Cox, E. Cernichiari and T.W. Clarkson. 2000. Secondary analysis from the Seychelles Child Development Study: the child behavior checklist. Environ. Res. 84: 12–9.

Nair, A., M. Jordan, S. Watkins, R. Washam, C. DuClos, S. Jones, J. Palcic, M. Pawlowicz and C. Blackmore. 2014. Fish consumption and hair mercury levels in women of childbearing age, Martin County, Florida. Matern. Child Health J. 18: 2352–61.

Nicholson, J.K., J.A. Timbrell and P.J. Sadler. 1985. Proton NMR spectra of urine as indicators of renal damage. Mercury-induced nephrotoxicity in rats. Mol. Pharmacol. 27: 644–651.

Nicholson, J.K., J.C. Lindon and E. Holmes. 1999. "Metabonomics": Understanding the metabolic responses of living systems to pathophysiological stimuli via multivariate statistical analysis of biological NMR spectroscopic data. Xenobiotica. 29: 1181–1189.

Ogra, Y. and Y. Anan. 2012. Selenometabolomics explored by speciation. Biol. Pharm. Bull. 35: 1863–1869.

Oken, E., R.O. Wright, K.P. Kleinman, D. Bellinger, C.J. Amarasiriwardena, H. Hu, J.W. Rich-Edwards and M.W. Gillman. 2005. Maternal fish consumption, hair mercury, and infant cognition in a U.S. Cohort. Environ. Health Perspect. 113: 1376–1380.

Okyere, H., R.B. Voegborlo and S.E. Agorku. 2015. Human exposure to mercury, lead and cadmium through consumption of canned mackerel, tuna, pilchard and sardine. Food Chem. 179: 331–5.

Ontario Ministry of Environment. Guide to eating Ontario Sport Fish 2011–2012. Toronto.

Orisakwe, O.E., O.J. Afonne, E. Nwobodo, L. Asomugha and C.E. Dioka. 2001. Low-dose mercury induces testicular damage protected by zinc in mice. Eur. J. Obstet. Gynecol. Reprod. Biol. 95: 92–96.

Oskarsson, A., A. Schütz, S. Skerfving, I.P. Hallén, B. Ohlin and B.J. Lagerkvist. 1996. Total and inorganic mercury in breast milk and blood in relation to fish consumption and amalgam fillings in lactating women. Arch. Environ. Health 51: 234–241.

Othman, M.S., G. Safwat, M. Aboulkhair and A.E. Abdel Moneim. 2014. The potential effect of berberine in mercury-induced hepatorenal toxicity in albino rats. Food Chem. Toxicol. 69: 175–81.

Palenzuela, B., L. Manganiello, A. Ríos and M. Valcárcel. 2004. Monitoring inorganic mercury and methylmercury species with liquid chromatography–piezoelectric detection. Anal. Chim. Acta 511: 289–294.

Pedrero, Z., L. Ouerdane, S. Mounicou, R. Lobinski, M. Monperrus and D. Amouroux. 2012. Identification of mercury and other metals complexes with metallothioneins in dolphin liver by hydrophilic interaction liquid chromatography with the parallel detection by ICP MS and electrospray hybrid linear/orbital trap MS/MS. Metallomics 4: 473–9.

Pena-Pereira, F., I. Lavilla and C. Bendicho. 2010. Liquid-phase microextraction techniques within the framework of green chemistry. TrAC Trends Anal. Chem. 29: 617–628.

Percy, A.J., M. Korbas, G.N. George and J. Gailer. 2007. Reversed-phase high-performance liquid chromatographic separation of inorganic mercury and methylmercury driven by their different coordination chemistry towards thiols. J. Chromatogr. A 1156: 331–9.

Pesch, A., M. Wilhelm, U. Rostek, N. Schmitz, M. Weishoff-Houben, U. Ranft and H. Idel. 2002. Mercury concentrations in urine, scalp hair, and saliva in children from Germany. J. Expo. Anal. Environ. Epidemiol. 12: 252–8.

Ramirez, G.B., M.C.V. Cruz, O. Pagulayan, E. Ostrea and C. Dalisay. 2000. The tagum study I: Analysis and clinical correlates of mercury in maternal and cord blood, breast milk, meconium, and infants' hair. Pediatrics 106: 774–781.

Rice, D.C. 2004. The US EPA reference dose for methylmercury: sources of uncertainty. Environ. Res. 95: 406–13.

Río Segade, S. and J.F. Tyson. 2003. Determination of inorganic mercury and total mercury in biological and environmental samples by flow injection-cold vapor-atomic absorption spectrometry using sodium borohydride as the sole reducing agent. Spectrochim. Acta Part B At. Spectrosc. 58: 797–807.

Roels, H.A., P. Hoet and D. Lison. 1999. Usefulness of biomarkers of exposure to inorganic mercury, lead, or cadmium in controlling occupational and environmental risks of nephrotoxicity, in: Renal Failure 21: 251–262.

Rose, M. and A. Fernandes. 2013. Persistent Organic Pollutants and Toxic Metals in Foods. Woodhead Publishing, Elsevier. 486 p.

Sakamoto, M., M. Kubota, X.J. Liu, K. Murata, K. Nakai and H. Satoh. 2004. Maternal and fetal mercury and n-3 polyunsaturated fatty acids as a risk and benefit of fish consumption to fetus. Environ. Sci. Technol. 38: 3860–3863.

Sakamoto, M., T. Kaneoka, K. Murata, K. Nakai, H. Satoh and H. Akagi. 2007. Correlations between mercury concentrations in umbilical cord tissue and other biomarkers of fetal exposure to methylmercury in the Japanese population. Environ. Res. 103: 106–11.

Sanz-Medel, A. 2008. Heteroatom(isotope)-tagged genomics and proteomics. Anal. Bioanal. Chem. 390: 1–2.

Soon, R., T.D. Dye, N.V. Ralston, M.J. Berry and L.M. Sauvage. 2014. Seafood consumption and umbilical cord blood mercury concentrations in a multiethnic maternal and child health cohort. BMC Pregnancy Childbirth 14: 209.

Spanopoulos-Zarco, P., J. Ruelas-Inzunza, M. Meza-Montenegro, K. Osuna-Sánchez and F. Amezcua-Martínez. 2014. Health risk assessment from mercury levels in bycatch fish species from the coasts of Guerrero, Mexico (Eastern Pacific). Bull. Environ. Contam. Toxicol. 93: 334–8.

Stern, A.H. and A.E. Smith. 2003. An assessment of the cord blood: maternal blood methylmercury ratio: implications for risk assessment. Environ. Health Perspect. 111: 1465–1470.

Strain, J.J., P.W. Davidson, M.P. Bonham, E.M. Duffy, A. Stokes-Riner, S.W. Thurston, J.M.W. Wallace, P.J. Robson, C.F. Shamlaye, L.A. Georger, J. Sloane-Reeves, E. Cernichiari, R.L. Canfield, C. Cox, L.S. Huang, J. Janciuras, G.J. Myers and T.W. Clarkson. 2008. Associations of maternal long-chain polyunsaturated fatty acids, methyl mercury, and infant development in the Seychelles Child Development Nutrition Study. Neurotoxicology 29: 776–82.

Szpunar, J. 2004. Metallomics: a new frontier in analytical chemistry. Anal. Bioanal. Chem. 378: 54–6.

Tainer, J.A., V.A. Roberts and E.D. Getzoff. 1991. Metal-binding sites in proteins. Curr. Opin. Biotechnol. 2: 582–591.

Takeuchi, T., K. Eto and H. Tokunaga. 1989. Mercury level and histochemical distribution in a human brain with Minamata disease following a long-term clinical course of twenty-six years. Neurotoxicology 10: 651–657.

Takeuchi, T., K. Eto, Y. Kinjo and H. Tokunaga. 1996. Human brain disturbance by methylmercury poisoning, focusing on the long-term effect on brain weight. Neurotoxicology 17: 187–190.

Tao, G. and S.N. Willie. 1998. Determination of total mercury in biological tissues by flow injection cold vapour generation atomic absorption spectrometry following tetramethylammonium hydroxide digestion. Analyst 123: 1215–1218.

Templeton, D.M., F. Ariese, R. Cornelis, L.G. Danielsson, H. Muntau, H.P. Van Leeuwen and R. Łobiński. 2000. Guidelines for terms related to chemical speciation and fractionation of elements. Definitions, structural aspects, and methodological approaches (IUPAC recommendations 2000). Pure Appl. Chem. 72: 1453–1470.

Thiry, C., A. Ruttens, L. De Temmerman, Y.-J. Schneider and L. Pussemier. 2012. Current knowledge in species-related bioavailability of selenium in food. Food Chem. 130: 767–784.

Toxicological effects of Methylmercury. 2000. National Academy Press, Washington, DC.

Van Wijngaarden, E., S.W. Thurston, G.J. Myers, J.J. Strain, B. Weiss, T. Zarcone, G.E. Watson, G. Zareba, E.M. McSorley, M.S. Mulhern, A.J. Yeates, J. Henderson, J. Gedeon, C.F. Shamlaye and P.W. Davidson. 2013. Prenatal methyl mercury exposure in relation to neurodevelopment and behavior at 19 years of age in the Seychelles Child Development Study. Neurotoxicol. Teratol. 39: 19–25.

Watras, C.J., N.S. Bloom, S.A. Claas, K.A. Morrison, C.C. Gilmour and S.R. Craig. 1995. Methylmercury production in the anoxic hypolimnion of a Dimictic Seepage Lake. Water, Air, Soil Pollut. 80: 735–745.

Wei, L., P. Liao, H. Wu, X. Li, F. Pei, W. Li and Y. Wu. 2008. Toxicological effects of cinnabar in rats by NMR-based metabolic profiling of urine and serum. Toxicol. Appl. Pharmacol. 227: 417–429.

Williams, R.J. 2001. Chemical selection of elements by cells. Coord. Chem. Rev. 216-217: 583–595.

Wu, H., X. Liu, J. Zhao and J. Yu. 2012. Toxicological responses in halophyte Suaeda salsa to mercury under environmentally relevant salinity. Ecotoxicol. Environ. Saf. 85: 64–71.

Zalups, R. 1995. Molecular interactions with mercury in the kidney. Toxicol. Appl. Pharmacol. 130: 121–131.

Zalups, R.K. and J. Koropatnick. 2000. Temporal changes in metallothionein gene transcription in rat kidney and liver: Relationship to content of mercury and metallothionein protein. J. Pharmacol. Exp. Ther. 295: 74–82.

Zayas, Z.P., L. Ouerdane, S. Mounicou, R. Lobinski, M. Monperrus and D. Amouroux. 2014. Hemoglobin as a major binding protein for methylmercury in white-sided dolphin liver. Anal. Bioanal. Chem. 406: 1121–9.

4

Selenium Speciation in Marine Environment

Madrid Yolanda

ABSTRACT

Selenium is an emerging global metalloid contaminant with potential ecological and human risk second only to mercury. Selenium in marine environments is biomagnified in food webs, with birds and fish being particularly susceptible to the effect of this metalloid. Primary producers, especially phytoplankton, generally concentrate selenium from 10^2- to 10^6-fold above ambient dissolved concentrations. In the marine environment, selenium and its compounds are submitted to physical, chemical, and biochemical transformations with or without modification of their oxidation states. Despite the fact that the total selenium content is a fundamental measure in assessing risk, it does not provide a complete picture on its bioavailability and mobility. These characteristics depend on the chemical properties of a specific selenium species. The purpose of this chapter is to summarize and to discuss the existing analytical methods for selenium speciation in marine environments, in order to provide easy access to this topic to all interested researchers. Advantages and limitations of the analytical procedures applied are critically discussed with respect to the sample preparation and speciation analysis. Very recent developments such as the use of nanomaterials as sorbents for the preconcentration of inorganic selenium in water samples are highlighted.

Introduction

Selenium (Se) has increasingly been recognized as an emerging global metalloid contaminant with potential ecological and human risk second only to mercury (Hu et al. 2009; Luoma and Rainbow 2008). Selenium achieves this distinction due to its non-point sources that are sensitive to geological and climatic conditions, the narrow margin of safety between its deficiency and toxicity, and its propensity to biomagnify through the food web.

Selenium is released into the aquatic environments through weathering and erosion of rocks and soils and through anthropogenic activities. Selenium is distributed globally but not uniformly in organic-rich marine sedimentary rocks (e.g., black shales, petroleum source rocks, phosphorite). The common range of Se in soils is 0.01–2 mg kg^{-1}, but the distribution varies from almost zero up to 1250 mg kg^{-1}

Departamento de Química Analítica, Facultad de Ciencias Químicas, Universidad Complutense de Madrid, Avda. Complutense s/n 28040 Madrid.
Email: ymadrid@quim.ucm.es

(Keskinen et al. 2009). Naturally high selenium concentrations can be found in some types of soils, called selenoferous soils, which are found in the arid and semi-arid areas of the world, including North America, Australia, Canada, and China. Selenium is also discharged into aquatic environments through anthropogenic activities such as metal smelting, mining (coal, hard rock, uranium, phosphate), power generation (coal-fired power plants, oil refineries) and agriculture activities (agricultural irrigation runoff and drainage to watersheds). Irrigation of semi-arid farmland in selenoferous regions is by far the most common driver of Se contamination particularly in the presence of a subsurface impermeable layer (e.g., clay), where leached Se can accumulate in toxic concentrations, a phenomenon that has been well documented in the San Joaquin Valley of California (Presser and Ohlendorf 1987). Furthermore, certain new technologies that use Se, such as nanotechnology, may have unpredicted impact in the environment.

Selenium Cycling in Marine Environment

As-mentioned above, selenium is released into the aquatic environments through weathering and erosion of rocks and soils and through industrial activities. Dissolved selenium in seawater is found as selenite Se(IV), selenite Se(VI), and dissolved organic selenides Se(-II), although thermodynamic equilibrium suggests only selenate should be present in seawater. However, selenate coexists with selenite in many aquatic systems suggesting that the oxidation state of selenium is not only controlled by thermodynamic equilibria, but also by other conditions and processes such as biochemical reactions and surface interactions. Elemental selenium is insoluble in water and is expected to accumulate in anaerobic sediments. The slow oxidation, by dissolved oxygen, of elemental selenium to selenite allows this element to be present also in oxic enviroments (oxigen- rich environments). Biotic transformations of selenium species are also numerous which could lead to a cocktail of different selenium species that can be found in marine environments. Several studies have shown that Se undergoes oxidation and reduction reactions mediated by microorganisms. Certain bacteria can reduce Se^0 to selenide (Herbel et al. 2003) or oxidize Se^0 to selenite and selenate (Dowdle and Oremland 1998). Red amorphous Se is produced by heterotrophic aerobic bacteria (Stolz et al. 2006), heterotrophic anaerobic bacteria (Siddique et al. 2006), and hemolithotrophic bacteria (Rauschenbach et al. 2011). It has been shown that biogenic Se remains in aqueous suspensions for prolonged times. This is due to the fact that, regardless of the microbial strain used, biogenic elemental Se is typically of nanoparticulate size (50–500 nm) (Buchs et al. 2013).

Similarly to other bioactive trace elements such as iron and zinc, selenite and selenate exhibit "nutrient-type" ocean profiles, indicating the utilization of inorganic selenium by phytoplankton in the surface ocean. In this line, total dissolved selenium, selenite, and selenate displayed surface water depletion and deep water enrichment when vertical profiles species were obtained in the Celebes Sea, the Sulu Sea, and the South China Sea (Nakaguchi 2004). The ocean is also an important source of releasing selenium into the atmosphere by means of an oceanic (byo) geochemical pathway similar to sulfur. Amouroux and Donard 1996, report the first assessment of at least three volatile selenium species (MeSeH, DMSe, DMDSe) in marine water and air. Dimethyl selenide (DMSe) was the main compound encountered and the total concentration of the volatile selenides was found to be correlated with the marine plankton biomass as measured through chlorophyll *a* content, suggesting the biogenic origin of the volatile compounds measured. Later, Amouroux et al. 2001, carried out measurements of volatile selenium species in samples from the North Atlantic sea. The dominant volatile forms were DMSe and the mixed S/Se compound dimethyl selenyl sulfide (DMSeS). The ocean-to-atmosphere vapor phase selenium flux has been estimated to be around of $5–8 \times 10^9$ g Se yr^{-1}. The geographical distribution of selenium-containing aerosol appears to be more closely related to the primary productivity of the surrounding waters rather than the transport of continentally derived material into the central ocean basin (Byard et al. 1987).

Concentrations of Se in organisms are classically biomagnified in food webs, with birds and fish being particularly susceptible to the effect of this metalloid. Primary producers, especially phytoplankton, generally concentrate selenium from 10^2- to 10^6-fold above ambient dissolved concentrations. Phytoplankton transformation of inorganic species of Se, such as selenite and selenate, to Se-containing amino acids is a critical step in the accumulation and magnification of selenium. Numerous studies have shown that dietary consumption of Se-laden microalgae by suspension- or deposit-feeding invertebrates can be responsible for > 90% of the total Se uptake in typical coastal waters (Schlenk et al. 2007). Uptake of Se by marine phytoplankton is rapid from solution, and there is preference for the selenite species, although selenate tends to be the predominant form in aqueous systems.

Nowadays, it is widely accepted that major selenium uptake route into fish is not accumulation from water, but rather via the food chain. Oviparous (egg-laying) vertebrates such as fish and waterbirds are the organisms most sensitive to Se out of those studied to date. Toxicity can result from maternal transfer of organic Se to eggs in oviparous vertebrates. The most sensitive diagnostic indicators of Se toxicity in vertebrates occur when developing embryos metabolize organic Se that is present in egg albumen or yolk. Toxicity endpoints include embryo mortality (which is the most sensitive endpoint in birds), and a characteristic suite of teratogenic deformities (such as skeletal, craniofacial, and fin deformities, and various forms of edema) that are the most useful indicators of Se toxicity in fish larvae (Holm et al. 2005; Tashjian et al. 2006). The threshold dietary Se toxicity concentration in water sturgeons juveniles has been estimated to be between 10 and 20 μg g^{-1} (Tashjian et al. 2007). The mechanism of selenium effect has not been fully elucidated yet. It has been postulated that certain metabolites of organic Se can become involved in oxidation–reduction cycling, generating reactive oxidized species that can cause oxidative stress and cellular dysfunction. Recently, Sivester et al. 2010, applied a proteomic approach for establishing variations in protein expression in larval surgeons exposed to heat stress and selenium (added as selenometionine). The most significant changes in abundance were observed in proteins involved in correct protein folding, protein synthesis, protein degradation, ATP supply, and structural proteins. Valosin-containing protein (VCP), a protein involved in aggresome formation and in protein quality control decreased more than 50% in response to selenium. Results suggest that selenium behaves similar to cadmium in inducing the formation of aggresomes, cellular organelle in which ubiquitinated and unfolded proteins accumulate when the capacity of the intracellular protein degradation machinery is exceeded.

Fish and shellfish are considered Se-rich organisms, although the levels of Se can vary depending on the species. Recently, Se content in seafood materials has been extensively reviewed by Yoshida 2012; Cabañero et al. 2005, determined total Se content in different highly consumed fish such as tuna, swordfish, and sardines, with values oscillating between 1.13 ± 0.04 μg g^{-1} and 1.58 ± 0.02 μg g^{-1}. Similar Se concentrations were found in Niboshi, a processed Japanese anchovy (*Engraulis japonicus*) (Yoshida et al. 2012). A higher content of Se was quantified in *Misgurnus anguillicaudatus* muscle tissue (6.280 μg g^{-1}) (Gong et al. 2013). Burger et al. 2001, found levels of Se in fish muscle tissues in the range of 0.21–0.64 μg g^{-1}.

Although Se content in marine animals is higher than in terrestrial ones, little is known about the chemical forms in which selenium is present. Proteinaceous Se species, such as selenomethionine (SeMet) and selenocysteine (SeCys) have been found in different types of seafood (Yoshida et al. 2011; Pedrero et al. 2011; Cabañero et al. 2005; Moreno et al. 2004). The selenium species distribution in the different tissues and organs is clearly related to the metabolic activity of the different tissues.SeMet is the major species detected in muscle, while SeCys is the main species in the liver, and it is also found in the kidney. The presence of these species, released from selenoproteins, in these metabolically active glandular visceral tissues, can be explained as a result of the conversion of the selenium species initially present in the supplemented diet (Pedrero et al. 2011).

However, a large percentage of Se can be associated to organic compounds other than amino acids or their derivatives. For instance, a novel selenometabolite, selenonine (2-selenyl-N,N,Ntrimethyl-L-histidine) was identified by Yamashita and Yamashita in 2010, in the blood blue tuna (*Thunus orientalis*). Selenonine appears to show a radical scavenging activity, although its biological function is still unknown. This selenometabolite has been also identified in sea turtles where it is preferentially accumulated in the liver. However, selenoneine content was below detection limit in terrestrial turtle suggesting that selenoneine is transferred among marine organisms via the food web (Anan et al. 2011).

Another biological significance of selenium is that it detoxifies inorganic mercury by directly interacting with it. Some authors has postulated that selenium is able to decrease Hg toxicity when the molar ratio of Se:Hg is greater than 1 (Sørmo et al. 2011; Burger et al. 2012). At equimolar ratio Hg and Se formed a complex (mercury selenide, tiemannite) that is accumulated in the liver of marine animals. The Hg–Se complex may be formed upon the degradation of proteins that bind inorganic mercury and selenium in the lysosomes (Ikemoto et al. 2004). The interaction selenium-mercury in marine environments is still an area of active research work. The studies include the evaluation of different parameters such as, measurement of selenium and mercury concentration in different fish specimens and biota (Kehrig et al. 2009), measurement of the metal detoxification potential by analyzing the hepatic metallothionein (MT) content (Siscar et al. 2014) and evaluation of the interaction of inorganic mercury and methlymercury with selenium by performing *in vivo* studies (Anan et al. 2011).

Selenoproteins in Marine Samples

Selenium is an essential micronutrient due to its requirement for biosynthesis and function of the 21st amino acid, selenocysteine (SeCys). The amino acid SeCys is found in the active sites of selenoproteins, most of which exhibit redox function, in all three domains of life. The best known selenoproteins are glutathione peroxidases, thioredoxin reductases, and iodothyroninedeiodinases (Lopez-Heras et al. 2011). Marine animals have more selenoproteins (around 30–37 selenoproteins) than terrestrial animals. Fish have several species-specific selenoproteins (fish 15 kDa selenoprotein-like protein (Fep15), selenoprotein J, and selenoprotein L) whose biological functions are still unclear. The Global Ocean Sampling Expedition Project detected more than 3,600 selenoprotein gene sequences belonging to 58 protein families in the microbial marine community (Zhang and Gladyshev 2008). In this study, samples were clustered in various groups based on geographic location, water temperature (tropical or temperate) and salinity (sea water, fresh water, estuaries, or hypersaline lakes). Geographic location had little influence on SeCys utilization as measured by selenoprotein variety and the number of selenoprotein genes detected; however, both higher temperature and marine (as opposed to freshwater and other aquatic) environment were associated with increased use of this amino acid. The study generated the largest selenoproteome reported to date and provide important insights into microbial Se utilization and its evolutionary trends in marine environments. Similarly, Seale et al. 2014 evaluated the influence of acclimation salinity on the selenoproteins expression (iodothyroninedeiodinases 1, 2 and 3 (Dio1, Dio2, Dio3), Fep15, glutathione peroxidase 2, selenoproteins J, K, L, M, P, S, and W) in eight tissues (brain, eye, gill, kidney, liver, pituitary, muscle, and intraperitoneal white adipose tissue) of the Mozambique tilapia, *Oreochromismossambicus*. Results show that selenoprotein transcripts are differentially regulated according to acclimation salinity suggesting selenoproteins and their synthesis factors as key points in osmoregulation processes.

In conclusion, the larger selenoproteomes found in aquatic organisms with respect to terrestrial animal may results from several marine-specific factors influencing the SeCys utilization, such as availability of selenium, gradients of temperature, pH, pressure, oxygen content, salinity, and chemical environment. In light of this, it is surprising that little information is available on the abundance and distribution of selenoproteins in fish species.

Selenium Speciation in Marine Environments—General Remarks

As was mentioned above, selenium is naturally found in the marine environment in a great variety of chemical forms: inorganic (i.e., selenate (Se(VI)) and selenite (Se(IV)) or organic (i.e., selenomethionine (SeMet), selenocysteine (SeCys), dimethylselenide (DMSe), and dimethyldiselenide (DMDSe)) forms, and is ultimately incorporated into selenoproteins as the amino acid selenocysteine (SeCys).

Knowledge about organic selenium compounds in the aquatic ecosystem is still very limited because most analytical work in this field was done for total selenium or for the inorganic selenium species. The organic bound selenium is an essential part of the natural selenium cycle which links the inorganic with the more complex organic selenium compounds. The coexistence of organic selenium and inorganic selenium in seawater suggests that selenium is involved in biological activity. The question regarding why selenium speciation should be studied in marine environments has a simple answer, and is that it is needed for a complete understanding of the complex chemistry and behavior of selenium in environment and organisms. In environmental studies the identification of the chemical forms of an element is still a challenge. Speciation analysis requires the application of several steps: sample treatment (extraction,

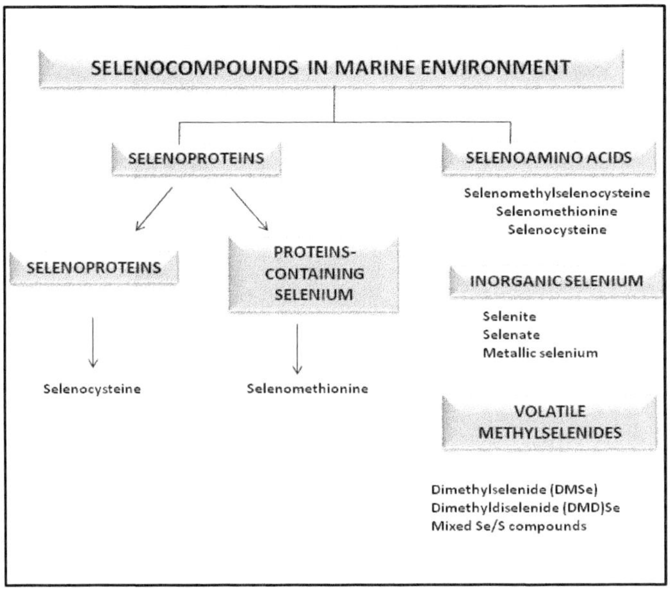

Fig. 1: Most relevant selenocompounds in the marine environment.

preconcentration, derivatization), separation of species, and identification. Generally, for selenium species determination, the coupling of chromatography techniques such as gas chromatography (GC), high performance liquid chromatography (HPLC), and capillary electrophoresis (CE) with sensitive and selective detectors such as, atomic absorption or fluorescence spectrometry (AAS or AFS) and inductively coupled plasma emission spectrometry (ICP-MS) has been widely used. Nowadays the coupling of HPLC-ICPMS is well implemented in most research groups focus in selenium speciation due to the simplicity of its interface and the high sensitivity of the ICPMS. However, it is still a costly technique requiring training personnel which makes difficult its implementation in routine laboratories. As an alternative, the coupling of HPLC HG-AFS offers very attractive features such as an inexpensive and affordable technique, simplicity of operation, and high sensitivity. The main drawback is that selenium species determination is not straightforward; it requires a previous analytical step for transforming both Se(IV) and selenium organic compounds into Se(VI). Identification of selenium species, once separated by HPLC is not an easy task. HPLC-ICP-MS usually produces a number of peaks whose identification should be considered with extreme care in order to avoid data misinterpretation. The assignment of analytes on the basis of retention times can often lead to wrong identification, therefore a proper identification of the species implies the application of at least two chromatographic columns with different separation mechanims. Unfortunately, the number of available Se-compounds standards is limited and therefore the aplicability of HPLC-ICPMS is inadecuate when Se unknown compounds need to be determined in environmental matrices. For this purpose, in the last few years, electrospray mass spectrometry (ES-MS, ES-MS-MS) or matrix-assisted laser desorption ionization time-of flight (MALDI-TOF) measurements has been incorporated to speciation studies. Selenocompounds cannot be directly analyzed by ES-MS because of the presence of the matrix composed of high molecular weight compounds and salts suppressing the signal. The identification of selenocompounds in biological matrices is usually based on the purification of the compounds of interest by multidimensional chromatography followed by the characterization of isolated compounds by mass spectrometry.

Besides the significant advances obtained in selenium speciation, sample treatment is still the Achilles' heel of selenium speciation. In this line, selenium species must be first quantitatively extracted from the matrix without changing the original species form and then unambiguously identified and correctly quantified. Selenium in biological samples are mainly presented as selenoamino acids, either or nor incorporated to proteins. Several approaches have been used for extracting selenoamino acids from samples: aqueous leaching, and acid, basic and enzymatic hydrolysis, the latter being the method of choice. In some cases, the extraction yield is far from being quantitative. Moreover, species transformation could take place during sample extraction (such as selenocysteine and selenomethionine oxidation).

Although there has been significant progress in the use of hyphenated techniques, it is important to keep in mind that there are a plethora of methods based in non-chromatography techniques (mainly solid phase extraction, SPE, and liquid-liquid microextraction) which could provide useful information about selenium chemical forms (mainly inorganic selenium) along with selenium preconcentration.

Sample treatment for selenium speciation in marine environments

Determination of inorganic selenium in water—Preconcentration procedures

As selenium concentration in environmental samples is very low, and those of each selenium species are much lower, the determination of inorganic selenium species in environmental samples, therefore, usually requires a preconcentration method. For this, analytes can be either retained on solid phase extraction (SPE) materials, extracted by dispersive liquid–liquid, or coprecipitated using insoluble metal hydroxides (Wake et al. 2004).

SPE is widely used in the preconcentration and separation of selenium. Recently, preconcentration of inorganic selenium has been extensively reviewed by Herrero-La Torre et al. 2013. These authors classified the SPE methodology according to the different behavior of inorganic selenium in relation to the SPE sorbent: (A) Selective SPE sorption of only one of the inorganic species, (B) SPE sorption of both inorganic species, and (C) selenium complexation and subsequent SPE sorption of chelated species.

Numerous substances have been used as adsorbent for solid phase extraction of selenium species, including different kinds of polymers and silica gel-based sorbents, metallicoxides, and hydroxides. Besides the typical SPE phases, nanometer size TiO_2 particles have been lately used for the preconcentration and separation of inorganic selenium species in waters. In all cases, selenium is retained on TiO_2 nanoparticles under acidic conditions because, in acidic media, TiO_2 has a high adsorptive capacity for Se(IV) and Se(VI). The mechanism of Se adsortion on TiO_2 nanoparticles has been discussed in the literature (Benedicto et al. 2013) and involves the strong adsorption of the anionic forms of Se on the TiO_2 surface (positively charged at low pH). The major advantage in using these nanoparticles is once they adsorbed selenium inorganic species from waters, the solid phase can be prepared to form a slurry which can be determined by GFAAS (Zhang et al. 2007). This methodology has been employed not only for the preconcentration of inorganic selenium species but also for separating organic and inorganic selenium. Huang et al. 2008 reported a speciation method for inorganic and organic selenium based on the use of both aluminum and titanium oxides in two SPE columns on line coupled to ICPMS. The first column was packed with the nanometer-sized Al_2O_3, where as the second contained TiO_2 chemically modified with dimercapto succinic acid (DMSAC). Due to the different absorption behavior of organic and inorganic selenium on the column, a selective elution of Se(IV) Se(VI), SeMet and SeCys was obtained. Another nice demonstration of the usefulness of nanomaterials as sorbents for selenium preconcentration is given by Kim and Lim 2013. The authors selectively extracted Se(IV) from natural waters by using TiO_2@SiO_2/Fe_3O_4 nanoparticles as sorbents. The mixture of nanoparticles has two noticeable capabilities: photocatalytic reduction of selenium cations to Se^0 atoms by TiO_2 shell and particle collection by Fe_3O_4 core. In addition to TiO_2 nanoparticles, other nanomaterials have been employed as sorbents for selenium speciation. In this line, Se(IV) was selectively adsorbed on Fe_3O_4 magnetic nanoparticles (within pH range of 1–10). The resulting Se(IV)-loaded nanoparticles were further separated from the solution by applying an external magnetic field and measured by ICPMS. A limit of detection (LOD) of 0.094 ng L^{-1} was achieved for Se(IV) with an enrichment factor of 287 (Huang et al. 2012). Wu et al. (2014) determined Se(IV) and Se(VI) in environmental water samples by using ZrO_2-modified coal cinder as sorbent and ICP-OES as detector. The separation mechanisms is based on the different pKa values of selenic acid and selenious acid and the isoelectric point of ZrO_2. An enrichment factor of 100 was obtained with 100 mL of water solution. The LOD were estimated to be 9 ng L^{-1} for Se(IV) (Wu et al. 2014).

Layered double hydroxides (LDHs), a class of nanostructured anionic clays, have been also used for preconcentrating selenium from water samples. Chen et al. (2012) extracted inorganic Se onto a mini-column packed with Mg–$FeCO_3$ LDHs coated cellulose fibre. The quantitative adsorption of Se(IV) and Se(VI) was performed at pH 3.8–8.0 and pH 5.8–7.0, respectively. Inorganic selenium was further eluted by applying 0.8% (m/v) NaOH as eluent, and measured by HG-AFS. An enrichment factor of 13.3 was obtained with a LOD of 11 ng L^{-1}. All these are examples of the capabilities of nanoparticles for preconcentrating selenium from water samples and open new perspectives in the use of nanomaterials as sorbents in the environmental speciation of selenium.

Regarding liquid phase microextraction (LPME) methods, single drop microextraction, hollow fiber liquid phase microextraction (HF- LPME), and dispersive liquid-liquid microextraction (DLLME) have been used for selenium species separation and preconcentration from water samples. All of them are miniaturized sample pretreatment techniques which have been usually employed in combination with electrothermal vaporization (ETV)-ICPMS and GFAAS, both being microamount sample introduction techniques. One of the techniques attracting special interest is DLLME. This procedure has been employed for extracting selenium in water samples using a mixture containing ethanol (disperser solvent), carbon tetrachloride (extraction solvent), and ammonium pyrrolidine dithiocarbamate (APDC, chelating agent). After centrifuging, the concentration of an enriched analyte in the carbon tetrachloride phase was determined by the iridium-modified pyrolitic tube GFAAS. The concentration of selenate was obtained as the difference between the concentration of selenite after and before pre-reduction of Se(VI) to Se(IV). Under the optimum conditions, an enrichment factor of 70 was obtained from only 5.00 mL of water sample with a detection limit of 0.05 µg L^{-1} (Bidari et al. 2007). The same procedure was used for selectively extracting Se(IV) by using 500 µL of ethanol (as disperser solvent) containing 70 µL chloroform (as extraction solvent) and 0.2 g L^{-1} Bismuthiol II (as chelating reagent) at pH = 2.0. After centrifugation, the complex of Se(IV) and Bismuthiol II concentrated in the extraction solvent was introduced into the ETV-ICP-MS for determination of Se(IV). Total selenium was determined after the reduction of Se(VI) to Se(IV) by using 5 mol L^{-1} HCl. This approach offers a LOD for Se of 0.047 ng mL^{-1} and an enrichment factor of 64.8 for only 5 mL of water sample (Zhang et al. 2013).

Ghasemi et al. 2010, employed HF-LPME for the simultaneous separation of selenium in environmental samples (sea water). The method involves the selective extraction of the Te(IV) and Se(IV) species by HF-LPME with the use of APDC as the chelating agent. The complex compounds were extracted into 10 µL of toluene and determined by GFAAS. Under optimal conditions, an enrichment factor of about 400 was achieved with limit of detection of 4 ng L^{-1}.

Regarding sensitivity, some of the latter methods indeed achieve low limits of detection (i.e., ng L^{-1} range). Still, diverse restrictions such as being labor intensive, time-consuming, costly and/or being species unspecific constrain their applicability for routine analysis needed in environmental studies.

Sample preparation for selenium speciation in fish and shellfish

As previously mentioned, enzymatic hydrolysis is the method of choice for selenium species extraction in food and biological samples. One of its advantages is that it allows selenium species extraction under mild operation conditions of temperature (37°C) and pH (7.0), preventing degradation of the original species. Following this approach, selenium species have been efficiently extracted from fish and other biological matrices using different methods such as controlled temperature incubation, microwave-assisted enzymatic extraction (MAEE), and sonication-assisted enzymatic hydrolysis through enzymatic hydrolysis using proteinases such as protease, subtilisin, trypsin, and proteinase N (Hinojosa et al. 2009; Siwek et al. 2005; Yang and Swami 2007; Lavilla et al. 2008; Pedrero et al. 2011).

A particular case of enzymatic extraction is establishing selenium bioaccessibility by using an *in vitro* gastrointestinal digestion method. There are several approaches that can be performed, the most widely applied method being the one developed by Luten et al. (1996). The procedure implies two steps: (1) gastric digestion, where the sample is submitted for a period of four hours at 37°C to the effect of gastric juice which consists of 6% of pepsine and (2) gastrointestinal digestion, where the solid residue from step one is treated with intestinal juice containing 1.5% pancreatine and 0.5% amylase. Cabañero et al. (2004) evaluated the bioaccessible fraction of selenium and mercury in different type of fish samples. Selenium solubility in the gastrointestinal supernatants was higher in swordfish and sardine (76 and 83%, respectively) than in tuna (50%). The results of speciation analysis (HPLC-ICPMS) of the gastrointestinal extracts show that SeMet was found to be the dominant Se species and its amount varied depending on the type of fish. Tuna and sardine had a higher bioaccessible SeMet concentration (0.290 ± 0.006 and 0.245 ± 0.006 μg g^{-1}, respectively) than swordfish (0.147 ± 0.005 μg g^{-1}). However, tuna, swordfish, and sardine provided similar percentages of bioaccessible SeMet (19, 14, and 16%, respectively) relative to the total Se content. Piñero et al. (2013) evaluated the *in vitro* bioavailability of selenium from seafood (white fish, cold water fish, and molluscs) by including dialysis of membranes during the simulated intestinal digestion. Selenium speciation in the resulting dializates evidenced that SeMet was the main selenium species. Bioavailability of selenium, calculated as the ratio between Se concentration in dialyzate extract and acid digestion extract, was very low and ranged from 4.0% to 13% of total selenium.

Although enzymatic hydrolysis followed by LC-ICPMS is widely implemented for Se speciation in seafood samples, it is important to point out that recoveries are often incomplete when analyzing fish sample extracts. Hinojosa et al. (2009) reported recoveries of 28, 35, 38, and 56% after enzymatic hydrolysis of tuna, shellfish, BCR62, and sharks, respectively. It has been demonstrated that during the enzymatic extraction from marine animal tissues, some Se compounds can remain bound in peptide form, depending on the cleavage specificity of the enzyme and the analyzed fish species. A clear example is shown in Fig. 2 representing the analysis of enzymatic extracts of fish tissue by SEC coupled to ICPMS. Se appeared to be associated to low molecular weight compounds (< 5 KDa). The low molecular mass range (Peak II) matched with a SeMet standard, represents around of 30% of the total selenium. The results obtained by SEC were in line with the data of SeMet quantification previously obtained by others authors when using either AE or RP separation. Furthermore, the fraction from peak I (6.5 min to 9.5 min) represented approximately a 60% of the total Se in the extract. Initially it was thought that the presence of this broad peak might be due to some Se compound that remained bound in peptide form. To investigate the characteristics of this fraction, the Se-containing peak I was collected and subjected to a second hydrolysis using two enzymes: protease and alcalase. Alcalase was chosen for being a highly efficient bacterial protease used to prepare protein hidrolyzates. This second hydrolysis with alcalase or protease did not significantly improve peak I hydrolysis (Fig. 2B). The chromatographic profile of the subsequent analysis of the fraction by RP-LC-ICPMS, shows a small peak at the void volume indicating that neither protease nor alcalase were able to hydrolyze the Se-compound present in Peak I.

The use of mixture of enzymes has been shown as very effective to quantitatively extract selenium from fish tissues. In this line, Zhang and Yang (2014) achieved acomplete hydrolysis of bay scallop tissues by the enzymes combination of Flavourzyme®500 L, carboxypeptidase Y, and trypsin (3 + 1 + 1), after a pre-treatment of the tissues with papain. Flavourzyme®500 L is a mixture of incision and excision enzymes, capable of breaking proteins into peptide fragments, where as carboxypeptidase Y is an excision enzyme, capable of cutting peptides into free amino acids. Trypsin and carboxypeptidase Y strengthened the ability of hydrolysis of the incision and excision enzymes in Flavourzyme®500 L, respectively.

Besides the presence of selenoamino acids in the extracts, a high percentage of Se can be associated to other lower molecular weight organic compounds different from selenoamino acids and their derivatives, and therefore resistants to enzymatic hydrolysis. For instance, selenium found in the water-extract from the Niboshi, a Japanese anchovy (*Engraulis japonicas*), is mostly ascribed to organoselenium compounds other than selenoamino acids with a molecular mass below 5 kDa and with anionic and/or amphoteric characteristics (Yoshida et al. 2012). Low-molecular mass Se species other than SeCys and SeMet were also detected in flat fish (Yoshida et al. 2011).

The development of new extraction procedures to quantitatively release selenium from fish tissues is an active field of research. Recently, subcritical water (formed from pure water under high temperature and high pressure) has been used for extracting selenium from fish tissues (Ohki et al. 2013). Subcritical water has a low dielectric constant and a high ionic product. The former leads to an increase in the solubility of organic compounds, while the latter produces the promotion of chemical reactions, such as hydrolysis. The experimental set up was very simple, the sample, a powdery dried fish tissue (tuna and mackerel), was subjected to the subcritical water system in an autoclave at 220°C in the liquid/solid (dry) ratio of

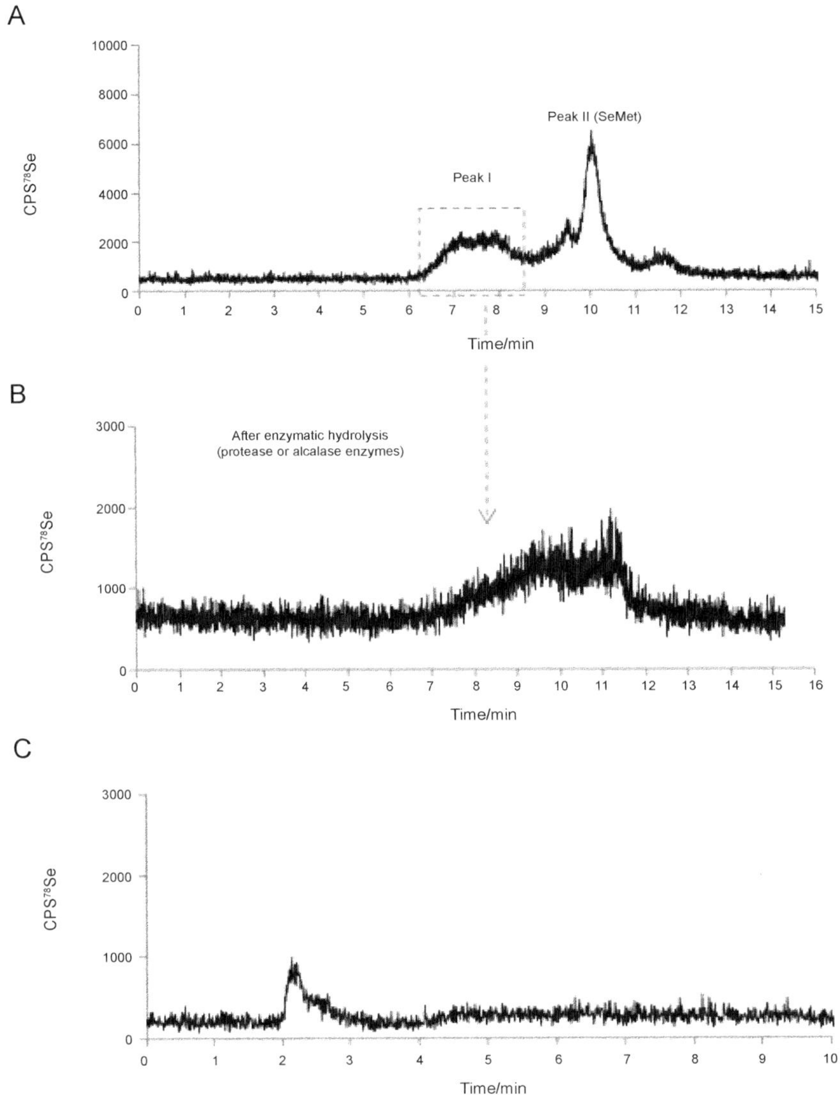

Fig. 2: (A) Chromatographic profile obtained by SEC-ICP-MS for ^{78}Se corresponding to enzymatically hydrolyzed extracts of lyophilized fish muscle. (B) Chromatographic profile obtained by SEC-ICP-MS for ^{78}Se after a second enzymatic extraction of Peak I by using 5:1 sample/alcalase mass ratio. (C) Chromatographic profile obtained by reverse phase LC-ICP-MS for ^{78}Se corresponding to a second enzymatic extraction of Peak I by using 5:1 sample/alcalase mass ratio (C).

20. Data from ICPMS reveal that selenium was quantitatively extracted from fish tissue when applying a sample treatment time of 120 min. The sample treatment time had a notable effect on the hydrolysis of selenoproteins which are gradually decomposed into lower molecular hydrophilic species, and dissolved in the aqueous phase as the sample treatment time increases.

Apart from low yield of extraction, species interconversion is another problem to take into account during sample treatment, especially when determining SeCys and SeMet. Selenocysteine is converted to dehydroalanine via oxidation to the selenoxide followed by β-elimination of seleninic acid. One approach to solve this problem is based on reduction with dithiotreitol (DTT) followed by acetylation with iodoacetamide (IAM). Pedrero et al. (2011) employed this procedure followed by enzymatic hydrolysis with protease to determine SeCys in fish organs by using HPLC-ICPMS. The method allowed authors to distinguish the contribution of selenoproteins (SeCys) and selenium-containing proteins (SeMet) as an alternative to Se proteins identification by ES-MS protocols.

Oxidation of selenomethionine to selenomethionine Se-oxide (SeMetO) during sample treatment is one of the main problems associated with the accurate determination of this species. Moreda-Piñeiro et al. (2013) confirmed SeMet oxidation during *in vitro* digestion procedure of sea food. The presence of SeMetO could lead to the overestimation of SeCys concentration by HPLC–ICP-MS when a Hamilton X-100 column is used. To detect the presence of SeMetO in the extracts, this selenocompound should be included among the working standard solutions of selenium. Selenomethionine Se-oxide (SeMetO) is not commercially available but it can be easily prepared in the laboratory by treating SeMet standard solution with 30% (w/v) H_2O_2 solution.

Analytical Techniques in Selenium Speciation in Marine Environments

Different instrumental techniques including atomic absorption spectrometry (AAS), atomic fluorescence spectrometry (AFS), inductively coupled plasma optical emission spectrometry (ICP-OES), and inductively coupled plasma mass spectrometry (ICP-MS) have been employed for selenium speciation. Among these techniques, ICP-MS has been frequently applied for determination of selenium due to its high sensitivity with a wide linear dynamic range and the capability of isotopic determination. Most of the reported procedures for selenium speciation in marine samples are based on liquid chromatography (LC) coupled to inductively coupled plasma mass spectrometry (ICP-MS). The separation of the selenium species can be accomplished by using reversed-phase columns in combination with an ion-pairing agent size-exclusion or ion exchange column (Pedrero and Madrid 1999). However, determination of selenium from ICPMS suffers from spectral, isobaric, and polyatomic (caused by atomic or molecular species that have the same mass of analyte of interest) and non spectral (caused by sample matrix) interferences. In selenium speciation, spectral interferences on their different isotopes have been the most studied. For instance, the most abundant selenium isotopes ^{78}Se (23.78%) and ^{80}Se (49.61%), overlap with argon dimers ($^{40}Ar,^{38}Ar^+$, and $^{40}Ar_2^+$, respectively) in conventional quadrupole ICP-MS, being the less abundant (8,7%) ^{82}Se isotope usually selected for selenium determination when a quadrupole ICP-MS is employed. One way of reducing spectral interferences is the use of ICP-MS equipped with collision/reaction cells. The gases most commonly used for overcoming interferences on Se determination are helium (collision, dissociation) and hydrogen (ion-molecules reactions). Despite the advantages of using collision/reaction cells, the procedure suffers from limitations. For instance, the use of hydrogen can lead to the production of new polyatomic interferences such as SeH^+ and BrH^+ which can be corrected by mathematical equations.

As alternative technique, the use of atomic fluorescence spectrometry (AFS), combined with hydride generation (HG), has gained wide use as an element-specific detector. The hydride-generation (HG) technique has proved to be an attractive interfacing technique for coupling HPLC and ICPMS or AFS. However, their application is often limited to total amount of selenium or to the inorganic form. Presently, sodium tetrahydroborate ($NaBH_4$) in acidic media is almost universally employed in HG applications. Considering the reduction kinetics of different Se species by $NaBH_4$, Se(VI) is not reducible to the hydride state. Consequently, as a mandatory protocol, inorganic speciation by HG techniques first involves determining Se(IV) and then transforming Se(VI) to Se(IV) before determining the total concentration. The most popular methodology involves the pre-reduction by halides such as chloride, bromide, or both in acidic medium followed by a heating process that will reach temperatures close to 100°. The selenate reduction has been also accomplished by using UV-light with both high and low pressure mercury lamps. Vilano et al. (2000) described an on-line method for the separation of Se(IV), Se(VI), SeMet, and SeCys using an anion exchange column followed by their reduction to Se(IV) through a 60-s exposure of the effluent to a 150-W high pressure mercury lamp. The method was validated by analyzing two water certified reference materials, in which only Se(VI) was detected. Similarly, Serra et al. (2012) develop a multisyringe flow injection analysis (MSFIA) system coupled to AFS for determining Se(IV), Se(VI), and SeMet. The method consists of three consecutive steps: (1) Se(IV) analysis by using HCL 1.2M and $NaBH_4$ 1% (w/v), (2) Se (VI) and SeMet analysis, as in 1, but now using UV light irradiation, and (3) total selenium determination, as in 1) but before reduction of Se(VI) to Se(IV) with KI 2.4M and NaOH 0.6M solution and using UV-light. The LODs for Se(IV), Se(VI), and SeMet were reported to be as low as 0.11, 0.12, and 0.13 μg L^{-1}, respectively.

Selenium species transformation has been also achieved by using other procedures avoiding the use of UV irradiation. Sanchez-Rodas et al. (2013) developed a method based on line coupling anion exchange HPCL-thermoreduction-HG-AFS for determining Se(IV), Se(VI), SeMet, SeCys, and selenomethylselenocysteine (MeSeCys). Selenium species transformation was performed by adding KBr during the heating step (150°C). Under optimized conditions (6M HCl, 1.5% $NaBH_4$, and 150°C), the method provides LODs at the low ppb range. The methodology was applied to selenium speciation in Se-rich algae after feeding them with Se(VI). SeMet was the main selenium specie found (3.4 mg kg^{-1}), followed by SeCys (1.8 mg kg^{-1}) and MetSeCys (1.0 mg kg^{-1}).

The reagent $NaBH_4$ is almost universally employed in hydride generation applications. However, its uses, when combined with ICP, may affect the stability of the Ar plasma because large quantities of hydrogen are delivered into the ICPMS. Therefore, it is of great interest to find new alternative vapor generation (VG) techniques. Recently, various VG techniques have been proposed, of which photoinduced VG has emerged as one of the most popular techniques. He et al. (2007) pioneered in the development of this technique for conversion of Se(VI) into gaseous compounds using a 15 W low pressure Hg lamp in the presence of formic, propionic, and malonic acid. Through systematic identification of the volatile photogenerated species using cryotrapping GC/MS or GC-ICP-MS, they found that inorganic Se(IV) was converted to SeH_2, SeCO, $(CH_3)_2Se$, and $(CH_3CH_2)_2Se$ in the presence of formic, acetic, propionic, and malonic acids, respectively. Nevertheless, insufficient efficiency of the pure photoinduced VG for conversion of Se(VI) into gaseous products remains an obstacle for the speciation of inorganic Se species. Sun et al. 2006, developed a novel UV/nano-TiO_2 photo-chemical vapor generation device to couple between LC and ICP-MS for the determination of Se(IV) and Se(VI). After HPLC separation and mixing with formic acid and nano-TiO_2 suspension, both Se(IV) and Se(VI) were transformed to volatile Se species under UV irradiation. Different systems, including UV/HCOOH, UV/ nanometricTiO_2-HCOOH, UV/ nanometricAg-TiO_2-HCOOH and UV/nanometric ZrO_2-HCOOH, were compared for HG of Se species. UV/nanometricZrO_2-HCOOH PCVG system as HPLC-ICP-MS interface provided the highest sensitivity for Se(VI) (0.016 ng mL^{-1}), Se(IV) (0.014 ng mL^{-1}), SeCys (0.018 ng mL^{-1}), and SeMet (0.007 ng mL^{-1}) determination (Li et al. 2012).

The capability of TiO_2 NPs to induce photocatalytic reduction of inorganic selenium into gaseous products was also explored by Shih et al. (2013), and by employing a microfluidic-based vapour generation (VG) system in conjunction with HPLC-ICPMS. Se(IV) and Se(VI) were first separated by an anion-exchange chromatography column and then vaporized in the microfluidic-based VG system and detected in the ICPMS. Both Se(IV) and Se(VI) were efficiently vaporised within 15s at pH = 3.0 in the presence of 25 mM HCOOH. The LODs were reported to be 0.043 and 0.042 μg L^{-1} for Se(IV) and Se(VI), respectively. The applicability of the method was examined by determining inorganic selenium species in the CRM NIST 1643e, artificial saline solution, and fortified irrigation water with recovery values close to 100%.

Finally, it is worth mentioning the determination of selenoproteins in marine samples by using MS-based techniques. The growing interest in the detection and identification of Se-containing proteins has derived into a progressive incorporation of the analytical tools commonly employed in proteomics into speciation studies. The classical proteomics methodology implies the following steps (1) proteins separation by 1D or 2D gel electrophoresis and (2) in gel digestion with trypsin followed by MALDI (if the sequence genome of the organism is known) and nano HPLC-ESI/MS/MS analysis. However the application of classical proteomics usually fails since very few of the proteins from the gel contain Se and concentration of selenoproteins in cells or the extract is very low (picomolar range). Besides the importance of selenoproteins in the marine environment, interest in determining selenoproteins or proteins–containing selenium in marine samples has been scarce. Most of the published studies have investigated the selenium distribution in the soluble protein fraction of fish species by using size exclusion chromatography (SEC) coupled to ICP-MS. Pedrero et al. (2009), performed the screening of selenium-containing proteins in the TRIS buffer soluble fraction of African catfish (*Clarias gariepinus*) muscle. The experimental set up includes the following: (1) separation of selenoproteins by 1D or 2D gel electrophoresis, (2) specific detection of a

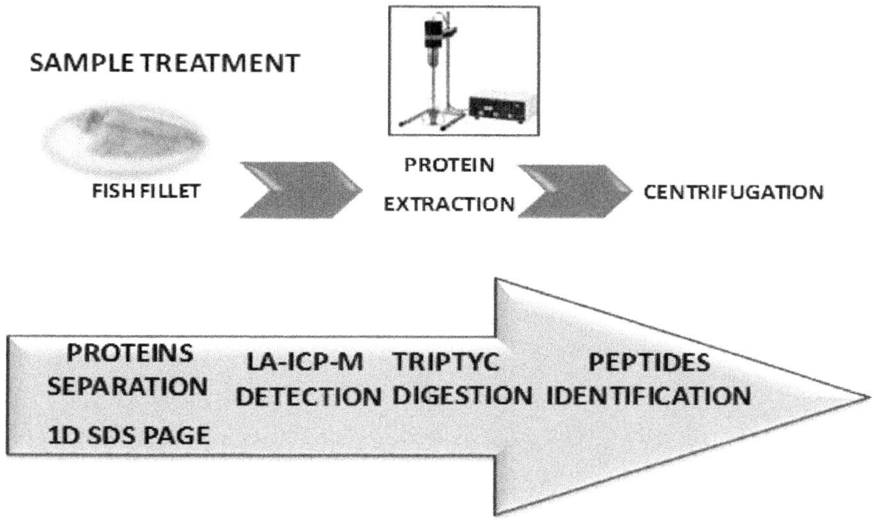

Fig. 3: Scheme of selenoproteins determination in fish muscle.

selenium-rich band or spots by laser ablation (LA)-ICPMS, (3) in-gel trypsin digestion of a selenium-rich band or spots, and (4) application of nano-HPLC-ICPMS for specific mapping of selenopeptides from gel bands or spots (Fig. 3).

Selenium was detected in more than 11 proteins spots in the range of 12 and 90 kDa. Database research revealed that only one protein was related to the African catfish (glyceraldehyde-3-phosphate dehydrogenase) and six proteins to the channel catfish, a close relative of the African fish. Unfortunately, the detection of selenoamino acids in the tryptic peptides was not possible. The reasons were attributable to the low concentration of the selenoprotein and/or selenium-containing protein extracted (often accompanied at least by their methionine analogues that are more abundant and preferentially ionized) and the suppression of the ionization by the concomitant species. The complexity of the task makes the determination of selenoproteins an almost impossible objective. Advances in mass spectrometers analyzers and tandem MS and MS^n will result in fast identification of selenoproteins. Lately, the use of Orbitrap MS/MS system allows detecting selenium at low ppb range (Bianga et al. 2013). Unfortunately, this methodology is not affordable for most laboratories working in selenium speciation. An alternative method to the identification of Se-proteins by using ES-MS protocols could be the determination of SeCys and SeMet to differentiate between the contribution of selenoproteins and selenium-containing proteins, respectively.

Final Remarks

In the marine environment, selenium poses its particular threat to wildlife. Selenium is a toxic or essential element depending on its concentration but more importantly, on its chemical forms. Several selenium species (organic and inorganic) can be simultaneously present in marine environments. New concerns are also rising regarding growing technologies that use selenium, such as nanotechnology. With the increasing use of nanotechnologies, it is expected that nanomaterials end up in natural aquatic systems, from freshwater to the sea. Some Quantum dots (QDs) contain a CdSe core generally coated with a ZnS shell and surrounded by an organic polymer for biological compatibility. The toxic effect of QDs on the marine environment is still an unresolved issue since it is not clear whether the toxic effects on the environment are due to nanoparticles *per se*, or due to the dissolved metals. In recent years, several papers have appeared on the toxicity of CdSe/ZnS quantum dots to aquatic model organisms, including marine diatom *Phaeodactylumtricornutum* and the green alga *Dunaliellatertiolecta* (Morelli et al. 2013; Morelli et al. 2012) as biological models in the marine environment. Therefore, future analytical developments need to be considered to fulfill the requirements of the coming environmental issues.

Although there has been significant evolution of analytical methodologies to determine selenium species in marine environments by using hyphenated techniques (HPLC-ICPMS) in combination with molecular characterization by MS, both speciation of selenium at very low Se levels and effective extraction of selenium species from marine samples still remain challenging.

Acknowledgments

The author thanks to the Spanish Commission of Science and Technology (CTQ2014-54801-C2-1-R) and the Community of Madrid/FEDER programme (S2013/ABI-3028, AVANSECAL-CM) for funding.

Keywords: Selenium, marine environment, sample treatment, preconcentraction, speciation

References

Amouroux, D. and O.F.X. Donard. 1996. Maritime emission of selenium to the atmosphere in Eastern Mediterranean seas. Geophyis. Res. Lett. 23(14): 1777–1780.

Amouroux, D., P.S. Liss, E. Tessier, M. Hamren-Larsson and O.F.X. Donard. 2001. Role of oceans as biogenic sources of selenium. Earth Planet. Sci. Lett. 189(3-4): 277–283.

Anan, Y., K. Ishiwata, N. Suzuki, S. Tanabe and Y. Ogra. 2011. Speciation and identification of low molecular weight selenium compounds in the liver of sea turtles. J. Anal. At. Spectrom. 26: 80–85.

Anan, Y., S. Tanabe and Y. Ogra. 2011. Comparison of selenonine found in marine organisms with selenite in the interaction with mercury compounds *in vitro*. J. Toxicol. Sci. 36(6): 725–731.

Benedicto, A., T. Missana and C. Degueldre. 2013. Predictions of TiO_2-driven migration of Se(IV) based on an integrated study of TiO2 colloid stability and Se(IV) surface adsorption. Sci. Total Environ. 449: 214–222.

Bianga, J., E. Govasmark and J. Szpunar. 2013. Characterization of selenium incorporation into wheat proteins by two-dimensional gel electrophoresis–laser ablation ICP MS followed by capillary HPLC–ICP MS and electrospray linear trap quadrupole Orbitrap MS. Anal. Chem. 85: 2037–2043.

Bidari, A., E. Jahromi, Y. Assadi, M. Reza and M. Hossei. 2007. Monitoring of selenium in water samples using dispersive liquid–liquid microextraction followed by iridium-modified tube graphite furnace atomic absorption spectrometry. Microchem. J. 87: 6–12.

Buchs, B., M.W. Evangelou, L.H.E. Winkel and M. Lenz. 2013. Colloidal properties of nanoparticular biogenic selenium govern environmental fate and bioremediation effectiveness. Environ. Sci. Technol. 47: 2401–2407.

Burger, J., K.F. Gaines, W.L. Stephens, C.S. Boring, I.L. Brisbin, L. Snodgrass, J. Peles, L. Bryan, M.H. Smith and M. Gochfeld. 2001. Mercury and selenium in fish from the Savannah River: species, trophic level, and location differences. Environ. Res. Section 87: 108–118.

Byard, W., M.R.A. Duce, J.M. Prospero and D.L. Savoie. 1987. Atmospheric selenium: Geographical distribution and ocean to atmosphere flux in the Pacific. Journal of Geophysical Research: Atmospheres (1984–2012) 92 (D11): 13277–13287.

Cabañero, A.I., Y. Madrid and C. Cámara. 2004. Selenium and mercury bioaccessibility in fish samples: an *in vitro* digestion method. Anal. Chim. Acta. 526: 51–61.

Cabañero, A.I., C. Carvalho, Y. Madrid, C. Batoréu and C. Cámara. 2005. Quantification and speciation of mercury and selenium in fish samples of high consumption in Spain and Portugal. Biol. Trace. Elem. Res. 103: 17–35.

Chen, M. and M. An. 2012. Selenium adsorption and speciation with Mg–$FeCO_3$ layered double hydroxides loaded cellulose fibre. Talanta 95: 31–35.

Dowdle, P.R. and R.S. Oremland. 1998. Microbial oxidation of elemental selenium in soil slurries and bacterial cultures. Environ. Sci. Technol. 32: 3749–3755.

Ghasemi, E., N.M. Najafi, F. Raofie and A. Ghassempou. 2010. Simultaneous speciation and preconcentration of ultra traces of inorganic tellurium and selenium in environmental samples by hollow fiber liquid phase microextraction prior toelectrothermal atomic absorption spectroscopy determination. J. Hazard Mat. 181: 491–496.

Gong, L., Q. Xu, C. Lee and H. Zhang. 2012. Selenium speciation analysis of *Misgurnusanguillicaudatus* selenoprotein by HPLC–ICP–MS and HPLC–ESI–MS/MS. Eur. Food Res. Technol. 235: 169–176.

He, Y., X. Hou, C. Zheng and R.E. Sturgeon. 2007. Critical evaluation of the application of photochemical vapor generation in analytical atomic spectrometry. Anal. Bioanal. Chem. 388: 769–774.

Herbel, M.J., J.S. Blum, R.S. Oremland and S.E. Borglin. 2003. Reduction of elemental selenium to selenide: experiments with anoxic sediments and bacteria that respire Se-oxyanions. Geomicrobiol. J. 20: 587–602.

Herrero-Latorre, C., J. Barciela-García, S. García-Martín and R.M. Peña-Crecente. 2013. Solid phase extraction for the speciation and preconcentration of inorganic selenium in water samples: A review. Anal. Chim. Acta 804: 37–49.

Hinojosa Reyes, L., J.L. Guzmán, G.M. Mizanur Rahman, B. Seybert, T. Fahrenholz and H.M. Skip Kingston. 2009. Simultaneous determination of arsenic and selenium species in fish tissues using microwave-assisted enzymatic extraction and ion chromatography–inductively coupled plasma mass spectrometry. Talanta 78: 983–990.

Holm, J., P.V. Siwik, P. Sterling, G. Evans and R. Baron. 2005. Developmental effects of bioaccumulated selenium in eggs and larvae of two salmonid species. Environ. Toxicol. Chem. 24: 2373–81.

Hu, X., F. Wang and M.L. Hanson. 2009. Selenium concentration, speciation and behavior in surface waters of the Canadian prairies. Sci. Total Environ. 407: 5869–5876.

Huang, C., B. HU, M. Ue and J. Duan. 2008. Organic and inorganic selenium speciation on environmental and biological samples by nanometers sized nanoparticles materials packed dual-column separation/preconcentration on line coupled with ICPMS. J. Mass Spectrometry 43: 336–345.

Huang, C.Z., W. Xie, X.C. Liu, J.P. Zhang, H.C. Xu, X. Li and Z.C. Liu. 2012. Highly sensitive method for speciation of inorganic selenium in environmental water by using mercapto-silica-Fe_3O_4 nanoparticles and ICP-MS. Anal. Methods 4(11): 3824–3829.

Ikemoto, T, T. Kunito, H. Tanaka, N. Baba, N. Miyazaki and s. Tanabe. 2004. Detoxification mechanism of heavy metals in marine mammals and sea birds: interaction of selenium with mercury, silver, copper, zinc and cadmium in liver. Arch. Environm. Contam. Toxicol. 3: 402–13.

Kehrig, H.A., G. Seixas, E.A. Palermo, A.P. Baêta, C.W. Castelo-Branco, O. Malm and I. Moreira. 2009. The relationships between mercury and selenium in plankton and fish from a tropical food web. Environ. Sci. Pollut. Res. 16: 10–24.

Keskinen, R., P. Ekholm, M. Yli-Halla and H. Hartikainen. 2009. Efficiency of different methods in extracting selenium from agricultural soils of Finland. Geoderma. 153: 87–93.

Kim, J. and H.B. Lim Bull. 2013. Separation of selenite from inorganic selenium ions using TiO_2 magnetic nanoparticles. Korean Chem. Soc. 34(1): 3362–3366.

Lavilla, I., P. Vilas and C. Bendicho. 2008. Fast determination of arsenic, selenium, nickel and vanadium in fish and shellfish by electrothermal atomic absorption spectrometry following ultrasound-assisted extraction. Food. Chem. 106: 403–409.

Li, H., Y. Luo, Z. Li, L. Yang and Q. Wang. 2012. Nanosemiconductor-based photocatalytic vapor generation systems for subsequent selenium determination and speciation with atomic fluorescence spectrometry and inductively coupled plasma mass spectrometry. Anal. Chem. 84: 2974–2981.

Lopez Heras, I., M. Palomo and Y. Madrid. 2011. Selenoproteins: the key factor in selenium essentiality. State of the art analytical techniques for selenoprotein studies. Anal. Bioanal. Chem. 400: 1717–1727.

Luoma, S.N. and P.S. Rainbow. 2008. Metal Contamination in Aquatic Environments. Cambridge, UK: Cambridge University Press.

Luten, J., H. Crews, A. Flynn, P. Van-Dael, P. Kastenmayer, R. Hurrell, H. Deelstra, L. Shen, S. Fairweather-Tait, K. Hickson, R. Farré, U. Schlemmer and W. Frohlich. 1996. Interlaboratory trial on the determination of the *in vitro* iron dialysability from food. J. Sci. Food Agric. 72(4): 415–424.

Moreda-Piñeiro, J., A. Moreda-Piñeiro, V. Romarís-Hortas, R. Domínguez-González, E. Alonso-Rodríguez, P. López-Mahía, S. Muniategui-Lorenzo, D. Prada-Rodríguez and P. Bermejo-Barrera. 2013. *In vitro* bioavailability of total selenium and selenium species from seafood. Food Chem. 139: 872–877.

Morelli, E., P. Cionia, M. Posarelli and E. Gabellieri. 2012. Chemical stability of CdSe quantum dots in seawater and their effects on a marine microalga. Aquat. Toxicol. 122-123: 153–162.

Morelli, E., E. Salvadori, R. Bizzarri, P. Cionia and E. Gabellieri. 2013. Interaction of CdSe/ZnS quantum dots with the marine diatom *Phaeodactylumtricornutum* and the green alga *Dunaliellatertiolecta*: A biophysical approach. Biophys. Chem. 182: 4–10.

Moreno, P., M.A. Quijano, A.M. Gutiérrez, M.C. Pérez-Conde and C. Cámara. 2004. Selenium species distribution in biological tissues by size exclusion and ionic exchange chromatography coupled to plasma spectroscopy. Anal. Chim. Acta 524: 315–327.

Nakaguchi, Y., M. Takey, H. Hattori, Y. Arii and Y. Yamaguchi. 2004. Dissolved selenium species in the Sulu Sea, the South China Sea and the Celebes Sea. Geochem. J. 38: 571–580.

Ohki, A., K. Hayashi, J. Ohsako, T. Nakajima and H. Takanashi. 2013. Analysis of mercury and selenium during subcritical water treatment of fish tissue by various atomic spectrometric methods. Microchem. J. 106: 357–36.

Pedrero, Z. and Y. Madrid. 2009. Novel approaches for selenium speciation in foodstuffs and biological specimens: A review. Anal. Chim. Acta 634: 135–152.

Pedrero, Z., Y. Madrid, C. Cámara, E. Schram, J.L. Lutten, I. Feldman and R. Jakuboswky. 2009. Screening of selenium containing proteins in the Tris-buffer soluble fraction of African catfish fillet by laser ablation-ICP-MS after SDS-Page and electroblotting onto membranes: J. Anal. At. Spectrom 24(6): 775–784.

Pedrero, Z., S. Murillo, C. Cámara, E. Schram, J.B. Luten, I. Feldmann, N. Jakubowskie and Y. Madrid. 2011. Selenium speciation in different organs of African catfish (Clarias gariepinus) enriched through a selenium-enriched garlic based diet. J. Anal. At. Spectrom. 26: 116–125.

Presser, T.S. and H.M. Ohlendorf. 1987. Biogeochemical cycling of selenium in the San Joaquin Valley, California, USA. Environmental Management 11(6): 805–821.

Rauschenbach, I., P. Narasingarao and M. Haggblom. 2011. *Desulfurispirillumindicum* sp. nov. A selenate and selenite respiring bacterium isolated from an estuarine canal in southern India. Int. J. Syst. Evol. Microbiol. 61: 654.

Sanchez-Rodas, D., F. Mellano, E. Morales and I. Giraldez. 2013. A simplified method for inorganic selenium and selenoamino acids speciation based on HPLC–TR–HG–AFS. Talanta 106: 298–304.

Schlenk, D., G. Batley, C. King, J. Stauber, M. Adams, S. Simpson, W. Maher and J.T. Oris. 2007. Effects of light on microalgae concentrations and selenium uptake in bivalves exposed to selenium-amended sediments. Arch. Environ. Contam. Toxicol. 53: 365–370.

Seale, L.A., C.L. Gilman, B.P. Moorman, M.J. Berry, E. Gordon, G. Andre and P. Seale. 2014. Effects of acclimation salinity on the expression of selenoproteins in the tilapia, *Oreochromismossambicus*. J. Trace Elem. Med. Biol. 28: 284–292.

Serra, A.M., J.M. Estela and V. Cerda. 2012. AMSFIA system for selenium speciation by atomic fluorescence spectrometry. J. Anal. At. Spectrom. 27: 1858–1862.

Shih, T., C. Lin, I. Hsu, J. Chen and Y. Sun. 2013. Development of a titanium dioxide-coated microfluidic-based photocatalyst-assisted reduction device to couple high-performance liquid chromatography with inductively coupled plasma-mass spectrometry for determination of inorganic selenium species. Anal. Chem. 85: 10091–10098.

Siddique, T., Y. Zang, B.C. Okeke and W.T. Frankenberger Jr. 2006. Characterization of sediment bacteria involved in selenium reduction. Bioresour. Technol. 97: 1041–1049.

Silvestre, F., J. Linares-Casenave, S.I. Doroshov and D. Kültz. 2010. A proteomic analysis of green and white sturgeon larvae exposed to heat stress and selenium. Sci. Total Environ. 408: 3176–31.

Siscar, R., S. Koenig, A. Torreblanca and M. Solé. 2014. The role of metallothionein and selenium in metal detoxification in the liver of deep-sea fish from the NW Mediterranean Sea. Sci. Total Environ. 466-467: 898–905.

Siwek, M., B. Galunsky and B. Niemeyer. 2005. Isolation of selenium organic species from Antarctic krill after enzymatic hydrolysis. Anal. Bioanal. Chem. 381: 737–741.

Sørmo, E.G., T.M. Ciesielski, I.B. Øverjordet, S. Lierhagen, G.S. Eggen and T. Berg. 2011. Selenium moderates mercury toxicity in free-ranging freshwater fish. Environ. Sci. Technol. 45: 6561–6.

Stolz, J.F., P. Basu, J.M. Santini and R.S. Oremland. 2006. Arsenic and selenium in microbial metabolism. Annu. Rev. Microbiol. 60: 107–130.

Sun, Y.C., Y.C. Chang and C.K. Su. 2006. On-line HPLC-UV/nano-TiO_2- ICPMS system for the determination of inorganic selenium species. Anal. Chem. 78: 2640–2645.

Tashjian, D., J. Cech and S. Hung. 2007. Influence of dietary selenomethionine exposure on the survival and osmoregulatory capacity of white sturgeon in fresh and brackish water. Fish Physiol. Biochem. 33: 109–19.

Tashjian, D.H., S.J. The, A. Sogomonyan and S.S.O. Hung. 2006. Bioaccumulation and chronic toxicity of dietary L-selenomethionine in juvenile white sturgeon (*Acipensertransmontanus*). Aquat. Toxicol. 79: 401–9.

Vilano, M. and R. Rubio. 2000. Liquid Chromatography-UV irradiation hydride generation atomic fluorescence spectrometry for selenium speciation. J. Anal. At. Spectrom. 15: 177–180.

Wake, B.D., A.R. Bowie, E.C.V. Butler and P.R. Haddad. 2004. Modern preconcentration methods for the determination of selenium species in environmental water samples. Trends Anal. Chem. 23(7): 491–500.

Wu, Y., L. Han, J. Guo and H. Sun. 2014. Speciation of inorganic selenium in environmental water samples by inductively coupled plasma optical emission spectrometry after preconcentration by using a mesoporous zirconia coating on coal cinder. J. Sep. Sci. 37: 2260–2267.

Yamashita, Y. and M. Yamashita. 2010. Identification of novel selenium compound, selenoneine as the predominant chemical form of organic selenium in the blood of blue fin tuna. J. Biol. Chem. 285: 18134–18138.

Yang, K.X. and K. Swami. 2007. Determination of metals in marine species by microwave digestion and inductively coupled plasma mass spectrometry analysis. Spectrochim. Acta Part B 62: 1177–1181.

Yoshida, S., M. Haratake, T. Fuchigami and M. Nakayama. 2011. Selenium in seafood materials. J. Health Sci. 57(3): 215–224.

Yoshida, S., M. Haratake, T. Fuchigami and M. Nakayama. 2012. Characterization of selenium species in extract from Niboshi (a processed Japanese anchovy). Chem. Pharm. Bull. 60: 348–353.

Zhang, Q. and G. Yang. 2014. Selenium speciation in bay scallops by high performance liquid chromatography separation and inductively coupled plasma mass spectrometry detection after complete enzymatic extraction. J. Chromatogr. A 1325: 83–91.

Zhang, Y. Morita, A. Sakuragawa and A. Isozaki. 2007. Inorganic speciation of As(III, V), Se(IV, VI) and Sb(III, V) in natural water with GF-AAS using solid phase extraction technology. Talanta 72: 723–729.

Zhang, Y. and V.N. Gladyshev. 2008. Trends in Selenium Utilization in Marine Microbial World Revealed through the Analysis of the Global Ocean Sampling (GOS) Project. PLoS Genet. 4(6): e1000095.

Zhang, Y., J. Duan, M. He, B. Chen and B. Hu. 2013. Dispersive liquid liquid microextraction combined with electrothermal vaporization inductively coupled plasma mass spectrometry for the speciation of inorganic selenium in environmental water samples. Talanta 115: 730–736.

5

Arsenic Occurrence in Marine Biota

The Analytical Approach

Angels Sahuquillo Estrugo,[a,*] *José Fermín López Sánchez,*[b] *Antoni Llorente Mirandes, Albert Pell Lorente, Roser Rubio Rovira* and *Mª José Ruiz Chancho*

ABSTRACT

Speciation studies of arsenic are of paramount importance as the toxicity of the element is strongly dependent on the chemical species. The development of reliable analytical methods for arsenic speciation will contribute to improve the knowledge about both the presence and the contents of arsenic species in a wide variety of marine samples and also will support the development of legislation for the sake of consumer health.

This chapter focuses on the occurrence of arsenic in marine biota focusing on the critical aspects to be taken into account during the analytical process to guarantee reliable results. Specific considerations for algae and seafood analysis are described.

Introduction: Arsenic in Marine Biota

Arsenic is present in the environment as a consequence of both anthropogenic and natural processes. It occurs in seawater mainly as inorganic arsenic compounds, in the trivalent (As(III)) and the pentavalent (As(V)) states, at the low μg L^{-1} concentration level (Naidu et al. 2006). In living marine organisms, although inorganic forms of arsenic can be also found, a large number of organoarsenical compounds have been identified. The structures of some of the most reported arsenocompounds found in marine biota are presented in Fig. 1.

In general, the presence of these compounds indicates that living marine organisms accumulate arsenic in such a way that the total arsenic content measured in them is several orders of magnitude greater than

Department of Analytical Chemistry, Faculty of Chemistry, University of Barcelona, Martí i Franquès, 1-11, 08028, Barcelona, Spain.

[a] Email: angels.sahuquillo@ub.edu

[b] Email: fermin.lopez@ub.edu

* Corresponding author

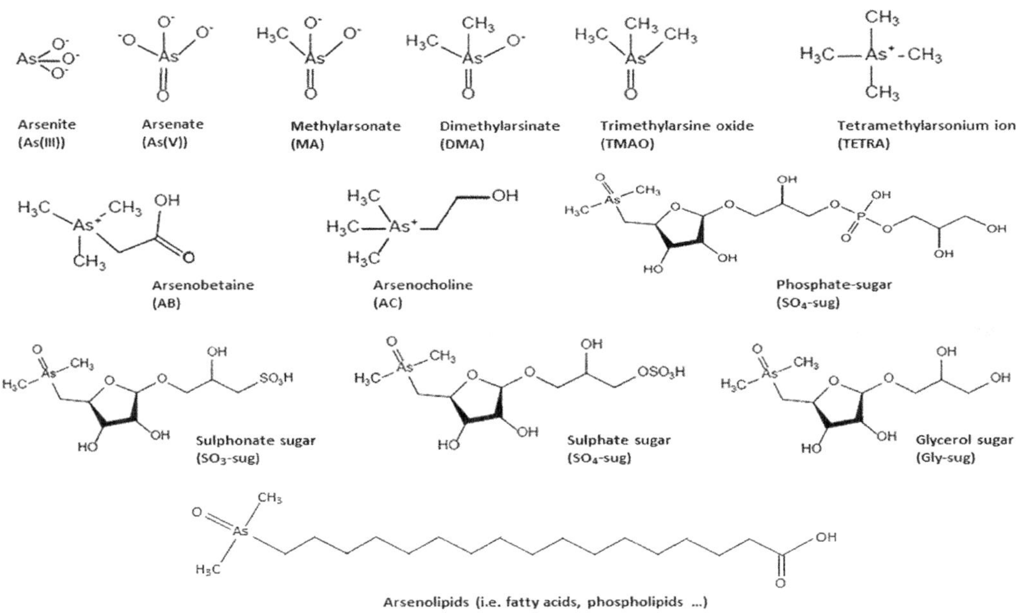

Fig. 1: The most reported arsenocompounds found in marine biota.

that present in the surrounding seawater (Edmonds et al. 2003). The biotransformation of assimilated arsenic in living marine organisms involves complex mechanisms. Several studies have proposed pathways involving animals, algae, and activity of the microbial community (Edmonds 2000; Traar and Francesconi 2006). For instance, algae can accumulate and biotransform inorganic arsenic to arsenosugars (derivatives of dimethylarsinoylribosides and trimethylarsinoribosides), and among them the derivatives of dimethylarsinoylribosides commonly named glycerol sugar (Gly-sug), phosphate sugar (PO_4-sug), sulfonate sugar (SO_3-sug), and sulfate sugar (SO_4-sug) are the most frequently occurring arsenic species in algae (Shibata et al. 1987; Francesconi 1997). As well as these compounds, other organo arsenicals like methylarsonate (MA), dimethylarsinate (DMA), arsenobetaine (AB) and arsenocholine (AC), and inorganic arsenic can also be found in marine algae, but generally in lower amounts than arsenosugars. Trimethylarsine oxide (TMAO) and tetramethylarsonium ion (TETRA) have also been reported in a very few studies (Hirata and Toshimitsu 2007; Thomson et al. 2007). As one of the latest steps in marine biotransformation is the production of AB, this is the predominant species in fish, bivalves, and crustaceans (Caumette et al. 2012).

The toxicity of arsenic in environmental and biological systems is strongly dependent on the chemical species (Irvin and Irgolic 1995). The inorganic forms of arsenic are highly toxic, whereas other arsenic species such as MA and DMA are less toxic to humans, with AB being considered non-toxic (Feldmann and Krupp 2011; Geng et al. 2009). Arsenosugars are also considered non-toxic although there are no reliable data (Niegel and Matysik 2010). The International Agency for Research on Cancer (IARC) considers inorganic arsenic compounds (arsenite and arsenate) highly toxic and classifies them as group I (human carcinogens) (Straif et al. 2009). The same considerations are reported by the Joint FAO/WHO Expert Committee on Food and Additives (JECFA/72/SC 2010) and by the European Food Safety Authority (EFSA 2005, 2009). Thus, the information obtained from total arsenic determination in a sample is not enough to assess the toxicological risk in environmental studies, analytical speciation of arsenic is necessary.

A large number of studies have addressed arsenic speciation in marine organisms, including algae, fish, bivalves, and crustaceans and a high number of publications have been generated in the last decades. Many are dedicated to examining new compounds and their behavior, transformations and metabolic pathways in other to understand the arsenic biogeochemistry. Furthermore, many marine organisms are used for human consumption, i.e., in many countries algae form part of the human diet, either through direct ingestion or as a food supplement, owing to their high contents of iodine, minerals, and vitamins. The increase in global consumption of seafood is associated with several benefits such as a reduction in risk of several diseases (Innis 2007). However, concerns about human health have arisen since more than 50 arsenic species have been detected in seafood (Francesconi 2010; Leufroy et al. 2011), reinforcing the importance of its chemical speciation, as the total amount of arsenic does not provide enough information about the toxicity of the analyzed sample.

To date, the European Union has not established a limit for total or speciated arsenic in fish and seafood in its legislation (Commission Regulation 2006). However countries such as New Zealand and Australia have developed legislation for the maximum levels of the more toxic inorganic arsenic species (iAs) in seafood and establishing a maximum level of inorganic arsenic of 2 mg kg^{-1} for crustaceans and fish, and 1 mg kg^{-1} for mollusks and seaweed (Australia New Zealand Food Authority 2013). The Republic of China establishes a maximum level of inorganic arsenic of 0.1 mg kg^{-1} (dry weight) for fish and 1.0 mg kg^{-1} (dry weight) for shells, shrimps, and crabs (Ministry of Health of China, 2005). On the other hand, the Brazilian authorities through the Ministry of Agriculture, Livestock and Food Supply (MAPA) establish a reference value (National Program for Residue 2012). Aware of this situation, the EFSA published (in 2009 and 2014) two reports about the dietary exposure to arsenic in the European population (EFSA 2009, 2014). Both reported the urgent need for further data on arsenic species, particularly iAs data, in particular in fish and seafood, and in other food groups, as rice and wheat based products, that provide a significant contribution to the dietary exposure to iAs to reduce the uncertainty of the exposure assessments of iAs. Thus, the need to produce specific legislation is becoming evident (EFSA 2009; Feldmann and Krupp 2011). Furthermore, the need to create certified reference materials for seafood and to develop arsenic speciation methods for a large range of samples and arsenic species was also emphasized. The increased focus on inorganic arsenic in food has led to several initiatives towards development of methods for selective determination of iAs in seafood. For this purpose, the Institute for Reference Materials and Measurements (IRMM) organized two proficiency tests (PT) in 2010 for measuring iAs, and trace metals in seafood (IMEP-109 and IMEP-30). One of the outcomes of these trials was that the determination of iAs in seafood test materials presented serious analytical problems. The expert laboratories were not able to agree on a value for the iAs within a reasonable degree of uncertainty and it was concluded that more research in extraction procedures and chromatographic separations was required for a reliable quantification of the iAs in seafood (Baer et al. 2011). The complexity of the seafood matrix requires accurate and robust procedures. However, the analytical procedures used to date do not comply with these requirements (Feldmann and Krupp 2011). Some authors reported inorganic arsenic values in several seafood CRMs collected from previously published studies (Leufroy et al. 2011; Pétursdóttir et al. 2012a,b, 2014). The observed differences found in the literature among the concentrations of iAs in several CRMs reinforce the need to develop reliable methodology in its determination. In this way Zmozinski et al. (2015) recently proposed a method for the determination of iAs as well as for AB, DMA, MA, AC, and TMAO species in seafood based on wet microwave assisted digestion with a solution containing 0.2% (w/v) of nitric acid and 1% (w/v) of hydrogen at temperature of 95°C. Regarding the claimed advantages of this method, the conversion of As(III) to As(V) which allows the quantification of iAs as As(V) is the most notable factor, as As(III) elutes near the void volume in the anion exchange column and it could co-elute with other cationic species usually found in seafood (specially AB). Therefore, the controlled oxidation of As(III) to As(V) allows the accurate determination of iAs as As(V) which is well separated from other As species. Such methodologies could be a valuable tool for food control laboratories which assessing the iAs in seafood samples and could be useful for regulatory institutions to assess dietary exposure to toxic iAs and for the development of further directives on arsenic in food commodities.

The goal of this chapter is to give to the reader an overview of arsenic occurrence in marine living organisms with the focus on the analytical techniques available to study arsenic speciation in such matrices, according to the experience of the authors.

Measuring As and its Species

When dealing with analytical speciation, reliable results are only obtained if all steps involved in the analytical process pursue sample integrity, including the distribution of species accounted in the original matrix. In the following sections, the most usual sample pretreatment procedures and the analytical techniques used for arsenic determination in marine biota are discussed.

Sample pretreatment

The relevance of sample handling in analytical processes is often underestimated, even more when analytical element-speciation analysis is undertaken. In this case, the removal of any interfering substances during sample pretreatment or possible loss or transformation of the native element species has to be carefully addressed.

Most of the sample pretreatment steps for marine biota involve a first cleaning of surfaces to be analyzed, a further drying, powdering, and homogenizing of samples, and an extraction step for speciation analysis or a digestion procedure for the analysis of total contents. Detailed information about the previous steps to extraction or digestion conditions is usually scarce in the literature, and in most studies, only the analytical procedures from the starting point of dried sample are referred to.

As far as cleaning is concerned, for marine samples such as algae and aquatic plants, one of the most critical steps is the removal of micro-organisms living in symbiosis, such as plant and animal epiphytic communities, as well as the potential presence of sediments on the surface of the bulk material. Small amounts of arsenic compounds, such as AB or some thio-arsenosugars, reported in algae, could be related to epiphytes or bacteria (Nischwitz and Pergantis 2005; Grotti et al. 2008). Even if there are not well-established procedures, the removal of epiphytes and epifauna is normally performed by scraping samples with razor blades previously cleaned with ethanol, or by using a nylon brush (Meier et al. 2005). A subsequent rinsing with sea water or a NaCl solution resembling sea composition (Thompson et al. 2007a; Slejkovec et al. 2006), tap water (Castelhouse et al. 2003; Raber et al. 2000; Zheng et al. 2003; Han et al. 2009; Van Hulle et al. 2002; Hirata et al. 2007), or deionized water (Meier et al. 2005; Thompson et al. 2007b; Madsen et al. 2000; Tukai et al. 2002), is then undertaken. As the use of aggressive chemical cleaning systems are not suitable for speciation analysis, the manual elimination of the epiphytes under a stereomicroscope before subsequent treatment is the most appropriate system to obtain representative results of arsenic species in algae (Rubio et al. 2010). For fish and seafood samples, an initial washing procedure with double deionized water is also advisable before cutting the tissues in pieces (Zmozinski et al. 2015a).

In general, samples are dried and ground to a fine powder before any subsequent extracting or digesting procedure. Freeze-drying or thermal treatments are used mainly for removing water from clean or dirty samples. For thermal drying, there is a wide range of proposals for temperature (from room temperature to 60°C) and time (18–48 h). For freeze drying, the process takes 24–48 h. Effects of sample processing have been systematically studied on arsenic speciation in marine macroalgae (Pell et al. 2013a). In this case, a pooled sample of *Cystoseira mediterranea* was split in portions, one portion being analyzed directly while fresh and two portions stored at –18°C and –80°C. Each of the last portions was defrosted at 1 and 45 days later and all portions were differently pretreated drying under an air current at room temperature (25°C), drying in an oven at 40°C, and freeze drying. Before drying, samples were equally cleaned removing epiphytes, washed with deionized water, and chopped into fine pieces. The level of total arsenic in all frozen samples was about 60% lower than that of the non-frozen samples, which could be attributed to the brake of algal cell structures during the freezing/defrosting process. Regarding arsenic compounds no changes in their relative contents were observed when frozen and non-frozen samples were compared.

A number of systems are described for powdering and homogenizing dried samples, including mechanical and automatic devices. While some studies working in biological sciences, highlight the risk of changes in the distribution of arsenic species when samples are powdered as a result of the destruction of tissues, others support powdering to small particles to ensure the rupture of cells and membranes and thereby guarantee both, the homogeneity of samples and the quantitative extraction of arsenic species (Kirby et al. 2004).

Once the sample is ready for wet analysis, sample characterization usually starts with total arsenic content determination. One problem that can be encountered, mainly in fish and marine species, is that the presence of especially stable arsenic species, such as AB, hinders the total element determination. In some cases, even with the use of strong oxidizing agents combined with strong acids at high temperatures, the complete degradation of AB is not achieved (Narukawa et al. 2005; Narukawa et al. 2006).

For the extraction of arsenic species in marine biota, a wide variety of techniques has been discussed and reviewed. Water is the extractant most recommended for the more polar or ionic arsenic species, such as arsenosugars, whereas mixtures of MeOH:water are also widely used for extracting less polar species, mainly associated with lipids and also with arsenic bound to cell components or proteins that are not extracted with water (Koch et al. 2000a; Koch et al. 2000b). Some studies propose a strategy that involves extracting only hydride-forming arsenic (inorganic arsenic and its methylated species) from marine tissues (Karadjova et al. 2007). In general, the extraction conditions can vary greatly depending on the extractant:sample ratio, the extracting system used, and the ranges of time and temperature. Extraction conditions influence not only the extraction efficiency but also the integrity of the native arsenic species during extraction. Even if it would be desirable to develop a well-defined protocol for extracting arsenic species in order to obtain comparable results, relevant differences between marine matrices, concerning both morphological variability and chemical composition, hinder this approach. Whatever the extraction system chosen, it is highly advisable to optimize the protocol to obtain reliable results on the basis of the extraction efficiency. Finally, it is crucial to pay special attention to the stability of arsenic species in the extracts. Even if it is widely recognized that the stability of species during the analytical process should be a priority for assessing the presence and identification of arsenic species, very little data are available on this issue (García Salgado et al. 2008).

Analytical techniques

Direct arsenic speciation methods

Direct speciation methods described in literature are mainly based on isotopic, electrochemical, or X-ray analytical techniques. Isotopic methods are mainly used for isotopic dilution quantification and mechanistic studies, taking advantage of a radiotracer which behaves in the same way as stable isotopes and a measurement which is mostly independent of physical and chemical influences (Caruso et al. 2003).

Electrochemical methods also offer possibilities to determine arsenic and arsenic compounds at low concentrations. However, direct measurements are only possible in relatively simple solutions free from interferences. Stripping techniques with gold and gold-film plated electrodes and HCl as support electrolyte have been used for arsenic determination. As(III) is deposited on the working gold electrode surface as elemental arsenic by electrochemical reduction. Elemental arsenic is then electrochemically oxidized (stripped) back to As(III). As(V) is determined after chemical or electrochemical reduction to As(III) determining total As. As(V) is calculated by difference. This approach has been successfully applied to seawater samples (Muñoz and Palmero 2005). A sequential injection system with anodic stripping voltammetric detection using a tubular electrochemical detection cell allowed inorganic arsenic speciation in waters (Rodríguez et al. 2012). A number of arsenic biosensors have been developed in the last decades from the coupling of biological engineering and electrochemical techniques. Some of these devices are based on whole-cell biosensors and others on cell-free (protein, DNA)-based biosensors. Available bacterial whole-cell biosensors are currently detecting only inorganic arsenic, and their performance can still be seriously compromised even in water samples by the presence of other ions or contaminants that might complex arsenic or affect the viability of the bacterial biosensor cells (Chen and Rosen 2014). Current developments in this field focus on the improvement of detection specificity and detection multiplexing, together with strategies for keeping the biosensor alive and active for prolonged periods of time.

There are a number of methods that permit *in situ* analysis of As and As speciation using X-rays, which fall under the broad term X-ray Absorption Spectroscopy (XAS). These methods include X-ray Absorption Near Edge Structure (XANES) and Extended X-Ray Absorption Fine Structure (EXAFS). The use of XAS spectra to understand As speciation in solids and waters, from natural or modeled environment, has dramatically increased in the past decade (Foster and Kim 2014), even in the case of *in-situ* analysis samples that are hydrated, anaerobic, or are living organisms. XAS has been particularly useful for studying changes in As speciation that accompany dynamic, often seasonal, redox conditions that can exist in aquatic sediments and in organic-rich matrices. With this technique, the knowledge on the dominant sorption complexes formed on the most relevant mineral phases in aquatic sediments, particularly ferrous and ferric phases with which As is strongly associated, has been improved.

Coupled-techniques

Taking advantage of the synergistic effects of different separation techniques for dealing with organic and inorganic compounds, and the capabilities of an element-selective analyzer (an atomic or mass spectrometer), the appearance of coupled-techniques has allowed the development of very powerful analytical methods for speciation purposes. The most common separation techniques used for this purpose, are gas chromatography (GC), high-performance liquid chromatography (HPLC), including ion chromatography (IC), capillary electrophoresis (CE), and field-flow fractionation (FFF), among others. Atomic absorption detectors (AAS), including cold vapor and hydride generation systems, atomic fluorescence spectrometry (AFS), atomic emission, and ICP-MS detection are those most used among the element specific detectors (Lobinski 1997; Caruso et al. 2003). The choice of the suitable separation technique will be determined by the properties of the species of interest, such as volatility, charge, and polarity, whereas the detection technique is determined by the expected concentration level in the sample. Arsenic analysis by a coupled technique will be often preceded by a more or less complicated wet chemical sample preparation as previously described.

Applications of coupled techniques in speciation analysis of arsenic have been comprehensively reviewed (Francesconi and Kuehnelt 2004; Michalski et al. 2012; Maher et al. 2012; Chen et al. 2014).

Assessing the Quality for As Speciation

Obtaining reliable results in any field, it requires the use of validated analytical methods with well-established performance and the implementation of quality control activities covering all steps of the analytical process. When dealing with arsenic speciation in marine biota, different problems have to be faced for assessing the quality of obtained results before drawing conclusions related to environmental or health protection problems. In this section, the main points to be considered when using analytical methods based on an extraction procedure followed by LC-ICPMS measurements will be reviewed but the situation will be comparable for other techniques.

For the identification of the obtained chromatographic peak, retention times are compared with those obtained for standards when available. When this is not the case, as for arsenosugars, which can be the arsenic predominant species in algae and freshwater plants, a brown seaweed *Fucus serratus* extract (Madsen et al. 2000) can be used for comparison, as a well-characterized sample extract. If any extract for the studied matrix is available, then the peak remains as unknown.

For quantification, when standards are available for the arsenic species of interest, the normal procedure consists of preparing fresh calibration solutions daily and to run the calibration curve at the beginning and at the end of the sample series being analysed, together with analysis of blanks. When no standards are available, as for arsenosugars, species can be quantified using the calibration curves of the nearest eluting standard compound (Llorente-Mirandes et al. 2011; Pell et al. 2013b,c) and the same procedure can be used when quantifying unknown arsenic compounds. Working in this way (Francesconi and Sperling 2005) implies that the element response of the ICPMS detection system is independent of the species, which is a controversial point as nebulization efficiency might be different for each compound (Polya et al. 2003; Entwisle and Hearn 2006).

The mass balance between total arsenic extracted and total arsenic content in the matrix provides an estimation of the extraction yield. Extraction efficiency values from 80–110%, calculated in this way, can be considered acceptable. However, it has to be taken into account that an optimized extraction procedure for arsenic speciation in a matrix (e.g., algae) can yield to low extraction efficiency values for an algae species (e.g., *Undaria pinnatifida*).

For quality assessment, column recovery must also be established to guarantee the correctness of the chromatographic separation. To this end, the ratio of the sum of the species eluted from the chromatographic column to the total arsenic in the extract injected into the column has to be assessed in sample replicates with good reproducibility. Depending on the combination matrix and arsenic species, column recovery values from 70 to 120% can be obtained. Low column recovery values could indicate the presence of several arsenic compounds in the sample that cannot be evaluated with the chromatographic separation used.

In general, certified reference materials (CRMs) are used to assess the accuracy and the reliability of results obtained. However, in speciation studies, the number of CRMs for element species contents is limited. For instance, CRM ERM-BC211 is a rice CRM where total As, dimethylarsinic acid, and the sum of arsenite and arsenate are certified, and in tuna fish tissue CRM (BCR-627) the contents of total As, AB, and dimethylarsinic acid are certified. Thus, the most commonly used practice comprises of performing analytical speciation on CRMs in which the total content of arsenic is certified. Then, several results found in the literature can be useful for comparison purposes. If considering algae and aquatic plants, only a few CRMs are available with certified values of a small number of arsenic species and most commonly, a number of CRMs of marine animals, mainly DORM 2 (currently replaced by DORM 3) are reported to assess the accuracy of results by comparison with the few certified arsenic species available or with several published results on non-certified arsenic species.

Application Examples

A great number of proposals regarding arsenic speciation procedures to be followed are reported in the literature and there are remarkable differences among research groups. The comparison and selection of the most suitable method is a very important task that ought to be performed for each matrix. As long as the nature of environmental samples is disparate, methods have been adapted consequently. Another source of method variation is the aim of the study, for instance: biodisponibility, detection of the more toxic species (iAs), transformation of species, identification of metabolites, among others. Specific considerations for two case studies concerning marine biota are described.

Algae

In these matrices, the extraction of the arsenicals is mandatory to perform the speciation, but the selection of the method is crucial. Using aggressive procedures and/or strong extracting reagents may increase the yield; still, the species' integrity is not ensured. From our experience, water is mainly selected for extraction of algae, since it is a soft extracting agent. Samples are shacked in an end-over-end rotator overnight, as long as a gentle method minimizes species inter-conversion. Table 1 summarizes some proposals as example of the used approaches.

Extraction efficiencies for algal samples ranged broadly between algal division, 28%–106% for ochrophyta (brown), 9%–48% for rhodophyta (red), and 14%–73% for chlorophyta (green), the brown algae having a higher fraction of extractable arsenic. To our mind, further comparisons might be highly speculative. Extraction efficiencies reported in the literature are in agreement with our values (see Table 1). Column recoveries for algal samples ranged broadly between algal division, 22%–102% for ochrophyta (brown), 11%–68% for rhodophyta (red), and 19%–105% for chlorophyta (green). The results obtained by the authors are in the range or lower than those reported in the literature but it is noticeable that several studies do not give this quality parameter.

A separation step preceding detection is required to quantify each compound. The most common technique used to analyze aqueous algae extracts is the HPLC coupled to an atomic detector. The following separation methodology is used in our laboratories for the majority of the studies on algae:

analytical anionic exchange column and using as mobile phase $NH_4H_2PO_4$ (20 mM, pH 5.8), and analytical cationic exchange column and using as mobile phase $C_5H_6N^+COO^-$ (20 mM, pH 2.6). The performance of the method is adequate for most samples and work properly with the ICP-MS. However, the lack of structural information provided by the ICP-MS is an important drawback of this approach, and we also use a chromatographic method compatible with a molecular mass spectrometer (MSD-TOF). The low volatility of $H_2PO_4^-$ discouraged the use of this electrolyte in coupling, so NH_4HCO_3 is selected due to its compatibility with the detector. So it is possible to work with a LC-MSD-TOF method that provides structural information for the identification of organoarsenical species. From our experience in this field three different detectors were used with different aim:

- ICP-MS: We use the coupling of LC and ICP-MS for speciation of samples because of the high specificity and sensitivity of this technique at trace level concentrations. However, identification of peaks relies on retention time matching with standards and standard addition.
- MSD-TOF: The coupling of LC and MSD-TOF allows checking the identity of a compound when matching of the retention time with available standards and the standard addition were not enough for confirming their identity.
- HG-AFS: For previous studies, i.e., the optimization of mobile phases is carried out using this detector due to the low operation costs and adequate detection limits. However, only those species that form the hydride can be detected which makes this detector unsuitable for algae speciation, since arsenosugars cannot be detected. HG-AFS was used in the quantification of total arsenic in samples, whose high concentration in chloride might cause interferences in the ICP-MS, but not in this technique.

Table 1: Some examples of extraction methods reported in the literature for arsenic speciation in algae.

Reference	Extraction method		Extraction efficiency (%)	Column recovery (%)
Hirata and Toshimitsu 2007	H_2O Ultrasound assisted (15 min)	Ochrophyta	9.3–94.6	51–72
Llorente-Mirandes et al. 2010	H_2O End-over-end (16 h)	Chlorophyta Rhodophyta Ochrophyta	45.6–69 77.4–86.9 55.6–88.6	78.7–53.4 77.4–86.9 72.0–83.2
Shaeffer et al. 2006	H_2O Ultrasound assisted (30s) End-over-end (16 h)	Chlorophyta[(a)] Aquatic plant[(a)]	10–38 26–53	90–97 83–98
Thomson et al. 2007	Sequential			
	Lipid fraction $CHCl_3$:MeOH (2:1 v/v) Vortex (30s) End-over-end (4h)	Chlorophyta Chlorophyta[(b)] Rhodophyta	19–44 5–21 21–24	Not reported
	Aqueous fraction H_2O 1 h 100°C	Chlorophyta Chlorophyta[(b)] Rhodophyta	23–35 17–30 45–46	Not reported
Miyashita et al. 2009	H_2O:MeOH (1:1 v/v) Ultrasound assisted (10 min)	Chlorophyta[(b)]	16	Not reported
This chapter	H_2O End-over-end (16 h)	Chlorophyta Rhodophyta Ochrophyta	14–73 9–48 28–106	19–105 11–68 22–102

(a) freshwater *(b) estuarine*

As stated before, one important drawback in elemental speciation is the lack of CRMs for species contents. Few attempts have been done by some CRMs producers and most of them empowered by the necessity of controlling food and feed. Regarding algae, one of the approaches to overcome this problem is to analyze the available reference materials to check accuracy, relying on the comparison of the obtained results with those reported in the literature. For instance, the CRM 279 Sea lettuce (*Ulva lactuca*) is used by several research groups as a control in speciation studies, even though arsenic species are not certified. The extraction efficiency obtained in our group using the water-soft method (57–59%) is in the order of the values reported in the literature using methods more aggressive (53–57%) as summarized in Table 2. Using assisted microwave extraction at higher temperature up to 84% extraction efficiency is reported (Foster et al. 2007; Hsieh and Jiang 2012) whereas the use of sonication yields up to 53% extraction efficiency (Caumette et al. 2011). The results obtained for the CRM speciation are shown in Table 3. Among the species extracted, there are the inorganic forms, arsenosugars, and methylated species. The reported methods produce similar results, except for the 2% HNO_3 extraction that had higher As(V)

Table 2: Comparison of extraction efficiencies for CRM 279 Sea lettuce (*Ulva lactuca*) reported in the literature.

Reference	Extraction efficiency	Extraction method
This chapter	57–59%	H_2O, room temperature 16 h End-over-end
Foster et al. 2007	77%	MeOH:H_2O (1:1, v/v) 70°C 15 min, MW assisted
Foster et al. 2007	57%	2% HNO_3 95°C 6 min, MW assisted
Caumette et al. 2011	53%	H_2O 60°C 4 h, Ultrasounds 15 min
Hsieh and Jiang 2012	84%	NH_4HCO_3 (0.5 mM, 1% MeOH, pH 8.5): MeOH (1:1 v/v) 60°C 20 min, MW assisted

Table 3: Reported data on arsenic species in the extracts of BCR CRM 279 Sea lettuce (*Ulva lactuca*). Values are given in mg As kg^{-1}.

References	As(V)	As(III)	DMA	MA	TETRA
Foster et al. 2007	0.52 ± 0.03		0.08 ± 0.01	0.07 ± 0.01	0.04 ± 0.01
Foster et al. 2007	1.2 ± 0.06		0.04 ± 0.01	0.03 ± 0.01	0.24 ± 0.02
Caumette et al. 2011	0.7		0.2		
Hsieh and Jiang 2012	0.67 ± 0.02	0.065	0.004	0.234	0.01
This chapter	0.53 ± 0.04	0.06 ± 0.03	0.06 ± 0.03	0.04 ± 0.01	
References	**Gly-sug**	**PO_4-sug**	**SO_3-sug**	**AB**	**Unknown**
Foster et al. 2007	0.05 ± 0.01	0.21 ± 0.03			0.01 ± 0.01
Foster et al. 2007					0.16 ± 0.02
Caumette et al. 2011	0.2		0.2	0.2	
Hsieh and Jiang 2012		0.302			0.008
This chapter	0.096 ± 0.004	0.08 ± 0.01		0.14 ± 0.01	0.07 ± 0.02

and no arsenosugars were found. On the other hand, the method used in our research group extracted more arsenosugars than the others and As(III) was quantified, which suggest that the method preserves the original species. It should be taken into account that this method is the simplest from an operational point of view, but it obtains results in the order of the most sophisticated methods.

A second approach, widely used, is the analysis of a extract from the brown seaweed *Fucus serratus* prepared by Madsen et al. (2000) to identify the arsenosugars present in seaweed samples. The authors identified the arsenosugars peaks in chromatograms by analyzing the extract by HPLC–ICP–MS and LC-ESI-MS. From that moment, there are a vast amount of data for this extract reported in the literature and they are used as reference values. In our research group the extract was analyzed several times, as reported in Table 4, and the obtained values are consistent with those reported in the literature. The use of *F. serratus* in which the most relevant arsenosugars occurring in algae are well identified allows assuring accuracy in arsenosugar identification in all the sample extracts by following the described procedure. As an example, the chromatograms obtained from cation and anion exchange chromatography are shown in Fig. 2. It shows the presence of phosphate, sulfonate, and sulphate sugar in the anion exchange chromatogram and Gly-sug in the cation exchange chromatogram.

Along several years the group participated in several projects to study arsenic occurrence in algae. In this context we have the opportunity to obtain algae samples collected in several countries such as Spain, Greece, Panama, Argentina, or Chile. An overall discussion on the obtained results is presented in the following paragraphs.

Considering total content, brown algae accumulates more arsenic than the rest of phyla (29 to 150 mg As kg^{-1}), with the exception of *Padina* spp. which had 7.7 to 11.5 mg As kg^{-1}, irrespective of species and sampling site. Red and green algae had 4.6 to 12.0 mg As kg^{-1} and 1.4 to 9.2 mg As kg^{-1} respectively, except for two samples of *C. fragile* which contained 23 to 30 mg As kg^{-1}. The sampling location seems to have no influence on the arsenic content. An ANOVA considering only brown algae ($\alpha = 0.05$) confirms this hypothesis. Arsenic content reported in the literature for algae from different families seems to be not related to the sampling site except from some specific areas such as the Venice Lagoon (Caliceti et al. 2002). Therefore, our values for Sargassaceae (4.2–150 mg As kg^{-1}), Laminareceae (68.0 ± 20.6 mg As kg^{-1}), Dictyotaceae (1.3–20.9 mg As kg^{-1}), Gracialariaceae (5.7–12.2 mg As kg^{-1}), Cystocloiniaceae (0.28–5.0 mg As kg^{-1}), Codiaceae (24 mg As kg^{-1}), and Ulvaceae (1.39–5.3 mg As kg^{-1}) are in the order of those reported.

Referring to arsenic speciation, the predominance of arsenosugars is overwhelming in most samples and, in some of them, they are the only detected species. Nonetheless, the arsenical distribution in Sargassaceae algae is at odds with the other families, since they present a large proportion of inorganic arsenic. Furthermore, *P. capitatus* and *H.* cf. *incressata* also had a high percentage of inorganic arsenic; however, only one sample was analysed, and biased conclusions might be drawn. In addition, *P. capitatus* did

Table 4: Species in the extract of *F. serratus*. Reported data in the literature. Values are given as μg in the extract.

References	DMA	Gly-sug	PO_4-sug	SO_3-sug	SO_4-sug
Madsen et al. 2000	0.005 (20%)	0.10 (3.6%) 0.088 (21%)	0.086 (2.9%) 0.075 (2.7%)	0.62 (3.8%) 0.57 (2.2%)	0.40 (3.1%) 0.10 (2.6%)
Almela et al. 2005		0.098 (1.6%)	0.084 (0.4%)	0.65 (3.4%)	0.39 (2.5%)
Schmeisser et al. 2004		0.098 (4.1%)	0.082 (6.5%)	0.58 (5.3%)	0.39 (5.9%)
Kohlmeyer et al. 2003		0.088 (4.7%)	0.089 (4.1%)	0.60 (3.2%)	0.40 (2.9%)
Ruiz-Chancho et al. 2010		0.10 ± 0.03	0.09 ± 0.01	0.69 ± 0.13	0.45 ± 0.08
Llorente-Mirandes et al. 2010	0.01 ± 0.01	0.10 ± 0.01	0.09 ± 0.01	0.64 ± 0.02	0.40 ± 0.01
Pell et al. 2013	0.01 ± 0.01	0.07 ± 0.01	0.07 ± 0.01	0.56 ± 0.04	0.37 ± 0.02

not have arsenosugars, which can be considered as an abnormal behaviour. The occurrence of arsenobetaine is not usual among samples, mostly due to the thorough cleaning of algae. Further discussion considering sampling points will be too speculative since not enough samples were analysed and it was not possible to sample all genus in every location.

In view of the fact that arsenosugars are the most abundant arsenic fraction, the occurrence of individual arsenosugar is sumarised in Table 5, with the exception of *P. capitatus*, that did not have arsenosugars, Gly-sug was present in all samples, being the most common arsenosugar. According to the metabolic routes proposed at the literature Gly-sug is an intermediate in the synthesis to the rest of arsenosugars (Edmonds et al. 1987; Caumette 2012). Moreover, the degradation of arsenosugars in marine environment converges in Gly-sug, which may also explain the presence of this arsenical in all samples (Pengprecha 2005). PO_4-sug and SO_3-sug majorly occur in brown algae, whilst few green algae had those compounds. Regarding red algae, some Gracilariaceae had PO_4-sug and SO_3-sug while Cystocloniaceae did not; however, only algae from two families and one location were analysed. Literature reported the occurrence of SO_4-sug in *Sargassum* spp. and *Fucus* spp. (Madsen 2000; Šlejkovec et al. 2006; García Salgado et al. 2008; Llorente-Mirandes et al. 2010) which agrees with our results. It should be stressed that the data collected from marine biota may contribute to further understanding of the role of arsenic in the environment.

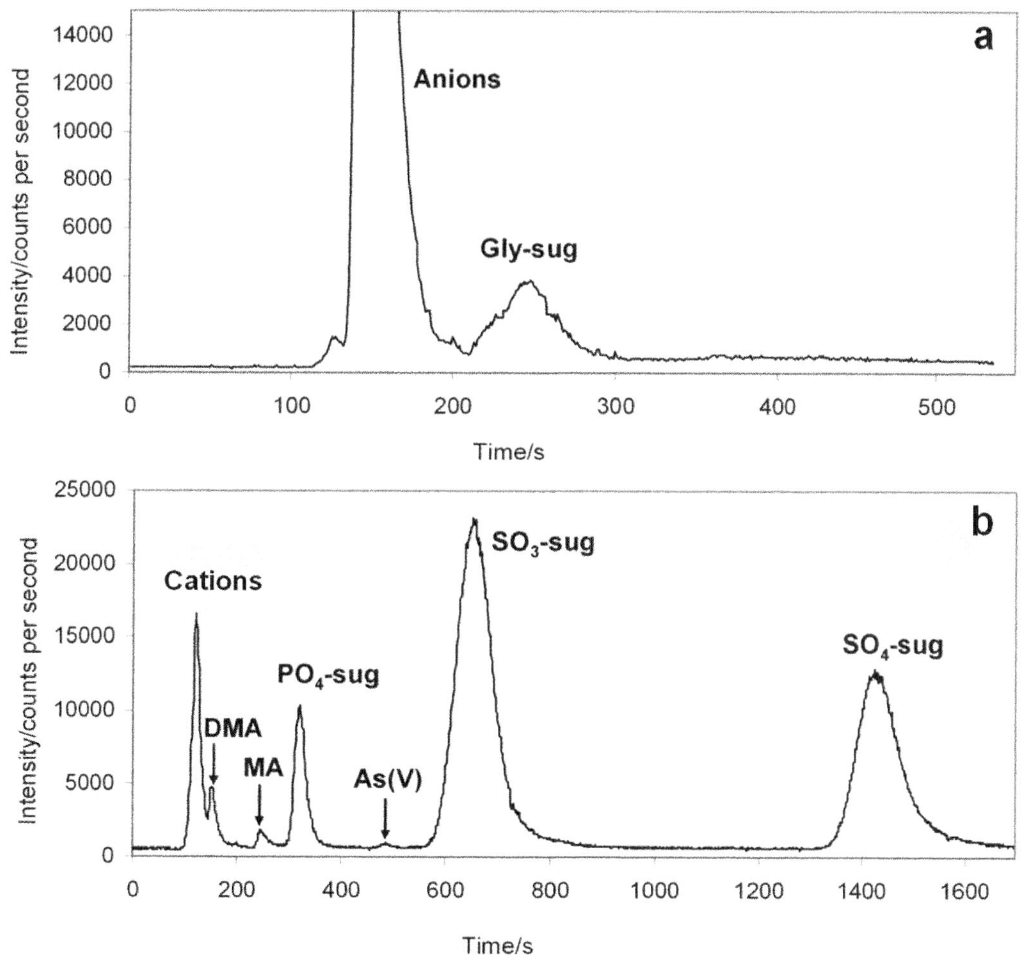

Fig. 2: Chromatograms of the *F. serratus* extract from cation exchange with the Zorbax SCX300 column (a) and anion exchange with the PRP-X100 column (b).

Table 5: Occurrence of arsenosugars in some studied marine algae.

Family	Species	Sampling Site	PO_4-sug	SO_3-sug	SO_4-sug	Gly-sug
■ Sargassaceae	*C. mediterranea*	Lloret de Mar (ES)		★		★
	C. mediterranea	Lloret de Mar (ES)	★	★		★
	C. barbata	Agia triada (GR)		★		★
	C. spinosa	Agia triada (GR)		★		★
	S. cf. *Acinarum*	Panama	★	★	★	★
	S. cf. *Acinarum*	Cayo Zapatilla (PA)	★	★	★	★
	S. cf. *Acinarum*	Istmito beach (PA)	★	★	★	★
■ Laminareceae	*M. pyrifera*	Fin del Mundo (AR)	★	★		★
	M. pyrifera	Est. Harberton (AR)	★	★		★
	U. pinnatifida	Rías Baixas (ES)	★			★
■ Dictyotaceae	*P. pavonica*	Lloret de Mar (ES)	★	★		★
	P. pavonica	Agia triada (GR)		★		★
	P. pavonica	Agia triada (GR)	★	★		★
	P. cf. *santae-crucis*	Cayo Zapatilla (PA)	★	★		★
■ Fucales	*F. vesiculosus*	Local market (ES)	★	★	★	★
▲ Gracilariaceae	*Gracilaria* sp.	Viamyl (GR)		★		★
	G. verrucosa	Viamyl (GR)	★		★	★
	G. verrucosa	Viamyl (GR)	★	★		★
	G. verrucosa	Kalohori (GR)	★	★		★
Family	**Species**	**Sampling Site**	**PO_4-sug**	**SO_3-sug**	**SO_4-sug**	**Gly-sug**
	G. verrucosa	Kalohori (GR)	★	★		★
	G. verrucosa	Viamyl (GR)				★
▲ Cystocloniaceae	*H. musciformis*	Agia Triada (GR)				★
	H. musciformis	Viamyl (GR)				★
♦ Codiaceae	*C. fragile*	Agia Triada (GR)				★
	C. fragile	Viamyl (GR)	★			★
	C. fragile	Viamyl (GR)				★

Table 5 contd....

Table 5 contd...

Family	Species	Sampling Site	PO_4-sug	SO_3-sug	SO_4-sug	Gly-sug
♦ Ulvaceae	*U. intestinalis*	Agia Triada (GR)				★
	U. intestinalis	Viamyl (GR)				★
	U. rigida	Agia Triada (GR)				★
	U. rigida	Kalohori (GR)				★
	U. lactuca	Viamyl (GR)				★
♦ Udoteaceae	*U.* cf. *flabellum*	Cayo Zapatilla (PA)	★	★		★
♦ Halimedaceae	*H.* cf. *incressata*	Cayo Zapatilla (PA)				★

Sample location: Spain (ES), Greece (GR), Argentina (AR) and Panama (PA)

Algae division: Ochrophyta (■), Rhodophyta (▲) and Chlorophyta (♦)

Seafood

Unlike the previous case study, when determining arsenic in seafood, the first problem to be faced is total element determination. In most of the fish and marine species, AB is present and, as its chemical decomposition is very difficult, as stated before, total arsenic content can be easily underestimated. Thus, it is necessary to be aware of this fact in order to select the most suitable digestion procedure to overcome this problem. For instance, to determine the total arsenic content in seafood we use a microwave digestion followed by ICP-MS measurement. 0.5–2 g samples are digested with 8 mL of concentrated HNO_3 and 2 mL of H_2O_2. Helium gas is used in the collision cell to avoid interferences in the ICP-MS measurements. To evaluate the accuracy of the method, several CRMs and one reference material were analysed: seafood CRMs (DOLT-4, DORM-4, BCR-627, TORT-2, TORT-3, SRM 2976, ERM CE278, SRM 1566b, BCR-279, NMIJ 7405-a, and NIES n9), and one RM (9th PT). The obtained results can be seen in Fig. 3. The comparison between each measured value of total As with its corresponding certified value showed no significant difference at a 95% confidence level when Student's t-test was applied.

For speciation studies, 0.2–1.0 g of samples were extracted with 10 mL of a solution containing 0.2% (w/v) HNO_3 and 1% (w/v) of H_2O_2 in a microwave system at 95°C. To evaluate the accuracy of the speciation method, four of the CRMs with certified species contents were analysed. The obtained results can be seen in Fig. 4. The comparison between each measured value with its corresponding certified value showed no significant difference at a 95% confidence level when Student's t-test was applied.

The results obtained in a selection of 22 seafood samples including crustaceans, bivalves, and fish are shown in Fig. 5. As expected, AB was found as the main arsenic species in all analysed samples, comprising 48% to 95% of the total arsenic. DMA was also detected as minority compounds in mussels, clams, and prawns. DMA was found in 73% of samples, and MA appeared in 36% of samples (prawns, shrimp, cockles, and oysters). DMA was found at higher levels than MA in fish samples. TMAO and AC were found in 50% and 18% of all samples respectively. An unknown compound with a retention time of 279 s was found using the cationic column (UC-A, ranged from 0.6% to 27% of total arsenic), along with a second unknown compound (UC-B, ranged from 0.3% to 6% of the total arsenic) with a retention time of 360s. These unknown cation species could be attributed to trimethylarsoniopropionate (TMAP) and tetramethylarsonium ion (TETRA), respectively, according to Kirby et al. (2004). However, the lack of appropriate standards made it impossible to corroborate this hypothesis.

In terms of anionic species, two unknown compounds, UA-A and UA-B, with a retention time of 148 and 251 s respectively, were found as minor species in crustacean and bivalve samples. These unknowns comprised 0.4% to 0.9% and 0.2% to 15% of the total arsenic, for UA-A and UA-B, respectively. These peaks could correspond to arsenosugar compounds such as Gly-sug and PO_4-sug, which were identified

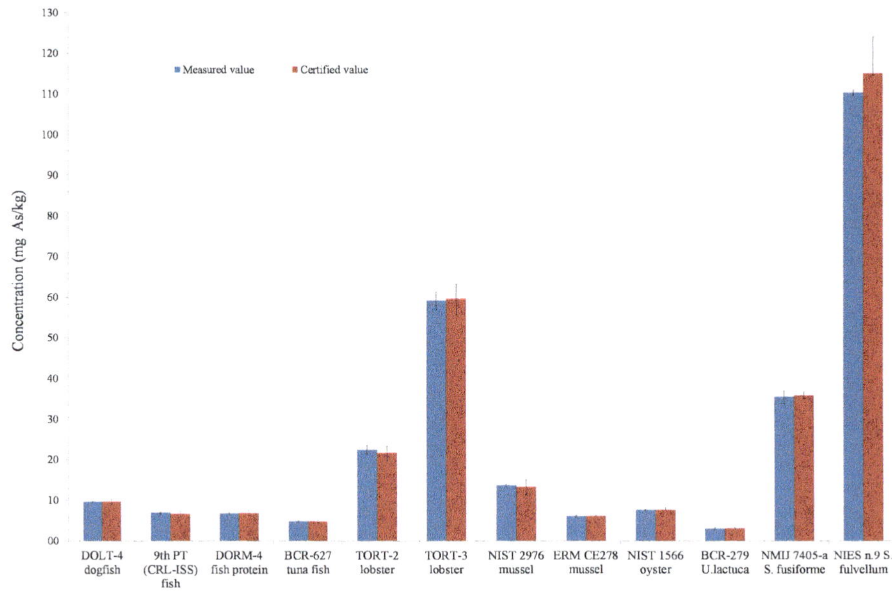

Fig. 3: Evaluation of the accuracy of the method for total As determination in seafood.

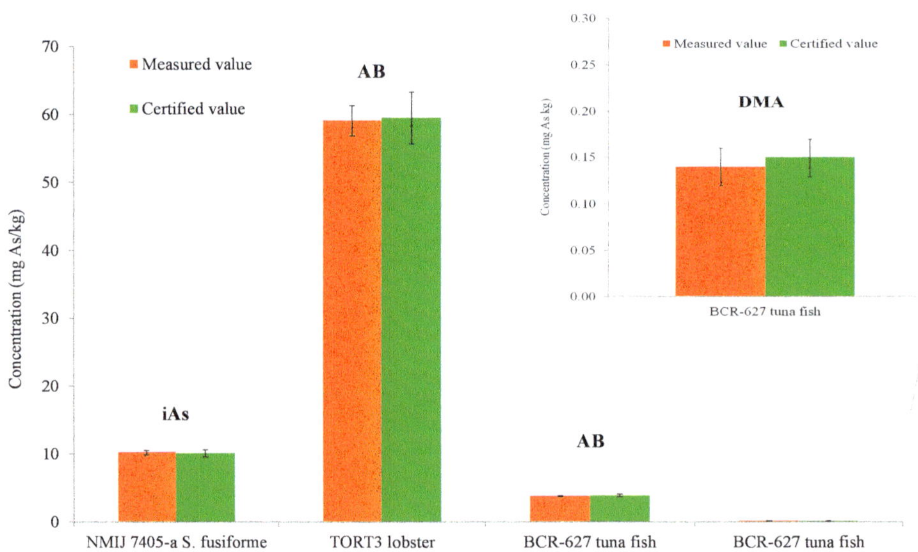

Fig. 4: Evaluation of the accuracy of the method for As speciation in seafood.

in fish and molluscs by Nischwitz et al. (2005). Due to the lack of appropriate standards, this attribution was not checked.

The iAs was extracted, identified, and quantified as As(V), and selectively separated from other arsenic compounds. It was found in 36% of all samples, always being below 3.3% of the total arsenic. For fish samples, the inorganic arsenic content is in all cases below the limit of detection. These results are in agreement with those found in the literature (Fontcuberta et al. 2011; Larsen et al. 2005; Leufroy et al. 2011). However, iAs was found in bivalves and crustaceans at concentrations of up to 0.35 mg kg^{-1} accounting for less than 3.3% of the total arsenic.

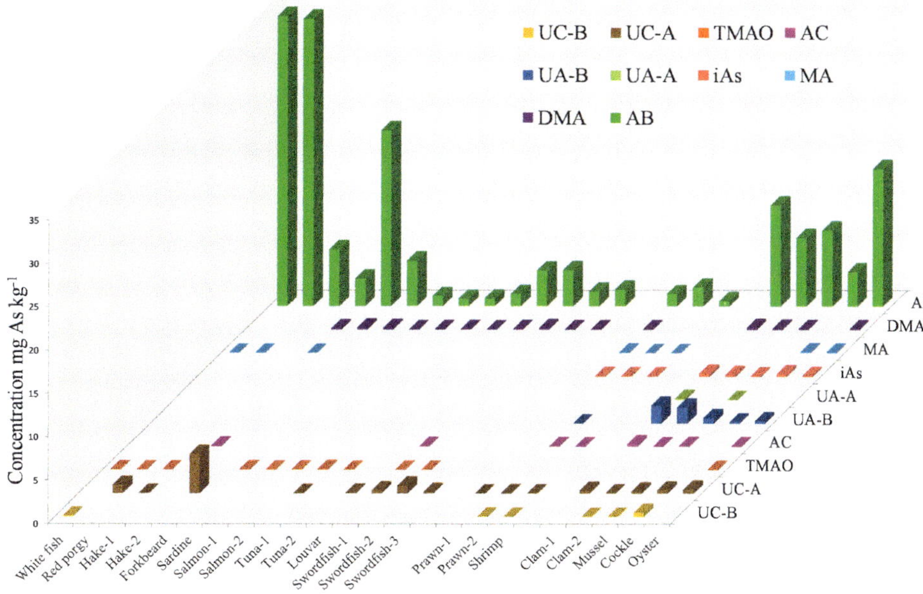

Fig. 5: Arsenic speciation of selected seafood samples.

As can be seen, when dealing with seafood samples, a wide variability in the arsenic species can be expected, highlighting the need to carry out speciation to discern the toxic from the non-toxic species.

Conclusions

Arsenic speciation studies for marine biota require analytical methods assuring the integrity of the distribution of species that are accounted for in the original matrix. The most used analytical methods involved coupled techniques and method validation is hindered by the lack of suitable standards and certified reference materials.

Both, total arsenic contents and the arsenic compounds present are highly dependent on type of matrices (algae or seafood) and can be different among families for the same type of matrix. For each specific study, the analytical conditions might be revised and optimized.

Acknowledgements

The authors thank the DGICYT (Project No. CTQ2010-15377) and the Grup de Recerca Consolidat (Project No. SGR2009-1188) for financial help received in support of this study. The authors also thank Dr. A. Padró (Centres Científics i Tecnològics de la Universitat de Barcelona, CCiTUB) for his valuable support with LC-ICP-MS measurements.

Keywords: Arsenic speciation, quality assurance, coupled techniques, seawater, algae, seafood, fish

References

Almela, C., J.M. Laparra, D. Vélez, R. Barberá, R. Farré and R. Montoro. 2005. Arsenosugars in raw and cooked edible seaweed: characterization and bioaccessibility. J. Agric. Food Chem. 53: 7344–51.

Australia New Zealand Food Standards Code. 2013. Standard 1.4.1—Contaminants and natural toxicants. Federal Register of Legislative Instruments F2013C00140. Issue: 139.

Baer, I., M. Baxter, V. Devesa, D. Vélez, G. Raber, R. Rubio et al. 2011. Performance of laboratories in speciation analysis in seafood–Case of methylmercury and inorganic arsenic. Food Control 22: 1928–1934.
Caliceti, M., E. Argese, A. Sfriso and B. Pavoni. 2002. Heavy metal contamination in the seaweeds of the Venice lagoon. Chemosphere 47: 443–454.
Caruso, J.A., B. Klaue, B. Michalke and D.M. Rocke. 2003. Group assessment: elemental speciation. Ecotoxicol. Environ. Safety 56: 32–44.
Castelhouse, H., C. Smith, A. Raab, C. Deacon, A.A. Meharg and J. Feldman. 2003. Biotransformation and accumulation of arsenic in soil amended with seaweed. Environ. Sci. Technol. 37: 951–957.
Caumette, G., I. Koch, E. Estrada and K.J. Reimer. 2011. Arsenic speciation in plankton organisms from contaminated lakes: Transformations at the base of the freshwater food chain. Environ. Sci. Technol. 45: 9917–9923.
Caumette, G., I. Koch and K.J. Reimer. 2012. Arsenobetaine formation in plankton: a review of studies at the base of the aquatic food chain. J. Environ. Monitor. 14: 2841–53.
Chen, J. and B.P. Rosen. 2014. Biosensors for inorganic and organic arsenicals. Biosensors 4: 494–512.
Chen, M.-L., L.-Y. Ma and X.-W. Chen. 2014. New procedures for arsenic speciation: A review. Talanta 125: 78–86.
Commission Regulation. 2006. Number 1881/2006 of 19 December 2006 setting maximum levels for certain contaminants in foodstuffs. Official Journal of the European Communities L364: 5–24.
Edmonds, J.S. and K.A. Francesconi. 1987. Transformations of arsenic in the marine environment. Experientia 43: 553–557.
Edmonds, J.S. 2000. Diastereoisomers of an "arsenomethionine"-based structure from *Sargassum lacerifolium*: the formation of the arsenic-carbon bond in arsenic-containing natural products. Bioorg. Med. Chem. Lett. 10: 1105–1108.
Edmonds, J.S. and K.A. Francesconi. 2003. *In*: P. Craig [ed.]. Organometallic Compounds in the Environment, Wiley, Chichester, West Sussex, UK.
EFSA. 2005. European Food Safety Authority. Opinion of the scientific panel on contaminants in the food chain on a request from the commission related to arsenic as undesirable substance in animal feed (Question EFSAQ-2003-031). EFSA J. 180: 1–35.
EFSA. 2009. European Food Safety Authority. Scientific opinion on arsenic in food. Panel on Contaminants in the Food Chain (CONTAM). EFSA Journal 7: 1351–1549.
EFSA. 2014. European Food Safety Authority. Dietary exposure to inorganic arsenic in the European population. EFSA Journal 12: 3597–3665.
Entwisle, J. and R. Hearn. 2006. Development of an accurate procedure for the determination of arsenic in fish tissues of marine origin by inductively coupled plasma mass spectrometry. Spectrochim. Acta Part B: At. Spectrosc. 61: 438–443.
Feldmann, J. and E.M. Krupp. 2011. Critical review or scientific opinion paper: Arsenosugars—A class of benign arsenic species or justification for developing partly speciated arsenic fractionation in foodstuffs? Anal. Bioanal. Chem. 399: 1735–1741.
Fontcuberta, M., J. Calderon, J.R. Villalbí, F. Centrich, S. Portaña and A. Espelt. 2011. Total and inorganic arsenic in marketed food and associated health risks for the Catalan (Spain) population. J. Agric. Food Chem. 59: 10013–10022.
Foster, A.L. and C.S. Kim. 2014. Arsenic speciation in solids using X-ray absorption spectroscopy. Rev. Mineral. Geochem. 79: 257–369.
Foster, S., W.A. Maher, F. Krikowa and S. Apte. 2007. A microwave-assisted sequential extraction of water and dilute acid soluble arsenic species from marine plant and animal tissues. Talanta 71: 537–549.
Francesconi, K.A. and J.S. Edmonds. 1997. Arsenic and marine organisms. Adv. Inorg. Chem. 44: 147–189.
Francesconi, K.A. and D. Kuehnelt. 2004. Determination of arsenic species: a critical review of methods and applications, 2000–2003. Analyst 129: 373–395.
Francesconi, K.A. and M. Sperling. 2005. Speciation analysis with HPLC–mass spectrometry: time to take stock. Analyst 130: 998–1001.
Francesconi, K.A. 2010. Arsenic species in seafood: origin and human health implications. Pure Appl. Chem. 82: 373–381.
García Salgado, S., M.A. Quijano Nieto and M.M. Bonilla Simón. 2008. Assessment of total arsenic and arsenic species stability in alga samples and their aqueous extracts. Talanta 75: 897–903.
Geng, W., R. Komine, T. Ohta, T. Nakajima, H. Takanashi and A. Ohki. 2009. Arsenic speciation in marine product samples: Comparison of extraction-HPLC method and digestion-cryogenic trap method. Talanta 79: 369–375.
Grotti, M., F. Soggia, C. Lagomarsino, W. Goessler and K.A. Francesconi. 2008. Arsenobetaine is a significant arsenical constituent of the red Antarctic alga *Phyllophora Antarctica*. Environ. Chem. 5: 171–175.
Han, C., X. Cao, J.-J. Yu, X.-R. Wang and Y. Shen. 2009. Arsenic speciation in *Sargassum fusiforme* by microwave-assisted extraction and LC-ICP-MS. Chromatographia 69: 587–591.
Hirata, S. and H. Toshimitsu. 2007. Determination of arsenic species and arsenosugars in marine samples by HPLC–ICP-MS. Appl. Organometal. Chem. 21: 447–454.
Hsieh, Y.-J. and S.J. Jiang. 2012. Application of HPLC-ICP-MS and HPLC-ESI-MS procedures for arsenic speciation in seaweeds. J. Agric. Food Chem. 60: 2083–2089.
Innis, S.M. 2007. Dietary (n-3) fatty acids and brain development. J. Nutr. 137: 855–859.
Irvin, T.R. and K.J. Irgolic. 1995. *In-vitro* prenatal toxicity of trimethylarsine, trimethylarsine oxide and trimethylarsine sulfide. Appl. Organometal. Chem. 9: 315–321.
JECFA/72/SC. 2010. Joint Food and Agricultural Organisation/World Health Organisation. Expert Committee on Food and Additives.

Karadjova, I.B., P.K. Petrov, I. Serafimovski, T. Stafilov and D.L. Tsalev. 2007 Arsenic in marine tissues – the challenging problems to electrothermal and hydride generation atomic absorption spectrometry. Spectrochim. Acta Part B 62: 258–268.

Kirby, J., W. Maher, M. Ellwood and F. Krikowa. 2004. Arsenic species determination in biological tissues by HPLC–ICP–MS and HPLC–HG–ICP–MS. Aust. J. Chem. 57: 957–966.

Koch, I., L. Wang, C.A. Ollson, W.R. Cullen and K.J. Reimer. 2000a. The predominance of inorganic arsenic species in plants from Yellowknife, Northwest Territories, Canada. Environ. Sci. Technol. 34: 22–26.

Koch, I., L. Wang, K.J. Reimer and W.R. Cullen. 2000b. Arsenic species in terrestrial fungi and lichens from Yellowknife, NWT, Canada. Appl. Organometal. Chem. 14: 245–252.

Kohlmeyer, U., E. Jantzen, J. Kuballa and S. Jakubik. 2003. Benefits of high resolution IC-ICP-MS for the routine analysis of inorganic and organic arsenic species in food products of marine and terrestrial origin. Anal. Bioanal. Chem. 377: 6–13.

Larsen, E., J. Engman, J. Sloth, M. Hansen and L. Jorhem. 2005. Determination of inorganic arsenic in white fish using microwave-assisted alkaline alcoholic sample dissolution and HPLC-ICP-MS. Anal. Bioanal. Chem. 381: 339–346.

Leufroy, A., L. Noël, V. Dufailly, D. Beauchemin and T. Guérin. 2011. Determination of seven arsenic species in seafood by ion exchange chromatography coupled to inductively coupled plasma-mass spectrometry following microwave assisted extraction: method validation and occurrence data. Talanta 83: 770–779.

Llorente-Mirandes, T., M.J. Ruíz Chancho, M. Barbero, R. Rubio and J.F. López-Sánchez. 2010. Measurement of arsenic compounds in littoral zone algae from the Western Mediterranean Sea. Occurrence of arsenobetaine. Chemosphere 81: 867–875.

Llorente-Mirandes, T., M.J. Ruiz-Chancho, M. Barbero, R. Rubio and J.F. López-Sánchez. 2011. Determination of water-soluble arsenic compounds in commercial edible seaweed by LC-ICPMS. J. Agric. Food Chem. 59: 12963–12968.

Lobinski, R. 1997. Elemental speciation and coupled techniques. Appl. Spectrosc. 51: 260A–278A.

Madsen, A.D., W. Goessler, S.N. Pedersen and K.A. Francesconi. 2000. Characterization of an algal extract by HPLC–ICP–MS and LC-electrospray MS for use in arsenosugar speciation studies. J. Anal. At. Spectr. 15: 657–662.

Maher, W., F. Krikowa, M. Ellwood, S. Foster, R. Jagtap and G. Raber. 2012. Overview of hyphenated techniques using an ICP-MS detector with an emphasis on extraction techniques for measurement of metalloids by HPLC–ICPMS. Microchem. J. 105: 15–31.

Meier, J., N. Kienzl, W. Goessler and K.A. Francesconi. 2005. The occurrence of thio-arsenosugars in some samples of marine algae. Environ. Chem. 2: 304–307.

MHC. 2005. Maximum levels of contaminants in foods, GB2762-2005. Beijing: Ministry of Health of China.

Michalski, R., S. Szopa, M. Jabłónska and A. Łyko. 2012. Application of hyphenated techniques in speciation analysis of arsenic, antimony, and thallium. Sci. World J. 2012: 1–17.

Miyashita, S., M. Shimoya, Y. Kamidate, T. Kuroiwa, O. Shikino, S. Fujiwara, K.A. Francesconi and T. Kaise. 2009. Rapid determination of arsenic species in freshwater organisms from the arsenic-rich Hayakawa River in Japan using HPLC-ICP-MS. Chemosphere 75: 1065–1073.

Muñoz, E. and Susana Palmero. 2005. Analysis and speciation of arsenic by stripping potentiometry: a review. Talanta 65: 613–620.

Naidu, R., E. Smith, G. Owens and P. Bhattacharya (eds.). 2006. Managing Arsenic in the Environment, CSIRO Publishing, Collingwood, Australia.

Narukawa, T., T. Kuroiwa, K. Inagaki, A. Takatsu and K. Chiba. 2005. Decomposition of organoarsenic compounds for total arsenic determination in marine organisms by the hydride generation technique. Appl. Organometal. Chem. 19: 239–245.

Narukawa, T., T. Kuroiwa, T. Yarita and K. Chiba. 2006. Analytical sensitivity of arsenobetaine on atomic spectrometric analysis and the purity of synthetic arsenobetaine. Appl. Organometal. Chem. 20: 565–572.

National Program for Residues. 2012. Brazil: Ministry of Agriculture, Livestock and Food Supply (MAPA).

Niegel, C. and F.M. Matysik. 2010. Analytical methods for the determination of arsenosugars – a review of recent trends and developments. Anal. Chim. Acta 657: 83–89.

Nischwitz, V. and S.A. Pergantis. 2005. First report on the detection and quantification of arsenobetaine in extracts of marine algae using HPLC-ES-MS/MS. Analyst 130: 1348–1350.

Pell, A., A. Márquez, R. Rubio and J.F. López-Sánchez. 2013a. Effects of sample processing on arsenic speciation in marine macroalgae. Anal. Methods 5: 2543–2550.

Pell, A., G. Kokkinis, P. Malea, S.A. Pergantis, R. Rubio and J. F. López-Sánchez. 2013b. LC–ICP–MS analysis of arsenic compounds in dominant seaweeds from the Thermaikos Gulf (Northern Aegean Sea, Greece). Chemosphere 93: 2187–2194.

Pell, A., A. Márquez, J.F. López-Sánchez, R. Rubio, M. Barbero, S. Stegen, F. Queirolo and P. Díaz-Palma. 2013c. Occurrence of arsenic species in algae and freshwater plants of an extreme arid region in northern Chile, the Loa River Basin. Chemosphere 90: 556–564.

Pengprecha, P., M. Wilson, A. Raab and J. Feldmann. 2005. Biodegradation of arsenosugars in marine sediment. Appl. Organometal. Chem. 19: 819–826.

Pétursdóttir, Á.H., H. Gunnlaugsdóttir, H. Jörundsdóttir, A. Mestrot, E.M. Krupp and J. Feldmann. 2012a. HPLC-HG-ICP-MS: A sensitive and selective method for inorganic arsenic in seafood. Anal. Bioanal. Chem. 404: 2185–2191.

Pétursdóttir, Á.H., H. Gunnlaugsdóttir, H. Jörundsdóttir, A. Raab, E.M. Krupp and J. Feldmann. 2012b. Determination of inorganic arsenic in seafood: Emphasizing the need for certified reference materials. Pure Appl. Chem. 84: 191–202.

Pétursdóttir, A.H., H. Gunnlaugsdóttir, E.M. Krupp and J. Feldmann. 2014. Inorganic arsenic in seafood: Does the extraction method matter? Food Chem. 150: 353–359.

Polya, D.A., P.R. Lythgoe, F. Abou-Shakra, A.G. Gault, J.R. Brydie, J.G. Webster, K.L. Brown, M.K. Nimfopoulos and K.M. Michailidis. 2003. IC–ICP–MS and IC–ICP–HEX–MS determination of arsenic speciation in surface and groundwaters: preservation and analytical issues. Mineral. Mag. 67: 247–261.

Raber, G., K.A. Francesconi, K.J. Irgolic and W. Goessler. 2000. Determination of 'arsenosugars' in algae with anion-exchange chromatography and an inductively coupled plasma mass spectrometer as element-specific detector. Fresenius J. Anal. Chem. 367: 181–188.

Rodríguez, J.A., E. Barrado, M. Vega, Y. Castrillejo and J.L.F.C. Lima. 2012. Sequential injection anodic stripping voltammetry at tubular gold electrodes for inorganic arsenic speciation. pp. 203–218. *In*: Y. Shao [ed.]. Electrochemical Cells - New Advances in Fundamental Researches and Applications. Publisher InTech, Croatia (available on-line).

Rubio, R., M.J. Ruiz-Chancho and J.F. López-Sánchez. 2010. Sample pre-treatment and extraction methods that are crucial to arsenic speciation algae and aquatic plants. Trends Anal. Chem. 29: 53–69.

Ruiz Chancho, M.J., J.F. López-Sánchez and R. Rubio. 2010. Occurrence of arsenic species in the seagrass Posidonia oceanica and in the marine algae Lessonia nigrescens and Durvillaea antarctica. J. Appl. Phycol. 22: 465–472.

Schaeffer, R., K.A. Francesconi, N. Kienzl, C. Soeroes, P. Fodor, L. Váradi, R. Raml, W. Goessler and D. Kuehnelt. 2006. Arsenic speciation in freshwater organisms from the river Danube in Hungary. Talanta 69: 856–865.

Shibata, Y., M. Morita and J.S. Edmonds. 1987. Purification and identification of arsenic-containing ribofuranosides from the edible brown seaweed, *Laminaria japonica* (MAKONBU). Agric. Biol. Chem. 51: 391–398.

Schmeisser, E., W. Goessler, N. Kienzl and K.A. Francesconi. 2004. Volatile analytes formed from arsenosugars: Determination by HPLC-HG-ICPMS and implications for arsenic speciation analyses. Anal. Chem. 76: 418–423.

Slejkovec, Z., E. Kápolna, I. Ipolyi and J.T. van Elteren. 2006. Arsenosugars and other arsenic compounds in littoral zone algae from the Adriatic Sea. Chemosphere 63: 1098–1105.

Straif, K., B. Benbrahim-Tallaa, R. Baan, Y. Grosse, B. Secretan, F. El Ghissassi, V. Bouvard, N. Guha, C. Freemas, L. Galuchet and V. Cogliano. 2009. A review of human carcinogens part C: metals, arsenic, dusts, and fibres. Lancet Oncol. 10: 453–454.

Thomson, D., W. Maher and S. Foster. 2007a. Arsenic and selected elements in intertidal and estuarine marine algae, south-east coast, NSW, Australia. Appl. Organometal. Chem. 21: 396–411.

Thomson, D., W. Maher and S. Foster. 2007b. Arsenic and selected elements in marine angiosperms, south-east coast, NSW, Australia. Appl. Organometal. Chem. 21: 381–395.

Traar, P. and K.A. Francesconi. 2006. Synthetic routes for naturally-occurring arsenic-containing ribosides. Tetrahedron Lett. 47: 5293–5296.

Tukai, R., W.A. Maher, I.J. McNaught and M.J. Ellwood. 2002. Measurement of arsenic species in marine macroalgae by microwave-assisted extraction and high performance liquid chromatography-inductively coupled plasma mass spectrometry. Anal. Chim. Acta 457: 173–185.

Van Hulle, M., C. Zhang, X. Zhang and R. Cornelis. 2002. Arsenic speciation in Chinese seaweeds using HPLC-ICP-MS and HPLC-ES-MS. Analyst 127: 634–640.

Zheng, J., H. Hintelmann, B. Dimock and M.S. Dzurko. 2003. Speciation of arsenic in water, sediment, and plants of the Moira watershed, Canada, using HPLC coupled to high resolution ICP–MS. Anal. Bioanal. Chem. 377: 14–24.

Zmozinski, A.V., T. Llorente-Mirandes, I.C.F. Damin, J.F. López-Sánchez, M.G.R. Vale, B. Welz and M.M. da Silva. 2015a. Direct solid sample analysis with graphite furnace atomic absorption spectrometry—A fast and reliable screening procedure for the determination of inorganic arsenic in fish and seafood. Talanta 134: 224–231.

Zmozinski, A.V., T. Llorente-Mirandes, J.F. López-Sánchez and M.M. da Silva. 2015b. Establishment of a method for determination of arsenic species in seafood by LC-ICP-MS. Food Chemistry 173: 1073–1082.

6

Biological Effects of Organotins in the Marine Environment

Teresa Neuparth,[1,a] *Luis Filipe C. Castro*[2,b] and *Miguel M. Santos*[2,c,*]

ABSTRACT

Organotin compounds (OTs) have been used for approximately 50 years as biocides in antifouling ship paints to prevent the settlement of marine organisms. Although initial evaluations suggested a negligible risk to non-target organisms, over the years the global concern regarding the ecological effects led to a phase out in 2008 on the use of organotins as antifoulants in ship paints. However, given the half-times in anoxic sediments, and the high toxicity to several metazoans, organotins are likely to persist for many years in sites with a high TBT legacy and impact sensitive *taxa*. Here, we aim at reviewing the biological effects of organotins in marine animals, focusing in particular on environmentally relevant low level effects and the associated molecular and biochemical mechanisms of disruption.

Introduction

Organotin compounds are a highly versatile group of organometallic chemicals characterized by a Sn atom covalently bound to one or more organic groups-R (such as alkyl groups, i.e., methyl, ethyl, propyl, butyl, or aryl groups) and to anionic species X (such as chloride, fluoride, oxide, hydroxy, carboxylate, and thiolate). These compounds are chemically represented by the general formula R_nSnX_{4-n} (Fig. 1).

According to the number of organic groups, organotins are grouped into mono-, di-, tri-, and tetra-substituted compounds. Generally, the toxicity of the organotins is influenced firstly by the number of organic groups attached to the tin moiety and secondly, by the nature of the organic groups. Inorganic tin compounds show a low toxicity. For most biological processes, tri-organotin compounds have been reported to be more toxic than di-organotin and mono-organotin. The organic group bound to tin also plays a role in the toxicity, butyl and phenyl being the most important groups.

[1] CIMAR/CIIMAR, Associated Laboratory, Interdisciplinary Centre for Marine and Environmental Research, University of Porto, Avenida General Norton de Matos, S/N, 4450-208 Matosinhos, Portugal.

[2] CIMAR/CIIMAR, Associated Laboratory, Interdisciplinary Centre for Marine and Environmental Research, University of Porto, Avenida General Norton de Matos, S/N 4450-208, Matosinhos, Portugal, University of Porto, FCUP, Faculty of Sciences, University of Porto, Department of Biology, Rua do Campo Alegre, 4169-007 Porto, Portugal.

[a] Email: tneuparth@ciimar.up.pt

[b] Email: filipe.castro@ciimar.up.pt

[c] Email: santos@ciimar.up.pt

* Corresponding author

Compound	Chemical structure	Molecular Formula
Tributyltin	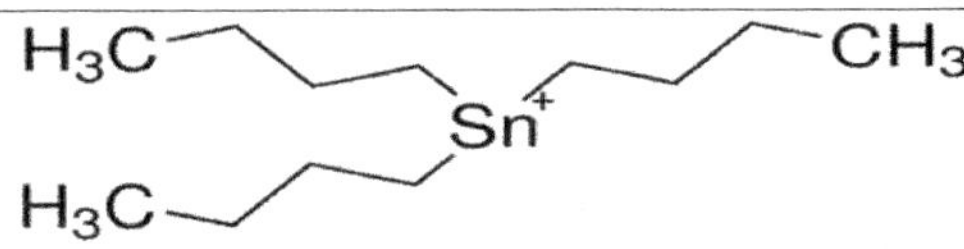	$C_{12}H_{27}Sn^{+}$
Dibutyltin	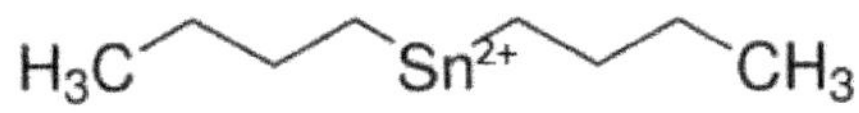	$C_8H_{18}Sn^{2+}$
Monobutyltin		$C_4H_9Sn^{3+}$
Triphenyltin	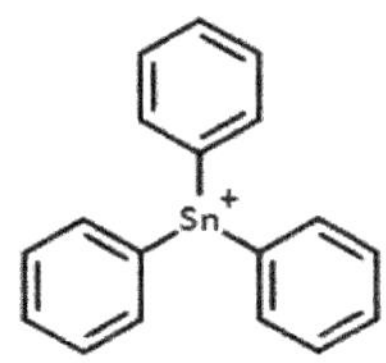	$C_{18}H_{15}Sn^{+}$
Diphenyltin	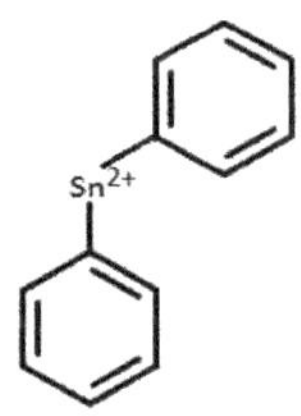	$C_{12}H_{10}Sn^{2+}$

Fig. 1: Chemical structure of organotins.

Although organotin compounds have been known since the 1850s, their commercial use started only in the 1940s when the plastic industry, particularly the production of polyvinyl chloride (PVC) began to expand (Hoch 2001). Even today, di-organotin compounds are used as heat and light stabilizer additives in PVC processing, since this addiction prevents the discoloration and embrittlement of the PVC polymer. The biocidal properties of mainly tri-organotin compounds were discovered in the late 1950s. Since then, tri-organotin compounds have had widespread use as biocides in agrochemicals (fungicides, insecticides, miticides, antifeedants), antifouling agents for ship bottoms and fishery and aquaculture structures, antifungal agent in textiles and industrial water systems, and as a general wood preservative (Furdek et al. 2011). Organotin compounds reach the marine environments mostly by their widespread use as pesticides and antifouling paints. Most attention was given by scientific comunity to tributylin (TBT), due to its wide used in marine antifouling paints to prevent the growth of organisms on the hull of ships, while the use of triphenyltin (TPT) was limited to some paints and always in a lower percentage in comparison with TBT. With respect to the TPT, its presence in the ecosystem is mainly associated to its use as pesticide. The intense development of TBT industrial production started in the 1960s, after the excellent and long lasting antifouling properties of TBT were discovered. TBT was considered a true revolution in the market of antifouling paints. Attachment and growth of barnacles, clams, and other organisms in hulls diminished ships' performance, increased fuel consumption, and its mechanic removal was costly and time consuming. TBT not only reduced the costs caused by fouling, but also the need for frequent repainting as its effects could last up to 4–5 years (Hall et al. 1987).

In the late 1970s, significant disturbances were observed in the oyster farms of the Arcachon Bay along the French Atlantic Coastline (Alzieu 2000). TBT was found to be responsible for reproductive failure and for abnormalities occurring in the shell calcification of adult oysters (*Crassostrea gigas*), leading to stunted growth (Alzieu 2000). In parallel, in the early 1970s, Blaber (1970) noticed the presence of a penis like outgrowth behind the right cephalic tentacle in spent females of the dog-whelk, *Nucella lapillus* females collected in Plymouth Sound, England. This was the first report of masculinization in female gastropods. The term imposex, however, was coined by Smith (1971) to describe the syndrome of a superimposition of male type genital structures, such as the penis and vas deferens, on female gastropods. But it was only in 1981 that Smith established that imposex was an abnormal phenomenon caused by exposure to components of antifouling paints used in ships' hulls, more specifically to TBT (Smith 1981a,b). In 1986, Gibbs and Bryan demonstrated experimentally that exposure to TBT at concentrations as low as 0.5 ng TBT Sn/L could induce imposex in *N. lapillus* and cause, at advanced stages, female reproductive failure resulting in population decline or even local extinction. Since then, numerous studies reported the occurrence of imposex in other species, particularly in the vicinity of harbors, marinas, and other places with high maritime traffic (Santos et al. 2000, 2002; Gómez-Ariza et al. 2006; Titley-O'Neal et al. 2011). As of 2004, over 150 gastropod species have been reported to be affected by imposex worldwide (Horiguchi 2009). Although mollusks are particularly sensitive to TBT, organotins are known today to have diverse consequences on a wide range of species, from bacteria to mammals, at concentrations in the low ng/L range (Fent 1996; Janer 2005; Santos et al. 2012).

The undesired impact of TBT on marine ecosystems was therefore far greater than initially anticipated. Thus, as opposed to the benefits for the shipping industry, environmental and economic consequences of the utilization of TBT as antifoulants had to be considered. Efforts were undertaken in order to find a global solution to this problem and legal requirements have been imposed to protect the aquatic environment. Thus, the use of TBT in vessels < 25 m was initially regulated in many countries since the mid-1980s (Konstantinou and Albanis 2004). Because of worrying side-effects of TBT on oysters, France was the first country to ban the use of organotin-based antifouling paints on vessels less than 25 m in 1982 (Alzieu et al. 1986). Comparable regulations came into effect a few years later in North America, UK, Australia, New Zealand, Hong Kong, and most European countries after 1987 (Antizar-Ladislao 2008). A directive with the same guidelines was published by the European Union, in 1989 (89/677/CEE) (Santos et al. 2002). However, whereas in some locations these measures were sufficient to allow full recovery of local mollusk populations, other studies reported unexpectedly slower recovery rates. Re-surveys in several locations revealed that imposex levels in *N. lapillus* had not decreased demonstrating that measures taken in the early 1990s were not being effective (Santos et al. 2000; Barroso and Moreira 2002; Santos et al. 2002). In addition, a first report highlighted the impact of TBT in open seas, as a result of intense shipping traffic (Hallers-Tjabbes et al. 1994).

Since shipping is a worldwide activity, control on antifouling system individually by each country was not effective enough to prevent organotin pollution. Therefore, the problem was brought to the International Maritime Organization (IMO) in order to develop and maintain a global regulatory framework for the organotins ban. It was in 2001 that the Convention on the Control of Harmful Anti-fouling Systems on Ships (AFS) was adopted by IMO. This convention established a compromise to ban the application or re-application of TBT containing antifouling paints in all ships from January 1st 2003 onwards (MEPC 2001), and to ban the presence of TBT in ship hulls from the 1st January 2008, unless covered by a coating that prevented organotin leaking. The convention entered into force on the 17th of September 2007, when 25 states (representing 38.09% of the world tonnage) ratified it (Gipperth 2009). Therefore, the total ban was effective 12 months later, 17th September 2008 (Santos et al. 2009). To promote the implementation of the AFS Convention in the member states, the European Union moved to the adoption of decision no. 2455/2001/EC and defines 11 priority hazardous substances, including TBT compounds, subject to cessation of emissions, discharges and losses into water. Additionally, the adoption of the Directive 2002/62/EC and Regulation 782/2003 established the total interdiction of organotins antifouling paints application on European Union ships after 1st July 2003, and the prohibition of organotins antifouling paints on ships' hulls from 1st January 2008. This prohibition was then extended to all ships entering European Union ports. Later the EC Regulation 1907/2006 restricted the manufacture, placing on the market, and use of organotin compounds. The most recent European Commission initiatives concerning organotins occurred in 2009 by the adoption of the Decision 2009/425/EC that deals with the restriction of these substances in consumer products (Sousa et al. 2014).

However, after the complete phase out of TBT-based antifouling paints, organotin levels in coastal waters are still a reason for concern. Although the use of TBT in antifouling paints has been forbidden, its low degradation rates in sediments will contribute for long persistence of TBT in aquatic ecosystems, particularly in estuarine areas (Santos et al. 2004). The Persistence of TBT in sediments is indicated by a half-time of between 2 and 30 years in temperate regions (Dowson et al. 1996; Maguire 2000; WHO 2006). Moreover, organotin compounds continues to be used in other industrial applications such as wood, textiles, cotton, and in PVC production (Janer 2005). In addition, although TBT has a half-life of 6 days to four months in superficial waters (Maguire 1987), it shows up to 360–770 days in surface sediments and to tens of years in anaerobic ones (Maguire 1987; Dowson et al. 1996). In fact, a maximum of 87 ± 17 years was described for deep anaerobic sediments in enclosed marine systems at sub-arctic temperatures (Viglino et al. 2004; Coray and Bard 2007). This behavior can explain the elevated levels of TBT still found today in some ecosystems, such as in the Venice lagoon (Castro and Fillmann 2012; Barroso et al. 2011). Recent data from several locations, in particular coastal areas, reported a clear decrease in TBT levels and associated effects such as imposex in marine snails (Smith et al. 2008; Santos et al. 2009; Barroso et al. 2011, 2015; Longstone et al. 2015).

Although many studies have evaluated the acute toxicity of organotins in a range of aquatic organisms, few have addressed the effects of chronic low level exposure, and even less studies have focused on full-life cycle exposures. Here, we aim at reviewing the biological effects of organotins in marine animals, focusing in particular on environmentally relevant low level effects.

Mollusks gastropods

As early as the 1980s, TBT was known to be highly toxic to some marine organisms (Smith 1981a,b), mollusks being the most sensitive group. Many studies, dealing with the toxic effects of TBT on sexual development and reproductive functions, have been focused on gastropods (Santos et al. 2012). The development of male sexual characteristics, notably a penis and a vas deferens, in gastropods females a phenomenon known as imposex, is the best-documented example (Gibbs and Bryan 1986).

Imposex is an irreversible condition in most species of gastropods suggesting that this sexual abnormal development can have potential long-term impacts on the fitness of gastropod populations (Foale 1993; Stroben et al. 1992). Severe stages of imposex at higher TBT concentrations can lead to female sterilization and death (Mensink et al. 1996). Indeed, some gastropod populations with a 100% incidence of imposex have disappeared in several locations in the UK (Gibbs and Bryan 1986), the Cork Harbour, Ireland (Minchin et al. 1996), among others.

TBT bioaccumulates in gastropods at levels up to 100,000-fold greater than those measured in the aqueous environment (Bryan et al. 1987; Bryan et al. 1989). Available data seems to indicate that gastropods biotransform TBT by the cytochrome P450 enzymes (Fent 1996). However, P450-mediated detoxification/elimination reactions appear to be limited in mollusks (Livingstone et al. 1989; Solé and Livingstone 2005) resulting in incomplete biotransformation of TBT and increased accumulation (Sternberg et al. 2010). This tendency to accumulate TBT may contribute to the high sensitivity of gastropods to TBT toxicity (Sternberg et al. 2010). Gibbs and Bryan (1986) demonstrated experimentally that imposex in the dogwhelk *N. lapillus* was induced by exposure to TBT concentrations as low as 0.5 ng/l. Due to their high sensitivity to very low concentrations of TBT in marine ecosystems, several snail species have been used worldwide as bioindicators of TBT contamination (Gibbs et al. 1987; Stroben et al. 1995; Oehlmann et al. 1996; Swennen et al. 1997; Sole et al. 1998; Santos et al. 2000; Sousa et al. 2009; Sternberg et al. 2010; Abidli et al. 2012) and the imposex phenomenon is the most used method to evaluate the degree of TBT pollution. Imposex is recognized as a truly global phenomenon, most extensively studied in *N. Lapillus*, but effects of TBT have been observed in over 195 species of gastropods worldwide (Sternberg et al. 2010). Most of the Neogastropod species belong to the families Muricidae, Buccinidae, Conidae, and Nassariidae (Horiguchi 2008).

Table 1: Overview of the main effects of organotins in aquatic animals at environmentally relevant concentrations.

Organisms		Field studies	Laboratory studies
Gastropods		- Imposex - Female sterilization - Male abnormal penis development - Population decline and disappearance of the species sensitive living close to TBT sources - Evidences of genotoxic damage - Evidences of altered steroid metabolism	- Imposex - Female sterilization - Male abnormal penis development - Oogenesis suppressed and supplanted by spermatogenesis - Evidences of changes on lipid metabolism/steroid
Bivalves	Oyster	- Calcification anomalies - Reproductive failure - Strong decline of populations	- Larval growth disturbed - Embryogenesis disturbed
	Mussel	-	- DNA damage
	Clam	Evidences of population decline and overall impoverishment of macrofauna community	- Intersex - Reduced growth and filtration rates - Sex ratio changes - Gill abnormalities
Crustaceans	Amphipods	-	- Reduction of reproductive success - Altered sex-ratio toward females - Delayed growth rate and molting - Maturation inhibition - Decreased oocyte viability - Decreased number of juveniles hatched - Increase of microsporidian parasites in females
	Copepods	-	- Development delay - Reduction of population growth - Altered sex-ratio toward males - Female gonads reduced - Male reproductive system stimulated
Annelids		Evidences of population decline and overall impoverishment of macrofauna community	- Changes on embryo development - Decreased growth - Changes on gonad development - Genotoxic and cytotoxic damage

Table 1 contd....

Table 1 contd...

Organisms	Field studies	Laboratory studies
Fish	Evidences of disruption of embryonic development	- Sex reversal - Altered sex ratio toward males/females - Female and male gonad development disturbed - Delayed oogenesis - Changes on steroid levels - changes on reproductive related signaling pathways - Increased the ratio of testosterone to 17β-estradiol - Increase of apoptotic ovarian follicular cells - Histological damage of testis - Decline in total sperm counts - Decreased sperm motility - Lack of sperm flagella - Decreased reproduction - Decreased reproductive frequency of mating groups - Reduced hatchability - Embryo malformations - Low fertility rate - Spawn fail - Decrease gonotosomatic index - Decrease sexual exhibitions, such as dancing - Alterations of swimming behavior - Swim up failure in the next generation - Alterations of neuroendocrine and thyroidal status - Osmoregulation perturbations - Evidences of changes on lipid metabolism
Marine Mammals		- Alterations of the immune system

Species-specific sensitivity for the development of imposex has been observed in gastropods. Whereas TBT induces imposex in some species at rather low levels, inhibiting breeding and resulting in population declines or even local extinction; others are less susceptible to TBT, showing effects only in the most polluted areas and in some cases with little interference in reproduction (Fent 1996; Mensink et al. 2002). Interestingly, some gastropod species do not develop imposex even in highly TBT contaminated areas (Matthiessen and Gibbs 1998; Mensink et al. 2002; Swennen et al. 2009).An association between imposex severity and DNA damage (micronucleus) was also established in field *N. lapillus* populations (Hagger et al. 2006). Imposex, however, is not the only example of the adverse effects of TBT. Some gonochoristic gastropods, such as the periwinkle *Littorina littorea* (Bauer et al. 1995), develop a similar disruption, termed intersex, which consists in the transformation of female pallial organs into male morphological structures. Intersex in the periwinkle has been used in several regions, alongside with imposex, to monitor TBT contamination, particularly in estuarine areas (Minchin et al. 1996, 1997; Oehlmann et al. 1996; Bauer et al. 1997).

TBT-induced imposex in gastropods has been regarded as one of the clearest examples of endocrine disruption caused by an environmental contaminant. Several endocrine-related hypotheses have been proposed to define the possible mechanism(s) by which TBT causes imposex. However, the lack of the complete understanding of the endocrinology in gastropods hampers a full clarification of the target pathways.

Bivalves

The deleterious effects of TBT in bivalves were first documented in the Arcachon Bay, France, in the late 1970s. The peculiar sensitivity of bivalves from this bay to TBT was due to the fact that it is surrounded by marinas and numerous seasonal moorings, representing, at the time, a total accommodation capacity of over 7500 pleasure boats (Alzieu 2000). From 1975 to 1982, the production of the pacific oysters (*Crassostrea gigas*) was severely affected by a complete lack of reproduction and the appearance of calcification anomalies, including shell thickening and formation protein-containing jelly, which were responsible for a strong decline of the oyster populations (Alzieu 1991). Shell thickening was also observed in additional regions including UK and others, but the sensitivity varies greatly among oyster species (Kefi et al. 2011). Subsequent investigations have shown that alteration of calcification processes occurs even at TBT concentrations as low as 2 ng/l, but the mechanism of action remains to be fully clarified (His and Robert 1985; Santos et al. 2012).

Laboratory and field experiments revealed the extreme toxicity of TBT towards the embryogenesis and larval development of oysters (*C. gigas*) (His and Robert 1985). Sublethal effects at histological level showed that TBT induces histological modifications on the cells of digestive gland, when oysters (*C. gigas*) were exposed to TBT concentrations as low as 2 ng/l. Therefore, at the time, safe TBT levels for mariculture waters were suggested to be lower than 2 ng/l (Chagot et al. 1990).

Most of the available studies regarding the chronic toxicity of TBT on bivalves focus on the pacific oyster *C. gigas*. Investigation on other bivalve species is particularly limited. However, following the serious declines in the English native oyster (*Ostrea edulis*) fishery studies were performed to investigate the interference of TBT with the reproduction of this bivalve mollusk. Thain and Waldock (1986) showed that *O. edulis* exposed to TBT (240 ng/L for 74 d) developed either as males or were undifferentiated. Furthermore, field data from TBT-contaminated populations of *O. edulis* showed that only 5 to 6% of individuals were producing larvae. However, although this effect of TBT was probably an example of endocrine disruption, it was not observed in all investigated oyster species (Matthiessen and Gibbs 1998). Additionally, Hagger et al. (2005) found that TBT is genotoxic to *Mytilus edulis*. A laboratory exposure to environmentally realistic concentrations of TBT, for seven days, resulted in a statistically significant increase in DNA damage, detected using the comet and micronucleus assays, at TBT concentrations as low as 0.5 µg/l. More recently, Park et al. (2012) reported an alteration of several important biological responses of the clam *Gomphina veneriformis* after a chronic laboratory exposure to TBT concentrations between 0.2 and 20 ug/l. Reduced growth and filtration rates, changes in the GI and sex ratio, induction of intersex gonads, and gill abnormalities were observed in

G. veneriformis clams after TBT exposure. Supporting these findings, in heavily TBT-contaminated sites, clams populations of several species declined during the 1980s before the enforcement of the partial ban on the use of TBT in antifouling paints (Ruiz et al. 1994; Langston et al. 1990, 1994, 2015). Taken together, these results indicate that bivalves are particularly sensitive to low levels of TBT.

In mollusks, other effects in addition to imposex and shell thickening have been attributed to TBT exposure: reduced rates of fertilization and development (Fent 1996); alterations in steroid and lipid metabolism (Santos et al. 2012; Janer et al. 2007; Lyssimachou et al. 2009), interference with the immune system (Gopalakrishnan et al. 2011), and changes on pheromone production (Straw and Rittscof 2004).

Annelids

Evidence derived from laboratory tests indicate that annelids are also highly sensitive to TBT at environmentally relevant concentrations. Exposure of juveniles of the opheliid polychaete, *Armandia brevis* to TBT through sediments indicates an impact in growths at 191 ng/g sediment dry wt., after 21 days exposure (Meador and Rice 2001). TBT was also shown to be genotoxic, cytotoxic and affects the ontogenetic (embryo-larval) development of the polychaeta *Platynereis dumerilii* (Hagger et al. 2002). Similarly, chronic exposure to TBT affected several ecologically relevant endpoints of the marine polychaete *Hydroides elegans*; survivorship, percentage of settlement and time to reach settlement were significantly reduced at 10 ng TBT/L. Only *H. elegans* exposed to TBT levels below 100 ng/L reached maturity and egg production, success in fertilization and egg development were all reduced at TBT levels above 100 ng/L. Taken together, the data gathered from several different polychaeta species indicate that this group is highly sensitive to TBT at environmentally relevant concentrations and may explain the overall impoverishment of macrofauna community in areas from TBT contaminated sediments. However, the molecular and biochemical mechanisms involved have yet to be investigated, although our own unpublished data indicates it could be associated with the modulation of the Retinoid X receptor.

Crustaceans

In crustaceans, a few works addressing the effects of organotins at environmentally relevant levels have been performed mainly in amphipods and copepods and the majority of the studies focused on TBT (Lewis and Ford 2012). Aono and Takeuchi (2008) reported a reduction in reproductive success of *Caprella danilevskii* (Crustacea: Amphipoda) exposed to TBT (1.1 and 10.7 ng/l) during 49 days. Ohji et al. (2002) examined the biological effects of TBT exposure (0, 10, 100, 1000, and 10,000 ng/l) on the same amphipod species (*Caprella danilevskii*) during the embryonic stage (5d) and observed that sex ratios changed dramatically in the hatched juvenile with increases in TBT concentrations; i.e., the proportion of females was found to increase to 55.6% at 10 ng/l, 85.7% at 100 ng/l, and 81.8% at 1000 ng/l. One year later, the same authors published a study where *C. danilevskii* were exposed to TBT (0, 10, 100, 1000, and 10,000 ng/l) over a generation after hatching. They observed marked delays in growth and molting in both 100 and 1000 ng/l TBT concentrations. Also, inhibition of maturation and reproduction, such as delaying in the time to reach maturity and a decrease in the number of juveniles hatched was apparent in 10 and 100 ng/l TBT concentrations. Furthermore, brood loss, and failure in egg formation and hatching were observed in 1000 and 10000 ng/l TBT concentrations. No significant changes in sex ratio were observed at any of the TBT exposure levels (Ohji et al. 2002).

In a recent study, Jacobson et al. 2011 exposed the benthic amphipod *Monoporeia affinis* to TBT, for five weeks in static microcosms with natural sediment (70 and 170 ng/g sediment d wt) during gonad maturation. TBT exposure resulted in a significant adverse effect on oocyte viability and an increase of the prevalence of microsporidian parasites in females. Significant effects were also observed in male sexual maturation in the 70 ng TBT/g d wt exposure and on ecdysteroid levels in the 170 ng/g sediment d wt exposure.

In comparison to TBT, very few studies have examined the effects of TPT in crustaceans. Yi et al. (2014) investigated the individual and population responses of the marine copepod, *Tigriopus japonicus* in a waterborne exposure to TPT (0.1, 0.5, and 1 ug/l) during a two generation assay. *T. japonicus* exposed to 1.0 ug/l TPT exhibited a delay in development and a significant reduction of population growth. At 0.1 ug/l TPT or above, the sex ratio of the second generation was significantly changed to a male-biased population. Watermann et al. (2013) exposed the marine copepod *Acartia tonsa* to TPT (1,4–8,8 ng/l) for 21 days and histologically observed that TPT stimulated the male reproductive system at 1,4 and 3,5 ng/l TPT, whereas it inhibited the female gonad at 8,8 ng/l TPT.

Fish

The dramatic impact of TBT in gastropods at the low ng/L ranged prompted full life-cycle studies with fish. In one of the early chronic studies, Shimasaki et al. (2003) reported the masculinization of genetically female Japanese flounder (*Paralichthys olivaceus*). Dietary exposure to TBT at concentrations of 0.1 and 1 μg of TBT/g during the sex differentiation period induced the sex reversal of female flounders into phenotypic males. Similarly, McAllister and Kime (2003) reported that zebrafish exposed to a concentration range from 0.1 to 100 ng/l of TBT, from hatching to 70 days, resulted in altered sex ratio toward males. Santos et al. (2006) also reported a masculinizing effect of dietary TBT exposure at 25 and 100 ng/g on zebrafish from embryos to day 120. This effect in sex ratio could be concentration related as a recent study by Lima et al. (2015) reported a feminizing effect of TBT in zebrafish after a full-life cycle dietary exposure to 1 μg of TBT/g of diet. Hence, evidence indicates that the impact of TBT on zebrafish sex ratio could follow a non-monotonic distribution. In contrast to zebrafish and Japanese flounder, TBT failed to affect sex ratio of other fish species such as the japanese medaka, following full-life exposure (Kuhl and Brouwer 2006). The reasons for these species-specific differences are as yet unknown. Philip et al. (2012) hypothesized it could be related with genetic differences namely in the portfolio of nuclear receptor RXR. It has been hypothesized that the disruption of fish sex ratio by TBT could be related with the modulation of P450 aromatase (at the transcription or enzymatic level). Indeed, aromatase inhibition is associated with increased levels of testosterone in mammals (O'Donnell et al. 2001) and development of testes in female chicken (Elbrecht and Smith 1992) and turtles (Richard-Mercier et al. 1995). Aromatase inhibition in fish was also related with the transformation of female fish into males (Uchida et al. 2004). However, the precise mechanism(s) of action of TBT in fish sex differentiation and reproduction have yet to be fully clarified.

Following the early studies of TBT in fish sex differentiation, a significant amount of works has been performed to evaluate the effects in additional endpoints. Several studies addressed the toxic effects of TBT in fish embryogenesis, most pointing to an impact at environmentally relevant levels, i.e., in the range of 1 and 50 ng/L (Zhang et al. 2011). Organotins also impact female and male gonad development at an environmentally relevant concentration. In fact, concentrations ranging from 1 to 100 ng/L affected the development of ovary in female cuvier (*Sebastiscus marmoratus*) and elevated testosterone levels, increased the ratio of testosterone to 17β-estradiol, delayed oogenesis, and a significant increase of apoptotic ovarian follicular cells was seen (Zhang et al. 2007). Similarly, Mochda et al. (2007) reported serious histological damage to the testis, including reduction in counts of spermatids and spermatozoa and malformation of somatic cells around the seminal duct in the marine fish *Fundulus heteroclitus* exposed to 5.8 μg/L TBT for 2 or 4 weeks. Reduced sperm counts in guppies (*Poecilia reticulata*) were also observed by Haubruge et al. (2000) following TBT exposure. These authors suggested that the damage of normal Sertoli cell function, which facilitates the transport of maturing sperm into the testicular deferent duct, could be the cause of sperm damage. McAllister and Kime (2003) examined sperm mobility in adults zebrafish exposed for 70 days to environmentally relevant levels of TBT and observed that the group exposed to 1 ng/l, the motility of sperm was significantly lower than that of control fish, while at 10 and 100 ng/l, all sperm lacked flagella. More recently, the effects of chronic TBT exposure on testicular development were investigated in the Asian marine fish, *Sebastiscus marmoratus* (Zhang et al. 2009). The exposure to environmentally relevant concentrations of TBT during 48 days produced a decrease in the gonotosomatic index, spermatogenesis suppression, an elevation of testosterone levels, and suppression of estradiol levels. These effects were accompanied by alterations in RXRs, PPARs, and estrogen receptor (ER) mRNA transcription levels in testes (Zhang 2009). Full-life cycle studies confirm that TBT, at environmentally relevant levels, decreases reproduction in at least two fish species (zebrafish and sheepshead minnow) (Manning et al. 1999; Lima et al. 2015).

In addition to the toxic effect of TBT on reproduction, several studies also addressed the effects of TBT on sexual behavior. Sexual exhibitions, such as dancing, decreased in TBT-exposed male Japanese medaka, *Oryzias latipes* (1 µg/g body weight for 3 weeks) (Nakayama et al. 2004). These treated fish also had low fertility rate and failed to spawn. TBT also reduced the reproductive frequency of mating groups of medaka, decreased hatchability and promoted swim up failure in the next generation (Nirmala et al. 1999). Given that preferential accumulation of TBT occurs in the nervous system, including the brain (Rouleau et al. 2003; Mochida and Fujii 2008), it is likely that the brain control in spawning behavior is damaged (Mochida and Fujii 2008). Alteration of swimming behavior from TBT exposure was also reported in rainbow trout (*Oncorhynchus mykiss*) (Triebskorn et al. 1994) and carp (*Cyprinus carpio*) (Schmidt et al. 2004, 2005). Syngnathidae larvae exposed to TBT single and in combination with EE2 also showed change in behavior (Sária et al. 2011).

Other adverse outcomes have been induced by TBT under laboratory conditions such as perturbation of brain and neuroendocrine status (Zang et al. 2008; Chen et al. 2011; Yu et al. 2013), alteration in thyroidal status (Zang et al. 2009), thymus atrophy (Grinwis et al. 2009), perturbation of osmoregulation (Pinkney et al. 1989; Hartl et al. 2001), and both TBT and TPT have been shown to be genotoxic at environmentally relevant levels (Micael et al. 2007).

More recently, organotins were shown to work as potent obesogens in mammalian animal models. Although data on fish is very limited (Santos et al. 2012), Meador et al. (2011) has recently reported an impact of TBT in the juvenile trout weight, and immature males and females of rockfish exposed to 100 ng/L TBT (as Sn) for 48 days showed increased total lipids and lipid droplets in the testis (Zang et al. 2009) or ovary (Zang et al. 2013). Lyssimachou et al. (2015) observed a massive fat accumulation and disruption of lipid homeostasis following chronic low level exposures (10 ng TBT Sn/L, nominal levels). This indicates a new unidentified target of organotins in fish and perhaps additional metazoans *taxa*.

In contrast to gastropods, where many detailed studies have clearly indicated the impact of TBT and TPT under field conditions, including local species extinction, there is an almost complete paucity of data addressing the effects of organotins in fish under field conditions. Given the array of negative impacts demonstrated under laboratory conditions over a range of environmentally relevant TBT/TPT concentrations, it is very likely that most sensitive species from moderate to polluted sites experienced fitness decrease. This hypothesis is fully supported by an elegant study performed with the Chinese sturgeon, *Acipenser sinensis*. In the field, a large proportion of embryomal formations were observed and directly correlated with TPT levels. In the laboratory, injection of the same range of TPT levels recapitulated the effects observed under field conditions thus supporting the role of TPT in impairment of embryo development (Hu et al. 2009).

Marine mammals

Before the global ban on the use of TBT as antifoulant was enforced, marine mammals found stranded in many coastal areas around the globe have been shown to be contaminated with high levels of butyltin compounds, suggesting that these compounds could be involve in decreased immunocompetence (Kannan et al. 1996). Despite these observations, potential effects of organotin exposure on marine mammals have been little investigated, most focusing on the immune system. Frouin et al. (2008) assessed the effects of tributyltin (TBT) and its dealkylated metabolites dibutyltin (DBT) and monobutyltin (MBT) on the immune responses of harbour seals (*Phoca vitulina*). Peripheral blood mononuclear cells isolated from pup and adult harbour seals were exposed *in vitro* to varying concentrations of organotins. DBT resulted in a significant decrease at 100 and 200 nM of phagocytotic activity and reduced significantly phagocytic efficiency at 200 nM in adult seals. No effects in phagocytosis were observed under TBT and MBT exposure. In pups, the highest concentration (200 nM) of DBT inhibited phagocytic efficiency. The immune functions were more affected by organotins exposure in adults than in pups.

Although earlier studies indicated a low toxicity of TBT in mammalian cells (Santillo 2001), subsequent evidence indicates that mammals are also sensitive to TBT exposure. Some of the described effects are similar to those observed in fish: interference with steroid metabolism (Janer 2005), decreased estradiol levels and impact in ERα and ERβ expression (Chen et al. 2008), and immunotoxicity (Chen et al. 2011). Moreover, recent data suggested the involvement of organotins in the development of obesity (Grün et al. 2006; Santos et al. 2012). Indeed, TBT has recently been considered an "environmental obsesogen" (Grün and Blumberg 2009) as it causes the induction of adipogenesis in both cell culture models and *in vivo*, increases adipose mass in frogs and mice (Grün et al. 2006), induces the differentiation of preadipocyte 3T3-L1 cells into adipocytes and expression of adipogenic gene markers (Kanayama et al. 2005; Li et al. 2011), increases triglyceride storage (Li et al. 2011), and ectopic fat cell production (Grün et al. 2006). Since effects in mammalian models are recorded at environmentally relevant concentrations, some authors argue that exposure to TBT could likely increase the adipose mass over time (Kirchner et al. 2010), and hence affect marine mammals and humans.

While the risk for humans is yet to be properly assessed, as few data exists for most countries, TBT's extremely high toxicity, worldwide contamination, and broad spectrum effects towards numerous species turn this compound into one of the most dangerous ever produced.

The possibility that TBT and its metabolites can accumulate throughout the food chain has been largely underestimated (Santos et al. 2009), although some authors detected its presence in top predators, such as bottlenose dolphins (Kannan et al. 1996; Kannan and Falandysz 1997) and sea otters (Kannan et al. 1998) and in marine organisms used for human consumption (Santos et al. 2009).

Other groups of organisms

Information regarding chronic low levels effects in the remaining *taxa* is particularly limited, and therefore will not be revised here.

Mechanism(s) of Action of Organotins

The mechanism(s) by which TBT disrupts several biological processes in most metazoans is still not fully understood. Historically, several hypotheses have been suggested to explain the effects of TBT in gastropods. The original work of Féral and Le Gall (1983) indicated that the disruption of the synthesis of neuroendocrine factors that are under the control of the pedal and cerebropleural ganglia were responsible by imposex induction. Oberdorster and McClellan-Green (2000) proposed that the neuropeptide APGW amide was one of the neuroendocrine factors involved in imposex induction in *Ilyanassa obsoleta*, although follow up studies failed to validate this hypothesis in other gastropod species (Santos et al. 2006; Castro et al. 2007). The disruption of steroid signaling and physiological balance has also been proposed as a potential driver for imposex development. Some studies indicate that the androgenic effects of TBT appear to be caused by interference with steroid biosynthesis. Elevation of testosterone and/or testosterone/estradiol ratio has been reported in snails exposed to TBT in the laboratory (Spooner et al. 1991; Schulte-Oehlmann et al. 1995; Bettin et al. 1996; Santos et al. 2005) and clams (Morcillo et al. 1998). This imbalance would be caused by TBT inhibition of aromatase (CYP19), an enzyme responsible for the aromatization of androgens to estrogens. Other hypotheses have been put forward to explain the steroid imbalance induced by TBT, these included decrease in the amount of testosterone sulphur-conjugates, though only observed at rather high TBT concentrations (Ronis and Mason 1996), and a decrease in the esterification of testosterone with fatty acids (Gooding et al. 2003; Santos et al. 2005; Abidli et al. 2013). However, a CYP19 ortholog or a functional equivalent has not been reported to date outside chordates; neither have orthologs of the vertebrate specific androgen receptor, which questions the physiological relevance of TBT impact over steroid pathways in gastropods.

In an attempt to identify the affinity of several EDCs towards human nuclear receptors (NRs) (Nishikawa et al. 1999; Nishikawa et al. 2004; Kanayama et al. 2005; Fig. 2), TBT was found to be a high affinity ligand for the retinoid X receptor (RXR) and the Peroxisome proliferator-activated receptor (PPAR). NRs form an important super-family of ligand-dependent and independent transcription factors that regulate numerous biological processes in vertebrates, including morphogenesis, differentiation, metabolism, and reproduction (Castro and Santos 2014). RXR seems to be particularly important because it serves and as a heterodimeric partner of many other NRs (Philips et al. 2012). Hence, based on these early findings, Nishikawa and co-workers (2004) and Castro et al. (2007) showed that 9-cis retinoic acid, the putative natural ligand of RXR, induced imposex at rather low concentrations. The same findings have been reported under injections of RXR synthetic agonists which further support the role of RXR in imposex induction. More recently, it has been hypothesized that the heterodimer PPAR/RXR could be directly involved in imposex induction (Santos et al. 2012; Pascoal et al. 2013). However, PPAR has not been functionally characterized in Lophotrochozoans (Santos et al. 2012; Castro and Santos 2014), and therefore our understanding is limited. In contrast, today it is well established that TBT-induced imposex operates through the modulation of RXR, although the full cascade of events leading to imposex induction still remains elusive (Lima et al. 2011; Gesto et al. 2013).

A limited number of studies addressed the mode of action of TBT in fish. Zhang et al. (2009) observed that TBT inhibits testes development concomitantly with changes in testicular RXR, PPARγ and ER expression in the rockfish *Sebastiscus marmoratus*. McGinnis and Crivello (2011) screened the transcription of genes typically associated with female and male sex differentiation and concluded that exposure to TBT in adulthood induced an overall masculinizing effect tempered by feminizing effects in both gonads and brain (McGinnis and Crivello 2011). More recently, Lima et al. (2015) showed that TBT significantly decreased fecundity at environmentally relevant levels and, in contrast to previous reports, altered zebrafish sex ratio towards females. These negative effects were observed concomitantly with the down regulation of brain mRNA levels of aromatase b (CYP19a1b) in females and of PPARγ in both males and females, suggesting an involvement of these pathways, i.e., CYP 19 and PPARγ/RXR in reproductive impairment associated with TBT. Altogether, in fish, these reports suggest that a complex mechanism(s) is involved in the reproductive toxicity induced by TBT and emphasize the need to perform more detailed studies.

RXR is an ancient member of the nuclear receptor superfamily, being evolutionary widely distributed (Fig. 2) (Bertrand et al. 2004; Bridgham et al. 2010). It has been reported or predicted in most Bilaterian lineages and it has been isolated in some Cnidarians and Placozoans. RXR sequences are highly conserved throughout evolution with approximately 90% of sequence similarity between mollusks and vertebrates,

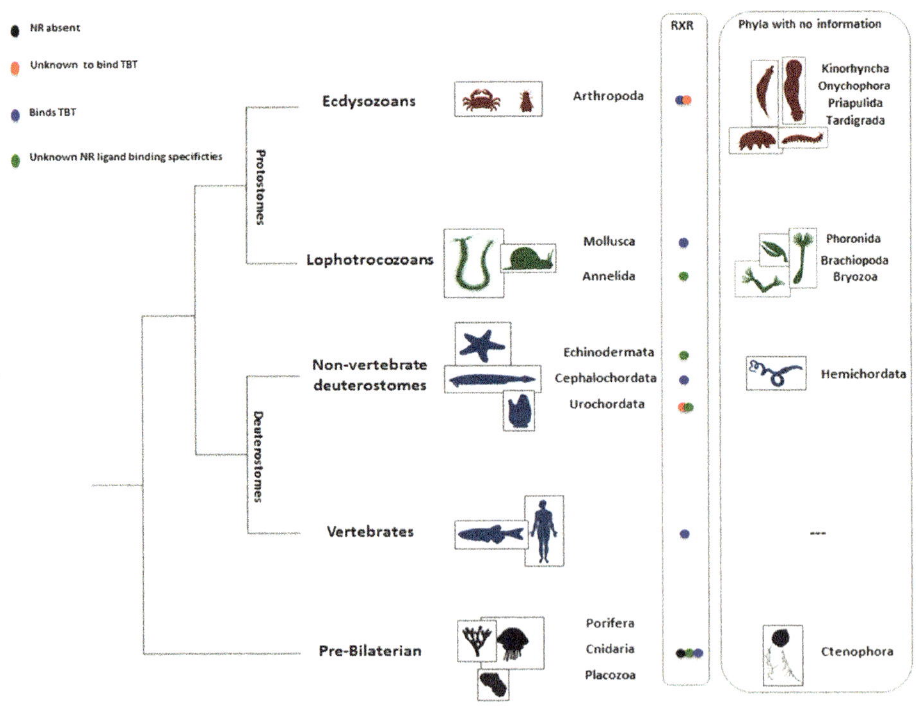

Fig. 2: Phylogenetic relationships between major extant Metazoan lineages. The capacity of TBT to bind/transactivate expression of RXR is identified. Examples of some unsampled phyla are also depicted and should also be included in a priority focused research approach towards understanding the full impact of EDCs. Modified from Castro and Santos (2014).

which have diverged over 600 million years ago. Importantly, available data points to a mostly conserved binding profile to either natural ligands and/or xenobiotics (e.g., TBT and TPT). Therefore, given the degree of RXR conservation in different *taxa*, the entire Metazoan ecosystems are likely to be impacted by this group of chemicals. This might explain the extremely high toxicity of TBT and TPT to most metazoan species study to date.

Final Remarks

Although organotins have been in use for over 50 years, our understanding of the full spectrum of affected marine *taxa* is still fragmentary. This is mostly related with the fact that for many metazoans, chronic data is absent, and no mechanistic studies have been carried out. Given the high conservation of RXR, it is likely that many *taxa* have been affected by organotins at environmentally relevant levels (André et al. 2014). This might explain many of the observed negative impacts such as the low macrofauna diversity reported in TBT-moderately contaminated sediments. The prohibition on the use of organotins as antifoulants for ships hulls has resulted in a clear decrease trend of organotins in different compartments in several rocky-shores and, though to a lesser extent, in estuarine sediments, reaching in several areas the Environmental Quality Standards (EQS) for water quality (2 ng TBT/L) and other established criteria for sediments and biota. This has resulted in a sharp decrease on imposex severity and recovery of macrofauna community (Smith et al. 2008; Santos et al. 2009; Sousa et al. 2009; Langstone et al. 2015). However, given the half-times in anoxic sediments, organotins are likely to persist for many years in sites with a high TBT legacy, with re-suspension occurring during extreme climate events or dredging (Santos et al. 2004), and therefore impacting the most sensitive *taxa*.

Acknowledgements

T. Neuparth was supported by Postdoctoral fellowships from the Portuguese Science and Technology Foundation (FCT), refs. SFRH/BPD/77912/2011. This research was supported by the FCT project PTDC/MAR/115199/2009 and by the European Regional Development Fund (ERDF) through the COMPETE—Operational Competitiveness Programme and national funds through FCT—Foundation for Science and Technology, under the project "PEst-C/MAR/LA0015/2013".

Keywords: Organotins, antifouling, low level effects, marine animals, risk assessment, regulation

References

Abidli, S., M.M. Santos, Y. Lahbiba, L.F.C. Castro, M.A. Reis-Henriques and N.T. El Menif. 2012. Tributyltin (TBT) effects on *Hexaplex trunculus* and *Bolinus brandaris* (Gastropoda: Muricidae): imposex induction and sex hormone levels insights. Ecol. Ind. 13: 13–21.

Abidli, S., L.F. Castro, Y. Lahbib, M.A. Reis-Henriques, T.N. El Menif and M.M. Santos. 2013. Imposex development in *Hexaplex trunculus* (Gastropoda: Caenogastropoda) involves changes in the transcription levels of the retinoid X receptor (RXR). Chemosphere 93: 1161–1167.

Alzieu, C.L., J. Sanjuan, J.P. Deltriel and M. Borel. 1986. Tin contamination in Arcachon Bay: Effects on oyster shell anomalies. Mar. Poll. Bull. 17: 494–498.

Alzieu, C. 1991. Environmental problems caused by TBT in France: assessment, regulation, prospects. Mar. Environ. Res. 3: 121–130.

Alzieu, C. 2000. Impact of tributyltin on marine invertebrates. Ecotoxicology 9: 71–76.

André, A., R. Ruivo, M. Gesto, L.F.C. Castro and M.M. Santos. 2014. Retinoid metabolism in invertebrates: When evolution meets endocrine disruption. Gen. Comp. Endocrinol. 208: 134–145.

Antizar-Ladislao, B. 2008. Environmental levels, toxicity and human exposure to tributyltin (TBT)-contaminated marine environment. A review. Environ. Int. 34: 292–308.

Aono, A. and I. Takeuchi. 2008. Effects of tributyltin at concentrations below ambient levels in seawater on *Caprella danilevskii* (Crustacea: Amphipoda: Caprellidae). Mar. Pollut. Bull. 57: 515–23.

Barroso, C.M. and M.H. Moreira. 2002. Spatial and temporal changes of TBT pollution along the Portuguese coast: inefficacy of the EEC directive 89/677. Mar. Pollut. Bull. 44: 480–486.

Barroso, C.M., M. Rato, A. Veríssimo, A. Sousa, J.A. Santos, S. Coelho, M.B. Gaspar, F. Maia and S. Galante-Oliveira. 2011. Combined use of *Nassarius reticulatus* imposex and statolith age determination for tracking temporal evolution of TBT pollution in the NW Portuguese continental shelf. J. Environ. Monit. 13: 3018–3025.

Bauer, B., P. Fioroni, I. Ide, S. Liebe, J. Oehlmann, E. Stroben and B. Watermann. 1995. TBT effects on the female genital system of *Littorina littorea*, a possible indicator of tributyltin pollution. Hydrobiologia 309: 15–27.

Bauer, B., P. Fioroni, U. Schulte-Oehlmann, J. Oehlmann and W. Kalbfus. 1997. The use of *Littorina littorea* for tributyltin (TBT) effect monitoring-results from the German TBT survey 1994/1995 and laboratory experiments. Environ. Pollut. 96: 299–309.

Bertrand, S., F.G. Brunet, H. Escriva, G. Parmentier, V. Laudet and M. Robinson-Rechavi. 2004. Evolutionary genomics of nuclear receptors: from twenty-five ancestral genes to derived endocrine systems. Mol. Biol. Evol. 21: 1923–1937.

Bettin, C., J. Oehlmann and E. Stroben. 1996. TBT-induced imposex in marine neogastropods is mediated by an increasing androgen level. Helgol. Mar. Res. 50: 299–317.

Blaber, S.J.M. 1970. The occurrence of a penis-like outgrowth being the right tentacle in spent females of *Nucella lapillus* (L.). Proc. Malacol. Soc. London 39: 231–233.

Bouton, D., H. Escriva, R.L. de Mendonca, C. Glineur, B. Bertin, C. Noel, M. Robinson-Rechavi, A. de Groot, J. Cornette, V. Laudet and R.J.A. Pierce. 2005. Conserved retinoid X receptor (RXR) from the mollusc *Biomphalaria glabrata* transactivates transcription in the presence of retinoids. J. Mol. Endocrinol. 34: 567–582.

Brande-Lavridsen, N., B. Korsgaard, I. Dahllöf, J. Strand, Z. Tairova and P. Bjerregaard. 2013. Abnormalities in eelpout *Zoarces viviparus* upon chemical exposure. Mar. Environ. Res. 92: 87–94.

Bridgham, J.T., G.N. Eick, C. Larroux, K. Deshpande, M.J. Harms, M.E. Gauthier, E.A. Ortlund, B.M. Degnan and J.W. Thornton. 2010. Protein evolution by molecular tinkering: diversification of the nuclear receptor superfamily from a ligand-dependent ancestor. PLoS Biol. 8: 1–13.

Castro, I.B. and G. Fillmann. 2012. High tributyltin and imposex levels in the commercial muricid *Thais chocolata* from two Peruvian harbor areas. Environ. Toxicol. Chem. 31: 955–60.

Castro, L.F.C., D. Lima and A. Machado. 2007. Imposex induction is mediated through the retinoid X receptor signalling pathway in the neogastropod *Nucella lapillus*. Aquat. Toxicol. 85: 57–66.

Castro, L.F.C. and M.M. Santos. 2014. To bind or not to bind: the taxonomic scope of nuclear receptor mediated endocrine disruption in invertebrate phyla. Environ. Sci. Technol. 48: 5361–5363.

Chagot, D., C. Alzieu, J. Sanjuan and H. Grizel. 1990. Sublethal and histopathological effects of trace levels of tributyltin fluoride on adult oysters *Crassotrea gigas*. Aquat. Living Resour. 3: 121–130.

Chen, J., C. Huang, L. Zheng, M. Simonich, C. Bai, R. Tanguay and Q. Dong. 2011. Trimethyltin chloride (TMT) neurobehavioral toxicity in embryonic zebrafish. Neurotoxicol. Teratol. 33: 721–726.

Chen, Q., Z. Zhang, R. Zhang, Y. Niu, X. Bian and Q. Zhang. 2011. Tributyltin chloride-induced immunotoxicity and thymocyte apoptosis are related to abnormal Fas expression. Int. J. Hyg. Envir. Heal. 214: 145–150.

Chen, Y., Z. Zuo, S. Chen, F. Yan, Y. Chen, Z. Yang and C. Wang. 2008. Reduction of spermatogenesis in nice after tributyltin administration. Toxicology 251: 21–27.

Coray, C. and S. Bard. 2007. Persistence of thributyltin-induced imposex in dogwhelks (*Nucella lapillus*) and intersex in periwinkles (*Littorina littorea*) in Atlantic Canada. Water Qual. Res. J. Can. 42: 111–122.

Dowson, P.H., J.M. Bubb and J.N. Lester. 1996. Persistence and degradation pathways of tributyltin in freshwater and estuarine sediments. Estuar. Coast. Shelf 42: 551–562.

Elbrecht, A. and R.C. Smith. 1992. Aromatase enzyme activity and sex determination in chickens. Science 24: 467–70.

Fent, K. 1996. Ecotoxicology of organotin compounds. Crit. Rev. Toxicol. 26: 3–117.

Féral, C. and S. Le Gall. 1983. The neuroendocrine mechanism responsible for penis differentiation in *Crepidula fornicata*. pp. 169–173. *In*: J. Lever and H.H. Boer [eds.]. Molluscan Neuroendocrinology. North Holland, Amsterdam.

Fioroni, P., J. Oehlmann and E. Stroben. 1991. The pseudohermaphoditism of prosobanchs; morfological aspects. Zoological. Anz. 286: 1–26.

Foale, S. 1993. An evaluation of the potential of gastropod imposex as a bioindicator of tributyltin pollution in Port Phillip Bay, Victoria. Mar. Pollut. Bull. 26: 546–552.

Frouin, H., M. Lebeuf, R. Saint-Louis, M. Hammill, E. Pelletier and M. Fournier. 2008. Toxic effects of tributyltin and its metabolites on harbour seal (*Phoca vitulina*) immune cells *in vitro*. Aquat. Toxicol. 21: 243–51.

Furdek, M., M. Vahcic, J. Šcancar, R. Milacic, G. Kniewald and N. Mikac. 2012. Organotin compounds in seawater and *Mytilus galloprovincialis* mussels along the Croatian Adriatic Coast. Mar. Pollut. Bull. 64: 189–199.

Gibbs, P.E. and G.W. Bryan. 1986. Reproductive failure in populations of the dog-whelk, *Nucella lapillus*, caused by imposex induced by tributyltin from antifouling paints. J. Mar. Biol. Assoc. UK 66: 767–777.

Gibbs, P.E., G.W. Bryan and P.L. Pascoe. 1987. The use of the dog-whelk, *Nucella lapillus*, as an indicator of tributyltin (TBT) contamination. J. Mar. Biol. Assoc. UK 67: 507–523.

Gipperth, L. 2009. The legal design of the international and European Union ban on tributyltin antifouling paint: direct and indirect effects. J. Environ. Manage. 90 Suppl. 1: S86–95.

Gooding, M.P., V.S. Wilson, L.C. Folmar, D.T. Marcovich and G.A. Leblanc. 2003. The biocide tributyltin reduces the accumulation of testosterone as fatty acid esters in the mud snail (*Ilyanassa obsoleta*). Environ. Health Persp. 111: 426–430.

Gómez-Ariza, J.L., M.M. Santos, E. Morales, I. Giráldez, D. Sánchez-Rodas, A. Velasco, N. Vieira, J.F. Kemp, J.P. Boon and C.C. Ten-Hallers-Tjabbes. 2006. Organotin contamination in open Atlantic Sea along the Iberian coast in relation to shipping. Chemosphere 64: 1100–1108.

Gopalakrishnan, S., W.-B. Huang, Q.W. Wang, M.L. Wu and J. Liu. 2011. Effects of tributyltin and benzo[a]pyrene on the immune-associated activities of hemocytes and recovery responses in the gastropod abalone, *Haliotis diversicolor*. Comp. Biochem. Phys. C 154: 120–128.

Grinwis, G.C., P.W. Wester and A.D. Vethaak. 2009. Histopathological effects of chronic aqueous exposure to bis(tri-n-butyltin) oxide (TBTO) to environmentally relevant concentrations reveal thymus atrophy in European flounder (*Platichthys flesus*). Environ. Pollut. 157: 2587–93.

Grün, F., H. Watanabe and Z. Zamanian. 2006. Endocrine-disrupting organotin compounds are potent inducers of adipogenesis in vertebrates. Mol. Endocrinol. 20: 2141–55.

Grün, F. and B. Blumberg, 2009. Minireview: the case for obesogens. Mol. Endocrinol. 23: 1127–34.

Hagger, J.A., A.S. Fisher, S.J. Hill, M.H. Depledge and A.N. Jha. 2002. Genotoxic, cytotoxic and ontogenetic effects of tri-n-butyltin on the marine worm, *Platynereis dumerilii* (Polychaeta: Nereidae). Aquat. Toxicol. 57: 243–255.

Hagger, J.A., M.H. Depledge and T.S. Galloway. 2005. Toxicity of tributyltin in the marine mollusc *Mytilus edulis*. Mar. Pollut. Bull. 51: 811–816.

Hagger, J.A., M.H. Depledge, J. Oehlmann, S. Jobling and T.S. Galloway. 2006. Is there a causal association between genotoxicity and the imposex effect? Environ. Health. Perspect. 114: 20–26.

Hall, L.W. Jr., M.J. Lenkevich, W.S. Hall, A.E. Pinkney and S.J. Bushong. 1987. Evaluation of butyltin compounds in Maryland waters of Chesapeake Bay. Mar. Pollut. Bull. 18: 78–83.

Hartl, M.G.J., S. Hutchinson and L.E. Hawkins. 2001. Organotin and osmoregulation: quantifying the effects of environmental concentrations of sediment-associated TBT and TPhT on the freshwater-adapted European flounder, *Platichthys flesus* (L.). J. Exp. Mar. Biol. Ecol. 256: 267–278.

Haubruge, E., F. Petit and M.J.G. Gage. 2000. Reduced sperm counts in guppies (*Poecilia reticulata*) following exposure to low levels of tributyltin and bisphenol A. Proc. R. Soc. Lond. B 267: 2333–2337.

His, Z. and R. Robert. 1985. Developpement des veligeres de *Crassotrea gigas* dans le bassin d'Arcachon. Etudes sur les mortalites larvaires. Re. Tra. Inst. Peches Marit. 47: 63–88.

Hoch, M. 2001. Organotin compounds in the environment: an overview. Appl. Geochem. 16: 719–743.

Horiguchi, T. 2009. Mechanism of imposex induced by organotins in gastropods. pp. 111–124. *In*: T. Arai et al. [eds.]. Ecotoxicology of Antifouling Biocides, Springer.

Hu, J., Z. Zhang, Q. Wei, H. Zhen, Y. Zhao, H. Peng, Y. Wan, J.P. Giesy, L. Li and B. Zhang. 2009. Malformations of the endangered Chinese sturgeon, *Acipenser sinensis*, and its causal agent. Proc. Natl. Acad. Sci. USA 106: 9339–9344.

Jacobson, T., B. Sundelin, G. Yang and A.T. Ford. 2011. Low dose TBT exposure decreases amphipod immunocompetence and reproductive fitness. Aquat. Toxicol. 17: 72–77.

Janer, G. 2005. Steroid levels, steroid metabolic pathways and their modulation by endocrine disruptors in invertebrates. PhD Thesis, Universitat Autonoma de Barcelona.

Janer, G., J.C. Navarro and C. Porte. 2007. Exposure to TBT increases accumulation of lipids and alters fatty acid homeostasis in the ramshorn snail *Marisa cornuarietis*. Comp. Biochem. Physiol. C Toxicol. Pharmacol. 146: 368–74.

Kanayama, T., N. Kobayashi and S. Mamiya. 2005. Organotin compounds promote adipocyte differentiation as agonists of the peroxisome proliferator-activated receptor gamma/retinoid X receptor pathway. Mol. Pharmacol. 67: 766–74.

Kannan, K., S. Corsolini, S. Focardi, S. Tanabe and R. Tatsukawa. 1996. Accumulation pattern of butyltin compounds in dolphin, tuna, and shark collected from Italian coastal waters. Arch. Environ. Contam. Toxicol. 31: 19–23.

Kannan, K. and J. Falandysz. 1997. Butyltin residues in sediment, fish, fish-eating birds, harbour porpoise and human tissues from the Polish coast of the Baltic Sea. Mar. Pollut. Bull. 34: 203–207.

Kannan, K., K.S. Guruge, N.J. Thomas, S. Tanabe and J.P. Giesy. 1998. Butyltin residues in southern sea otters (*Enhydra lutris nereis*) found dead along California coastal waters. Environ. Sci. technol. 32: 1169–1175.

Kefi, F.J., Y. Lahbib, L.G. Abdallah and N.T. Menif. 2011. Shell disturbances and butyltins burden in commercial bivalves collected from the Bizerta lagoon (northern Tunisia). Environ. Monit. Assess. 184: 6869–6876.

Kirchner, S., T. Kieu, C. Chow, S. Casey and B. Blumberg. 2010. Prenatal exposure to the environmental obesogen tributyltin predisposes multipotent stem cells to become adipocytes. Mol. Endocrinol. 24: 526–539.

Konstantinou, I.K. and T.A. Albanis. 2004. Worldwide occurrence and effects of antifouling paint booster biocides in the aquatic environment: a review. Environ. Int. 30: 235–48.

Kuhl, A.J. and M. Brouwer. 2006. Antiestrogens inhibit xenoestrogen-induced brain aromatase activity but do not prevent xenoestrogen-induced feminization in Japanese medaka (*Oryzias latipes*). Environ. Health Persp. 114: 500–506.

Langston, W.J., N.D. Pope, M. Davey, K.M. Langston, S.C. O' Hara, P.E. Gibbs and P.L. Pascoe. 2015. Recovery from TBT pollution in English Channel environments: A problem solved? Mar. Pollut. Bull. 95: 551–564.

Lau, M.C., K.M. Chan, K.M. Leung, T.G. Luan, M.S. Yang and J.W. Qiu. 2007. Acute and chronic toxicities of tributyltin to various life stages of the marine polychaete *Hydroides elegans*. Chemosphere 69: 135–144.

Lewis, C. and A.T. Ford. 2012. Infertility in male aquatic invertebrates: a review. Aquat. Toxicol. 120-121: 79–89.

Li, X., J. Ycaza and B. Blumberg. 2011. The environmental obesogen tributyltin chloride acts via peroxisome proliferator activated receptor gamma to induce adipogenesis in murine 3T3-L1 preadipocytes. J. Steroid. Biochem. Mol. Biol. 127: 9–15.

Lima, D., M.A. Reis-Henriques, R. Silva, A.I. Santos, L.F.C. Castro and M.M. Santos. 2011. Tributyltin-induced imposex in marine gastropods involves tissue-specific modulation of the retinoid X receptor. Aquat. Toxicol. 101: 221–227.

Lima, D., L.F. Castro, I. Coelho, R. Lacerda, M. Gesto, J. Soares, A. André, R. Capela, T. Torres, A.P. Carvalho and M.M. Santos. 2015. Effects of Tributyltin and other retinoid receptor agonists in reproductive-related endpoints in the Zebrafish (*Danio rerio*). J. Toxicol. Environ. Health A 12: 747–760.

Lyssimachou, A., J.C. Navarro, J. Bachmann and C. Porte. 2009. Triphenyltin alters lipid homeostasis in females of the ramshorn snail *Marisa cornuarietis*. Environ. Pollut. 157: 1714–1720.

Lyssimachou, A., J.G. Santos, A. André, J. Soares, D. Lima, L. Guimarães, C.M.R. Almeida, C. Teixeira, L.F.C. Castro and M.M. Santos. 2015. The Mammalian "Obesogen" Tributyltin Targets Hepatic Triglyceride Accumulation and the Transcriptional Regulation of Lipid Metabolism in the Liver and Brain of Zebrafish. PLoS ONE 10(12): e0143911.

Manning, C.S., T.F. Lytle, W.W. Walker and J.S. Lytle. 1999. Life-cycle toxicity of Bis(Tributyltin) oxide to the sheepshead minnow (*Cyprinodon variegatus*). Arch. Environm. Cont. Tox. 37: 258–266.

Maguire, R.J. 1987. Environmental aspects of tributyltin. App. Organomet. Chem. 1: 475–498.

Maguire, R.J. 2000. Review of the persistence, bioaccumulation and toxicity of tributyltin in aquatic environments in relation to Canada's toxic substances management policy. Water Qual. Res. J. Can. 35: 633–679.

Matthiessen, P. and P.E. Gibbs. 1998. Critical appraisal of the evidence for tributyltin-mediated endocrine disruption in mollusks. Environ. Toxicol. Chem. 17: 37–43.

McAllister, B.G. and D.E. Kime. 2003. Early life exposure to environmental levels of the aromatase inhibitor tributyltin causes masculinisation and irreversible sperm damage in zebrafish (*Danio rerio*). Aquat. Toxicol. 65: 309–316.

McGinnis, C.L. and J.F. Crivello. 2011. Elucidating the mechanism of action of tributyltin (TBT) in zebrafish. Aquat. Toxicol. 103: 25–31.

Meador, J.P. and C.A. Rice. 2001. Impaired growth in the polychaete *Armandia brevis* exposed to tributyltin in sediment. Mar. Environ. Res. 51: 113–129.

Meador, J.P., F.C. Sommers, K.A. Cooper and G. Yanagida. 2011. Tributyltin and the obesogen metabolic syndrome in a salmonid. Environ. Res. 111: 50–56.

Mensink, B., H. Everaats, C. Kralt, C. Hallers-Tjabbes and J.P. Boon. 1996. Tributyltin exposure in early life stages induces the development of male sexual characteristics in the common whelk, *Buccinum undatum*. Mar. Environ. Res. 42: 151–154.

Mensink, B.P., H. Kralt, A.D. Vethaak, C.C. Ten Hallers-Tjabbes, J.H. Koeman, B. van Hattum and J.P. Boon. 2002. Imposex induction in laboratory reared juvenile *Buccinum undatum* by tributyltin (TBT). Environ. Toxicol. Pharmacol. 11: 49–65.

MEPC. 2001. International Convention on the Control of Harmful Anti-fouling Systems on Ships, IMO, London, 5 October 2001.

Micael, J., M.A. Reis-Henriques, A.P. Carvalho and M.M. Santos. 2007. Genotoxic effects of binary mixtures of xenoandrogens (tributyltin, triphenyltin) and a xenoestrogen (ethinylestradiol) in a partial life-cycle test with Zebrafish (Danio rerio). Environ. Int. 33:1035–1039.

Minchin, D., E. Stroben, J. Oehlmann, B. Bauer, C.B. Duggan and M. Keatinge. 1996. Biological indicators used to map organotin contamination in Cork Harbour, Ireland. Mar. Pollut. Bull. 32: 188–195.

Mochida, K., K. Ito, K. Kono, T. Onduka, A. Kakuno and K. Fujii. 2007. Molecular and histological evaluation of tributyltin toxicity on spermatogenesis in a marine fish, the mummichog (*Fundulus heteroclitus*). Aquat. Toxicol. 83: 73–83.

Mochida, K. and K. Fujii. 2008. Toxicity for aquatic organisms. pp. 149–160. *In*: T. Arai et al. [eds.]. Ecotoxicology of Antifouling Biocides. Japan, Springer.

Nakanishi, T., J.I. Nishikawa, Y. Hiromori, H. Yokoyama, M. Koyanagi, S. Takasuga, J.I. Ishizaki, M. Watanabe, S.I. Isa, N. Utoguchi, N. Itoh, Y. Kono, T. Nishihara and K. Tanaka. 2005. Trialkyltin compounds bind retinoid X receptor to alter human placental endocrine functions. Mol. Endocrinol. 19: 2502–2516.

Nakanishi, T., J. Nishikawa and K. Tanaka. 2006. Molecular targets of organotin compounds in endocrine disruption: do organotin compounds function as aromatase inhibitors in mammals? Environ. Sci. 13: 89–100.

Nakanishi, T. 2008. Endocrine disruption induced by organotin compounds; organotins function as a powerful agonist for nuclear receptors rather than an aromatase inhibitor. J. Toxicol. Sci. 33: 269–76.

Nakayama, K., Y. Oshima, T. Yamaguchi, Y. Tsuruda, I.J. Kang, M. Kobayashi, N. Imada and T. Honjo. 2004. Fertilization success and sexual behavior in male medaka, *Oryzias latipes*, exposed to tributyltin. Chemosphere 55: 1331–1337.

Niederreither, K. and P. Dollé. 2008. Retinoic acid in development: towards an integrated view. Nat. Genet. 9: 541–553.

Nirmala, K., Y. Oshima, R. Lee, N. Imada, T. Honjo and K. Kobayashi. 1999. Transgenerational toxicity of tributyltin and its combined effects with polychlorinated biphenyls on reproductive processes in Japanese medaka (*Oryzias latipes*). Environ. Toxicol. Chem. 18: 717–721.

Nishikawa, J., K. Saito, J. Goto, F. Dakeyama, M. Matsuo and T. Nishihara. 1999. New screening methods for chemicals with hormonal activities using interaction of nuclear hormone receptor with coactivator. Toxicol. Appl. Pharmacol. 154: 76–83.

Nishikawa, J.I., S. Mamiya, T. Kanayama, T. Nishikawa, F. Shiraishi and T. Horiguchi. 2004. Involvement of the retinoid X receptor in the development of imposex caused by organotins in gastropods. Environ. Sci. Technol. 38: 6271–6276.

Oberdörster, E. and P. McClellan-Green. 2000. The neuropeptide APGW amide induces imposex in the mud snail, *Ilyanassa obsoleta*. Peptides 21: 1323–1330.

O'Donnell, L., K.M. Robertson, M.E. Jones and E.R. Simpson. 2001. Estrogen and spermatogenesis. Endocrinol. Rev. 22: 289–318.

Oehlmann, J., P. Fioroni, E. Stroben and B. Market. 1996. Tributyltin (TBT) effects on *Ocinebrina aciculata* (Gastropoda: Muricidae): imposex development, sterilization, sex change and population decline. Sci. Total. Environ. 188: 205–223.

Ohji, M., T. Arai and N. Miyazaki. 2002. Effects of tributyltin exposure in the embryonic stage on sex ratio and survival rate in the caprellid amphipod *Caprella danilevskii*. Mar. Ecol. Prog. Ser. 235: 171–176.

Ohji, M., T. Arai and N. Miyazaki. 2003. Chronic effects of tributyltin on the caprellid amphipod *Caprella danilevskii*. Mar. Pollut. Bull. 46: 1263–72.

Park, K., R. Kim, J.J. Park, H.C. Shin, J.C. Lee, H.S. Cho, Y.G. Lee, J. Kim and I.S. Kwak. 2012. Ecotoxicological evaluation of tributyltin toxicity to the equilateral venus clam, *Gomphina veneriformis* (Bivalvia: Veneridae). Fish Shellfish Immun. 32: 426–433.

Pascoal, S., G. Carvalho, O. Vasieva, R. Hughes, A. Cossins, Y. Fang, K. Ashelford, L. Olohan, C. Barroso, S. Mendo and S. Creer. 2013. Transcriptomics and *in vivo* tests reveal novel mechanisms underlying endocrine disruption in an ecological sentinel, *Nucella lapillus*. Mol. Ecol. 22: 1589–1608.

Philip, S., L.C.F. Castro, R.R. da Fonseca, M.A. Reis-Henriques, V. Vasconcelos, M.M. Santos and A. Antunes. 2012. Adaptive evolution of the Retinoid X receptor in vertebrates. Genomics 99: 81–89.

Pinkney, A.E., D.A. Wright, M.A. Jepson and D.W. Towle. 1989. Effects of tributyltin compounds on ionic regulation and gill ATPase activity in estuarine fish. Comp. Biochem. Physiol. 92C: 125–129.

Richard-Mercier, N., M. Dorizzi, G. Desvages, M. Girondot and C. Pieau. 1995. Endocrine sex reversal of gonads by the aromatase inhibitor Letrozole (CGS 20267) in *Emys orbicularis*, a turtle with temperature-dependent sex determination. Gen. Comp. Endorc. 100: 314–326.

Ronis, M.J.J. and A.Z. Mason. 1996. The metabolism of testosterone by periwinkle (*Littorina littorea*) *in vitro* and *in vivo*: effects of tributyltin. Mar. Environ. Res. 42: 161–166.

Rouleau, C., Z.H. Xiong and G. Pacepavicius. 2003. Uptake of waterborne tributyltin in the brain of fish: axonal transport as a proposed mechanism. Environ. Sci. Technol. 37: 3298–3302.

Ruiz, J.M., G.W. Bryan and P.E. Gibbs. 1994. Chronic toxicity of water tributyltin (TBT) and copper to spat of the bivalve *Scrobicularia plana*: ecological implications. Mar. Ecol. Prog. Ser. 113: 105–117.

Santillo, P.J. 2001. Chapter 13. Tributyltin (TBT) antifoulants: a tale of ships, snails and imposex. In Late lessons from early warning: the precautionary principle 1896-2000. EEA, Environmetal issue report nº 22/2001 135–148.

Santos, M.M., N. Vieira and A.M. Santos. 2000. Imposex in the dog-welk *Nucella lapillus* (L.) along the Portuguese coast. Mar. Pollut. Bull. 40: 643–646.

Santos, M.M., C.C. Ten Hallers-Tjabbes, A.M. Santos and N. Vieira. 2002. Imposex in *Nucella lapillus*, a bioindicator for TBT contamination: re-survey along the Portuguese coast to monitor the effectiveness of EU regulation. J. Sea Res. 48: 217–223.

Santos, M.M., J. Micael, A.P. Carvalho, R. Morabito, P. Booy, P. Massanisso, M. Lamoree and M.A. Reis-Henriques. 2006. Estrogens counteract the masculinizing effect of tributyltin in zebrafish. Comp. Biochem. Physiol. C 142: 151–155.

Santos, M.M., P. Enes, M.A. Reis-Henriques, J. Kuballa, L.F.C. Castro and M.N. Vieira. 2009. Organotin levels in seafood from Portuguese markets and the risk for consumers. Chemosphere 75: 661–666.

Santos, M.M., M.A., Reis-Henriques and L.F.C., Castro. 2012. Lipid homeostasis perturbation by organotins: Effects on vertebrates and invertebrates. pp. 83–96. *In*: A. Pagliarani et al. [eds.]. Biochemical and Physiological Effects of Organotins. Bentham Science Publishers, Italy.

Sárria, M.P., M.M. Santos, M.A. Reis-Henriques, N.M. Vieira and N.M. Monteiro. 2011. Drifting towards the surface: a shift in newborn pipefish's vertical distribution when exposed to the synthetic steroid ethinylestradiol. Chemosphere 85: 618–624.

Schmidt, K., C.E. Steinberg, S. Pflugmacher and G.B. Staaks. 2004. Xenobiotic substances such as PCB mixtures (Aroclor 1254) and TBT can influence swimming behavior and biotransformation activity (GST) of carp (*Cyprinus carpio*). Environ. Toxicol. 19: 460–470.

Schmidt, K., G.B. Staaks, S. Pflugmacher and C.E. Steinberg. 2005. Impact of PCB mixture (Aroclor 1254) and TBT and a mixture of both on swimming behavior, body growth and enzymatic biotransformation activities (GST) of young carp (*Cyprinus carpio*). Aquat. Toxicol. 71: 49–59.

Schulte-Oehlmann, U., C. Bettin, P. Fioroni, J. Oehlmann and E. Stroben. 1995. *Marisa cornuarietis* (Gastropoda, Prosobranchia): A potential TBT bioindicator for freshwater environments. Ecotoxicology 4: 372–384.

Shimasaki, Y., T. Kitano and Y. Oshima. 2003. Tributyltin causes masculinization in fish. Environ. Toxicol. Chem. 22: 141–144.

Simões-Costa, M., A.P. Azambuja and J. Xavier-Neto. 2008. The search for non-chordate retinoic acid signaling: lessons from chordates. J. Exp. Zool. 310B: 54–72.

Smith, B.S. 1971. Sexuality in the American mud snail, *Nassarius obsoletus*. J. Molluscan Stud. 39: 377–378.

Smith, B.S. 1981a. Male characteristics on female mud snails caused by antifouling bottom paints. J. Appl. Toxicol. 1: 22–25.

Smith, B.S. 1981b. Tributyltin compounds induce male characteristics on female mud snails *Nassarius obsletus* (*Iiyanassa obsoleta*). J. Appl. Toxicol. 1: 141–144.

Smith, R., S.G. Bolam, H.L. Rees and C. Mason. 2008. Macrofaunal recovery following TBT ban; long-term recovery of subtidal macrofaunal communities in relation to declining levels of TBT contamination. Environ. Monit. Assess. 136: 245–256.

Sole, M., Y. Morcillo and C. Porte. 1998. Imposex in the commercial snail Bolinus brandaris in the Northwestern Mediterranean. Environ. Pollut. 99: 241–246.

Solé, M. and D.R. Livingstone. 2005. Components of the cytochrome P450-dependent monooxygenase system and 'NADPH-independent benzo[a]pyrene hydroxylase' activity in a wide range of marine invertebrate species. Comp. Biochem. Physiol. C Toxicol. Pharmacol. 141: 20–31.

Sonoda, J., L. Pei and R.M. Evans. 2008. Nuclear receptors: decoding metabolic disease. FEBS Letters 582: 2–9.

Sousa, A., F. Laranjeiro, S. Takahashi, S. Tanabe and C.M. Barroso. 2009. Imposex and organotin prevalence in a European post-legislative scenario: temporal trends from 2003 to 2008. Chemosphere 77: 566–573.
Sousa, A.C.A., M.R. Pastorinho, S. Takahashi and S. Tanabe. 2014. History on organotin compounds, from snails to humans. Environ. Chem. Lett. 12: 117–137.
Spooner, N., P.E. Gibbs, G.W. Bryan and L.J. Goad. 1991. The effect of tributyltin upon steroid titers in the female dogwhelk, *Nucella lapillus*, and the development of imposex. Mar. Environ. Res. 2: 37–49.
Sternberg, R.M., A.K. Hotchkiss and G.A. LeBlanc. 2008. The contribution of steroidal androgens and estrogens to reproductive maturation of the eastern mud snail Ilyanassa obsoleta. Gen. Comp. Endocrinol. 156: 15–26.
Sternberg, R.M., M.P. Gooding, A.K. Hotchkiss and G.A. LeBlanc. 2010. Environmental-endocrine control of reproductive maturation in gastropods: implications for the mechanism of tributyltin-induced imposex in prosobranchs. Ecotoxicology 19: 4–23.
Straw, J. and D. Rittschof. 2004. Responses of mud snails from low and high imposex sites to sex pheromones. Mar. Pollut. Bull. 48: 1048–1054.
Stroben, E., J. Oehlmann and P. Fioroni. 1992. *Hinia reticulata* and *Nucella lapillus* - comparison of two gastropod tributyltin bioindicators. Mar. Biol. 114: 289–296.
Swennen, C., U. Sampantarak and N. Ruttanadakul. 2009. TBT-pollution in the Gulf of Thailand: a re-inspection of imposex incidence after 10 years. Mar. Pollut. Bull. 58: 526–532.
Ten Hallers-Tjabbes, C.C., J.F. Kemp and J.P. Boon. 1994. Imposex in whelks (*Buccinum undatum*) from the open North Sea: relation to shipping traffic intensities. Mar. Pollut. Bull. 28: 311–313.
Thain, J.E. and M.J. Waldock. 1986. The impact of tributyltin, TBT antifouling paints on molluscan fisheries. Water Sci. Technol. 18: 193–202.
Thornton, J.W. 2003. Nuclear receptor diversity: phylogeny, evolution and endocrine disruption. Pure Appl. Chem. 75: 1827–1839.
Titley-O'Neal, C.P., K.R. Munkittrick and B.A. MacDonald. 2011. The effects of organotin on female gastropods. J. Environ. Monit. 13: 2360–2388.
Triebskorn, R., H. Kohler, J. Flemming, T. Braunbeck, R. Negele and H. Rahmann. 1994. Evaluation of bis(tri-n-butyltin) oxide (TBTO) neurotoxicity in rainbow trout (*Oncorhynchus mykiss*). I. Behaviour, weight increase, and tin content. Aquat. Toxicol. 30: 189–197.
Uchida, D., M. Yamashita, T. Kitano and T. Iguchi. 2004. An aromatase inhibitor or high water temperature induce oocyte apoptosis and depletion of P450 aromatase activity in the gonads of genetic female zebrafish during sex-reversal. Comp. Biochem. Physiol. A Mol. Integr. Physiol. 137: 11–20.
Viglino, L., E. Pelletier and R. St-Louis. 2004. Highly persistent butyltins in the northern marine sediments: a long-term threat for the Saguenary Fjord (Canada). Environ. Toxicol. Chem. 23: 2673–2681.
Yi, A.X., J. Han, J.S. Lee and K.M. Leung. 2014. Ecotoxicity of triphenyltin on the marine copepod *Tigriopus japonicus* at various biological organisations: from molecular to population-level effects. Ecotoxicology 23: 1314–25.
Yu, A., X. Wang, Z. Zuo, J. Cai and C. Wang. 2013. Tributyltin exposure influences predatory behavior, neurotransmitter content and receptor expression in *Sebastiscus marmoratus*. Aquat. Toxicol. 128-129: 158–162.
Watermann, B.T., T.A. Albanis, T. Dagnac, K. Gnass, K. Ole Kusk. V.A. Sakkas and L. Wollenberger. 2013. Effects of methyltestosterone, letrozole, triphenyltin and fenarimol on histology of reproductive organs of the copepod *Acartia tonsa*. Chemosphere 92: 544–54.
White, R.J., Q. Nie, A.D. Lander and T.F. Schilling. 2007. Complex regulation of cyp26a1 creates a robust retinoic acid gradient in the zebrafish embryo. PLoS Biology 5: 12.
Zhang, J., Z.H. Zuo, Y.X. Chen, Y. Zhao, S. Hu, C.G. Wang. 2007. Effects of tributyltin on the development of ovary in female cuvier (*Sebastiscus marmoratus*). Aquat. Toxicol. 83: 174–179.
Zhang, J., Z.H. Zuo, R. Chen, Y.X. Chen and C. Wang. 2008. Tributyltin exposure causes brain damage in *Sebastiscus marmoratus*. Chemosphere 73: 337–343.
Zhang, J., Z.H. Zuo, C.Y. He, J.L. Cai, Y.Q. Wang, Y.X. Chen and C.G. Wang. 2009. Effect of tributyltin on testicular development in *Sebastiscus marmoratus* and the mechanism involved. Environ. Toxicol. Chem. 28: 1528–1535.
Zhang, J., Z.H. Zuo, Y.Q. Wang, A. Yu, Y.X. Chen and C.G. Wang. 2011. Tributyltin chloride results in dorsal curvature in embryo development of *Sebastiscus marmoratus* via apoptosis pathway. Chemosphere 82: 437–442.
Zhang, J., Z. Zuo, J. Xiong, P. Sun, Y. Chen and C.G. Wang. 2013. Tributyltin exposure causes lipotoxicity responses in the ovaries of rockfish, *Sebastiscus marmoratus*. Chemosphere 90: 1294–1299.

7

Functional Genomics Approaches in Marine Pollution and Aquaculture

Nieves Abril,[a,*] *María José Prieto-Álamo*[b] and *Carmen Pueyo*[c]

ABSTRACT

Anthropogenic pollution affects the health of all creatures in the ocean and also affects the health of humans. Aquaculture is a solution to the increased food demand of a growing human population. However, there are several challenges that must be met to make aquaculture productive, feasible, and sustainable, including minimizing the aggravation of marine pollution. Functional genomics provides information about the biomolecules involved in biological defense mechanisms as well as in the restoration of processes altered by contamination. We present here recent research results on the application of different omics methodologies in two important species, *Procambarus clarkii,* the dominant North American commercial crayfish, and the Senegalese sole (*Solea senegalensis*), a flatfish species with a high potential for use in marine aquaculture diversification. These studies will help to improve a sustainable aquaculture and identify novel potential biomarkers useful for monitoring aquatic environments and assessing the health of the marine ecosystem.

Introduction

The marine environment is a sink for potentially hazardous chemical pollutants emitted from industrial and domestic sources. For decades, humans have acted as though the oceans were so vast and so full of life that any amount of pollution could be tolerated without ocean life being affected. Contaminants were dumped into the deep sea as though they would simply disappear forever. Nothing has ever been less true. Human activities have overloaded marine and coastal ecosystems with large quantities of nitrogen, phosphorus, heavy metals, pesticides, drugs, and many other environmental toxins. The consequences affect the health of all creatures in the ocean and also affect human health. Aquaculture is a solution to the inability of wild sources of fish and other aquatic species to keep up with the increased demand of

Department of Biochemistry & Molecular Biology, University of Cordoba, Campus de Rabanales, 14071-Cordoba, Spain.
[a] Email: bb1abdim@uco.es
[b] Email: bb2pralm@uco.es
[c] Email: bb1pucuc@uco.es
* Corresponding author

a growing human population. However, aquaculture must solve three main challenges to be productive, feasible and sustainable: (1) diversification of diets, (2) diseases, stresses, and/or deterioration of culturing conditions, and (3) introduction of new species (Prieto-Alamo et al. 2012b). Large-scale genome projects have changed the paradigm of biological experimentation. During the late 1980s, structural genomics allowed the generation and analysis of information about genes and genomes. Currently, the mass of genome data is being converted into gene-function data. Functional genomics attempts to make use of the vast wealth of data produced by genome projects (such as genome sequencing projects) to describe gene (and protein) functions and interactions. Functional genomics permits monitoring large numbers of genes for their expression under a range of different conditions at the transcript, protein, or metabolite levels. Consequently, this type of study will undoubtedly identify potential candidate genes that have large phenotypic effects and may be of commercial importance in aquaculture.

Expansion of aquaculture can result in detrimental effects on the environment and aggravation of pollution of the marine environment. The study of environmental stress episodes and of environmental risk monitoring requires the evaluation of biological responses to contaminants. These responses are difficult to decipher because they are modulated by numerous factors related to synergistic/antagonistic interactions between contaminants and other substances present in the ecosystem. Only the integrated use of the powerful analytical tools offered by functional genomics provides simultaneous information about the biomolecules involved in biological defense mechanisms as well as in the restoration of processes altered by contamination. Nevertheless, the application of transcriptomic, proteomic, or any of the many "omics" approaches in aquaculture or environmental monitoring assessment is currently limited. One important reason is that the organisms of interest are usually non-model organisms, and there is thus very limited genomic information available in public databases, as well as a complete lack of commercial resources.

We present here recent research results on the application of different omics methodologies to improve sustainable aquaculture and to identify novel potential biomarkers useful for monitoring aquatic environments and assessing the health of their ecosystem, which are essential to protect the environment, human health, and sustainable development. Focusing on two important species, *Procambarus clarkii* (the dominant North American commercial crayfish) and the Senegalese sole (*Solea senegalensis*, a flatfish species with a high potential for use in marine aquaculture diversification), we have used transcriptomic, proteomic, and metabolomic approaches to gain deep insight into the appraisal and interpretation of their biological responses.

Procambarus clarkii, a Valuable Source of Food and an Environmentally Devastating Species

Crustaceans occupy a central place in the food web and are highly sensitive to environmental stressors. Accordingly, crayfish have been used as bioindicators for monitoring pollution in aquatic environments (Alcorlo et al. 2006, and references herein). The Decapod crustacean *Procambarus clarkii* (red swamp crayfish, freshwater crayfish, or American crayfish) is a prolific species of crayfish that has a long life cycle, has a relatively sedentary life style, is able to tolerate extreme and polluted environments, and accumulates heavy metals and toxicants in its tissues (Gherardi 2006, and references herein). These features contribute to make *P. clarkii* an effective bioindicator species to assess the effect of contaminants under controlled conditions (e.g., Vioque-Fernández et al. 2009b) or in real environments (e.g., Vioque-Fernández et al. 2009a). Currently, transcriptomics in crayfish, and more specifically in *P. clarkii*, is mainly used to identify genes related to the response to infections by pathogens that cause serious economic losses in the crustacean farming industry, in particular, the white spot syndrome virus (WSSV), one of the most devastating viruses to *penaeid* shrimp (Ou et al. 2013; Shi et al. 2010; Zeng and Lu 2009). Nevertheless, there are relatively few sequenced and well-defined *P. clarkii* genes in the open access databases.

Procambarus clarkii was introduced to Spain through the Lower Basin of the Guadalquivir River in 1974. Currently, it is both a widespread and non-native species (and therefore collectable with minimal ecosystem disturbance) in the Doñana National Park (DNP) (SW Spain), a wildlife reserve North of Guadalquivir River. Therefore, it has been used in recent years as a bioindicator species in environmental studies to monitor the contamination of water by pesticides and metals in the DNP. This park, which represents one of the largest and most important remaining wetlands in Europe, was declared a World Heritage Site by UNESCO in 1994. It is constituted by a marshy area, shallow streams, and sand dunes. It includes a great variety of aquatic ecosystems and constitutes a critical stopover for millions of migrating birds mainly at its core, the Doñana Biological Reserve (DBR) (Oñate et al. 2003). In spite of its environmental relevance, biogeochemical and ecological research at this park was begun only three decades ago (Serrano et al. 2006). DNP conservation is threatened by intense and multi-sector economic activity, mainly due to intensive agriculture in its surroundings, marsh drainage, irrigation projects, industrial settlements at Huelva, and nearby pyrite mining activity (as exemplified by the Aznalcollar spill in 1998) (Oñate et al. 2003).

Procambarus clarkii was first used as bioindicator through the use of biochemical biomarkers (Vioque-Fernandez et al. 2007) and next in a global proteomics analysis (Vioque-Fernández et al. 2009a,b). In this second study, an altered PES (protein expression signature) was used as a multimarker, though the altered proteins were not identified because *P. clarkii* was absent from the public sequence databases (Vioque-Fernández et al. 2009a,b). More recently, 1st, 2nd, and 3rd generation proteomic approaches have been used to quantify changes in the protein profiles caused in the crayfish inhabiting highly polluted aquatic systems. These studies have identified proteins involved in immune and stress responses and other proteins whose altered abundances indicate an imbalance of homeostasis in the polluted crayfish. With respect to the environmental transcriptomic analyses, *Suppression Subtractive Hybridization* (SSH) methodology was employed for the first time in a field study using the fresh water crayfish *P. clarkii* as a bioindicator to identify differentially expressed genes in aquatic ecosystems close to the DNP with different pollution statuses (Osuna-Jimenez et al. 2014), as discussed below.

Transcriptomic studies in P. clarkii

Environmental transcriptomic analyses in *P. clarkii* (Osuna-Jimenez et al. 2014) have mainly focused on the hepatopancreas, a central metabolic organ in crayfish that executes intestinal, hepatic, and pancreatic functions and also participates in detoxifying heavy metals and organic xenobiotics (Schram and Vaupel Klein 2012). Animals were captured at six sites in areas within and around the DNP (Fig. 1). "Luciodel Palacio" (LDP) is at DRB, a clean site, while the other five sites are close to areas with intensive agricultural use. The "Partido" (PAR) and "Ajolí" (AJO) sites, the upstream part and downstream sites, respectively, of the Partido stream, are under the influence of citrus fruit and grape fields and receives effluents from Almonte, a medium-size town, and many horticultural activities. The "Rocina" (ROC)

Fig. 1: Location of sampling areas. The Iberian Peninsula (A) with the rectangle enlarged on the right showing the Doñana National Park (DNP) and surroundings (B). The problem areas PAR and AJO are located along the course of Partido stream, ROC and BER along the course of Rocina stream, and MAT is near de course of Guadiamar River. The control site, LDP, is within the Doñana Biological Reserve at the core of DNP.

and "Bernabe" (BER) sites, which are the upstream and downstream parts, respectively, of the Rocina stream, are near strawberry, citrus fruit, and grape fields. The "Matochal" (MAT) site is located next to the Guadiamar stream and is affected by rice fields. The pollutant load sustained by the crayfish was evaluated using the concentrations of four elements (Cu, Zn, As, and Cd) in the hepatopancreas of the crayfish as indicators of pollution. The concentrations of these elements were significantly different between the sites. In general, the lowest concentrations were found in the LDP population, highlighting its usefulness as a negative control. Similarly, the metal concentration confirmed MAT as the most polluted zone, as previously published (Vioque-Fernandez et al. 2009a; Vioque-Fernandez et al. 2007). Overall, the specimens collected at the upper parts of the Partido and Rocina streams (PAR and ROC) exhibited lower metal levels than the specimens collected in the downstream locations (AJO and BER).

Construction of SSH libraries to identify differentially expressed genes in P. clarkii specimens from sites of different pollution status

Suppression Subtractive Hybridization (SSH) is an efficient technique to identify genes that are differentially expressed between two samples at the mRNA level (Diatchenko et al. 1996) in the absence of sequence information, which allows researchers to identify novel genes involved in a particular function or response in non-model organisms. An additional benefit of SSH methodology is that a number of expressed sequence tags (EST) may be obtained and incorporated into public databases. The construction of subtractive libraries in *P. clarkii* allowed the identification of differentially expressed genes in crayfish from moderately (PAR) and highly polluted (MAT) zones compared to crayfish from the control site in the Doñana Biological reserve (LDP). For both the moderately and the highly polluted zones, forward (F) and reverse (R) libraries were designed to obtain clones of genes that were up- and down-regulated, respectively, in response to environmental pollutants. A total of 606 clones were sequenced (Table 1), and the ESTs were identified by comparison with entries in open access databases. A total of 43 unique sequences were isolated, and 26 were identified as relating to major physiological functions. The relatively high proportion of un-identified ESTs is not surprising given that only three crustacean genomes had been completed at the time of this study, and, moreover, they were from species distantly related to *P. clarkii* (Pagani et al. 2012). Most ESTs represented redundant sequences, and it is remarkable that only one transcript (identified as HC2) accounted for 124 out of a total of 289 ESTs in the two F libraries. Similarly, in the two R libraries, one gene (CTSZ) accounted for 154 out of 275 ESTs, consistent with previous results obtained in hemocytes of *P. clarkii* infected with the WSSV (Zeng and Lu 2009).

Overall, the SSH libraries showed that the crayfish from sites with different amounts of pollution displayed different transcriptional expression patterns. The differentially expressed genes encode proteins involved in multiple physiological functions, many of them dealing with different aspects of the immune response. The proteins encoded by the differentially expressed genes included metabolic enzymes, proteases and their inhibitors, stress response-related proteins, intracellular components potentially related to signaling cascades, transcription factors, cytoskeletal proteins, and ribosomal proteins (Table 2).

Table 1: Characteristics of the *P. clarkii* SSH libraries (Osuna-Jimenez et al. 2014).

	PAR	MAT
Number of sequenced clones	336	270
Number of analyzed clones	332	244
Number of unique ESTs	36	14
ESTs Up-regulated	20	6
Identified ESTs	11	5
Non-identified ESTs	9	1
ESTs Down-regulated	19	8
Identified ESTs	12	6
Non-identified ESTs	7	2

Determination of transcriptional profiles by quantitative RT-PCR

The differential gene expression was further validated by quantifying the transcript copy number of nine selected genes by real-time qRT-PCR. Given that the immune system is one of the essential physiological systems at the interface between organism and environment (Chapman 2001) and given the evidence of immunotoxicity in aquatic organisms exposed to pollutants (Snape et al. 2004), priority was given to genes that were putatively related to the immune response, and the specificity of their responses was taken into consideration (Table 3).

To precisely assess the differential expression of the selected genes, a rigorous approach based on the absolute quantification of transcripts (exact number of molecules of each transcript in each sample) at an individual level was carried out (Osuna-Jimenez et al. 2014). Absolute quantification provided valuable additional information on the profiles of the selected transcripts, allowing the determination of their steady-state levels in *P. clarkii* hepatopancreas. As expected, substantial differences in abundance were found depending on the transcript examined (Table 4). Of note are the outstanding high copy numbers of HC2 and CTSL, which could most likely explain their aforementioned extreme redundancy in the R and F libraries, respectively. The results also indicate that the study was not particularly vulnerable to misinterpretation due to inter-individual differences in crayfish transcript levels or inter-individual susceptibility to the effects of living in a particular area. In general, and irrespective of the different steady-state levels, the qRT-PCR analyses confirmed the SSH results (Tables 2 and 4). They corroborated the lack of differential mRNA expression for AST in the F and R libraries from the PAR area and confirmed the up-regulated (BTF3, HC, HC2, SPINK4) and down-regulated (CTSL, CHIT, ALP, and ferritin) mRNAs. Moreover, qRT-PCR analyses were consistent with the site-specific expression detected in the forward SSH libraries (Tables 2 and 4).

The absolute quantification of these transcripts allowed us to calculate a statistically significant positive correlation between the hepatopancreatic content of copper and the HC and HC2 levels. This relationship was specific for this metal and was not limited to the animals from the location used to obtain the SSH libraries because a similar correlation was found when considering the six sites that were initially included in the study. Consequently, the determination of the levels of HC and HC2 transcripts in the hepatopancreas of *P. clarkii* could be a useful tool for monitoring the environmental contamination caused by copper.

Single nucleotide polymorphisms (SNP) can be identified from EST studies when several individuals are used and the redundancy of the libraries is high enough. During the aforementioned SSH analyses (Osuna-Jimenez et al. 2014), a substantial number of SNPs were detected in the ESTs. A more detailed analysis of the SNP profiles of the selected transcripts was performed by sequencing cDNA segments of these transcripts from a total of 21 animals from the three sampling sites. The SNP frequency correlated markedly with the pollution degree of the sampling sites (MAT > PAR > LDP). On the other hand, SNP densities can also differ greatly between genes (Osuna-Jimenez et al. 2014). These results support the use of EST-derived SNPs as markers to identify ecologically responsive genes (Orsini et al. 2011) and constitute an initial and very promising approach to identify point mutations that could arise in an adaptive response to environmental stressors in *P. clarkii*.

Proteomic studies in P. clarkii

Transcription and translation have a far from linear and simple relationship, and many studies indicate that the correlation between mRNA and protein abundances in the cell is notoriously poor. The cause is the modulation of protein expression at different levels from transcription to maturation of the polypeptides produced by translation of mature mRNAs. Additional post-translational modifications are of key importance as they yield to multiple protein products from a single gene, each of which may have different functions. Proteomics addresses the post-genomic challenge of examining the entire set of proteins (proteome) expressed by a genome in a cell, tissue, or organ at a given time under defined conditions (Pueyo et al. 2011). Proteins are initially separated by two-dimensional electrophoresis (2-DE, (Wilkins et al. 1996)) and identified by mass spectrometry analysis of their peptide mass fingerprint (MALDI-TOF-PMF) or by *de novo* sequencing of some peptides (nESI-MS/MS), then comparing the results with entries in public databases (Simpson 2003). Fluorescent labeling gave rise to difference gel electrophoresis (DIGE), which

Table 2: *P. clarkii* forward and reverse SSH libraries (Osuna-Jimenez et al. 2009).

Protein	No. of clones[a]	
	PAR	MAT
Forward library		
Acetyl-CoA C-acyltransferase-like	13 (1)[b]	–
CHH-like protein precursor, alternatively spliced	8 (1)	–
Cytochrome b	–	2 (1)
Endo beta-1,4-glucanase	4 (1)	5 (1)
Hemocyanin	6 (2)	3 (1)
Hemocyanin 2	56 (2)	68 (4)
Kazal-type serine proteinase inhibitor 4	–	4 (1)
14-3-3-like protein	3 (1)	–
Ribosomal protein	6 (1)	–
Ribosomal protein S7	3 (1)	–
Ribosomal protein S17	2 (1)	–
Transcription factor BTF3	20 (1)	–
Zinc proteinase (astacin gene)	18 (1)	–
Unknown function and novel sequences		
Clone h9_A10 mRNA sequence	20 (1)	–
Predicted protein (NEMVEDRAFT_v1g223528)	16 (1)	–
EST1	–	1 (1)
EST2	11 (1)	–
EST3	9 (1)	–
EST4	4 (1)	–
EST5	2 (1)	–
EST6	2 (1)	–
EST7	2 (1)	–
EST8	1 (1)	–
Reverse library		
Actin 2	3 (1)[b]	–
Alkaline phosphatase-like	–	3 (1)
Arginine kinase	3 (1)	–
Chitinase	3 (1)	–
Cysteine proteinase preproenzyme	69 (3)	85 (2)
Ferritin	18 (1)	28 (1)
Hepatopancreas trypsin	2 (1)	–
Pseudo hemocyanin	14 (2)	5 (1)
Ribosomal protein L13	1 (1)	–
Ribosomal protein l44	2 (1)	–
Ribosomal protein s24	–	8 (1)
25 kDsynaptosome-associated protein-like mRNA	1 (1)	6 (1)
Vitelline	2 (1)	–

Table 2 cont....

Table 2 cont...

Protein	No. of clones[a]	
	PAR	MAT
Reverse library		
Zinc proteinase (astacin gene)	1 (1)	–
Unknown function and novel sequences		
clone cherax_161 mRNA sequence	1 (1)	–
clone cherax_169 mRNA sequence	1 (1)	3 (1)
EST3	1 (1)	–
EST9	–	13 (1)
EST10	1 (1)	–
EST11	1 (1)	–
EST12	1 (1)	–
EST13	1 (1)	–

[a]Numbers are the total ESTs in both up- and down-regulated libraries.
[b]The number of contigs are given in parentheses.

Table 3: Selected transcripts to be validated by real-time qRT-PCR (Osuna-Jimenez et al. 2014).

	SSH[a]			
	Forward		Reverse	
Transcript (Symbol)[b]	PAR	MAT	PAR	MAT
Zinc proteinase (astacin gene) (AST)	+	–	+	–
Transcription factor BTF3	+	–	–	–
Hemocyanin (HC)	+	+	–	–
Hemocyanin 2 (HC2)	+	+	–	–
Kazal-type serine proteinase inhibitor 4 (SPINK4)	–	+	–	–
Alkaline phosphatase-like (ALP)	–	–	–	+
Chitinase (CHIT)	–	–	+	–
Cysteine proteinase preproenzyme (CSTL)	–	–	+	+
Ferritin	–	–	+	+

[a]SSH library where each selected transcript was identified (+) or not (–).
[b]Gene symbols are according to the NCBI Gene database.

analyzes, also by bidimensional electrophoresis, two samples in a single gel, facilitating a quantitative assessment of protein abundance (Ünlü et al. 1997) as the measurements are standardized. Lately, shotgun methodologies are gaining popularity as they allow the analysis of complex protein mixtures after full digestion, by multidimensional separation coupled with tandem liquid chromatography (LC/LC) and MS/MS (Washburn et al. 2001).

Proteomic studies in *P. clarkii* outside of our group are relatively scarce. In the aquaculture field, there are some studies focused on finding clues for reproductive genetic breeding of this species. Shui et al. (2012) used the 2-DE patterns of total proteins to identify those functionally involved in ovarian development in *P. clarkii*. In the environmental ambit, Xu et al. (2014a) have recently reported the development of a non-denaturing 2-DE protocol for screening *in vivo* uranium-protein targets in *P. clarkii*. However, the application of proteomic technology in environmental studies discloses multiple changes in protein abundance that may be associated with contamination. Thus, proteins involved in toxicological responses never described before may be revealed. The lack of genomic information in

Table 4: Absolute transcript levels in crayfish hepatopancreas determined by real-time qRT-PCR (Osuna-Jimenez et al. 2014).

	mRNA molecules/pg of total RNA[a]		
Transcript[b]	**LDP**	**PAR**	**MAT**
AST	709 ± 89	607 ± 75	698 ± 93
BTF3	48 ± 5.3	86 ± 8.1**	43 ± 8.0
HC	688 ± 170	2407 ± 605	3946 ± 1146
HC2	11.116 ± 1.480	25.171 ± 4.844**	39.901 ± 7.365**
SPINK4	91 ± 11	102 ± 12	154 ± 24**
ALP	75 ± 21	15 ± 3.4*	23 ± 6.1*
CHIT	3.412 ± 360	731 ± 79***	2198 ± 329*
CSTL	20.936 ± 1.130	3.787 ± 905***	9383 ± 1.929***
Ferritin	2.461 ± 191	1.104 ± 172**	1711 ± 156*

[a]Data are means ± SEM (n = 7). Statistical significance with respect to LDP is expressed as: ***P < 0.001; **P < 0.01; *P < 0.05.
[b]Gene symbols are according to the NCBI Gene database.

public databases for most non-model sentinel organisms makes it difficult but not impossible to identify differentially abundant proteins. After key proteins indicating exposure or effect are identified, proteomics can be used in risk assessment.

2-DE proteomic patterns to monitor aquatic ecosystems with P. clarkii

In environmental proteomics studies, the difficulty of protein identification in non-model organisms was initially circumvented by a pattern-only approach that focused on recognizing proteomic pattern signatures (PES) and not on protein identification. This approach resulted in sets of proteins useful as markers for the general state of the organism. Only the appearance of altered patterns was required to demonstrate the presence of adverse effects from pollution; there was no need to identify the proteins with altered expression levels. As a complement to the transcriptomic studies, we carried out the monitoring of aquatic ecosystems of the DNP by holistically analyzing the biological effects on *P. clarkii* at the proteomic level. We used male crayfish that we captured along the course of four campaigns carried out from 2003 to 2004 (Vioque-Fernandez et al. 2009a) at the reference and polluted sites described above. Two-dimensional gel electrophoresis resolved more than 2000 gill protein spots, and image analysis detected 35 spots with significant intensity differences between the reference and the other studied sites. The extent of up-/down-regulation of the 35 differentially expressed proteins differentiated the LDP site, located within the DBR, from the other capture sites (Fig. 2).

The higher proteomic responses found at the upstream "Rocina" (ROC) and "Partido" (PAR) sites indicate that non-persistent agrochemicals are mainly used in the Doñana surroundings. The highest responses corresponded to rice-growing areas (MAT) located between the Guadiamar stream and the Guadalquivir River, corresponding to the extended and intensive use of agrochemicals in such areas.

2D-DIGE, a second generation proteomic approach in aquaculture and environmental studies

The 2D-DIGE methodology has become a method of choice for quantitative proteomics as it provides more accurate and reliable quantification of proteins than simple 2-DE. In this method, the samples to be compared and an internal standard, each labeled with different fluorescent dyes, are run together on the same gel, eliminating potential gel-to-gel variation (Kondo 2008) (Fig. 3). This strategy reduces the

Capture site	Protein number																													
	1	2	3	4	5	7	8	9	10	11	12	14	15	16	18	19	20	21	22	23	24	25	26	28	29	30	31	32	33	34	35
PAR		3												15	-4		2	3	-3		12								-6	3	-24
MAT	-8	3	3	3	-9			5	6	2	-13		7		-5		3	3	-7	-5	13	-13	-12	47	-17	-12	-17	-41		5	
ROC	-3	4	5	4		-3	3	3	4	9		-67	2	6	-4					-3				12	-28		-19	-4	-7	3	-119

Fig. 2: Fold-number variations of *P. clarkii* gill protein spots intensities referred to the LDP control sample. Decreases in abundance are indicated in red and the increases in green. Adapted from (Vioque-Fernández et al. 2009a).

complexity of spot pattern comparison and provides a reliable method that can be applied to ecotoxicology studies to detect altered protein abundances.

The adaptation of 2D-DIGE methodology to our particular and complex tissues, the digestive gland and gills of *P. clarkii*, allowed us to resolve over 2000 protein spots in each gel and to differentiate the pollution level of the studied LDP, PAR, and MAT sites. After filtering data (volume ratios of $\geq$ 2- and $\leq$ 2-fold differences in expression in at least one of the two problem sites), 29 spots from the digestive gland and 15 spots from the gills were identified, corresponding to a total of 31 unique proteins. The proteomes from the digestive gland and gills shared only three of these unique proteins. The greatest changes were found in animals from the PAR site, the changes being lower in crayfish from the MAT rice fields.

A functional analysis indicated that proteins related to stress and immune responses and those related to energy generation greatly increased in abundance, presumably so that the animal could cope with the adverse environmental conditions. In contrast, pollution caused a diminution of other proteins related to the immune system and to xenobiotics biotransformation.

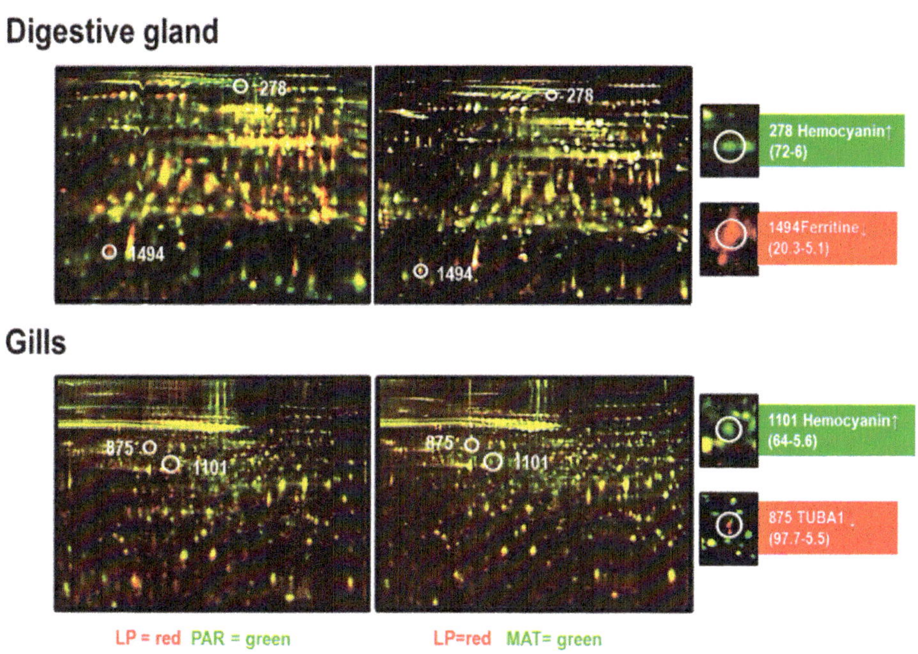

Fig. 3: Virtual two-dimensional differential in gel electrophoresis (2D-DIGE) images for comparison of LDP and PAR/MAT *P. clarkii* proteomes. For digestive gland or gills, equal amounts of Cy2 (IS, internal standard with equally mixed samples), Cy5 (LDP, reference crayfish), and Cy3 (PAR/MAT animals) labeled samples were mixed and then separated on analytical 2D-DIGE. Gels were scanned and a set of Cy5, Cy3, and Cy2 (A) images were obtained from each gel. An overlay of three dye scan-images was also obtained. The spot intensities and the relative expression ratio were computed using the DeCyder 6.5 software (Amersham Biosciences). Statistical significances were determined with the Student's t-test. As an example, circles on gels mark some spots whose intensities increased (green) or decreased (red) in relation to the IS because of the crayfish origin; for these four spots, the number assigned, the name of the protein, the direction of the change in intensity, the Mw and the pI, are given in brackets).

In summary, by using a 2D-DIGE proteomic approach, we separated the complex protein mixtures isolated from digestive gland and gill and identified the metabolic alterations and possible cellular damage in *P. clarkii* crayfish living in polluted sites.

Redox proteomics

Several types of pollutants, such as metals and oxidative organic chemicals, generate reactive oxygen and nitrogen species (ROS/RNS) that induce oxidative stress (Braconi et al. 2011; Lopez-Barea 1995). Proteins are major targets of ROS/RNS; many post-translational modifications (PTMs) have been described as being induced by ROS/RNS (Butterfield and Dalle-Donne 2012; Cabiscol and Ros 2006; Davies 2005). PTMs affect protein structure, biological activity, cellular localization, and/or interactions with other proteins (Cabiscol and Ros 2006; Hwang et al. 2009; Lee et al. 2009; Yamakura and Kawasaki 2010). Pollutant-promoted oxidative modifications of proteins, including carbonylation (Braconi et al. 2011) and oxidation of –SH groups are globally known as redox proteomics. The utility of redox proteomics as a novel biomarker was evaluated in aquatic ecosystems of the DNP and its surroundings using *P. clarkii* as abioindicator (Fernandez-Cisnal et al. 2014). In this study, after fluorescently labeling reversibly oxidized cysteine residues, proteins of the digestive gland and gills were separated by 2-DE electrophoresis (Fig. 4).

The total density of proteins with reversibly oxidized thiols was found to be much higher in animals from the MAT and ROC sites, while no difference was found in crayfish from PAR and those from the reference site LDP (Fernandez-Cisnal et al. 2014). Nineteen out of 35 proteins with significant differences in thiol oxidation were identified via MALDI-TOF/TOF. Among these, ferritin showed higher oxidation levels in ROC, while many others (superoxide dismutase, protein disulfide isomerase, actin, nucleoside diphosphate kinase, fructose-biphosphate aldolase, fatty acid-binding protein, and phosphopyruvate hydratase) were over oxidized in MAT. For most of the identified proteins, spots corresponding to different oxidized forms of cysteine were detected; spots corresponding to the native forms of many of the proteins were also found (Fernandez-Cisnal et al. 2014). The identified thiol-oxidized proteins provide information about the physiological processes affected by the oxidative stress caused by pollutants.

Gel-free/label-free proteomic analysis, the third-generation proteomic approach

Technological improvements in liquid chromatography and mass spectrometry have made possible the development of so called "gel-free proteomics" in which proteins are trypsinized and the resulting peptides are separated via high resolution chromatography and identified by tandem mass spectrometry. The gel-free and label-free technique has higher sensitivity than gel-based approaches, and can be easily automated to provide better reproducibility and lower influence of intrinsic protein characteristics (pI, molecular weight, etc.). Nevertheless, 2-DE also presents some advantages over a gel-free approach, such as the ability to detect protein isoforms, which is still complicated with gel-free approaches.

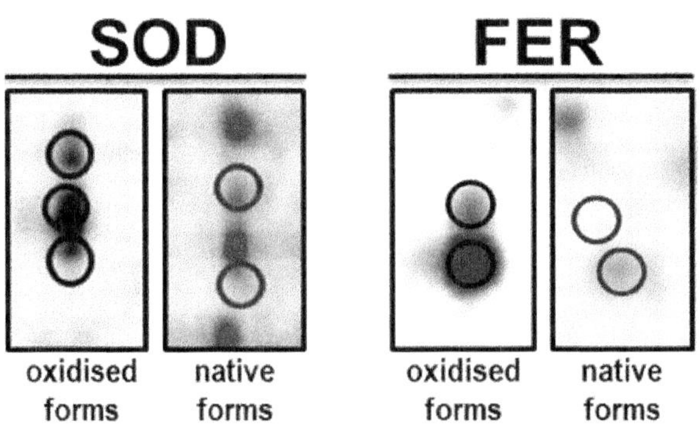

Fig. 4: Enlarged representative regions for SOD and FER protein spots. IAF labeled oxidized thiol proteins (in red) of *P. clarkii* (MAT) digestive gland were separated by 2-DE were separated, the fluorescence image captured, and the gels re-stained with SYPRO Ruby-stained and the image captured again (Adapted from Fernandez-Cisnal et al. 2014).

A preliminary study using a gel-free approach has been carried out to assess the effects of environmental pollution on the central nervous tissue of *P. clarkii*. In this study, male crayfish from PAR, MAT, and ROC were compared to LDP reference animals. Sixty-six different proteins with > ± 1.5-fold-change differences were identified. A similar number of deregulated proteins were found in each of the three polluted sites, but while in PAR samples predominated the increases in protein abundances, in the ROC samples, the changes were mainly reductions in the amounts of protein (Fig. 5A). Less than 20% of the deregulated proteins were common to the three polluted sites, indicating that singular pollutants were present in PAR, MAT, and ROC and that these pollutants were differentially affecting the nervous tissue metabolism and signaling pathways in the nervous tissue of *P. clarkii* (Fig. 5B). Though the data are still under analysis, an increase in glycolytic enzymes was detected, accompanied by an increase in enzymes involved in fermentative pathways. These changes suggest that a metabolic switch known as the Warburg effect (Vander Heiden et al. 2009; Bartrons and Caro 2007; Vatrinet et al. 2015), which has been described in animals exposed to several pollutants (Xu et al. 2014b), might have occurred. Proteins involved in Ca^{2+} metabolism were also more abundant in polluted crayfish. Ca^{2+} in the cytoplasm of a neuron plays a pivotal role in controlling universal cellular events, such as hormone and peptide secretion, gene transcription, and cell proliferation. Some pollutants increase Ca^{2+} uptake, causing neuronal excitotoxicity (Chi et al. 2012). These preliminary results are highly promising and suggest that it will be possible to elucidate the neuronal alterations caused by exposure to pollution.

Metabolomics in P. clarkii

Metabolomics is based on the study of the complete set of endogenous low-molecular-weight compounds in living organisms at a specified time under specific environmental conditions. This approach is a good tool to understand biological processes in complex systems because metabolites may be considered the final product of interactions between genes and proteins and the cellular environment. Different platforms involving gas chromatography (GC) or liquid chromatography (LC) coupled with mass spectrometry analyses (MS) are used to detect metabolites.

The use of metabolomic analysis in aquaculture is helping to develop methods for both better farming of the species and better processing techniques that guarantee the quality of the product. Metabolic studies have assessed the suitability of different dietary regimes (Schock et al. 2012), the use of biomarkers for fish storage time (Savorani et al. 2010), and the metabolic consequences of fish handling (Karakach et al. 2009). However, the use of metabolomic analyses on aquatic organisms has been dominated by environmental studies. One pioneering environmental metabolomics study characterized organismal responses to toxic stress and discovered novel biomarkers for use in environmental diagnostics (Bundy et al. 2009).

A recent study (Gago-Tinoco et al. 2014) determined the responses induced by contaminants in *P. clarkii* specimens from the DNP and surroundings. For this purpose, a metabolomic analysis was carried out by direct infusion into a triple quadrupole time-of-flight mass spectrometer with an electrospray ionization source (DI-ESI-QqQ-TOF) to investigate the relationships between metal levels and the metabolic response. Altered levels of several metabolites were discovered as potential biomarkers of pollution, such as decreased levels of carnosine, alanine, niacinamide, acetoacetate,

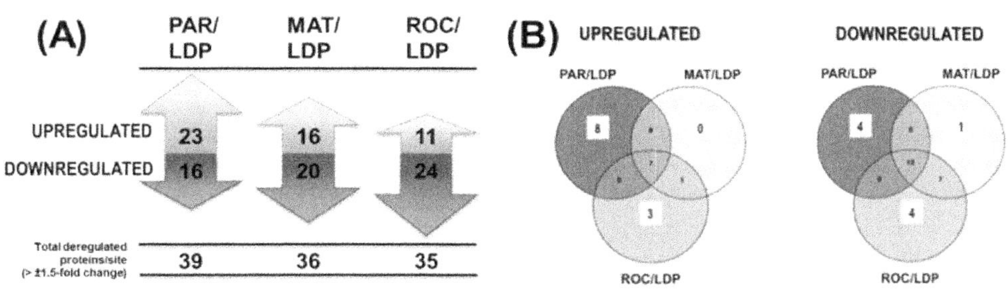

Fig. 5: (A) Numbers of up- and down-regulated proteins in the problem sites studied in relation to the LDP reference site. (B) Venn diagram showing the overlap of the various differentially expressed proteins in the three problem sites.

pantothenic acid, ascorbate, glucose-6-phosphate, arginine, glucose, lactate, phospholipids, and triglycerides, as well as elevated levels of acetyl carnitine, phosphocholine, choline, and uric acid. In this way, metal-induced toxicity could be related to metabolic impairments, principally oxidative stress, metabolic dysfunction, and dyslipidemia. The data demonstrated the strong impact of pollutants, chiefly metals, in the metabolism of *P. clarkii*, which was used as a bioindicator (Gago-Tinoco et al. 2014). New metabolomic analyses are being carried out in our labaimed at better understanding the responses in specific tissues of crayfish exposed to complex environmental pollution and also to single metal or other contaminants, under controlled laboratory exposures. Our analyses are also aimed at quantifying the metabolites identified as biomarkers. The integration of these results with the proteomic and transcriptomic data obtained from both the field and the lab studies will allow us to obtain a complete vision of the metabolic and signaling pathways altered by pollution in the crayfish, of the intensity of the changes, and of the association between individual pollutants and the triggered responses. In addition, it will be possible to define a battery of biomarkers to easily and efficiently monitor environmental risks to the ecosystem.

Solea senegalensis, an Interesting Fish in Environmental and Aquaculture Studies

Fish diseases are indicators of marine ecosystem health because they provide a biological end-point of historical exposure to stressors. Flatfish are good candidates as sentinels for the biological effect of aquatic pollution as they are benthic organisms that bury and feed in sediments and therefore experience significant exposure to sediment-associated toxicants. Furthermore, a number of flatfish are important food resources, with a high commercial interest due to the elevated cost of their white flesh. For these reasons, the aquaculture of several of these species has been enhanced over the last few years. Nonetheless, the limited knowledge of the complex biology of flatfish has limited the development of an efficient aquaculture industry. In recent years, an important effort has been made to employ functional genomic and proteomic approaches to better understand the molecular mechanisms underlying different physiological processes, including immunological and toxicological processes, in several flatfish species (Cerda et al. 2010; Cerda and Manchado 2013; Forne et al. 2010; Prieto-Alamo et al. 2012a; Williams et al. 2014). Nevertheless, because most of these species are not considered model organisms, their genomic resources are still very limited, and it has been necessary to devise experimental approaches to overcome the problem of limited sequence coverage.

In Southern latitudes of Europe, the flatfish *Solea senegalensis* has been chosen as a sentinel in several field and laboratory studies assessing pollution (Sole et al. 2014 and references herein). It is also an important candidate for diversification of Southern Europe aquaculture due to its high market value and fast growth rates (Benzekri et al. 2014). However, the aquaculture of *S. senegalensis* is hindered by its high sensitivity to different stresses and infectious diseases, which results in its high mortality in the exploitations. Consequently, development of genomic resources is a priority to facilitate the characterization of the Senegalese sole's biological responses to different stress situations, infections, and pollutants.

Transcriptomic studies in Solea senegalensis

Senegalese sole genes whose transcription is altered in the liver and/or head kidney in response to toxicants (copper and LPS) were identified using a combination of different and complementary experimental and methodological approaches, such as SSH libraries, heterologous DNA microarrays, and quantitative real-time RT-PCR (Osuna-Jimenez et al. 2009; Prieto-Alamo et al. 2009; Prieto-Alamo et al. 2012a). Although copper is an essential element, its overload is potentially toxic to most organisms. Moreover, the biological response of the Senegalese sole to copper exposure is of interest also because $CuSO_4$ is used in aquaculture as a bactericide and algaecide (Han et al. 2001). LPS is a complex molecule composed of a lipo-polysaccharide chain and a toxic lipid moiety responsible for its immunostimulatory properties (Swain et al. 2008). In most fish species the liver is the main metabolic organ involved in detoxifying both organic and inorganic chemicals and plays a central role in the defense system. The head kidney is an important hematopoietic organ in teleosts and also serves as a secondary lymphoid organ that induces and strengthens the induction and elaboration of immune response.

Construction of SSH libraries to identify differentially expressed genes in S. senegalensis exposed to copper and LPS

Four subtractive libraries in *S. senegalensis* were obtained to identify genes whose expression levels in response to LPS in the head kidney or to $CuSO_4$ in the liver differed relative to their levels in response to PBS controls (Prieto-Alamo et al. 2009). In both cases, forward (F) and reverse (R) libraries were designed to identify genes that were up- or down-regulated, respectively. To circumvent inter-individual variations and temporal differences in the responses, the libraries were constructed with total RNA from pooled head kidney or liver ($\geq$ 10 fish/condition) of soles treated with LPS or $CuSO_4$ for 6 and 24 h. Four hundred sixty clones were sequenced, and ESTs compared with the open access databases. A total of 222 unique sequences were detected, and 185 were identified as being associated with genes involved in central physiological functions (Table 5).

Consistent with the immunological role of the head kidney and the immunostimulatory properties of LPS, the putative products of an elevated number of the identified ESTs were related to immune responses (Fig. 6). Interestingly, the number of transcripts involved in immunity was even higher in the liver. These transcripts code essentially for acute phase proteins (lysozyme, coagulation factors, proteinase inhibitors, complement components, Fe transport/homeostasis proteins, etc.), according to previous studies that point to the liver as an important source of immune transcripts (Ewart et al. 2005) that mediate a powerful acute phase response (Bayne and Gerwick 2001). Furthermore, these results are also consistent with those of a study that showed that copper induces up-regulation of the cytokine TGF-β in the striped bass (Geist et al. 2007), supporting the capacity of metals (including copper) to alter immunological competence.

Other genes identified in the libraries from the livers of $CuSO_4$-treated soles encoded products involved in osmoregulation and nitrogen excretion (e.g., liver angiotensinogen, sodium potassium ATPase beta subunit, kininogen 1, angiotensin I converting enzyme 1, and alanine-glyoxylate aminotransferase 2-like), consistent with the previously reported effects of copper on osmoregulation, acid-base balance, and nitrogen excretion (See, e.g., Blanchard and Grosell 2006; Evans et al. 2005).

DNA microarrays

DNA microarrays permit, in a single experiment, the analysis of the levels of thousands of transcripts, making them a high-throughput methodology in functional genomics studies. Moreover, heterologous hybridization allows the use of microarrays manufactured from transcripts of one species to probe gene expression in another related species. The microarray developed for the European flounder *Platichthys flessus* (GENIPOL platform, (Williams et al. 2006)) has been used to assess cross-species hybridization to the transcriptome of many species of fish (including several flatfish), confirming the suitability of heterologous microarray analyses between closely related species (Cohen et al. 2007). The GENIPOL microarray is being successfully used in flatfish to identify gene expression changes in response to environmental toxicants or as a result of genetic adaptation (Nakayama et al. 2008; Williams et al. 2014; Williams et al. 2006). Along this line, this platform turned out to be a very useful tool for analyzing transcriptional expression in *S. senegalensis*. In a previous study (Osuna-Jimenez et al. 2009), changes in gene expression were analyzed in pooled livers of soles treated with $CuSO_4$ or LPS for 6 or 24 h. A total of 405 genes were differentially expressed (172 up-regulated and 233 down-regulated) after copper treatment for 6 h, 468 genes were differentially expressed (251 up-regulated and 217 down-regulated) after copper treatment for 24 h, 271 genes were differentially expressed (101 up-regulated and 170 down-regulated) after LPS treatment for 6 h, and 664 genes were differentially expressed (341 up-regulated and 323 down-regulated) after LPS treatment for 24 h. A gene ontology analysis (Blast 2GO software) identified genes specifically responsive to $CuSO_4$ (cell junction and cell signaling genes), to LPS (glutathione transferase and immune response genes) or to both treatments (immune response, digestive enzyme, unfolded protein binding, intracellular transport

Table 5: Characteristics of the *S. senegalensis* SSH libraries (Prieto-Alamo et al. 2009; Prieto-Alamo et al. 2012a).

	LPS (head kidney)	$CuSO_4$ (liver)
Number of sequenced clones	231	229
Number of analyzed clones	222	226
Number of unique ESTs	133	89
ESTs Up-regulated	62	48
Identified ESTs	49	44
Non-identified ESTs	13	4
ESTs Down-regulated	71	41
Identified ESTs	58	34
Non-identified ESTs	13	7

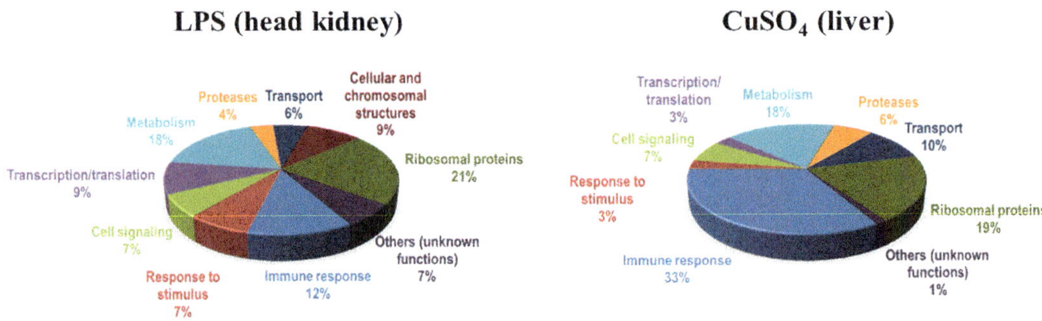

Fig. 6: Functional classification of ESTs obtained in the SSH libraries from LPS and CuSO4-treated. *S. senegalensis* individuals (Prieto-Alamo et al. 2009).

and secretion, and proteasome genes) (Table 6). For instance, among the genes specifically induced by copper in the cell junction category are those related to cellular adhesion, such as the claudins (CLDN26) and also genes related to cell signaling, such as GIT2, which encodes G-protein-coupled receptor kinase interactor 2, consistent with the ability of copper to alter the permeability of tight junctions in human intestinal mucosa (Ferruzza et al. 2002)? Glutathione-S-transferases (GSTs) were identified among the transcripts down-regulated in response to LPS treatment for 24 h, consistent with the ability of LPS to down-regulate biotransformation activities such as GSTs in several fish species (Reynaud et al. 2008). On the other hand, genes related to the immune response were specifically induced by LPS, including genes coding for the antimicrobial peptide hepcidin (HAMP), TNFα-induced protein 9 (TNFAIP9), cytokines (IL8, IL25) and chemotaxins (LECT2).

Conversely, other genes involved in immunity were induced by both LPS and $CuSO_4$ treatments (Table 6). This was the case for acute phase proteins haptoglobin (HP) or the C7 component of the complement system (Bayne and Gerwick 2001). The up-regulation of C7 by copper is consistent with the novel biological functions of complement proteins, distinct from their role in innate immunity (Mastellos et al. 2005). The GO analysis revealed that although the fish were fasting during the experiment, copper and LPS treatments down-regulated genes encoding digestive enzymes, such as trypsin (PRSS2), chymotrypsin (CTRB), elastase (ELA4), and carboxypetidase A (CPA1) and B (CPA2); this down-regulation might be due to a general stress caused by the treatments (Auslander et al. 2008). LPS and copper treatments also stimulated the expression of transcripts encoding proteins that bind to unfolded proteins (which are induced in fish tissues in response to different stressors (Basu et al. 2002)), proteasomal proteins (consistent with previous results in mammalian cells (Fernandes et al. 2006; Qureshi et al. 2003)), or proteins related to intracellular transport and secretion, to accommodate

Table 6: Selected genes significantly (FDR < 0.05) differentially expressed during copper and/or LPS treatments in the liver of the *S. senegalensis* (Osuna-Jimenez et al. 2009; Prieto-Alamo et al. 2012a).

		$CuSO_4$		LPS	
Biological function	**Gene**	**6 h**	**24 h**	**6 h**	**24 h**
Cell junctions	CLDN26, GIT2	↑[a]	–	–	–
Glutathione-S-transferases	GST-A, GST1, GST3	–	–	–	↓
Immune response	HAMP, TNFAIP9, IL8, IL25, LECT2	–	–	↑	↑
	C7, HP	↑	↑	↑	↑
Digestive enzymes	PRSS2, CTRB, ELA4, CPA1, CPA2	↓	↓	↓	↓
Unfolded protein binding	GP96, HSP70	–	↑	–	↑
Proteasome	PMSD3, MSUG1	–	↑	–	↑
Intracellular transport/secretion	ARF5, TMED7, SEC22	–	↑	↑	↑

[a]↑ means up-regulations and ↓ down-regulations.

the rapid onset of cytokine secretion and to facilitate the membrane traffic associated with phenotypic changes accompanying immune activation (Pagan et al. 2003).

We also analyzed the suitability of the GENIPOL microarray platform, which was constructed from ESTs derived from the flounder liver and which has mainly been used in studies on hepatic expression (Cohen et al. 2007; Williams et al. 2008; Williams et al. 2006; Williams et al. 2007) to analyze the transcriptional response in other fish tissues. To this end, basal expression in head kidney and liver in *S. senegalensis* were compared. From a quantitative point of view, although the microarray was constructed with hepatic ESTs, the number of genes identified as overexpressed was similar in both organs, though slightly higher in the head kidney than in the liver: 418 transcripts were expressed at higher levels in the liver than in the head kidney, and 586 transcripts were expressed at higher levels in the head kidney than in the liver.

A more detailed functional survey revealed that the most represented biological processes amongst the genes up-regulated in the liver compared to the head kidney were innate immune response, digestion, lipid transport, and monooxygenase activity (Table 7). As previously discussed, the liver is the main source of plasmatic proteins of the acute phase response, therefore, the category innate immune response grouped genes coding for these proteins. The term "monooxygenase activity" includes cytochrome P450 family genes, according to the detoxifying capability and the biotransformation activity of fish liver (Thorgaard et al. 2002). The functional categories associated with the transcripts more abundant in head kidney than in liver were predominantly related to cellular division and protein turnover (proteins involved in protein degradation, the proteasome, and ribosomal proteins) (Table 7).

These results are in consonance with the above mentioned role of the head kidney as a major hematopoietic organ in teleosts and with its function as a lymphoid organ involved in clearing soluble and particulate antigens from circulation (Whyte 2007). Surprisingly, in both organs, two sequences of the same gene (GAPDH; glyceraldehyde 3-phosphate dehydrogenase) were identified as up-regulated, although both sequences showed different tissue expression patterns. One of them was more abundant in the liver, and the second sequence was more abundant in the head kidney. *Solea senegalensis* possesses two different paralogous GAPDH genes that exhibit different tissue expression profiles, with GAPDH1 being more abundant in the liver and GAPDH2 isoform in the head kidney (Manchado et al. 2007). A more detailed analysis of the sequences detected in the microarrays revealed that they matched the described isoforms.

Overall, these results demonstrate the suitability of the GENIPOL microarray platform for analyzing the transcriptional response in the head kidney of *S. senegalensis*, even discriminating between genes encoding transcripts with specific tissue expression patterns. Therefore, these microarrays were used to evaluate the response to LPS in the head kidney. After 24 h of LPS treatment, 224 genes were significantly differentially expressed in the head kidney (117 up-regulated and 107 down-regulated). The functional analysis revealed

Table 7: Selected genes statistically (FDR < 0.05) differentially expressed in liver vs. head kidney in *S. senegalensis* (Osuna-Jimenez et al. 2009; Prieto-Alamo et al. 2012a).

Genes up-regulated in liver
Innate immune response
alpha-1-antitrypsin, alpha-2-macroglobulin, anticoagulant protein C precursor, coagulation factor VIIc, chemotaxin, complement component C3, complement component C8, complement component C9, complement regulatory plasma protein, fibrinogen alpha, fibrinogen beta chain precursor, fibrinogen gamma chain precursor, haptoglobin, hepcidin precursor, interleukin 8 precursor, kininogen 1, plasma protease C1 inhibitor precursor, prothrombin precursor, putative complement factor, transferrin
Digestive enzymes
chymotrypsinogen 1, chymotrypsinogen 2, trypsinogen 2 precursor
Lipid transport
apolipoprotein A-I, apolipoprotein A-IV, apolipoprotein C-I precursor, apolipoprotein E, apolipoprotein H, 14 kDa apolipoprotein, fatty acid-binding protein
Monooxygenase activity
cytochrome P450 2F2, cytochrome P450 2X, cytochrome P450 3A, cytochrome P450 3A45, cytochrome P450 8B1, cytochrome P450 monooxygenase
Genes up-regulated in head kidney
Cellular division/cytoskeleton
alpha-tubulin, actin-related protein 3 homolog, actin related protein 2/3 complex subunit 4, beta-actin, cofilin 2, coronin 1A, lamin B1, microtubule-based motor protein, mitotic spindle assembly checkpoint protein, myosin regulatory light chain 2, nuclear movement protein PNUDC, thymosin beta-4
Protein degradation/proteasome
polyubiquitin, proteasome alpha 1 subunit isoform 2, proteasome (prosome, macropain) subunit alpha type 7, proteasome beta-subunit C5, proteasome subunit beta type 3, proteasome (prosome, macropain) subunit beta type 5, proteasome 26S ATPase subunit 5, proteasome subunit N3, ubiquitin carboxyl-terminal hydrolase isozyme L, ubiquitin specific protease 9
Ribosomal proteins
40S ribosomal protein S3a, 40S ribosomal protein S4, 60S ribosomal protein L3, 60S ribosomal protein L4, 60S ribosomal protein L13

that the LPS treatment affected similar biological processes in the head kidney and liver. The functional groups that were most affected among the up-regulated genes were immune response, unfolded protein binding, intracellular transport/secretion, and proteasome. Among the down-regulated genes, the functional group that was most affected was digestive enzymes. On the contrary, the GSTs category, which was down-regulated in response to LPS treatment in the liver, was not altered in the head kidney.

Recently, two species-specific oligo-DNA microarray platforms have been described for *S. senegalensis*. An earlier version representing 5087 unigenes (Cerda et al. 2008) was designed against the 3'-end sequenced ESTs from a normalized cDNA library obtained from six tissues and five different developmental stages. This microarray was validated in a preliminary larval development survey and has been used to obtain information on the molecular basis of ovarian development (Tingaud-Sequeira et al. 2009) and spermatogenesis (Forne et al. 2011). An improved second version representing 30,119 non-redundant transcripts (Benzekri et al. 2014) has been obtained using transcriptome information produced by NGS (*New-Generation Sequencing*) technologies covering a large number of developmental stages. This microarray has been tested with larvae incubated at two salinities (10 and 36 ppt) (Benzekri et al. 2014), making it a useful tool for future large-scale gene expression studies in *S. senegalensis*.

Determination of transcriptional profiles by quantitative RT-PCR

As stated above, the determination of absolute transcript abundance (the exact number of molecules of a transcript in all samples of the study) is the only adequate procedure to accurately assess the expression of a gene (Prieto-Alamo et al. 2003) but these types of studies are infrequent because of the experimental difficulties that impair these analyses. The absolute expression of a set of more than 20 genes was assessed in the liver and/or head kidney of $CuSO_4$- or LPS-treated soles. The transcripts were selected based on the magnitude and specificity of their responses in the SSH (Prieto-Alamo et al. 2009) or in the heterologous microarray surveys (Osuna-Jimenez et al. 2009) and on the physiological relevance of their products: complement component C3, complement component C7, transferrin (TF), haptoglobin (HP), ferritin M, natural killer enhancing factor (NKEF), tumor necrosis factor alpha-induced protein 9 (TNFAIP9), hepcidin (HAMP), nonspecific cytotoxic cell receptor protein-1 (NCCRP1), angiotensinogen, sequestosome 1 (SQSTM1), and tumor necrosis factor receptor associated factor (TRAF3) in immune response; CCAAT/enhancer binding protein (CEBPB), cold inducible RNA binding protein (CIRBP), DNA-damage-inducible transcript 4-like (DDIT4L), and NHP2 non-histone chromosome protein (HMGB2) in stress response; glyceraldehyde-3P-dehydrogenasese (GAPDH), transketolase (TKT), and NADH-dehydrogenase 1 alpha subcomplex 4 (NDUFA4) in energy metabolism; asparaginil-tRNAsynthetase (NARS), heat shock protein GP96, proteasome 26S non-ATPase subunit 3 (PSMD3), and cathepsin Z (CTSZ) in protein synthesis, folding, and degradation; and α-globin in transport (Osuna-Jimenez et al. 2009; Prieto-Alamo et al. 2009).

In general, the results obtained with the subtractive libraries and the DNA microarrays were confirmed by qRT-PCR quantification. Nevertheless, while microarray and qRT-PCR data highly correlated quantitatively, the absolute quantification of transcripts was more sensitive because in most cases, the magnitude of the changes detected by qRT-PCR was greater. Additionally, a high proportion (~50%) of the genes isolated by SSH (Prieto-Alamo et al. 2009) were also identified as differentially expressed by microarray analyses (Osuna-Jimenez et al. 2009), providing further confidence in the data. To analyze in depth the transcriptional profiles of the selected transcripts, the absolute quantification was not limited to the samples used for the transcriptomic analyses but rather extended to samples from different individuals, organs, treatments and exposure times, validating at an individual level a substantial part of the changes in transcript copy number previously quantified in pooled samples. Therefore, these analyses provided valuable additional information on transcriptional expression profiles of *S. senegalensis* in response to toxicants. Individual quantification is also mandatory to prevent biased interpretations of specimens with abnormal expression levels. In this context, the absolute real-time qRT-PCR analyses on individual samples demonstrated that inter-individual variations of most of the examined transcripts in the treated soles were in the range of those of the control fish, indicating similar susceptibility to LPS or $CuSO_4$ challenge between individuals. In general, different expression patterns were distinguished based on both the specificity of the stressor and course response, always as a function of the organ analyzed.

Proteomic studies in Solea senegalensis

Proteomics is a powerful comparative tool and has therefore been increasingly used over the last decades to address different questions in aquaculture, regarding welfare, nutrition, health, quality, and safety. Most proteomic studies have been carried out in Atlantic salmon (*Salmo salar*), rainbow trout (*Oncorhynchus mykiss*), catfish (*Ictalurus punctatus*), and cod (*Gadusmorhua*), mainly because of their commercial importance. Studies have focused on stress related to farming conditions, muscle quality, the effects of administering probiotics, and the responses to common infections because infection and concomitant death are one of the main causes of heavy economic losses in marine fish cultures (Rodrigues et al. 2012, and references herein).

An interesting study analyzed the changes in the proteome of *S. senegalensis* during larval development. The study identified 23 proteins grouped in five functional categories including structural proteins, metabolism, transport, stress response, organization of chromatin, and gene expression control (Fig. 7).

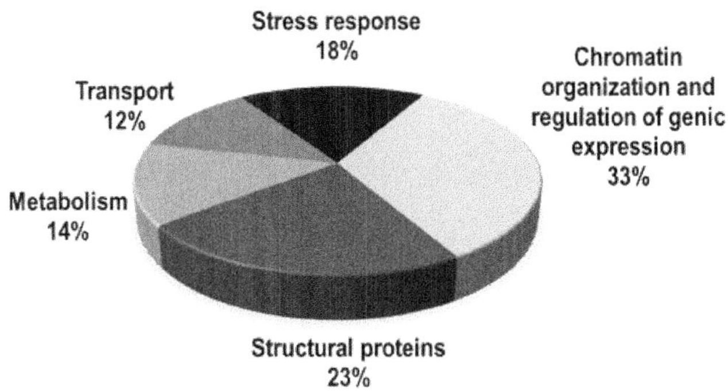

Fig. 7: Functional categories grouping the identified proteins with differential abundance along the *S. senegalensis* larvae metamorphosis.

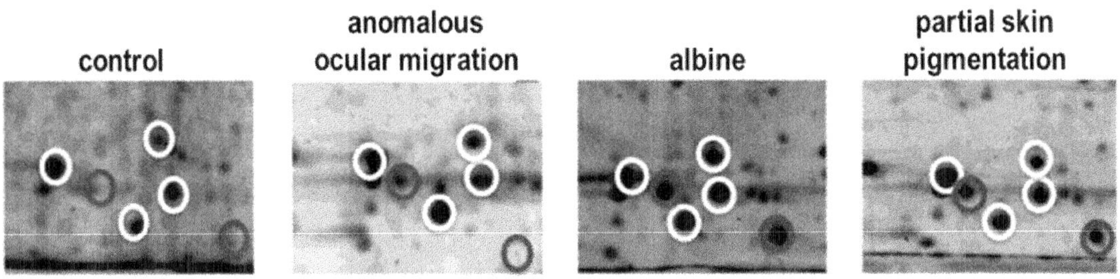

Fig. 8: Representative gel images showing changes in abundance of some proteins linked to anomalous ocular migration and skin pigmentation anomalies.

These proteome profiles were used as references in subsequent studies analyzing alterations linked to erroneous ocular migration or skin pigmentation anomalies (Fig. 8).

Evaluation of ecological risk assessment of aquatic sediments by determining Solea senegalensis responses at the proteomic level

The ecological risk assessment of aquatic sediments is extremely complex because of the presence of mixtures of contaminants and the intrinsic sediment characteristics that determine their bioavailability (Costa et al. 2009).

The benthic fish *S. senegalensis* has been successfully employed in bioassays with contaminated sediments (Costa et al. 2009; Jiménez-Tenorio 2007; Costa et al. 2008). However, little research has been able to relate toxicity to responses to specific classes of mixed sediment contaminants. To understand better the biological mechanisms underlying exposure to complex mixtures of contaminants, as in natural sediments, juvenile Senegalese soles were exposed to contaminated sediments collected from the Sado Estuary (West Portugal), an ecosystem impacted by humans (Fig. 9). The west bank of the Sado Estuary is classified as a natural reserve. However, this large coastal area is also subjected to urbanistic, industrial, agricultural, aquacultural, and touristic activities.

Sediment characterization confirmed that the two contaminated sediments contained a combination of metals, polycyclic aromatic hydrocarbons (PAHs), and organochlorines. All the pollutants shown in Fig. 10 were present at levels exceeding those described in the literature as frequently causing effects (Costa et al. 2012).

Changes in liver cytosolic protein profiles were determined by 2-DE, and deregulated proteins were identified by *de novo* sequencing by tandem mass spectrometry (Fig. 11) (Costa et al. 2012). Forty-one

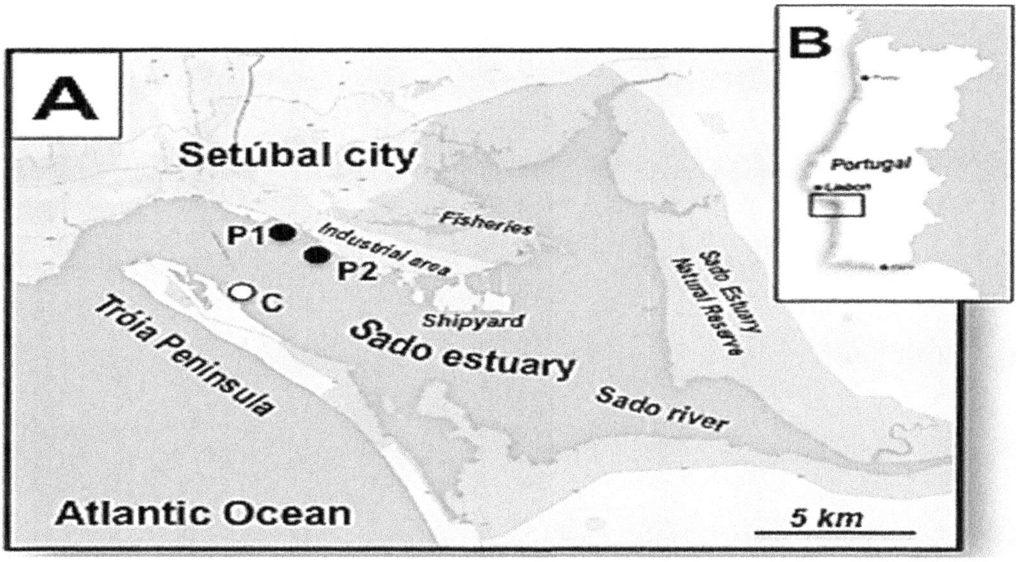

Fig. 9: Location of sampling sites. (A) The Sado estuary around the city of Setúbal, showing the areas where polluted (P1, P2) and reference samples were collected (C). (B) Location of the studied area at Portugal (SW Europe).

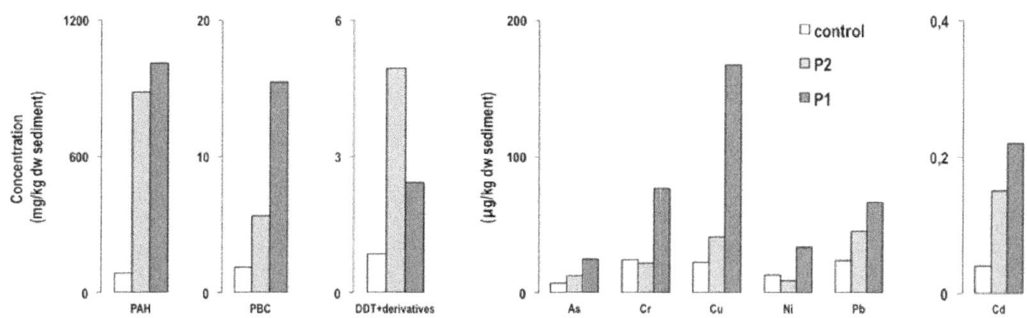

Fig. 10: Characterization of PAHs, organochlorines and metals in tested sediments collected at the control C and the two polluted P1 and P2 sites. Metals were quantified by ICP-MS using a Thermo Elemental X-Series equipment according to (Caetano et al. 2007). Sediment PAHs were quantified by GC–MS after Soxhlet extraction with an acetone/hexane (Martins et al. 2008). Organochlorines, DDT and its main metabolites pp'DDE and pp'DDD were determined by GC-ECD after Soxhlet extraction with n-hexane and column fractioning (Ferreira et al. 2003) (Adapted from (Costa et al. 2012)).

spot proteins with statistically significantly differences in abundance between control and polluted fish were gel-excised and sequenced, and 19 proteins were identified.

The functional analysis of these proteins indicated that exposure to polluted sediments affected multiple cellular processes such as anti-oxidative defense, energy production, proteolysis, and xenobiotic catabolism (especially oxidoreductase enzymes). Soles exposed in the laboratory to contaminated sediments failed to induce, or even markedly down-regulated, many proteins (58%) when compared to reference fish. The few proteins with increased abundance in polluted fish were peroxiredoxin, an anti-oxidant enzyme; IST1, a MAPK kinase involved in cell signaling, and a few others. Down-regulation of basal metabolism enzymes, mainly related to energy production and gene transcription, in fish exposed in the laboratory to contaminated sediment, may be linked to sediment-bound contaminants and likely compromised the organisms' ability to arrange adequate responses against situations of stress (Costa et al. 2012).

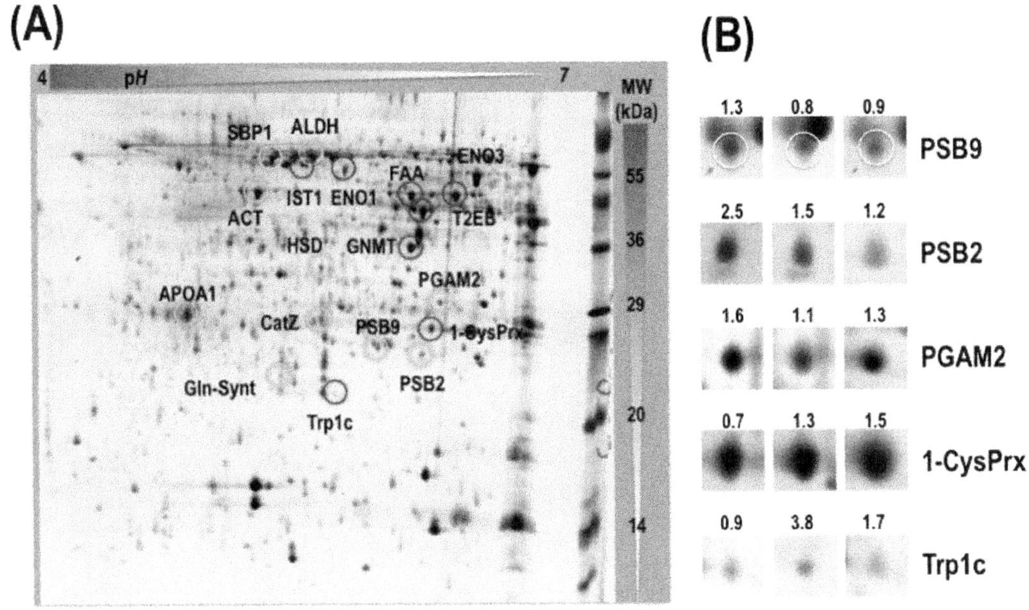

Fig. 11: 2-DE analysis of changes in the protein profiles of Senegalese sole exposed to sediments collected at C (reference), P1 and P2 (two polluted sites at the Sado estuary). (A) Master gel combining spots images of the three samples. The relative *Mr* is given on the left and the p*I* is given at the top of the figure. Protein spots with a significant change in abundance between C and P1/P2 are indicated with circles. (B) Representative spots showing changes in protein abundance.

Responses in *Solea senegalensis* to metals and PAHs

Organic enriched sediments and/or suspended particles act as sinks for metals, such as cadmium, and hydrophobic organic xenobiotics, such as polycyclic aromatic hydrocarbons (PAHs). Cadmium is a natural constituent of ocean water, with average levels ranging from 5 to 20 ng/L in the open seas (Kremling and Streu 2001; OSPAR 2002). However, due to the wide use of Cd in industrial processes (anticorrosive agent, stabilizer in PVC products, color pigment, neutron-absorber in nuclear power plants, fabrication of nickel-cadmium batteries) higher levels of this metal, reaching 73.8 mg/L, have been reported in polluted coastal waters (Yilmaz and Sadikoglu 2011; Beiras et al. 2003). Cd ions are highly toxic even at low doses (Yilmaz and Sadikoglu 2011; Beiras et al. 2003), but the exact mechanisms of Cd toxicity are not yet fully understood. It is thought that Cd may displace iron and zinc from metallothioneins, proteins such as copper-zinc superoxide dismutase (CuZn SOD), and zinc-finger class proteins (Lopez-Barea and Gomez-Ariza 2006). Exposure to this metal is known to induce depletion of reduced glutathione (GSH), inhibition of antioxidant enzymes and of energy metabolism, and enhanced production of reactive oxygen species (ROS) (Zhai et al. 2013). Thus, increased lipid peroxidation and oxidative DNA damage, inflammatory processes, apoptosis, and necrosis have been described among the mechanisms of Cd-induced liver injury (Templeton and Liu 2010; Satarug et al. 2010; Rana 2008; Murugavel and Pari 2007; Moniuszko-Jakoniuk et al. 2005; Matovic et al. 2012; Koyu et al. 2006; Jurczuk et al. 2004; Jurczuk et al. 2003; Jihen et al. 2010; Filipič 2012; Brzoska et al. 2011; Brzoska and Rogalska 2013; Asara et al. 2013; Afolabi et al. 2012). Benzo(a)pyrene (B[a]P) is a PAH arising as a byproduct of incomplete combustion or pyrolysis of organic material. B[a]P is metabolized to form a number of metabolites (PAH quinones or the highly genotoxicdiol epoxides) that may be toxic because of their ability to bind to DNA. The formation of B[a]P-DNA adducts can interfere with or alter DNA replication and have been associated with an increased risk of several forms of cancer.

Many studies have been carried out to evaluate Cd and B[a]P toxicity in aquatic and terrestrial environments (Olatunji et al. 2015; Osorio-Yanez et al. 2012; Friesen et al. 2008; Nordberg 2010, 2009). Both xenobiotics have also been widely employed as model toxicants in *in vivo* and *in vitro* studies. Such studies revealed the existence of antagonistic and synergistic effects of Cd and B[a]P mixtures in fish (Sandvik et al. 1997). Lewinska et al. (2007) did not find any interaction between Cd and B[a]P in inducing DNA damage in polychromatic erythrocytes in mouse bonemarrow. In contrast, others have reported that Cd enhances DNA damage induced by a B[a]P epoxide metabolite, most likely by impairment of DNA repair in human HeLa cell extracts (Mukherjee et al. 2004). This synergism was speculated to be caused by Cd-induced reduction of CYP1A activity and B[a]P-induced inactivation of metallothionein (Hurk et al. 1988), which is consistent with the findings on hepatic MT and CYP1A responses in *S. senegalensis* exposed to contaminated sediments (Costa et al. 2009). However, the effects of co-exposure to metals and PAHs on cells still remain largely unknown. To better understand the immediate mechanisms underlying the consequences of Cd and B[a]P co-exposure in the liver, the hepatic proteome profiles of soles (*S. senegalensis*) injected with subacute doses of cadmium (Cd) and/or benzo[a]pyrene were screened for alterations in cytosolic protein abundances (Costa et al. 2010). This proteomics approach indicated that different biochemical pathways were affected by exposure to Cd, B[a]P, or the combination of both chemicals (Costa et al. 2010) (Fig. 12).

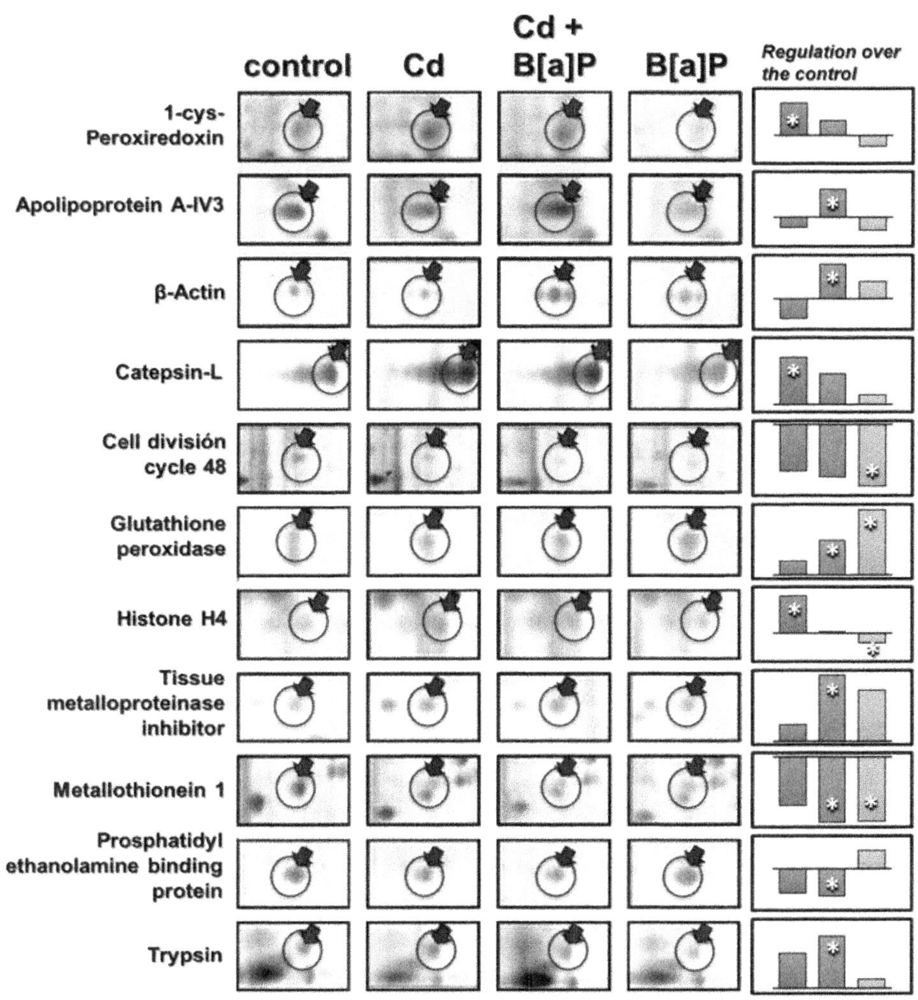

Fig. 12: 2-DE gel images showing the differential expression of eleven identified hepatic proteins in *S. senegalensis* treated with Cd, B[a]P, or Cd+B[a]P. Protein spots indicating changes > ± 1.5-fold in volume in each experimental group are shown. In addition, in the right of the figure the average protein regulation factors (expressed by an arbitrary unit relatively to control) are given. [*] significant differences of spot intensities relatively to control gels (Data adapted from (Costa et al. 2010).

Functional analysis of differentially abundant hepatic proteins in fish revealed their relationships with oxidative stress and inflammation/apoptosis. Both 1-cysperoxiredoxine (1-cysPrx) and glutathione peroxidase (GPX) are known to protect cells from apoptosis by scavenging oxidative radicals, thereby reducing oxidative stress (Gouaze et al. 2002; Manevich et al. 2002; Pak et al. 2002). As shown in Fig. 12, Cd specifically promoted 1-cysPrx up-regulation, and B[a]P specifically induced up-regulation of GPX. The combination of both xenobiotics, however, reduced the induction of both antioxidant enzymes, increasing the sensitivity of cells to oxidative damage.

A concurrent diminution of cell division cycle 48 (CDC48) and increase in abundance of cathepsin L (CatL) and trypsin caused by Cd administration clearly indicate induction of apoptosis, most likely linked to Cd-induced generation of ROS. The hexameric AAA-ATPase, CDC48, catalyzes an array of cellular activities, including endoplasmic reticulum (ER)-associated degradation (ERAD), ER/Golgi membrane dynamics, and DNA replication. Depletion of CDC48 has been demonstrated to affect continuance of the cell cycle by impairing DNA synthesis (Mouysset et al. 2008). Although DNA replication and induction of cell division by mitogens counterbalance apoptosis, the more specific role of CDC48 in the process remains unclear. CDC48 abundance was reduced by exposure to Cd and B[a]P together in a similar way. It is likely that down-regulation of CDC48 reflects a general failure of hepatic cell metabolism caused by both chemicals.

Cathepsin L and trypsin are proteases specialized in the degradation of cellular and intercellular structural proteins. In addition to their role in generalized lysosomal proteolysis, both proteases have been implicated in a variety of pathologies such as cancer and uncontrolled inflammation, and increased levels of CATL mRNA and protein have been linked to elevated apoptosis (Wei et al. 2013, and references herein). The effect of B[a]P on the expression of these proteases was small. However, the simultaneous administration of Cd and B[a]P enhanced the abundance of trypsin but diminished the amount of CatL. Although the 2-DE procedure failed to detect caspases or their precursors, our findings suggest that CatL may be involved in Cd-induced apoptosis and that Cd and B[a]P had, in this study, an antagonist effect on the regulation of this enzyme.

An imbalance between proteinases and their inhibitors has been demonstrated to be a key factor in many human diseases and pathological processes, including tumorigenesis. TIMPs are known apoptosis inhibitors, and TIMP-2 has been shown to suppress tumor growth and metastatic potential in many cell model systems, activities associated with inhibition of metalloproteinases, and the apoptotic process (Valente et al. 1998). TIMP2 is mainly induced by B[a]P, though treatment with both compounds exerts a synergic effect on its abundance. The reduced hepatocyte apoptosis observed in individuals cotreated with Cd and B[a]P (Costa et al. 2010) may thus be partially explained by TIMP up-regulation. Additionally, the up-regulation of beta-actin could contribute to the protection of hepatic cells.

Apolipoprotein ApoA-IV3 is a structural component of lipoprotein complexes that transport lipids. Recently, Apo-IV has been described as reducing oxidative stress-driven apoptosis by modulating the glutathione redox status (Spaulding et al. 2006), which may contribute to the reduced induction of apoptosis in fish exposed to both Cd and B[a]P.

Metallothioneins (MTs) are small (less than 10 kDa), cysteine-rich, metal-binding proteins that protect against metal toxicity. MT expression has been correlated with resistance to Cd and other toxic metal ions and chemicals in many studies, and the induction of MT proteins is believed to be the key mediator of induced cellular protection. It has also been reported that MT acts as an endogenous defensive factor against B[a]P-induced DNA damage (Takaishi et al. 2009). However, several experiments suggest that MT is not always the critical induced resistance factor and, more importantly, that MT overexpression may not provide additional protection beyond that afforded by low, basal MT levels (Kennette et al. 2005). In addition, it has been reported that the high expression of MT in certain tumors may block apoptosis. Cd alone or combined with B[a]P contributed to a more pronounced MT down regulation. One of the most important mechanisms of Cd toxicity is related to its ability to replace zinc (Zn) in several proteins including MTs themselves (Lopez-Barea and Gomez-Ariza 2006) and may be responsible for the observed effect. Interestingly, apoMT has been found to be rapidly degraded by cathepsins (including CatL) (Min et al. 1992). It is suspected that MT levels in Cd-treated livers may not reflect MT expression alone but also that of cysteine proteinases.

The present findings results (Costa et al. 2010) confirmed that the biochemistry of metal/PAH interactions in the vertebrate liver are complex and may involve a broad range of protein responses with multiple consequences at the histological and cytological levels.

The Utility of Omics Studies to Improve Animal Health under Cultivation

Omics approaches (genomics, transcriptomics, proteomics, and metabolomics) can assist the identification of suitable markers to monitor the welfare of cultured fish and the quality of aquaculture products. Therefore, these methodologies are being applied to better understand the molecular mechanisms underlying reproduction, development, nutrition, disease resistance, and the response to environmental stresses (revised in (Cerda et al. 2010; Rodrigues et al. 2012)). These studies will contribute to achieving an equilibrium between productivity and higher disease and stress resistance or a better feed efficiency to lower environment impact, which are critical for sustainable development of the fish farming industry.

So far, the EST databases obtained by Sanger sequencing of SSH libraries have been successfully used not only as a tool for holistic transcriptomic analyses but also as a source of cDNA sequences allowing the investigation of the expression of individual genes detected as responsive in specific conditions that could affect animal health and culture procedures. Along this line, the transcriptomic studies in *S. senegalensis* referred above (Osuna-Jimenez et al. 2009; Prieto-Alamo et al. 2009; Prieto-Alamo et al. 2012b) allowed the establishment of a panel of genes whose expression have been evaluated to elucidate the immunological response of *S. senegalensis* specimens under high stocking density and receiving the dietary probiotic *Shewanella putrefaciens* Pdp11, as well as the effect that these factors exert on disease resistance (Tapia-Paniagua et al. 2014). Overall, the data demonstrated that social stressors may provoke an immunological depression because the animals farmed under crowding conditions showed decreased levels of transcripts that have been previously described as relevant in the innate immune system of *S. senegalensis* (Osuna-Jimenez et al. 2009; Prieto-Alamo et al. 2009; Prieto-Alamo et al. 2012b), notably those involved in the acute phase response (C7, HP, and NARS) or that play a direct role against pathogens (G-LYZ). On the other hand, a group of fish exposed to high stocking density that presented a microbial infection showed up-regulation of a majority of the immune related genes studied. Interestingly, the levels of these transcripts returned to normal when the animals were fed the Pdp11 diet, indicating that the probiotic assisted the recovery of the specimens after infection and could improve stress tolerance of Senegalese soles to high stocking densities by modulating the expression of important immune related genes.

Currently, the use of NGS technologies allows a complete evaluation of the transcriptome in a given tissue under specific experimental conditions, even in species that lack a reference genome. Along this line, using RNA-seq, the reference transcriptomes of several species of interest in fish farming have been reported (e.g. (Liu et al. 2014; Robledo et al. 2014)), including that of *S. senegalensis* (Benzekri et al. 2014). This study has not only greatly increased the number of available DNA sequences but also permitted the identification of SNPs and simple-sequence repeats (SSRs), which constitute very useful molecular markers. The transcriptome information has been hosted in a public database, the SoleaDB, which will facilitate functional genomics studies in this important fish species at both transcriptomic and proteomic levels. However, a major gap in proteomics studies is that the proteomes of species commonly used in aquaculture are not annotated. Despite this limitation, the study of different aspects of aquaculture related to welfare and health, nutrition, or environmental sources of stress, have also benefited from proteomics technologies (Rodrigues et al. 2012). The application of a combination of transcriptomic and proteomic analyses constitutes a comprehensive approach (Forne et al. 2010) that has been used, for instance, to study the process of spermatogenesis in the Senegalese sole (Forne et al. 2011).

To conclude, it can be stated that the integration of complementary omic methodologies (transcriptomics and proteomics) provides a wider perspective of biological responses. This comprehensive approach will be very useful in identifying potentially relevant biomarkers to address future challenges in aquatic biology research.

Acknowledgements

The authors would like to thanks Prof. LópezBarea for his invaluable comments and suggestions, which have been of significant help for improving the manuscript.

This work was funded by grants CTM2009-12858-02, AGL2011-30381-C03-03 and CTM2012-38720-C03-02 (Spanish Ministry of Economy and Competitiveness), CYCE-00516, P08-CVI-03829 (Innovation and Science Agency, Andalusian Government).

Keywords: Functional genomics, transcriptomics, proteomics, metabolomics, qRT-PCR, *Solea senegalensis*, *Procambarus clarkii*, environmental pollution, aquaculture, biomarkers

References

Afolabi, O.K., E.B. Oyewo, A.S. Adekunle, O.T. Adedosu and A.L. Adedeji. 2012. Impaired lipid levels and inflammatory response in rats exposed to cadmium. EXCLI Journal 11: 677–687.

Alcorlo, P., M. Otero, M. Crehuet, A. Baltanas and C. Montes. 2006. The use of the red swamp crayfish (*Procambarus clarkii*, Girard) as indicator of the bioavailability of heavy metals in environmental monitoring in the River Guadiamar (SW, Spain). Sci. Total Environ. 366: 380–390.

Asara, Y., J.A. Marchal, E. Carrasco, H. Boulaiz, G. Solinas, P. Bandiera, M.A. Garcia, C. Farace, A. Montella and R. Madeddu. 2013. Cadmium modifies the cell cycle and apoptotic profiles of human breast cancer cells treated with 5-fluorouracil. Int. J. Mol. Sci. 14: 16600–16616.

Auslander, M., Y. Yudkovski, V. Chalifa-Caspi, B. Herut, R. Ophir, R. Reinhardt, P.M. Neumann and M. Tom. 2008. Pollution-affected fish hepatic transcriptome and its expression patterns on exposure to cadmium. Mar. Biotechnol. (NY) 10: 250–261.

Bartrons, R. and J. Caro. 2007. Hypoxia, glucose metabolism and the Warburg's effect. J. Bioenerg. Biomemb. 39: 223–229.

Basu, N., A.E. Todgham, P.A. Ackerman, M.R. Bibeau, K. Nakano, P.M. Schulte and G.K. Iwama. 2002. Heat shock protein genes and their functional significance in fish. Gene 295: 173–183.

Bayne, C.J. and L. Gerwick. 2001. The acute phase response and innate immunity of fish. Dev. Comp. Immunol. 25(8-9): 725–743.

Beiras, R., J. Bellas, N. Fernandez, J.I. Lorenzo and A. Cobelo-Garcia. 2003. Assessment of coastal marine pollution in Galicia (NW Iberian Peninsula); metal concentrations in seawater, sediments and mussels (*Mytilus galloprovincialis*) versus embryo-larval bioassays using *Paracentrotus lividus* and *Ciona intestinalis*. Mar. Environ. Res. 56: 531–553.

Benzekri, H., P. Armesto, X. Cousin, M. Rovira, D. Crespo, M.A. Merlo, D. Mazurais, R. Bautista, D. Guerrero-Fernandez, N. Fernandez-Pozo, M. Ponce, C. Infante, J.L. Zambonino, S. Nidelet, M. Gut, L. Rebordinos, J.V. Planas, M.L. Begout, M.G. Claros and M. Manchado. 2014. *De novo* assembly, characterization and functional annotation of Senegalese sole (*Solea senegalensis*) and common sole (*Solea solea*) transcriptomes: integration in a database and design of a microarray. BMC Genomics 15: 952.

Blanchard, J. and M. Grosell. 2006. Copper toxicity across salinities from freshwater to seawater in the euryhaline fish Fundulus heteroclitus: is copper an ionoregulatory toxicant in high salinities? Aquat. Toxicol. 80: 131–139.

Braconi, D., G. Bernardini and A. Santucci. 2011. Linking protein oxidation to environmental pollutants: redox proteomic approaches. J. Proteomics 74: 2324–2337.

Brzoska, M.M., J. Rogalska and E. Kupraszewicz. 2011. The involvement of oxidative stress in the mechanisms of damaging cadmium action in bone tissue: a study in a rat model of moderate and relatively high human exposure. Toxicol. Appl. Pharmacol. 250: 327–335.

Brzoska, M.M. and J. Rogalska. 2013. Protective effect of zinc supplementation against cadmium-induced oxidative stress and the RANK/RANKL/OPG system imbalance in the bone tissue of rats. Toxicol. Appl. Pharmacol. 272: 208–220.

Bundy, J., M. Davey and M. Viant. 2009. Environmental metabolomics: a critical review and future perspectives. Metabolomics 5: 3–21.

Butterfield, D.A. and I. Dalle-Donne. 2012. Redox proteomics. Antioxid. Redox. Signal 17: 1487–1489.

Cabiscol, E. and J. Ros. 2006. Oxidative damage to proteins: structural modifications and consequences in cell function. pp. 399–471. *In*: I. Dalle-Donne, A. Scaloni, D.A. Butterfield, D.M. Desiderio and N.M. Nibbering [eds.]. Redox Proteomics: From Protein Modifications to Cellular Dysfunction and Diseases. John Wiley & Sons, Inc. Hoboken, New Jersey, USA.

Caetano, M., N. Fonseca and R. Cesario Carlos Vale. 2007. Mobility of Pb in salt marshes recorded by total content and stable isotopic signature. Sci. Total Environ. 380: 84–92.

Cerda, J., J. Mercade, J.J. Lozano, M. Manchado, A. Tingaud-Sequeira, A. Astola, C. Infante, S. Halm, J. Vinas, B. Castellana, E. Asensio, P. Cañavate, G. Martinez-Rodriguez, F. Piferrer, J.V. Planas, F. Prat, M. Yufera, O. Durany, F. Subirada, E. Rosell and T. Maes. 2008. Genomic resources for a commercial flatfish, the Senegalese sole (*Solea senegalensis*): EST sequencing, oligo microarray design, and development of the Soleamold bioinformatic platform. BMC Genomics 9: 508.

Cerda, J., S. Douglas and M. Reith. 2010. Genomic resources for flatfish research and their applications. J. Fish Biol. 77: 1045–1070.

Cerda, J. and M. Manchado. 2013. Advances in genomics for flatfish aquaculture. Genes Nutr. 8: 5–17.

Chapman, R.W. 2001. EcoGenomics—a consilience for comparative immunology? Dev. Comp. Immunol. 25: 549–551.

Chi, Y.N., X. Zhang, J. Cai, F.Y. Liu, G.G. Xing and Y. Wan. 2012. Formaldehyde increases intracellular calcium concentration in primary cultured hippocampal neurons partly through NMDA receptors and T-type calcium channels. Neurosci. Bull. 28: 715–722.

Cohen, R., V. Chalifa-Caspi, T.D. Williams, M. Auslander, S.G. George, J.K. Chipman and M. Tom. 2007. Estimating the efficiency of fish cross-species cDNA microarray hybridization. Mar. Biotechnol. (NY) 9: 491–499.

Costa, P.M., J. Lobo, S. Caeiro, M. Martins, A.M. Ferreira, M. Caetano, C. Vale, T.A. DelValls and M.H. Costa. 2008. Genotoxic damage in *Solea senegalensis* exposed to sediments from the Sado Estuary (Portugal): effects of metallic and organic contaminants. Mutat. Res. 654: 29–37.

Costa, P.M., S. Caeiro, M.S. Diniz, J. Lobo, M. Martins, A.M. Ferreira, M. Caetano, C. Vale, T.A. DelValls and M.H. Costa. 2009. Biochemical endpoints on juvenile *Solea senegalensis* exposed to estuarine sediments: the effect of contaminant mixtures on metallothionein and CYP1A induction. Ecotoxicology 18: 988–1000.

Costa, P.M., E. Chicano-Galvez, J. Lopez Barea, T.A. DelValls and M.H. Costa. 2010. Alterations to proteome and tissue recovery responses in fish liver caused by a short-term combination treatment with cadmium and benzo[*a*]pyrene. Environ. Pollut. 158: 3338–3346.

Costa, P.M., E. Chicano-Galvez, S. Caeiro, J. Lobo, M. Martins, A.M. Ferreira, M. Caetano, C. Vale, J. Alhama-Carmona, J. Lopez-Barea, T.A. DelValls and M.H. Costa. 2012. Hepatic proteome changes in *Solea senegalensis* exposed to contaminated estuarine sediments: a laboratory and *in situ* survey. Ecotoxicology 21: 1194–1207.

Davies, M.J. 2005. The oxidative environment and protein damage. Biochim. Biophys. Acta 1703: 93–109.

Diatchenko, L., Y.F. Lau, A.P. Campbell, A. Chenchik, F. Moqadam, B. Huang, S. Lukyanov, K. Lukyanov, N. Gurskaya, E.D. Sverdlov and P.D. Siebert. 1996. Suppression subtractive hybridization: a method for generating differentially regulated or tissue-specific cDNA probes and libraries. Proc. Natl. Acad. Sci. U S A 93: 6025–6030.

Evans, D.H., P.M. Piermarini and K.P. Choe. 2005. The multifunctional fish gill: dominant site of gas exchange, osmoregulation, acid-base regulation, and excretion of nitrogenous waste. Physiol. Rev. 85: 97–177.

Ewart, K.V., J.C. Belanger, J. Williams, T. Karakach, S. Penny, S.C. Tsoi, R.C. Richards and S.E. Douglas. 2005. Identification of genes differentially expressed in Atlantic salmon (*Salmo salar*) in response to infection by *Aeromonas salmonicida* using cDNA microarray technology. Dev. Comp. Immunol. 29: 333–347.

Fernandes, R., J. Ramalho and P. Pereira. 2006. Oxidative stress upregulates ubiquitin proteasome pathway in retinal endothelial cells. Mol. Vis. 12: 1526–1535.

Fernandez-Cisnal, R., J. Alhama, N. Abril, C. Pueyo and J. Lopez-Barea. 2014. Redox proteomics as biomarker for assessing the biological effects of contaminants in crayfish from Donana National Park. Sci. Total Environ. 490: 121–133.

Ferreira, A.M., M. Martins and C. Vale. 2003. Influence of diffuse sources on levels and distribution of polychlorinated biphenyls in the Guadiana River estuary, Portugal. Mar. Chem. 89: 175–184.

Ferruzza, S., M. Scacchi, M.L. Scarino and Y. Sambuy. 2002. Iron and copper alter tight junction permeability in human intestinal Caco-2 cells by distinct mechanisms. Toxicol. *In Vitro* 16: 399–404.

Filipič, M. 2012. Mechanisms of cadmium induced genomic instability. Mutat. Res. 733: 69–77.

Forne, I., J. Abian and J. Cerda. 2010. Fish proteome analysis: model organisms and non-sequenced species. Proteomics 10: 858–872.

Forne, I., B. Castellana, R. Marin-Juez, J. Cerda, J. Abian and J.V. Planas. 2011. Transcriptional and proteomic profiling of flatfish (*Solea senegalensis*) spermatogenesis. Proteomics 11: 2195–2211.

Friesen, M.C., P.A. Demers, J.J. Spinelli and N.D. Le. 2008. Adequacy of benzo(*a*)pyrene and benzene soluble materials as indicators of exposure to polycyclic aromatic hydrocarbons in a Soderberg aluminum smelter. J. Occup. Environ. Hyg. 5: 6–14.

Gago-Tinoco, A., R. González-Domínguez, T. García-Barrera, J. Blasco-Moreno, M.J. Bebianno and J.-L. Gómez-Ariza. 2014. Metabolic signatures associated with environmental pollution by metals in Doñana National Park using *P. clarkii* as bioindicator. Environ. Sci. Pollut. Res. 21: 13315–13323.

Geist, J., I. Werner, K.J. Eder and C.M. Leutenegger. 2007. Comparisons of tissue-specific transcription of stress response genes with whole animal endpoints of adverse effect in striped bass (*Morone saxatilis*) following treatment with copper and esfenvalerate. Aquat. Toxicol. 85: 28–39.

Gherardi, F. 2006. Crayfish invading Europe: the case study of *Procambarus clarkii*. Mar. Freshw. Behav. Physiol. 39(3): 175–191.

Gouaze, V., N. Andrieu-Abadie, O. Cuvillier, S. Malagarie-Cazenave, M.F. Frisach, M.E. Mirault and T. Levade. 2002. Glutathione peroxidase-1 protects from CD95-induced apoptosis. J. Biol. Chem. 277: 42867–42874.

Han, F.X., J.A. Hargreaves, W.L. Kingery, D.B. Huggett and D.K. Schlenk. 2001. Accumulation, distribution, and toxicity of copper in sediments of catfish ponds receiving periodic copper sulfate applications. J. Environ. Qual. 30: 912–919.

Hurk, P.V.D., M. Faisal and M.H. Roberts, Jr. 1988. Interaction of cadmium and benzo[*a*]pyrene in mummichog (*Fundulus heteroclitus*): effects on acute mortality. Mar. Environ. Res. 46: 525–528.

Hwang, N.R., S.H. Yim, Y.M. Kim, J. Jeong, E.J. Song, Y. Lee, J.H. Lee, S. Choi and K.J. Lee. 2009. Oxidative modifications of glyceraldehyde-3-phosphate dehydrogenase play a key role in its multiple cellular functions. Biochem. J. 423: 253–264.

Jihen el, H., H. Fatima, A. Nouha, T. Baati, M. Imed and K. Abdelhamid. 2010. Cadmium retention increase: a probable key mechanism of the protective effect of zinc on cadmium-induced toxicity in the kidney. Toxicol. Lett. 196: 104–109.

Jiménez-Tenorio, N., C. Morales-Caselles, J. Kalman, M.J. Salamanca, M.L. González de Canales, C. Sarasquete and T.A. DelValls. 2007. Determining sediment quality for regulatory proposes using fish chronic bioassays. Environ. Int. 33: 474–480.

Jurczuk, M., M.M. Brzoska, J. Rogalska and J. Moniuszko-Jakoniuk. 2003. Iron body status of rats chronically exposed to cadmium and ethanol. Alcohol Alcohol 38: 202–207.

Jurczuk, M., M.M. Brzóska, J. Moniuszko-Jakoniuk, M. Gałażyn-Sidorczuk and E. Kulikowska-Karpińska. 2004. Antioxidant enzymes activity and lipid peroxidation in liver and kidney of rats exposed to cadmium and ethanol. Food Chem. Toxicol. 42: 429–438.

Karakach, T., E. Huenupi, E. Soo, J. Walter and L.B. Afonso. 2009. 1H-NMR and mass spectrometric characterization of the metabolic response of juvenile Atlantic salmon (*Salmo salar*) to long-term handling stress. Metabolomics 5: 123–137.

Kennette, W., O.M. Collins, R.K. Zalups and J. Koropatnick. 2005. Basal and zinc-induced metallothionein in resistance to cadmium, cisplatin, zinc, and tertbutyl hydroperoxide: studies using MT knockout and antisense-downregulated MT in mammalian cells. Toxicol. Sci. 88: 602–613.

Kondo, T. 2008. Tissue proteomics for cancer biomarker development: laser microdissection and 2D-DIGE. BMB Reports 41: 626–634.

Koyu, A., A. Gokcimen, F. Ozguner, D.S. Bayram and A. Kocak. 2006. Evaluation of the effects of cadmium on rat liver. Mol. Cell. Biochem. 284: 81–85.

Kremling, K. and P. Streu. 2001. The behaviour of dissolved Cd, Co, Zn, and Pb in North Atlantic near-surface waters (30°N/60°W–60°N/2°W). Deep. Sea Res. Part I: Oceanogr. Res. Pap. 48: 2541–2567.

Lee, E., J. Jeong, S.E. Kim, E.J. Song, S.W. Kang and K.J. Lee. 2009. Multiple functions of Nm23-H1 are regulated by oxido-reduction system. PLoS One 4: e7949.

Lewinska, D., J. Arkusz, M. Stanczyk, J. Palus, E. Dziubaltowska and M. Stepnik. 2007. Comparison of the effects of arsenic and cadmium on benzo(*a*)pyrene-induced micronuclei in mouse bone-marrow. Mutat. Res. 632: 37–43.

Liu, S., G. Gao, Y. Palti, B.M. Cleveland, G.M. Weber and C.E. Rexroad, 3rd. 2014. RNA-seq analysis of early hepatic response to handling and confinement stress in rainbow trout. PLoS One 9: e88492.

Lopez-Barea, J. 1995. Biomarkers in ecotoxicology: an overview. Arch. Toxicol. Suppl. 17: 57–79.

Lopez-Barea, J. and J.L. Gomez-Ariza. 2006. Environmental proteomics and metallomics. Proteomics 6 Suppl. 1: S51–62.

Manchado, M., C. Infante, E. Asensio and J.P. Cañavate. 2007. Differential gene expression and dependence on thyroid hormones of two glyceraldehyde-3-phosphate dehydrogenases in the flatfish Senegalese sole (Solea senegalensis Kaup). Gene 400: 1–8.

Manevich, Y., T. Sweitzer, J.H. Pak, S.I. Feinstein, V. Muzykantov and A.B. Fisher. 2002. 1-Cys peroxiredoxin overexpression protects cells against phospholipid peroxidation-mediated membrane damage. Proc. Natl. Acad. Sci. USA 99: 11599–11604.

Martins, M., A.M. Ferreira and C. Vale. 2008. The influence of *Sarcocornia fruticosa* on retention of PAHs in salt marshes sediments (Sado estuary, Portugal). Chemosphere 71: 1599–1606.

Mastellos, D., C. Andronis, A. Persidis and J.D. Lambris. 2005. Novel biological networks modulated by complement. Clin. Immunol. 115: 225–235.

Matovic, V., A. Buha, Z. Bulat, D. Ethukic-Cosic, M. Miljkovic, J. Ivanisevic and J. Kotur-Stevuljevic. 2012. Route-dependent effects of cadmium/cadmium and magnesium acute treatment on parameters of oxidative stress in rat liver. Food Chem. Toxicol. 50: 552–557.

Min, K.S., T. Nakatsubo, Y. Fujita, S. Onosaka and K. Tanaka. 1992. Degradation of cadmium metallothionein *in vitro* by lysosomal proteases. Toxicol. Appl. Pharmacol. 113: 299–305.

Moniuszko-Jakoniuk, J., M. Jurczuk, M.M. Brzoska, J. Rogalska and M. Galazyn-Sidorczuk. 2005. Involvement of some low-molecular thiols in the destructive mechanism of cadmium and ethanol action on rat livers and kidneys. Pol. Environ. Stud. 14: 483–489.

Mouysset, J., A. Deichsel, S. Moser, C. Hoege, A.A. Hyman, A. Gartner and T. Hoppe. 2008. Cell cycle progression requires the CDC-48UFD-1/NPL-4 complex for efficient DNA replication. Proc. Natl. Acad. Sci. USA 105: 12879–12884.

Mukherjee, J.J., S.K. Gupta, S. Kumar and H.C. Sikka. 2004. Effects of cadmium(II) on (+/–)-anti-benzo[*a*]pyrene-7,8-diol-9,10-epoxide-induced DNA damage response in human fibroblasts and DNA repair: a possible mechanism of cadmium's cogenotoxicity. Chem. Res. Toxicol. 17: 287–293.

Murugavel, P. and L. Pari. 2007. Effects of diallyl tetrasulfide on cadmium-induced oxidative damage in the liver of rats. Hum. Exp. Toxicol. 26: 527–534.

Nakayama, K., S. Kitamura, Y. Murakami, J.Y. Song, S.J. Jung, M.J. Oh, H. Iwata and S. Tanabe. 2008. Toxicogenomic analysis of immune system-related genes in Japanese flounder (*Paralichthys olivaceus*) exposed to heavy oil. Mar. Pollut. Bull. 57: 445–452.

Nordberg, G.F. 2009. Historical perspectives on cadmium toxicology. Toxicol. Appl. Pharmacol. 238: 192–200.

Nordberg, G.F. 2010. Biomarkers of exposure, effects and susceptibility in humans and their application in studies of interactions among metals in China. Toxicol. Lett. 192: 45–49.

Olatunji, O.S., O.S. Fatoki, B.O. Opeolu and B.J. Ximba. 2015. Benzo[*a*]pyrene and Benzo[*k*]fluoranthene in some processed fish and fish products. Int. J. Environ. Res. Public Health 12: 940–951.

Oñate, J.J., D. Pereira and F. Suarez. 2003. Strategic environmental assessment of the effects of European Union's regional development plans in Donana National Park (Spain). Environ. Manage. 31: 642–655.

Orsini, L., M. Jansen, E.L. Souche, S. Geldof and L. De Meester. 2011. Single nucleotide polymorphism discovery from expressed sequence tags in the waterflea *Daphnia magna*. BMC Genomics 12: 309.

Osorio-Yanez, C., J.L. Garcia-Tavera, M.T. Perez-Nunez, I. Poblete-Naredo, B. Munoz, B.S. Barron-Vivanco, S.J. Rothenberg, O. Zapata-Perez and A. Albores. 2012. Benzo(*a*)pyrene induces hepatic AKR1A1 mRNA expression in tilapia fish (*Oreochromis niloticus*). Toxicol. Mech. Methods 22: 438–444.

OSPAR. 2002. Cadmium, Hazardous Substances. OSPAR Commission.

Osuna-Jimenez, I., T.D. Williams, M.J. Prieto-Alamo, N. Abril, J.K. Chipman and C. Pueyo. 2009. Immune- and stress-related transcriptomic responses of *Solea senegalensis* stimulated with lipopolysaccharide and copper sulphate using heterologous cDNA microarrays. Fish Shellfish Immunol. 26: 699–706.

Osuna-Jimenez, I., N. Abril, A. Vioque-Fernandez, J.L. Gomez-Ariza, M.J. Prieto-Alamo and C. Pueyo. 2014. The environmental quality of Donana surrounding areas affects the immune transcriptional profile of inhabitant crayfish *Procambarus clarkii*. Fish Shellfish Immunol. 40: 136–145.

Ou, J., Y. Li, Z. Ding, Y. Xiu, T. Wu, J. Du, W. Li, H. Zhu, Q. Ren, W. Gu and W. Wang. 2013. Transcriptome-wide identification and characterization of the *Procambarus clarkii* microRNAs potentially related to immunity against *Spiroplasma eriocheiris* infection. Fish Shellfish Immunol. 35: 607–617.

Pagan, J.K., F.G. Wylie, S. Joseph, C. Widberg, N.J. Bryant, D.E. James and J.L. Stow. 2003. The t-SNARE syntaxin 4 is regulated during macrophage activation to function in membrane traffic and cytokine secretion. Curr. Biol. 13: 156–160.

Pagani, I., K. Liolios, J. Jansson, I.M. Chen, T. Smirnova, B. Nosrat, V.M. Markowitz and N.C. Kyrpides. 2012. The Genomes OnLine Database (GOLD) v.4: status of genomic and metagenomic projects and their associated metadata. Nucleic Acids Res. 40(Database issue): D571–579.

Pak, J.H., Y. Manevich, H.S. Kim, S.I. Feinstein and A.B. Fisher. 2002. An antisense oligonucleotide to 1-cys peroxiredoxin causes lipid peroxidation and apoptosis in lung epithelial cells. J. Biol. Chem. 277: 49927–49934.

Prieto-Alamo, M.J., J.M. Cabrera-Luque and C. Pueyo. 2003. Absolute quantitation of normal and ROS-induced patterns of gene expression: an *in vivo* real-time PCR study in mice. Gene Expr. 11: 23–34.

Prieto-Alamo, M.J., N. Abril, I. Osuna-Jimenez and C. Pueyo. 2009. *Solea senegalensis* genes responding to lipopolysaccharide and copper sulphate challenges: large-scale identification by suppression subtractive hybridization and absolute quantification of transcriptional profiles by real-time RT-PCR. Aquat. Toxicol. 91: 312–319.

Prieto-Alamo, M.J., I. Osuna-Jimenez, N. Abril, J. Alhama, C. Pueyo and J. López-Barea. 2012. Omics methodologies: new tools in aquaculture studies. pp. 361–390. *In*: Z. Muchlisin [ed.]. Aquaculture. InTech, Rijeka, Croatia.

Pueyo, C., J.L. Gómez-Ariza, M.A. Bello-López, R. Fernández-Torres, N. Abril, J. Alhama, T. García-Barrera and J. López-Barea. 2011. New methodologies for assessing the presence and ecological effects of pesticides in Doñana National Park (SW Spain). pp. 165–196. *In*: M. Stoytcheva [ed.]. Pesticides in the Modern World—Trends in Pesticides Analysis. InTech, Rijeka, Croatia.

Qureshi, N., P.Y. Perera, J. Shen, G. Zhang, A. Lenschat, G. Splitter, D.C. Morrison and S.N. Vogel. 2003. The proteasome as a lipopolysaccharide-binding protein in macrophages: differential effects of proteasome inhibition on lipopolysaccharide-induced signaling events. J. Immunol. 171: 1515–1525.

Rana, S.V. 2008. Metals and apoptosis: recent developments. J. Trace Elem. Med. Biol. 22: 262–284.

Reynaud, S., M. Raveton and P. Ravanel. 2008. Interactions between immune and biotransformation systems in fish: a review. Aquat. Toxicol. 87: 139–145.

Robledo, D., P. Ronza, P.W. Harrison, A.P. Losada, R. Bermudez, B.G. Pardo, M.J. Redondo, A. Sitja-Bobadilla, M.I. Quiroga and P. Martinez. 2014. RNA-seq analysis reveals significant transcriptome changes in turbot (*Scophthalmus maximus*) suffering severe enteromyxosis. BMC Genomics 15: 1149.

Rodrigues, P.M., T.S. Silva, J. Dias and F. Jessen. 2012. PROTEOMICS in aquaculture: Applications and trends. J. Proteomics 75: 4325–4345.

Sandvik, M., J. Beyer, A. Goksoyr, K. Hyll, E. Eliann and S. Janneche-Utne. 1997. Interaction of benzo[*a*]pyrene, 2,3,3',4,4',5 hexachlorobiphenyl (PCB 156) and cadmium on biomarker responses in flounder (*Platichthys flesus* L.). Biomarkers 2: 153–160.

Satarug, S., S.H. Garrett, M.A. Sens and D.A. Sens. 2010. Cadmium, environmental exposure, and health outcomes. Environ. Health Perspect. 118: 182–190.

Savorani, F., G. Picone, A. Badiani, P. Fagioli, F. Capozzi and S.B. Engelsen. 2010. Metabolic profiling and aquaculture differentiation of gilthead sea bream by 1H NMR metabonomics. Food Chemistry 120: 907–914.

Schock, T.B., S. Newton, K. Brenkert, J. Leffler and D.W. Bearden. 2012. An NMR-based metabolomic assessment of cultured cobia health in response to dietary manipulation. Food Chemistry 133: 90–101.

Schram, F. and C. Vaupel Klein. 2012. Treatise on Zoology—Anatomy, Taxonomy, Biology. The Crustacea. Vomen 9 Part B: Decapoda: Astacidea P.P. (Enoplometopoidea, Nephropoidea), Glypheidea, Axiidea, Gebiidea, and Anomura. BRILL,Leiden, The Netherlands.

Serrano, L., M. Reina, G. Martín, I. Reyes, A. Arechederra, D. León and J. Toja. 2006. The aquatic systems of Doñana (SW Spain): watersheds and frontiers. Limnetica 25: 11–32.

Shi, X.Z., X.C. Li, S. Wang, X.F. Zhao and J.X. Wang. 2010. Transcriptome analysis of hemocytes and hepatopancreas in red swamp crayfish, *Procambarus clarkii*, challenged with white spot syndrome virus. Invertebrate Surviv. J. 7: 119–131.

Shui, Y., Z.-B. Guan, Z.-H. Xu, C.-Y. Zhao, D.-X. Liu and X. Zhou. 2012. Proteomic identification of proteins relevant to ovarian development in the red swamp crayfish *Procambarus clarkii*. Aquaculture 370-371: 14–18.

Simpson, R.J. 2003. Proteins and Proteomics. A Laboratory Manual, Cold Spring Harbor Laboratory Press, New York, USA.

Snape, J.R., S.J. Maund, D.B. Pickford and T.H. Hutchinson. 2004. Ecotoxicogenomics: the challenge of integrating genomics into aquatic and terrestrial ecotoxicology. Aquat. Toxicol. 67: 143–154.

Sole, M., A. Fortuny and E. Mananos. 2014. Effects of selected xenobiotics on hepatic and plasmatic biomarkers in juveniles of *Solea senegalensis*. Environ. Res. 135: 227–235.

Spaulding, H.L., F. Saijo, R.H. Turnage, J.S. Alexander, T.Y. Aw and T.J. Kalogeris. 2006. Apolipoprotein A-IV attenuates oxidant-induced apoptosis in mitotic competent, undifferentiated cells by modulating intracellular glutathione redox balance. Am. J. Physiol. Cell. Physiol. 290: C95–C103.

Swain, P., S.K. Nayak, P.K. Nanda and S. Dash. 2008. Biological effects of bacterial lipopolysaccharide (endotoxin) in fish: A review. Fish Shellfish Immunol. 25: 191–201.

Takaishi, M., M. Sawada, A. Shimada, J.S. Suzuki, M. Satoh and H. Nagase. 2009. Protective role of metallothionein in benzo[*a*]pyrene-induced DNA damage. J. Toxicol. Sci. 34: 449–458.

Tapia-Paniagua, S.T., S. Vidal, C. Lobo, M.J. Prieto-Alamo, J. Jurado, H. Cordero, R. Cerezuela, I. Garcia de la Banda, M.A. Esteban, M.C. Balebona and M.A. Moriñigo. 2014. The treatment with the probiotic *Shewanella putrefaciens* Pdp11 of specimens of *Solea senegalensis* exposed to high stocking densities to enhance their resistance to disease. Fish Shellfish Immunol. 41: 209–221.

Templeton, D.M. and Y. Liu. 2010. Multiple roles of cadmium in cell death and survival. Chem. Biol. Interact. 188: 267–275.

Thorgaard, G.H., G.S. Bailey, D. Williams, D.R. Buhler, S.L. Kaattari, S.S. Ristow, J.D. Hansen, J.R. Winton, J.L. Bartholomew, J.J. Nagler, P.J. Walsh, M.M. Vijayan, R.H. Devlin, R.W. Hardy, K.E. Overturf, W.P. Young, B.D. Robison, C. Rexroad and Y. Palti. 2002. Status and opportunities for genomics research with rainbow trout. Comp. Biochem. Physiol. B Biochem. Mol. Biol. 133: 609–646.

Tingaud-Sequeira, A., F. Chauvigne, J. Lozano, M.J. Agulleiro, E. Asensio and J. Cerda. 2009. New insights into molecular pathways associated with flatfish ovarian development and atresia revealed by transcriptional analysis. BMC Genomics 10: 434.

Ünlü, M., M.E. Morgan and J.S. Minden. 1997. Difference gel electrophoresis. A single gel method for detecting changes in protein extracts. Electrophoresis 18: 2071–2077.

Valente, P., G. Fassina, A. Melchiori, L. Masiello, M. Cilli, A. Vacca, M. Onisto, L. Santi, W.G. Stetler-Stevenson and A. Albini. 1998. TIMP-2 over-expression reduces invasion and angiogenesis and protects B16F10 melanoma cells from apoptosis. Int. J. Cancer 75: 246–253.

Vander Heiden, M.G., L.C. Cantley and C.B. Thompson. 2009. Understanding the Warburg effect: the metabolic requirements of cell proliferation. Science (NY) 324: 1029–1033.

Vatrinet, R., L. Iommarini, I. Kurelac, M. DeLuise, G. Gasparre and A.M. Porcelli. 2015. Targeting respiratory complex I to prevent the Warburg effect. Internat. J. Biochem. Cell Biol. 63: 41–45.

Vioque-Fernandez, A., E. Alves de Almeida, J. Ballesteros, T. Garcia-Barrera, J.L. Gomez-Ariza and J. Lopez-Barea. 2007. Doñana National Park survey using crayfish (*Procambarus clarkii*) as bioindicator: esterase inhibition and pollutant levels. Toxicol. Lett. 168: 260–268.

Vioque-Fernandez, A., E. Alves de Almeida and J. Lopez-Barea. 2009a. Assessment of Doñana National Park contamination in *Procambarus clarkii*: integration of conventional biomarkers and proteomic approaches. Sci. Total Environ. 407: 1784–1797.

Vioque-Fernandez, A., E. Alves de Almeida and J. Lopez-Barea. 2009b. Biochemical and proteomic effects in *Procambarus clarkii* after chlorpyrifos or carbaryl exposure under sublethal conditions. Biomarkers 14: 299–310.

Washburn, M.P., D. Wolters and J.R. Yates. 2001. Large-scale analysis of the yeast proteome by multidimensional protein identification technology. Nature Biotechnol. 19(3): 242–247.

Wei, D.H., X.Y. Jia, Y.H. Liu, F.X. Guo, Z.H. Tang, X.H. Li, Z. Wang, L.S. Liu, G.X. Wang, Z.S. Jian and C.G. Ruan. 2013. Cathepsin L stimulates autophagy and inhibits apoptosis of ox-LDL-induced endothelial cells: potential role in atherosclerosis. Int. J. Mol. Med. 31: 400–406.

Whyte, S.K. 2007. The innate immune response of finfish—a review of current knowledge. Fish Shellfish Immunol. 23: 1127–1151.

Wilkins, M.R., C. Pasquali, R.D. Appel, K. Ou, O. Golaz, J.C. Sanchez, J.X. Yan, A.A. Gooley, G. Hughes, I. Humphery-Smith, K.L. Williams and D.F. Hochstrasser. 1996. From proteins to proteomes: large scale protein identification by two-dimensional electrophoresis and amino acid analysis. Biotechnology (NY) 14: 61–65.

Williams, T.D., A.M. Diab, S.G. George, R.E. Godfrey, V. Sabine, A. Conesa, S.D. Minchin, P.C. Watts and J.K. Chipman. 2006. Development of the GENIPOL European flounder (*Platichthys flesus*) microarray and determination of temporal transcriptional responses to cadmium at low dose. Environ. Sci. Technol. 40: 6479–6488.

Williams, T.D., A.M. Diab, S.G. George, V. Sabine and J.K. Chipman. 2007. Gene expression responses of European flounder (*Platichthys flesus*) to 17-beta estradiol. Toxicol. Lett. 168: 236–248.

Williams, T.D., A. Diab, F. Ortega, V.S. Sabine, R.E. Godfrey, F. Falciani, J.K. Chipman and S.G. George. 2008. Transcriptomic responses of European flounder (*Platichthys flesus*) to model toxicants. Aquat. Toxicol. 90: 83–91.

Williams, T.D., I.M. Davies, H. Wu, A.M. Diab, L. Webster, M.R. Viant, J.K. Chipman, M.J. Leaver, S.G. George, C.F. Moffat and C.D. Robinson. 2014. Molecular responses of European flounder (*Platichthys flesus*) chronically exposed to contaminated estuarine sediments. Chemosphere 108: 152–158.

Xu, M., S. Frelon, O. Simon, R. Lobinski and S. Mounicou. 2014. Development of a non-denaturing 2D gel electrophoresis protocol for screening *in vivo* uranium-protein targets in *Procambarus clarkii* with laser ablation ICP MS followed by protein identification by HPLC-Orbitrap MS. Talanta 128: 187–195.

Xu, T., J. Zhao, P. Hu, Z. Dong, J. Li, H. Zhang, D. Yin and Q. Zhao. 2014. Pentachlorophenol exposure causes Warburg-like effects in zebrafish embryos at gastrulation stage. Toxicol. Appl. Pharmacol. 277: 183–191.

Yamakura, F. and H. Kawasaki. 2010. Post-translational modifications of superoxide dismutase. Biochim. Biophys. Acta 1804: 318–325.

Yilmaz, S. and M. Sadikoglu. 2011. Study of heavy metal pollution in seawater of Kepez harbor of Canakkale (Turkey). Environ. Monit. Assess. 173: 899–904.

Zeng, Y. and C.P. Lu. 2009. Identification of differentially expressed genes in haemocytes of the crayfish (*Procambarus clarkii*) infected with white spot syndrome virus by suppression subtractive hybridization and cDNA microarrays. Fish Shellfish Immunol. 26: 646–650.

Zhai, Q., G. Wang, J. Zhao, X. Liu, F. Tian, H. Zhang and W. Chen. 2013. Protective effects of *Lactobacillus plantarum* CCFM8610 against acute cadmium toxicity in mice. Appl. Environ. Microbiol. 79: 1508–1515.

8

New Trends in Aquatic Pollution Monitoring

From Conventional Biomarkers to Environmental Proteomics

José Alhama Carmona,[a] *Carmen Michán Doña*[b] and *Juan López-Barea*[c,*]

ABSTRACT

Aquatic environments receive increasing amount of contaminants, which could affect the ecosystems, endanger living-beings' homeostasis, and risk human health. Besides pollution from anthropogenic sources, nowadays there is a growing concern about the effects of human activities on climate change. Environmental quality can be assessed by many methods (chemical, biological, ecological) that, if used individually, have limitations. Organisms and populations respond to stressors by changing different parameters (biomarkers) at biochemical, histological, immunological, physiological, or organismic levels. Biomonitoring requires the use of biomarker batteries, since contaminants induce many responses that are not necessarily correlated. In contrast to the so-called conventional or "classic" biomarkers, the *omic* approaches are becoming a powerful multidisciplinary strategy in environmental studies, since they monitor many biological molecules in an unbiased and high-throughput manner, providing a general appraisal of the biological responses altered by contaminants. Methodological limitations, mainly related to sample preparation for 2DE analysis, have been gradually overcome in the application of environmental proteomics. The relevance of redox proteomics is increasing as oxidative stress is one of the key modes of action of pollutant-induced toxicity. Since organisms dwelling in polluted environments modulate their responses via epigenetic changes, epigenetic "foot-printing" could identify chemical types to which an organism has been exposed throughout its lifetime.

Department of Biochemistry and Molecular Biology, Faculty of Sciences, University of Córdoba, Campus de Rabanales, 14071 Córdoba, Spain.

[a] Email: bb2alcaj@uco.es

[b] Email: bb2midoc@uco.es

[c] Email: bb1lobaj@uco.es

* Corresponding author

Introduction

Environmental quality can be assessed by several methods (chemical, biological, and ecological) that, if used individually, have multiple limitations. In fact, it is convenient to combine different approaches to understand the status of any ecosystem. The appearance of pollutants in the organisms and their biological effects has to be detected, in addition to their presence in the environment, before their possible risks can be evaluated. Chemical analysis of pollutants requires highly sophisticated tools due to the growing number of xenobiotics, the subsequent biotransformation of these parent compounds, the need to distinguish between different chemical forms, and the interest in characterizing the metabolites derived from them (Pueyo et al. 2011).

The environment receives contaminants of various types; nowadays over 150,000 are released in quantities that could affect the ecosystems. Environmental pollutants can represent a major concern to human health since they significantly contribute to many diseased states with major effects on public health (Braconi et al. 2011). Many pollutants are organics, such as petroleum-derived and polycyclic aromatic hydrocarbons (PAHs), polychlorinated (or poly-Br) biphenyls (PCBs, PBBs), or those derived from agriculture (fertilizers, herbicides, pesticides, etc.), transport (fuel and petroleum spills, biocides released from ships, etc.), industries (linear hydrocarbons, PAHs, PCBs, raw materials, intermediaries, final products, etc.), and therapeutic drugs. There are also transition metals, nitroaromatics, quinones, and other electrophiles. Some pollutants induce oxidative stress in organisms by a redox cycling process, when a pollutant or a chemical derived from it is reduced and passes electrons to O_2/N_2 generating reactive oxygen or nitrogen species (RO/NS). These RO/NS promote oxidative stress when they can not be properly quenched by cellular defense mechanisms (López-Barea 1995; López-Barea and Gómez-Ariza 2006). Recently, further concern has been raised about new emerging pollutants, such as pharmaceuticals, alkylphenols, linear alkylsulphonates, etc., that, while found in the ecosystems at very low concentrations, are ecologically very relevant, even as endocrine disruptors, deleterious for the reproduction of many species (Pueyo et al. 2011). Anti-inflammatory drugs, antibiotics, anti-contraceptives, etc. are used in human and animal health care. The amount of pharmaceuticals reaching the environment depends on animal and human consumption-excretion that reaches the sewage. Effluents of wastewater plants are the main source for these chemicals in the aquatic environment, followed by the release of outdated medicines down household (Ruhoy and Daughton 2007) and pharmaceutical industry waste (Larsson et al. 2007).

Genes exert their functions at the protein level, but the genetic responses to stress are often regulated at the transcriptional level. The concept of homeostasis describes the constant state of the internal medium of living beings. Homeostatic regulation allows them to function effectively in a broad range of conditions; pollution seriously endangers this situation. The cellular genome is in a dynamic equilibrium between processes that damage and those that maintain its integrity. Prior to death or sickness, organisms and populations respond to natural or anthropogenic-derived stressors by changing different parameters (so-called biomarkers) at biochemical, histological, immunological, physiological, or organismic levels. Biomarkers are defided as: "*Measurement of body fluids, cells, tissues or whole animals that indicate in biochemical, cellular, physiological, behavioral or energetic terms the presence of contaminants or the extent of the host response*" (López-Barea 1995). Biomarkers that reflect the health status of organisms at lower organization levels (molecular, cellular) (i) respond rapidly to stress, (ii) have high toxicological relevance, and (iii) are useful as early-warning indicators of environmental alterations before irreversible damages to the ecosystem occur. Those reflecting health at higher organizational levels (populations, communities, ecosystems) respond slowly and have lower toxicological relevance but are more ecologically relevant, although they often detect damages at an irreversible stage (López-Barea 1995).

Pollutant-exposed organisms display several responses, including induction of biotransforming enzymes. Those of phase I, such as cytochromes P450, add polar groups to the parent compounds, and those of phase II, such as glutathione-*S*-transferases (GSTs), conjugate their products to endogenous metabolites. Other inducible proteins, such as heat shock proteins (HSPs) and metallothioneins (MTs), refold denatured proteins or protect the organisms from metal toxicity. Primary antioxidative enzymes (e.g., superoxide dismutases (SODs), glutathione peroxidases (GSHPxs), and peroxiredoxins (PRDXs)), secondary enzymes (e.g., glucose-6-phosphate dehydrogenase (G6PDH) and glutathione reductase), and antioxidant proteins (e.g., thioredoxins and glutaredoxins), are induced upon exposure to many pollutants. Finally, low Mr antioxidants, including reduced glutathione (GSH) and vitamins C or E, also increase in contaminated organisms (López-Barea 1995; López-Barea and Gómez-Ariza 2006). Key biomolecules are damaged if toxic chemicals exceed pollutant-elicited defenses. Oxidized lipids generate fragments, such as malondialdehyde (MDA) and 4-hydroxy-2-nonenal. Some damages also arise in nucleic acids, including oxidized bases, abasic sites, and strand breaks. In proteins, disulfide bridges are formed and Met or His are oxidized to Met-sulfoxide or 2-oxoHis, respectively. Glutathione is oxidized to GSSG that forms mixed disulfides with protein thiols. Some of the molecular damages are repared by inducible enzyme systems (López-Barea 1995; López-Barea and Gómez-Ariza 2006). Many pollutant-elicited responses, mentioned above, are widely used as "conventional" biomarkers, although they require a profound previous knowledge of their toxic mechanisms. Thus, these biomarkers are only partially applicable, since they focus on known proteins but exclude other biomolecules that could also be altered, although their relationship to pollution is yet unknown.

Omic Approaches for Environmental Issues

After completion of the Human Genome Project, an increasing range of post-genomic techniques allow researchers to assess the complete content of genes, proteins, or metabolites in a cell under a particular situation. In contrast to the conventional biomarkers, these *omic* approaches, complemented with *in vitro* and *in silico* methods, are becoming a powerful multidisciplinary strategy in environmental studies. Nonetheless, their application is still at an early stage, mainly since most bioindicators are poorly represented in gene/protein sequence databases (Ruiz-Laguna et al. 2006). At least, the *omic* sciences include: (i) genomics for study of DNA variations, (ii) transcriptomics for genome-wide characterization of gene expression via the measurement of mRNAs, (iii) proteomics for measuring cell and tissue-wide expression of proteins, and (iv) metabolomics for global assessment of metabolite concentrations. These technologies give detailed molecular information that help to identify toxicity pathways and to define pollutant mechanisms and modes of action without requiring previous knowledge of pollutant actions. Another chapter of the present book reviews the utility of functional genomics approaches in "Marine Pollution and Aquaculture" (Abril et al. 2017).

Proteomics addresses the post-genomic challenge of examining the entire complement of proteins (proteome) expressed in a cell, tissue, or organ at a given time under defined conditions (James 1997). Protein expression is modulated at different levels from transcription to maturation of the polypeptides produced by translation of mRNAs. Proteins were initially separated by two-dimensional gel electrophoresis (2DE) (O'Farrell 1975), and their expression analyzed via 2D software. Proteins were then identified by mass spectrometry (MS) analysis of their peptide mass fingerprint (MALDI-TOF-PMF) or *de novo* sequencing of some peptides (nESI-MS/MS), contrasting the results with public databases (Simpson 2003). The 2DE technique, labor intensive and of a low reproducibility, requires a large amount of sample, and its narrow dynamic range is problematic with proteins of extreme Mr/pI. To overcome this limitation, fluorescent labeling gave rise to difference in-gel electrophoresis (DIGE), which analyzes two samples in one single gel, facilitating a quantitative assessment of expression (Ünlü et al. 1997). Alternatively, shotgun methods allow the analysis of complex protein mixtures after full digestion, by multidimensional separation coupling tandem liquid chromatography (LC/LC) and MS/MS (Washburn et al. 2001).

Environmental Proteomics: Looking for New Pollution Biomarkers

Environmental monitoring through conventional biomarkers has proven to have limited success since these techniques require previous knowledge of the mechanisms of action of pollutants. As mentioned above, this approach is intrinsically biased in assessing a limited number of well-known biomolecules while excluding many others because their relationship with contamination is not known. Biomonitoring requires the use of sets of different biomarkers, as environmental contaminants induce multiple responses in organisms that are not necessarily correlated. *Omic* technologies have been proposed as an alternative to conventional biomarkers since they quantitatively monitor many biomolecules in a high-throughput manner, thus providing a general appraisal of biological responses altered by exposure to contaminants (Gómez-Ariza et al. 2008; Van Aggelen et al. 2010; Abril et al. 2011; Pueyo et al. 2011; Ge et al. 2013). Within the growing body of proteomics, issues addressing Ecotoxicology aspects are relatively new (López-Barea and Gómez-Ariza 2006); yet such studies are increasing and diversifying their approaches, applications, techniques, and solutions being sought (Monsinjon and Knigge 2007). Proteomics can be a valuable tool in risk assessment in environmental studies with the potential to identify novel biomarkers, provide insight into toxicity mechanisms, and assess health risks (Rodriguez-Ortega et al. 2003; Romero-Ruiz et al. 2006; Monsinjon and Knigge 2007; Montes-Nieto et al. 2007; Vioque-Fernández et al. 2009a; Lemos et al. 2010; Montes Nieto et al. 2010; Ge et al. 2013; Fernández-Cisnal et al. 2014).

Recently developed proteomic approaches can also identify proteins that are significantly altered after pollutant exposure, and once validated, can be used as new advanced biomarkers. Furthermore, they may also help to establish the toxic mechanisms of pollutants (López-Barea and Gómez-Ariza 2006). However, the use of proteomic approaches for environmental studies (Environmental Proteomics, EP) is not free of limitations. Thus, in principle, they have to be applied in non-model sentinel organisms, whose sequences are scarcely present or totally absent from databases, which makes it difficult to identify proteins. Actually, MALDI-TOF is almost useless for identification of deregulated proteins in EP, which usually has to be carried out via *de novo* sequencing by LC-MS/MS (Barrett et al. 2005; López-Barea and Gómez-Ariza 2006).

Early Studies: Protein Expression Signatures

The initial approaches of EP were carried out by exposing molluscs or fish to model pollutants. The first study was published in 2000 by the Bradley group, in Baltimore, together with the Tedengren group, in Stockholm. Mussels (*Mytilus edulis*) were exposed for 7 days to Cu(II) and Aroclor 1248, a PCB mixture, and to decreased salinity. The difficulty in identifying proteins prompted the use of the protein expression signatures (PES), defined as the particular set of proteins observed in 2-D images as state markers of physical, chemical, or biological stressors; a different PES was found for each stressor with different sets of deregulated proteins (Shepard et al. 2000). Shepard and Bradley reported the progressive alteration of PES in gills of mussels exposed to increasing doses of Cu(II) (Shepard and Bradley 2000). In 2002, the group of Bradley extended the PES approach to rainbow trout (*Onchorrynchus mykiss*); PES typical of the endocrine disruptors were also detected in fish exposed to mixed sewage effluents (Bradley et al. 2002). The same year, the Hogstrand group, in London, used surface-excision laser-desorption ionisation (SELDI) proteomics to show the appearance of seven peptides in rainbow trout exposed to sublethal Zn doses for six days, the disappearance of another four peptides, and significant alterations in another 11 peptides (Hogstrand et al. 2002).

An early study showing the utility of proteomics in Marine Biology was published in 2001 by the group of J.L. López in Galicia (NW Spain) (López et al. 2001), that was subsequently applied to distinguish two mussel species, *M. edulis* and *M. galloprovincialis* (López et al. 2002). In 2003, our group at Córdoba, with that of Goksoyr in Bergen, showed that the expression of over 15 proteins was altered in *Chamelea gallina* clams exposed to increasing concentrations of Aroclor 1254, Cu(II), tributyltin (TBT), and arsenite. Many proteins showed parallel alterations, but two spots showed opposite changes depending on the pollutant. In contrast to the studies mentioned above, the identification of altered proteins was undertaken (Rodriguez-Ortega et al. 2003). MALDI-TOF and/or nESI-LC-MS/MS analysis allowed us to identify four of these proteins. Aroclor 1254 and Cu(II) induced isoforms of tropomyosin and of myosin light chain. Actin was repressed by Aroclor/Cu(II) but induced by TBT/As(III), while a truncated actin form had the opposite behavior. It was proposed that the sole identification of cytoskeletal proteins could be due to their abundance, their prevalence in mollusc databases, or their role as major targets of pollutant-related oxidative stress (Rodriguez-Ortega et al. 2003).

In contrast to the early start of studies in molluscs exposed to selected pollutants, in the next years few proteomic studies were developed in animals from natural ecosystems. The first time that

EP was applied to real ecosystems was in 2004, when the Andersen group in Stavanger used SELDI-TOF to identify peptides induced in mussels from Baltic Sea areas polluted by PAHs and metals. Animals from the polluted sites, showed an increase of over fivefold in a peptide of 3268 Da plus rises in the levels of other 18 peptides. The authors proposed that, even without identifying the proteins to which the peptides pertain, they could be used as new biomarkers (Knigge et al. 2004). Also in 2004, Olsson et al. reported alterations of seven stress proteins or HSPs in control *M. edulis* mussels and in those exposed to Aroclor 1248 or the PAHs and PCBs extracted from sediments taken at polluted sites of Baltic and Bosnian seas; correlating the induction levels of each particular protein with the site and type of contaminants (Olsson et al. 2004).

Proteomic Studies in Doñana National Park (SW Spain)

The viability of EP to assess the global contamination status was carried out in a very pristine ecosystem, the Doñana National Park (DNP, SW Spain, Fig. 1), a model and highly studied area of regional size. DNP was settled in 1969 and declared a World Heritage Site in 1981. This area of marsh, shallow streams, and sand dunes, in the mouth of the Guadalquivir River, has a biodiversity unique in Europe, contains different types of ecosystems, and shelters wildlife including millions of migratory birds, and endangered species such as the Imperial Eagle and Iberian Linx (Grimalt et al. 1999). Doñana is watered by the Rocina (ROC), Partido (PAR), and Guadiamar streams, and by the Guadalquivir River. DNP is surrounded by areas of intense agricultural activities and threatened by industries located 40 Km West, at the Huelva Estuary, formed by Odiel and Tinto rivers, that also carry metals from mining activity located further north. In addition, the Domingo Rubio stream discharges to this Estuary refuses from a petrochemical plant and pesticides from nearby strawberry crops (Montes-Nieto et al. 2007, 2010). Finally, in 1998, metals released during the accident of the tailings dam of Aznacóllar pyrite mine, 60 Km North, threatened DNP via Guadiamar stream (Grimalt et al. 1999).

Since the early studies, assessment of contaminants at Doñana area focused on the Guadiamar stream (Cabrera et al. 1984; Albaiges et al. 1987; Lopez-Pamo et al. 1999; Manzano et al. 1999). The high metal

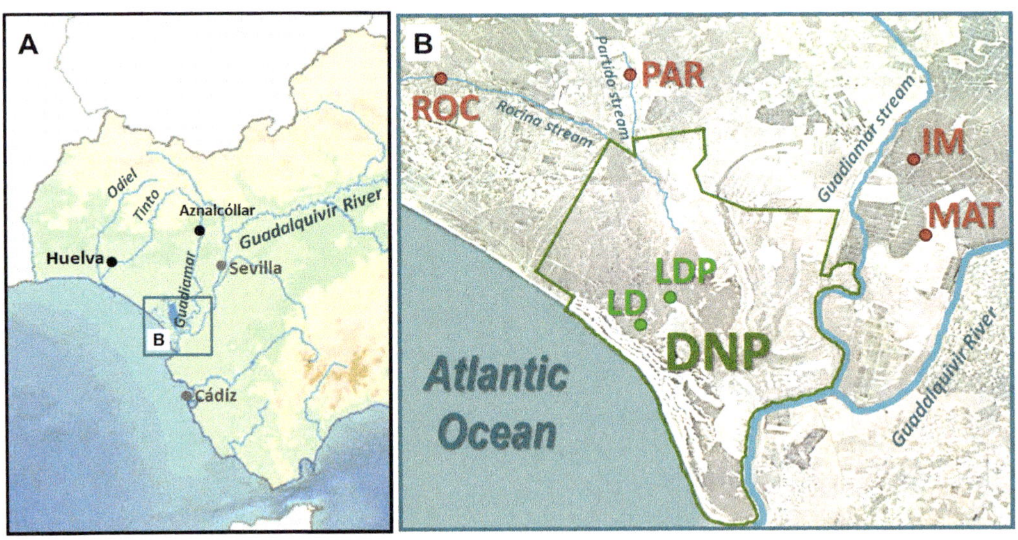

Fig. 1: Localization of the areas studied in Doñana National Park (DNP). (A) Map of Western Andalusia (SW Spain) with the courses of Guadalquivir River and streams, highlighting Huelva Estuary, and Aznalcóllar pyrite mine. Square B is enlarged. (B) DNP and its surroundings. Localizations of four problem sites are indicated; two along the course of the *Rocina* (ROC) and the *Partido* (PAR) streams, and two near the course of the Guadiamar Stream: *Isla Mayor* (IM) and *el Matochal* (MAT). Two reference sites within the Doñana Biological Reserve, the *Laguna Dulce* (LD) and the *Lucio del Palacio* (LDP), are also indicated. Blue lines show the courses of the three streams and of the Guadalquivir River, and green line the limits of DNP.

levels and low pH detected in its upper course were due to acid-mine draining from Aznalcóllar pyrite mine; metals and herbicides were also found at the rice growing fields East of DNP (Cabrera et al. 1984). The first global evaluation of DNP status was made by the Scientific Research Council (Albaiges et al. 1987). Streams feeding Doñana marshes were considered as the main source of contaminants. In addition to metals and pesticides, Guadiamar stream contributed with organic matter from urban/agro-industrial sources; petroleum hydrocarbons also entered DNP from Guadalquivir River at the South. In addition to water sources, the uniform distribution of PAHs and PCBs at stream-isolated lagoons suggested an aeolian transport from industries located at Huelva Estuary (Albaiges et al. 1987). After the partial collapse of the tailings dam of Aznalcóllar pyrite mine, the effect of this spill in DNP was studied by several groups (Lopez-Pamo et al. 1999; Manzano et al. 1999), including ours. Although the spill did not have a dramatic effect within Doñana, increasing concern was raised about the presence of contaminants at DNP core, not directly related to this overflow but possibly derived from the agrochemicals used in nearby areas (Ruíz-Laguna et al. 2001; Bonilla-Valverde et al. 2004).

The utility of proteomics to assess pollutant response of clams from three sites of Guadalquivir Estuary at the southern end of DNP was studied also by 2DE in *Scrobicularia plana* gills (Romero-Ruiz et al. 2006). Nearly 2000 well-resolved spots were detected in silver-stained gels in the 4.0–6.5 pH range. Different protein expression signatures were found at each site, with the highest number of resolved spots in animals with the highest metal content. The nineteen more intense spots were analysed by nESI-LC-MS/MS, *de novo* sequencing, and bioinformatics. Although sequence tags of sixteen of these spots were obtained, including several proteins induced by pollutant exposure of model organisms, only two were unambiguously identified: hypoxanthine guanine phosphoribosyl transferase and glyceraldehyde-3-P dehydrogenase. Both enzymes increased significantly in animals with top metal contents. The highest number of upregulated protein spots was found in animals from sites of Guadalquivir Estuary near its mouth in the Atlantic Ocean, corresponding to higher levels of several redox-active elements. Taken together, these results suggested that metals detected at Guadalquivir Estuary did not originate from the Aznalcóllar spill, but were carried out by the Guadalquivir River and deposited near its mouth at high tide conditions (Romero-Ruiz et al. 2006).

Freshwater aquatic DNP ecosystems were monitored using *Procambarus clarkii* crayfish in four campaigns from 2003 to 2004 (Vioque-Fernández et al. 2009a). In the conventional approach, 12 biomarkers responsive to pesticides, organics, and prooxidants were used. Low activity of catalase (CAT), glucose-6-P dehydrogenase (G6PDH), carboxyl esterase (CbE), and acetylcholinesterase (AChE) existed at ROC, PAR, MAT, and IM, plus high MDA, MT, and GSSG levels, suggesting that metals and other prooxidants were present at sites potentially polluted by agrochemicals. In contrast, high CAT, CbE, and AChE activities, plus increased GSH levels, and low MDA, MT, and GSSG contents existed at LDP. A parallel proteomic approach was also carried out: 2DE resolved > 2500 gill spots, and 35 proteins were deregulated (Vioque-Fernández et al. 2009a). The fold-number of up-/down regulation separated different PES: (i) animals captured within or close to Doñana Biological Reserve (LDP, LD, ROC, and PAR) showed clean/low polluted PES, with 32 proteins of unaltered intensity compared to LDP; (ii) Site PAR had a moderately polluted PES, with 7 proteins up- and other 6 down-regulated; (iii) crayfish from the upper ROC course showed a polluted PES, with 13 proteins up- and other 13 down-regulated compared to LDP; (iv) the higher proteomic responses at the upper ROC and PAR courses suggest that agrochemicals are used in Doñana surroundings; (v) the highest responses correspond to crayfish from the rice growing areas at MAT, with 30 of the 35 proteins being significantly altered, according to the extended and intensive use of agrochemicals in such areas. Both the conventional biomarker and the EP approaches indicate that sites within Doñana Biological Reserve are scarcely polluted, while the agricultural areas around DNP are highly polluted.

In vivo effects of two model pesticides, chlorpyrifos and carbaryl, were studied under controlled exposure conditions in *P. clarkii* (Vioque-Fernández et al. 2009b) to corroborate our results in natural ecosystems. Chlorpyrifos inhibited CbE in a concentration-dependent mode, AChE being less sensitive; in contrast, the effects of carbaryl were less clear. Chlorpyrifos decreased the levels of EROD, CAT, and GSSG but raised the GST activity, while carbaryl raised EROD, CAT, and GST, but lowered the GSHPx and GSH levels. The effects on global protein expression were studied also by 2DE (Vioque-Fernández et al.

2009b). In gill and nervous tissue, about 2000 spots were resolved, with quite different expression patterns. Chlorpyrifos deregulated 72 proteins, mostly in nervous tissue, and carbaryl 35, distributed evenly between organs. Several spots were selected as specific PES for chlorpyrifos or carbaryl exposure in both organs.

Methodological Limitations to Environmental Proteomics

Comparative proteomic analyses in Ecotoxicology/EP require reproducible display of as many proteins as possible. Within the proteomic workflow, sample preparation merits particular attention. For 2DE analysis, accurate and reproducible protein extraction is a critical step that must also be compatible with protein identification via MS. Previous studies with different organisms have underlined the need to adapt sample preparation to their specific characteristics (Knigge et al. 2013; Panchout et al. 2013; Wu et al. 2013). Usually, precipitation steps are required to remove contaminating compounds which may interfere with isoelectric focusing (IEF) separation, such as lipids or polysaccharides. Treatments with trichloroacetic acid (TCA)-acetone, TRIzol reagent or 2-D Clean-Up kit (GE Healthcare) have been applied in marine EP to extract proteins from sentinel animals (Knigge et al. 2013; Panchout et al. 2013; Wu et al. 2013; Jebali et al. 2014; Ghedira et al. 2015). As TRIzol allows the joint extraction of nucleic acids and proteins, it has been suggested for joint transcriptomic and proteomic analyses (Wu et al. 2013). In *Carcinus maenas*, gills represent a key challenge for protein extraction since their chitin endocuticule may interfere. Grinding of gills in liquid N_2 and TCA-acetone precipitation provides good 2DE patterns with a correct reproducibility level (Panchout et al. 2013). We have also successfully used the 2-D Clean-Up kit for proper resolution after 2DE (Fernández-Cisnal 2008; Ghedira 2011). Figure 2 compares the results obtained by applying a straightforward lysis buffer (Montes Nieto et al. 2010) and precipitation with the 2-D Clean-Up kit (Ghedira et al. 2011) to soluble gill proteins of *C. maenas*.

For EP, the identification of differentially expressed proteins is required to: (i) understand the action mechanisms of toxic compounds, and (ii) discover new molecular biomarkers which may overcome the limitations attributed to conventional biomarkers (López-Barea and Gómez-Ariza 2006; Monsinjon and Knigge 2007; Montes Nieto et al. 2010; Fernández-Cisnal et al. 2014; Ghedira et al. 2015). This is highly challenging in EP studies, in which the non-model-organisms used as bioindicators are poorly represented in protein/gene sequence databases, seriously hindering protein identification (López-Barea and Gómez-Ariza 2006; Romero-Ruiz et al. 2006; Vioque-Fernández et al. 2009a; Vioque-Fernández et al. 2009b). In fact, high-throughput MALDI-TOF-PMF analysis is almost useless to identify deregulated proteins, which usually has to be carried out by *de novo* nESI-LC-MS/MS

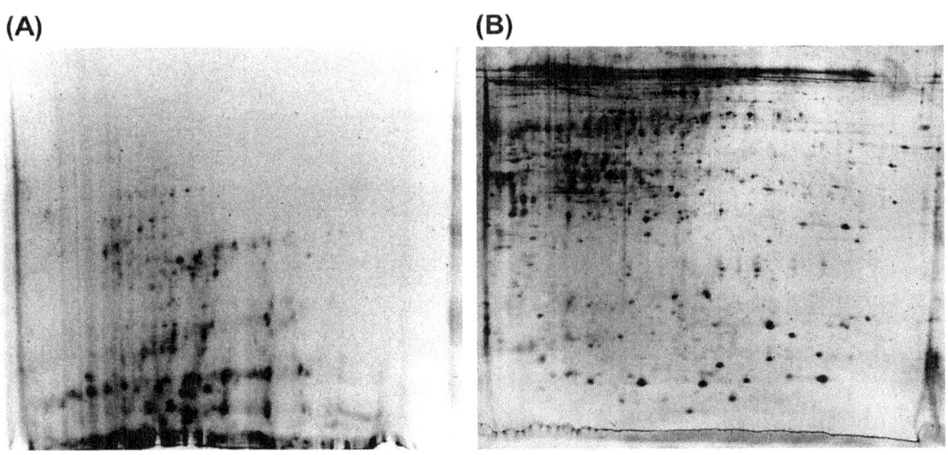

Fig. 2: 2DE gel images of soluble gill proteins of *Carcinusmaenas* extracted using a straightforward lysis buffer (Montes Nieto et al. 2010) (A) and precipitated with the 2-D Clean-Up kit (GE Healthcare) (Ghedira et al. 2011) (B). Proteins were separated on a 4–7 linear pH gradient in the first dimension and the gels (18 cm) were stained with SYPRO Ruby® (Bio-Rad) for protein visualization.

and cross-species identification (Rodriguez-Ortega et al. 2003; Romero-Ruiz et al. 2006; Costa et al. 2010; Montes Nieto et al. 2010; Costa et al. 2012; Fernández-Cisnal et al. 2014; Jebali et al. 2014; Ghedira et al. 2015); on the other hand, this technique is quite expensive, time-consuming and highly complex for routine biomonitoring.

To illustrate this point, an extensive characterization of proteins was carried out to increase knowledge of the *M. edulis* gill proteome. Around 700 spots were resolved by 2DE and over 300 were submitted to identification by nLC-MS/MS. Sixty-five percent of them were identified in line with the general yield of protein identification for *Mytilus* species. Sequence information on molluscs steadily increases, now accounting over 70,000 *Mytilidae* expressed sequence tags. However, since over one third of the selected spots remained unidentified, genomic information for this non-model organism is still scarce (Rocher et al. 2015). To change this, fully sequenced genomes of EP-relevant organisms are needed (López-Barea and Gómez-Ariza 2006; Monsinjon and Knigge 2007). Protein identification is absolutely needed to disclose the toxicity mechanisms of contaminants (Monsinjon and Knigge 2007; Fernández-Cisnal et al. 2014; Ghedira et al. 2015).

Even when not relying on bioinformatic approaches, useful information may be obtained from proteomic analysis alone, without MS identification of the altered proteins (Monsinjon et al. 2006; Monsinjon and Knigge 2007). Aiming to distill a set of biologically meaningful proteins from proteomics data, which may constitute a biomarker itself, new terms such as "key subset", "protein biomarker signatures", "expression pattern", or "pattern-only approach" were proposed in addition to PES (Monsinjon and Knigge 2007). Cluster analysis can be carried out to plot and analyze the protein patterns as similar or dissimilar groupings, and has been successfully used to assess the biological effects of contaminants in *P. clarkii* from aquatic ecosystems of DNP and its surroundings. After 2DE separation, a hierarchical cluster analysis was carried out to visually quantify the alterations in oxidized thiol levels of 35 selected spots, 19 of which were identified via MALDI-TOF/TOF. Proteins and sites were grouped based on similarities in their expression patterns according to pollution levels (Fernández-Cisnal et al. 2014). More recently, using *C. maenas* as bioindicator, the biological effects of heavy metals have been assessed along the Tunisian coast. A proteomic study was carried out to identify changes in the expression of proteins that might potentially be used as a biomarker battery. To quantify expression differences among the 23 spots selected from the digestive gland and the 16 from gills, a hierarchical cluster analysis was carried out. Besides this visual quantitative analysis, 10 proteins were identified in digestive glands and 8 in gills which are related to molting, oxidative stress and inflammation, innate immune response, and proteolysis. All of them can be used in a "multimarker approach" for EP studies and shed new light on the toxicity mechanisms of contaminants (Ghedira et al. 2015).

Another limitation of the *omic* approach is the recurrent identification of highly abundant, multifunctional, and ubiquitous proteins, thus proteins are identified by their abundance, sequence conservation and prevalence in the databases, rather than by their specific relation to a certain condition. The study of subproteomes and of low abundance proteins, by prefractionation and/or enrichment, was proposed to solve this problem (Monsinjon and Knigge 2007). Hence, to characterize GST activity and its subunit pattern in *Sparus aurata* under laboratory and in-field conditions, different hepatic GST isoforms were purified by GSH-affinity chromatography followed by a proteomic analysis. Nineteen highly reproducible GSTs were resolved by 2DE, some of them with different and specific expression patterns in response to contaminants. This proved that proteomic analysis of purified GST subunits can be a reliable tool for EP research (Jebali et al. 2013).

Redox Proteomics

Several lines of evidence confirm the key role of oxidative stress in pollutant toxicity (Braconi et al. 2011). Although RO/NS are normally produced in many natural contexts, they are greatly increased by environmental pollutants (Braconi et al. 2011). Several compounds, including metals and oxidative organic chemicals, generate RO/NS. If this overproduction is not counter balanced by antioxidative defenses, an oxidative stress develops with deleterious consequences (López-Barea 1995; Braconi et al. 2011; Fernández-Cisnal et al. 2014). Its main effect is oxidative damage to biomolecules, a key event in toxicity, leading to cell death and related to various pathologies (Levine 2002; Dalle-Donne et al. 2006; Eaton 2006; Roberts et al. 2009). Nonetheless, RO/NS should not be considered only as deleterious, since at low levels they also play a key role in signal transduction of stress-response pathways (Eaton 2006; Tell 2006; Oktyabrsky and Smirnova 2007; Janssen-Heininger et al. 2008; Winterbourn and Hampton 2008; Hansen et al. 2009).

Due to their abundance and high rate reaction constants, proteins are major targets for oxidative stress since they absorb 70% of RO/NS (Davies 2005). Many post-translational modifications of proteins, including reversible/irreversible modifications of their side-chains have been described and shown to have key roles in cellular localization, protein-protein interactions, and their structural or biological activity (Davies 2005; Cabiscol and Ros 2006; Hwang et al. 2009; Butterfield and Dalle-Donne 2012). RO/NS complicate the proteome by altering: (i) the levels of individual proteins, via gene expression changes and altered degradation of modified proteins, and (ii) the covalent structure of proteins. Advances in proteomics have led to an increasing number of reports assessing changes in the redox state of proteins in oxidative conditions. Redox proteomics aims to detect and analyse redox-based changes within the proteome both in redox signalling and in oxidative stress (Dalle-Donne et al. 2006; Janssen-Heininger et al. 2008; Sheehan et al. 2010; Braconi et al. 2011; Butterfield and Dalle-Donne 2012; Company et al. 2012; Fernández-Cisnal et al. 2014). In EP, redox proteomics is of growing interest, since oxidative stress is one of the key modes of action of pollutant-induced toxicity (Braconi et al. 2011; Rainville et al. 2015).

Protein Carbonylation

Amino acid side-chains can be irreversibly modified into carbonyl groups leading to protein aggregation, inactivation, or degradation. Carbonylation by several RO/NS and by-products of lipid oxidation, is a major form of protein oxidation commonly used as an indicator of oxidative stress (Levine 2002; Yan and Forster 2011; Rainville et al. 2014). Several sensitive methods have been developed to detect protein carbonyls in gel-based analysis (Yan and Forster 2011). Treatment with 2,4-dinitrophenylhydrazyne (DNPH) and with the fluorescein-5-thiosemicarbazide (FTC) probe are the most widely used in EP studies.

In early studies, after derivatization with DNPH and gel electrophoresis, carbonyls associated to specific bands/spots were detected on Western blots using commercially available anti-DNPH antibodies. Using this approach, increased specific carbonylation of proteins was shown via 1DE and 2DE in gills, digestive glands, and mantle of mussels exposed to environmental pollutants, such as Cd (Chora et al. 2008), gold nanoparticles (Tedesco et al. 2008), or nonylphenol (Chora et al. 2010), even in samples from a polluted Cork Harbour site, compared to a reference site (McDonagh et al. 2005). Ubiquitination, determined on membranes incubated with anti-ubiquitin, was studied in parallel to carbonylation, confirming its potential as sensitive and specific markers of oxidative stress inflicted by several environmental stressors (Chora et al. 2008; Tedesco et al. 2008; Chora et al. 2010). Distinct ubiquitination and carbonylation patterns were found, thus showing that they are independent processes, and that carbonylated proteins in clams are not strongly targeted for ubiquitination. Immunoblots also showed tissue-specific differences in protein ubiquitination and carbonylation (Chora et al. 2008; Tedesco et al. 2008; Chora et al. 2010).

Recently, protein carbonyls were quantified by labelling oxidized proteins with FTC followed by separating FTC-labelled proteins (Hu et al. 2014; Rainville et al. 2014; Sellami et al. 2014b; Sellami et al. 2015b; Sellami et al. 2015a). Protein redox status for EP has been evaluated in bivalves by detecting protein carbonylation with this methodology and total protein thiols, labelled with 5-iodoacetamido fluorescein (IAF) (Sellami et al. 2014b; Sellami et al. 2015a).

Reversibly Oxidized Thiols

Although Cys residues are rare, about 1–3% of total residues, total protein thiols (10–30 mM) are more abundant than glutathione (1–10 mM) and play major roles in intracellular defense against oxidative stress (Hansen et al. 2009; Company et al. 2012). Cys are key residues involved in protein catalysis, oxidative folding/trafficking, redox signalling, and regulation. However, its unique redox properties renders Cys vulnerable to many electrophiles, especially RO/NS, leading to redox modification and alteration of protein structure, function, or redox signalling (Ying et al. 2007; Fuentes-Almagro et al. 2012). Advances in proteomic technologies have led to an increasing number of reports assessing changes in the redox state of protein thiols in oxidatively stressed cells. Methods for detection, quantification, and identification of oxidant-sensitive thiol proteins have already been reviewed (Bonetto and Ghezzi 2006; Eaton 2006; Chiappetta et al. 2010; Chouchani et al. 2011; Charles et al. 2013). Since thiol-containing proteins are not abundant, enrichment methods are required. Activated thiol-sepharose (ATS) was used to concentrate the thiol-containing subproteome in different organisms (Hu et al. 2010; Company et al. 2012; Tedesco et al. 2012; Karlsen et al. 2014). 2DE profiles of ATS-selected proteins revealed that some proteins present in controls were absent in oxidant-treated extracts, which was attributed to thiol oxidation (Hu et al. 2010). The thiol subproteome of *Bathymodiolus azoricus* mussels from different vent sites showed differential expression of thiol-containing proteins and site-specific susceptibility to oxidative stress (Company et al. 2012). Alterations in the hepatic thiol-proteome of Atlantic cod (*Gadus morhua*) after methyl-mercury (MeHg) exposure were studied by 2DE and MS analyses for protein identification. Proteins involved in processes that are well-known targets of MeHg were revealed, but novel proteins not previously associated with MeHg-induced toxicity were also identified (Karlsen et al. 2014).

As an alternative to enrichment, very sensitive and specific thiol-tagging methods have been developed (Hill et al. 2009; Charles et al. 2013), including alkylation with iodoacetamide (IAM) or *N*-ethylmaleimide (NEM) (Sheehan et al. 2010; Charles et al. 2013). Fluorescent or radioactive reporters have also been combined with these thiol reactive compounds resulting in highly sensitive and specific tags. Thus, IAF specifically reacts with reduced thiols (-SH) but not with the oxidized variants, such as sulphenic acids (-SOH) or disulphides (-S-S), formed by oxidative stress (Baty et al. 2002), and the conjugates are visualized as fluorescent bands/spots in electrophoresis (Charles et al. 2013). Thus, 1DE separation of IAF-labelled proteins decreased in total thiol-containing proteins in *M. edulis* from polluted sites near Cork Harbour compared with the out-harbour control site; 2DE profiles of the thiol-containing proteins trapped on ATS also revealed differences between sites (Tedesco et al. 2012). Decrease in protein thiol groups has also been found in bivalves exposed to diclofenac (Jaafar et al. 2015), PAHs and insecticides (Sellami et al. 2014a; Sellami et al. 2015a), and Cu-nanoparticules (Hu et al. 2014). The tag-based method described above detects Cys oxidation by loss of labelling due to thiol modification. However, identifying an oxidant-induced loss of signal is challenging, particularly against a high background signal (Fuentes-Almagro et al. 2012; Charles et al. 2013). To tackle this problem, a method for the identification of modified thiols by a gain of signal has been developed (Charles et al. 2013).

Glutathionylation, via formation of reversible mixed disulphides between GSH and Cys-thiols, can be shown by Western blotting using anti-GSH antibodies. Increased glutathionylation occurred in gills and digestive gland of *M. edulis* in response to oxidative stress (McDonagh et al. 2005). Our group has also used IAM to block native reduced thiols, thus avoiding oxidation during the extraction, later reducing the native disulphide bonds and fluorescence labelling with IAF to aid with identification (Fernández-Cisnal et al. 2014). This method marks native disulfides, i.e., inter-/intramolecular protein disulphides and protein-glutathione mixed disulphides (Fuentes-Almagro et al. 2012; Charles et al. 2013; Fernández-Cisnal et al. 2014). Furthermore, we have recently shown that the total density of proteins with reversibly oxidized thiols increased in *P. clarkii* collected from areas surrounding DNP with high levels of pesticides and metals, compared to those from the control site. Figure 3 shows representative gel images of fluorescent labelled oxidized thiol proteins of the digestive gland of crayfish from the different sites. The total density of proteins with reversibly oxidized thiols was much higher in animals from MAT followed by ROC, while no difference was found in crayfish from PAR compared to those from the DNP core at LDP. After the hierarchical cluster analysis, proteins and sites were grouped based on similarities in their expression patterns. The oxidized protein spots were grouped into three clusters: cluster 1 included 4 spots that were exclusively up-regulated at ROC, cluster 2 included 17 proteins with particularly high fluorescence in MAT and that were also high at ROC, and cluster 3 included 14 spots that were up-regulated only at MAT. Principal component analysis (PCA) shows the distribution among the first two components of the three replicates of the four sampling sites in DNP. These first two components explained 93.86% of the variation in the dataset. PAR is close to LDP, the reference site, in the upper right of the graph. The other two sampling sites separate quite well in the 2D space: component 1 highly separates MAT which is localized in the upper left, while ROC is in the down center, separated from LDP by component 2. Nineteen proteins were identified via MALDI-TOF/TOF providing information about the metabolic pathways and/or physiological processes affected by pollutant-elicited oxidative stress. MS analysis identified specific thiol modifications on vulnerable Cys, important for understanding the response to stress and cell signalling (Fernández-Cisnal et al. 2014).

Identification of specific (carbonylation or thiol) modification sites is of paramount importance for an improved understanding of the consequences of protein oxidation (Eaton 2006; Braconi et al. 2011; Charles et al. 2013; Fernández-Cisnal et al. 2014). In summary, redox proteomics allows assessing the environmental quality of ecosystems. If the changes can be directly linked to exposure to a particular group of chemicals, they could be markers of exposure, which would be highly useful in biomonitoring programs.

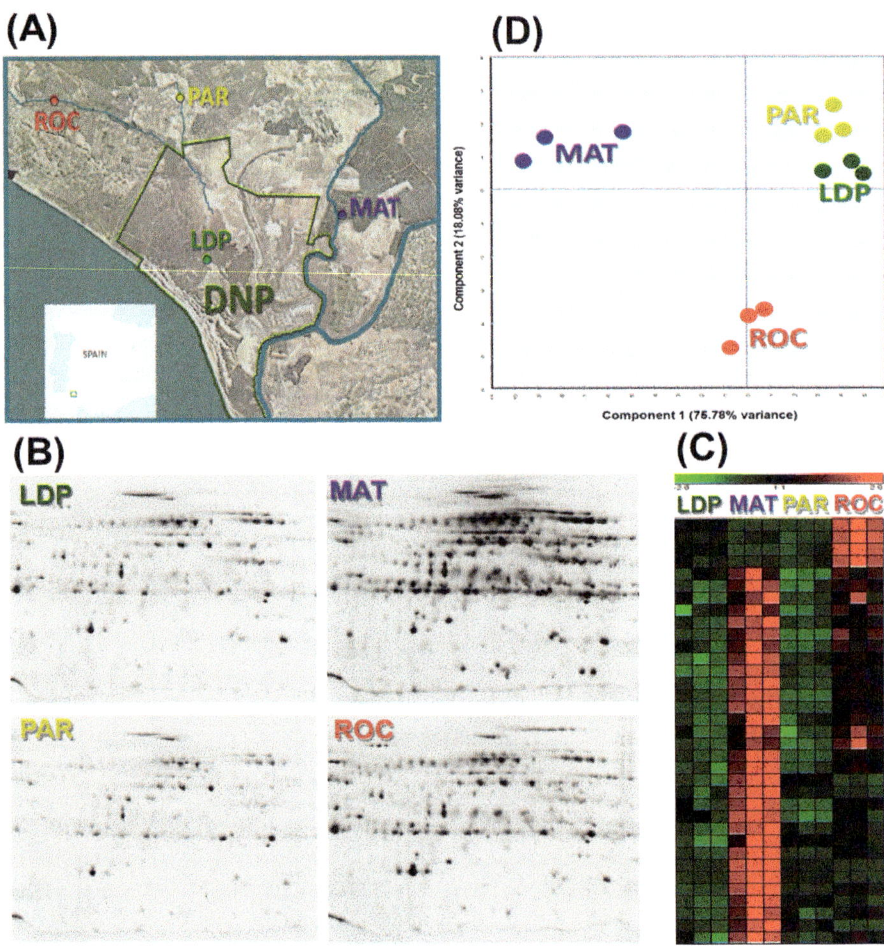

Fig. 3: Use of redox proteomics for assessing the biological effects of contaminants in DNP crayfish (A). Representative gel images of fluorescence-labeled oxidized thiol proteins of the digestive gland of crayfish from the different sites are shown (B). A hierarchical cluster analysis was carried out using the Genesis package to visually quantify the alterations in oxidized thiol levels among the 35 spots which presented differences in intensity (> 2 fold, $p < 0.05$) compared to the LDP reference site (C). Principal component analysis shows the distribution among the first two components of the three replicates of the four sampling sites in DNP and its surroundings where crayfish were collected (D) (modified from Fernández-Cisnal et al. 2014).

Integrating Omic Technologies for Ecological Risk Assessment

Toxicogenomics aims to determine the transcriptional responses of an organism to toxicants and to interpret these data to infer information on their mode of action and effects, including adverse outcomes (Williams et al. 2014a). As part of the GENIPOL project, a high density 13,270-clone cDNA microarray was prepared for the sentinel fish European flounder (*Platichthys flesus*) by combining clones obtained via suppressive subtractive hybridization and a liver cDNA library. Using this array, the global response to $CdCl_2$ doses relevant for environmental exposures were initially studied (Williams et al. 2006). This technology was later combined with data at molecular-, cellular-, tissue-, individual-, and population-level in an international consortium on Fish Toxicogenomics, aiming to integrate omics methods for aquatic Ecological Risk Assessment and Environmental Monitoring (Van Aggelen et al. 2010). Using this approach, hepatic gene expression was assessed in fish exposed to multiply polluted estuarine sediments, showing the lack of conventional exposure biomarker responses in contrast with the induction of inflammatory, innate immune, and apoptotic pathways (Leaver et al. 2010). In addition, *P. flexus* fish exposed to a brominated flame retardant mixture showed that the hepatic transcriptional changes were much more sensitive as exposure biomarkers than the enzyme activities and hormone levels measured previously for this aim (Williams et al. 2013). The molecular responses to contaminated sediments made by mixing material from the Forth (high organics) and Tyne (high metals and TBT) estuaries were also assessed in this organism (Williams et al. 2014b). This toxicogenomic approach, including transcriptomic and metabolomic analysis, was extended to stickleback fish (*Gasterostedus aculeatus*) exposed to copper (Santos et al. 2010) or ethinyl-estradiol (Katsiadaki et al. 2010).

The same approach was applied to the water flea *Daphnia magna*, by combining the Comet assay and transcriptomic analysis using Agilent microarrays, to assess the relative susceptibility to a variety of genotoxicants (David et al. 2011). More recently, transcriptomic and metabolomics approaches jointly applied to the freshwater alga *Chlamydomonas reinhardtii* showed that molecular toxicity of cerium oxide nanoparticles is obtained at supra environmental exposure concentrations, but not at predicted environmental levels (Taylor et al. 2015). Lately, 2nd-generation proteomics has been added to transcriptomics and metabolomics to study the toxicological effects of exposing *Mus musculus* mice to $CdCl_2$ (García-Sevillano et al. 2015).

Epigenetic Effects of Environmental Pollutants

Epigenetics studies the heritable changes of gene expression that are not derived from direct changes of DNA sequences. These changes include, at least, methylation of DNA (at a cytosine in a CpG site), different types of histone modification (acetylation, methylation, etc.), and microRNAs expression. Epigenetic changes affect most steps of the information flow, including: (i) interaction of DNA and transcription factors, (ii) transcript stability, (iii) DNA folding, (iv) nucleosome positioning, (v) chromatin compaction, and, (vi) higher-order organization of DNA. CpG dinucleotides are abundant in promoters, and their hypermethylation usually silences transcription, via a decreased affinity for transcription factors and an increased affinity for methylated DNA-binding proteins (MBDs), plus via recruiting histone deacetylases and other corepressors (Ho 2010). While methylated/silent promoters contain repressive histone marks, such as H3 trimethylated at Lys9 and Lys27, unmethylated/active promoters have lower affinity for MBDs, and display activated histone marks, such as H3 trimethylated at Lys4 and acetylated at Lys9. While DNA methylation mediates gene silencing in the long run, histone modifications promote faster responses (Ho 2010).

Recently, the epigenetic effects of pollutants has generated an increasing interest, following indications that organisms living in polluted environments modulate their responses via epigenetic changes (Williams et al. 2014a). Compared to mammals, fish have a high level of DNA methylation, with up to 7% of methylated cytosines. Despite these differences, there are similarities between zebrafish and mouse DNA methylation patterns (Feng et al. 2010), and between types of genes differentially methylated in human and zebrafish liver cancers (Mirbahai et al. 2011). Integration of high-throughput screening, omics and bioinformatics with high-throughput testing of zebrafish embryos, would lead to the discovery of adverse outcome pathways (AOP), and subsequently ensure the safety of chemicals in the environment (Williams et al. 2014a). Although far from the scope of this review, we have recently reported significant alterations of seven proteins involved in DNA and histone modification in terrestrial animals from DNP areas with different contaminant levels (Abril et al. 2015).

A recent review shows that epigenetic factors are key mechanisms of adaptation and response to environmental changes which, if persistent, could reflect previous stress exposures (Mirbahai and Chipman 2014). For this reason, epigenetic "foot-printing" could identify classes of chemicals to which organisms have been exposed throughout their lifetime. One of the most concerning effects of epigenetic modulation by environmental toxicants is their potential to modulate germ line programming, causing a transgenerational "memory" (Mirbahai and Chipman 2014). Common dabs (*Limanda limanda*) from polluted UK sites and high prevalence of liver tumors have been studied with a dab-specific microarray and high-throughput DNA sequencing. In hepatocellular adenomas, genes involved in cancer-related pathways, including apoptosis, wnt/β-catenin signaling, and estrogen responses, showed both altered levels of methylation and transcription (Mirbahai et al. 2015). These authors hypothesized that chronic exposure to a mixture of environmental contaminants contributes to a global hypomethylation followed by further epigenetic and genomic changes. These findings suggest a link between environment, epigenetics, and cancer in fish tumors in the wild, and show the utility of this methodology for studies in non-model organisms (Mirbahai et al. 2015), in agreement with our results in free-living animals from Doñana National Park (Abril et al. 2011; Abril et al. 2015).

The so-called AOP concept provides a framework for organizing knowledge about the progression of toxicity events across scales of biological organization; this would lead to adverse outcomes relevant for risk assessment. AOPs are aimed to improve both our understanding of chronic toxicity, including delayed toxicity and epigenetic or transgenerational effects of chemicals, and our ability to predict adverse outcomes. The requirements and challenges in application of AOP for chemical and site-specific risk assessment have been extended across species (Groh et al. 2015b). Recent AOP studies have been developed for pyrethroids, serotonin reuptake inhibitors, and Cd, focusing in growth impairment in fish, an apical endpoint commonly assessed in chronic toxicity tests for which a replacement is desirable (Groh et al. 2015a).

Overall Assessment of the Environmental Quality of Waters

Assessment of freshwater quality is usually based on cultivation-based methods. Express diagnosis in suspected faecal-contaminated samples can be gained looking at inhibition of pollutant-sensitive enzymatic activities (cholinesterases, luciferases, DNase, phosphatases, trypsin, amilase, etc.). These sensitive and inexpensive approaches could be useful for an initial screening of the integral pollution of waters (Fiksdal et al. 1994; Farnleitner et al. 2002; Fiksdal and Tryland 2008; Menzorova et al. 2014). An automated methodology based on the β-D-glucuronidase assay has been developed to test microbial or biochemical parameters of waters (Ryzinska-Paier et al. 2014). Inhibitive enzymatic assays for monitoring marine aquatic environments are not so simple, as enzymes are usually very sensitive to salts. The use of a salt-resistant DNase from eggs of the sea urchin *Strongylocentrotus intermedius* has been proposed for both marine- and fresh-waters (Menzorova et al. 2014). Yet, most of these bioassays cannot determine the toxins involved, although the inhibition of proteases has been specifically related to heavy metal pollution (Shukor et al. 2008; Shukor et al. 2009; Baskaran et al. 2013), and further molecular studies (e.g., EP) are needed to determine the contaminating chemical.

Sea water contains a huge number of microbes (protists, bacteria, viruses, etc.), mainly belonging to the plankton group, which play essential roles in regulating the flux of biomass and nutrients in oceans, and by extension, in the whole planet. When studying these organisms, the first problem faced by the researchers is that many of them have not been isolated or even identified. Since the start of 2000s, The Global Ocean Sampling Expedition (http://www.jcvi.org/cms/research/projects/gos/overview), and other metagenomic studies have analysed and sequenced the microorganisms in marine environments all around the world showing their vast taxonomic diversity and huge metabolic potential (Heidelberg et al. 2010; Wang et al.

2014). These studies provided the base for subsequent metaproteomic approaches, and in the last years several researchers started studying the proteome profiles from different sea regions (e.g., see (Siggins et al. 2012; Wang et al. 2014)) or the seasonal variations in an specific location (Georges et al. 2014).

Obviously, the presence of pollutants in sea spots should alter their microbial communities. Thus, changes in the metaproteomic profiles of aquatic areas, due to fluctuations in the species profile or to changes in their metabolisms, could in theory be used as pollution biomarkers. The use of microbes will overcome a key problem of multicellular bioindicators: the inconsistency in biomarkers due to intraclass biological variability—interindividual, age, sex and genetic background variations, etc.(Williams et al. 2013; Williams et al. 2014a). Similarly, the proteomes of microbial soil communities are now being used as pollution biomarkers (Siggins et al. 2012; Bastida et al. 2014). Nonetheless, metagenomic and metaproteomic microbiome studies from aquatic and terrestrial samples are methodologically difficult as proteins and/or DNAs are scarce in these media and their isolation produce poor yields, due to the presence of compounds that interfere in the extraction, including salts, humic acids, phenolic compounds, etc. (Becher et al. 2007; Benndorf et al. 2007; Pierre-Alain et al. 2007; Bastida et al. 2009; Taylor and Williams 2010; Keiblinger et al. 2012; Siggins et al. 2012; Bastida et al. 2014). Recently, commercial kits have become available to isolate nucleic acids from these environmental matrices, while those for protein isolation are not so developed yet. Our group is currently working on new methods to isolate proteins from the microbes present in soils, mud, and waters, in search of novel molecular biomarkers for early detection of polluted areas. Preliminary results show clearly different protein profiles from soils exposed to several sources of pollution in the surroundings of DNP (Rodríguez-Carmona 2015).

Proteomics in Climate Change Studies

As this revision has described, the EP approach is a powerful tool for understanding how the environment affects the biology of marine organisms. Thus, proteomic analyses have advanced towards the characterization of the biological effects of pollutants and the identification of comprehensive and pollutant-specific biomarker sets. Besides anthropogenic pollution, nowadays there is growing concern about the effects of human activities on climate change. This involves changes in a number of physical factors (e.g., temperature, salinity, ocean pH, hypoxia, etc.) that could significantly affect marine organisms, which must adapt. A chapter of the present book (Diniz et al. 2017) specifically reviews the proteomic approaches used to assess the effects of climate change in marine organisms. Tomanek has reviewed his previous work showing how thermal and osmotic stress, anaerobic conditions, and low pH due to ocean acidification, change the proteome. His work provides a framework of the common pathways that affect organisms under these environmental stresses, and identifies functionally adaptive variations (Tomanek 2011, 2012, 2014). A global analysis led to the proposal that oxidative stress may be an important factor affecting the physiological costs of elevated physical stress (Tomanek 2011; Tomanek et al. 2011, Tomanek 2014), similarly to pollution-promoted oxidative stress. Cytoskeletal elements were shown to be common and highly responsive stress targets. Stressors also disrupt protein homeostasis, as indicated by the induction of molecular chaperones and proteasome isoforms. Aerobic energy metabolism is in general depressed by the above-mentioned stressors, while other metabolic pathways are activated to switch among metabolic fuels, and to respond to the increased oxidative stress by scavenging RO/NS (Tomanek 2011, 2012, 2014).

Oxidative stress was also the main symptom in juveniles of the Senegal sole (*Solea senegalensis*) affected by gas bubble disease (GBD) in hyper-oxygenated ponds under aquaculture conditions (Fig. 4). Physical-chemical conditions inducing hyperoxia, include radiation, temperature, and high levels of dissolved oxygen. Exophthalmia, subcutaneous emphysemas, obstruction of gill lamellae, haemorrhages, and anomalous swimming accompanied by loss of orientation, near lethargy status and individual isolation were the main effects of oxygen super-saturation (Fig. 4A). A proteomic study was carried out in search of protein alteration patterns that might be used as potential new and unbiased biomarkers of hyperoxic stress. Protein expression profiles were studied by 2DE in cytosolic fractions of gills and livers. A total of 1,525 and 1,632 spots were detected in the four gill and liver gels, respectively. Figure 4B shows the master gels of cytosolic gill (upper) and liver (lower) proteins. A total of 205 protein spots were differentially expressed in the gills and 498 in the liver in each health status. Figure 4C shows the number of spots which are present only, absent, increased, or diminished in each of the studied conditions, and the total number of changes found. A significantly higher number of differentially expressed spots were found in GBD-affected soles. Of these, 25 gill spots and 23 in liver were selected for identification using nESI-IT MS/MS, *de novo* sequencing, and bioinformatics search. To quantify the expression differences among these spots, a hierarchical cluster analysis was carried out using the Genesis package (Fig. 4D). In gills, most of the spots with altered expression were upregulated in GBD-affected soles and only three were downregulated when compared with healthy fish. In the liver, 14 spots were upregulated and nine downregulated in GBD-affected fish. Sequence tags were obtained from nine (gills) and five (liver) of the selected spots, resulting in a total of 14 identified in the GBD status, which are indicated in the right of each cluster (Fig. 4C). Proteins identified in gills of GBD-affected fish were related to oxidative alteration of cytoskeleton structure/function, motility, or regulatory pathways, reflecting the central role of gills in oxygen exchange. Hepatic proteins were related to protein oxidative damages, protection from oxidative stress, and inflammatory response, in agreement with the predominant metabolic role of liver. The identified proteins were: MLC1, myosin light chain; α-TMP, alpha-tropomyosin; β-ACT, beta-actin; RKIP, Raf kinase inhibitor protein; CaM, calmodulin; GAPDH, glyceraldehyde-3-phosphate dehydrogenase; β-TBB, beta-tubulin; FABP, fatty acid-binding protein; GNMT, glycine N-methyltransferase; C3, complement component C3; DCXR, Dicarbonyl L-xylulose reductase; HBB, beta-globin. Some of these proteins might be good hyperoxia stress biomarkers and could be used in the early warning detection of GBD outbreaks (Salas-Leiton et al. 2009).

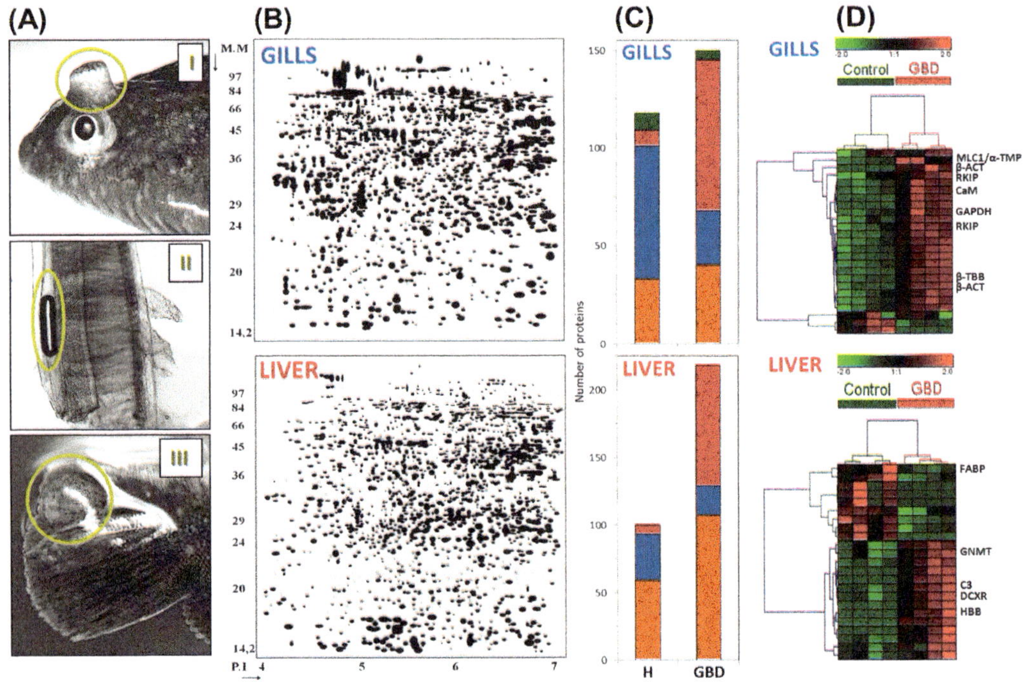

Fig. 4: (A) Visible symptoms in GBD-affected soles, including: exophthalmia via retrobulbar bubbles (I), a bubble obstructing a gill lamella (II), and a large bubble at caudal fin (III). (B) Master gels of GBD-affected soles after 2DE of cytosolic gill and liver. (C) Number of proteins showing distinct protein expression patterns in healthy (H) and GBD-affected fish. Bars represent the number of protein spots present only (■), absent (■), increased (■), and diminished (■) in each of the conditions. (D) Cluster analysis of the spots selected for identification by MS analysis. Each row represents one spot, showing successfully identified proteins on the right. Columns represent the average expression profile at each health condition. Green rectangles indicate samples with lower intensity for one particular protein spot relative to other samples, and red rectangles the samples where the spot is more intense. The color intensity is proportional to the fold change as represented by the scale (modified from Salas-Leiton et al. 2009).

Final Remarks

This chapter examines the progression of the methods used for aquatic pollution monitoring, since the initial assessment of conventional biomarkers, using a small number of end points, until the present, in which the use of *omic* methodologies is allowing the massive assessment of biological responses to pollutants. Initial biomarkers focused in the level of one particular enzymatic activity, or, after the development of genomic procedures, on the expression level of a certain gene/mRNA. After the appearance of massive sequencing techniques, this field entered the exciting era of postgenomic biology, that lead to an expanding range of new methodologies which could be used for environmental monitoring and risk assessment. *Omic* sciences include, among others, genomics for the study of DNA variations, transcriptomics for the genome-wide characterization of gene expression via the measurement of RNAs, epigenomics/epigenetics for the identification of specific transcriptomic regulating factors not directly coded in the genomic sequence, proteomics for measuring cell and tissue-wide expression of proteins, and more recently, metabolomics for the assessment of metabolites. Meaningful, proteomics provides additional and complementary information to genomic and transcriptomic analysis since, after transcription, mRNAs are differentially processed during the splicing route and, in addition, proteins may be modified after synthesis or translation, in a process known as post-translational modification. Changes in proteins caused by RO/NS have the potential to be very sensitive global markers of environmental stress. The incorporation of the different types of biomarkers should be reflected in the number of publications including different terms. Figure 5 shows the time-course of the number of references for "Biomarkers + Marine Organisms", found in the Science Direct database. The papers, grouped in biannual periods since 1998, were further tagged to include the

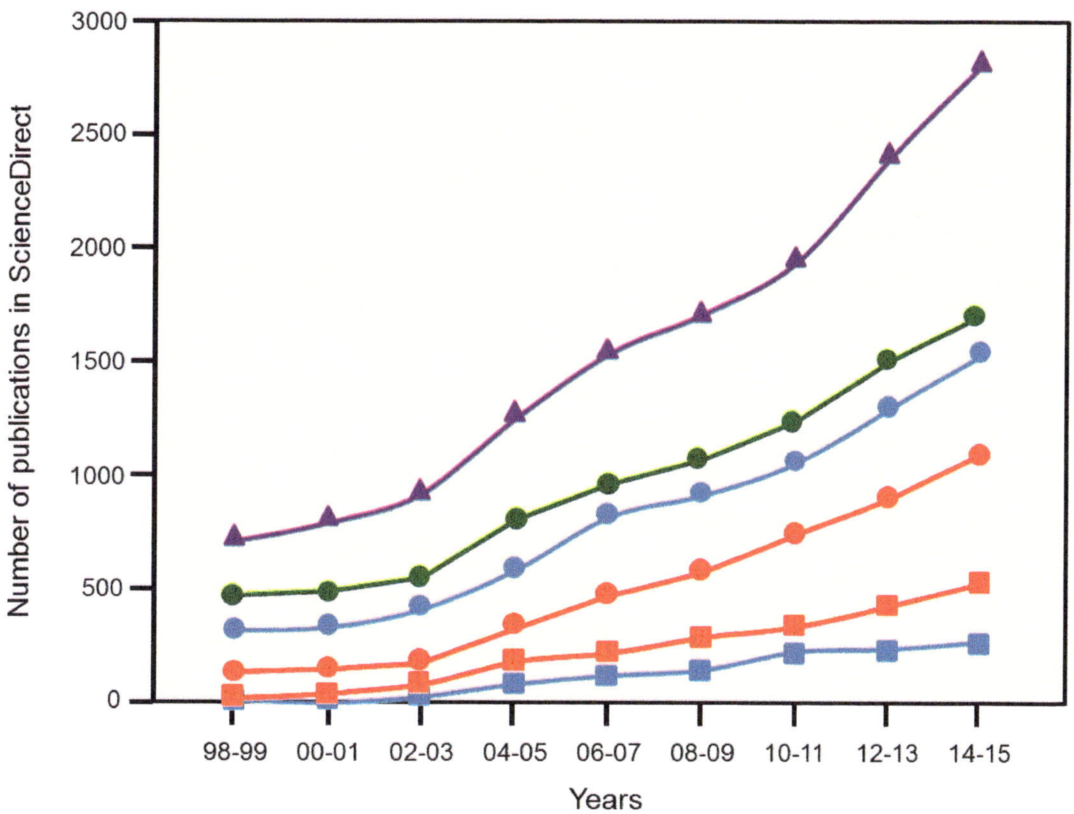

Fig. 5: The number of scientific outputs was determined from the Science Direct database. The terms used for the analysis were: "Marine+Biomarker" (Total, ▲), plus adding "Molecular" (●), "Protein" (●), "Gene" (●), "Genomic" (■), or "Proteomic" (■) to the search.

following terms: "Molecular", "Protein", "Gene", "Genomic" and "Proteomic". Around 2002, a drastic increase was observed in the number of publications about Marine Biomarkers. Close to two thirds of these papers included the term "Molecular", most of which also included "Protein", thus reflecting the early contributions of enzymatic activities. Although clearly retarded, the application of *omic* technologies is steadily increasing; yet, the "Genomic" approaches usually double the "Proteomic" studies, probably due to the methodological pitfalls mentioned above that should be overcome before their general application, plus the initial lack of fully-sequenced genomes of aquatic organisms.

In aquatic environmental studies, *omic* technologies have emerged as a powerful alternative to conventional or "classic" biomarkers since these techniques allow quantitatively monitoring many biological molecules in a high-throughput manner, and provide a general appraisal of the biological responses altered by exposure to contaminants and other stresses (e.g., temperature, salinity, pH, hypoxia/hyperoxia, etc.). It is becoming clear that any single *omic* approach is not sufficient to characterize the complexity of ecosystems. Future research for aquatic ecological risk assessment and environmental monitoring should focus towards the integration of different *omic* (and also conventional) technologies. This integration will provide a holistic information about specific biological responses, which combined with the results from contaminant analysis, will allow defining a non-biased growing set of specific biomarker patterns for the global biomonitoring of any ecosystem, thus contributing to the required technological renewal.

Acknowledgements

Authors thank the following organisations that supported the research included: (i) Spanish Ministry of Economy and Competitiveness (projects CTM2006-08960-C02-02, CTM2009-12858.C02-02, and CTM2012-38720.C03-02), (ii) European Regional Development Fund (INTERREG 3B, project 091-AAAG), (iii) Spanish Agency of International Cooperation (Spain-Tunisia Program), (iv) Agency of Economy, Competitiveness, Science and Employment of the Andalusian Government (projects AGR-516 and P08-CVI-03829, groups BIO151 and BIO187).

Keywords: Biomarkers, climate change, cluster analysis, Doñana National Park, environmental proteomics, environmental quality of waters, epigenetic changes, ecological risk assessment, mass spectrometry, oxidative stress, *Omic* approaches, protein carbonylation, protein expression signatures, redox proteomics, reversibly oxidized thiols

References

Abril, N., J. Ruiz-Laguna, I. Osuna-Jiménez, A. Vioque-Fernández, R. Fernández-Cisnal, E. Chicano-Gálvez, J. Alhama, J. López-Barea and C. Pueyo. 2011. Omic approaches in environmental issues. J. Toxicol. Environ. Health A 74: 1001–1019.

Abril, N., E. Chicano-Gálvez, C. Michan, C. Pueyo and J. López-Barea. 2015. iTRAQ analysis of hepatic proteins in free-living *Mus spretus* mice to assess the contamination status of areas surrounding Doñana National Park (SW Spain). Sci. Total Environ. 523: 16–27.

Abril, N., M.J. Prieto-Álamo and C. Pueyo. 2017. Functional genomics approaches in marine pollution and aquaculture. *In*: T. García-Barrera and J.L. Gómez-Ariza [eds.]. Environmental Problems in Marine Biology: Methodological Aspects and Applications. Science Publishers Inc., Enfield, New Hampshire, USA.

Albaiges, J., J. Algaba, P. Arambarri, F. Cabrera, G. Baluja, L.M. Hernández and J. Castroviejo. 1987. Budget of organic and inorganic pollutants in the Doñana National Park (Spain). Sci. Total Environ. 63: 13–28.

Barrett, J., P.M. Brophy and J.V. Hamilton. 2005. Analysing proteomic data. Int. J. Parasitol. 35: 543–553.

Baskaran, G., N.A. Masdor, M.A. Syed and M.Y. Shukor. 2013. An inhibitive enzyme assay to detect mercury and zinc using protease from *Coriandrum sativum*. Sci. World J. 2013: 678356.

Bastida, F., J.L. Moreno, C. Nicolás, T. Hernández and C. García. 2009. Soil metaproteomics: a review of an emerging environmental science. Significance, methodology and perspectives. Eur. J. Soil Sci. 60: 845–859.

Bastida, F., T. Hernández and C. García. 2014. Metaproteomics of soils from semiarid environment: functional and phylogenetic information obtained with different protein extraction methods. J. Proteomics 101: 31–42.

Baty, J.W., M.B. Hampton and C.C. Winterbourn. 2002. Detection of oxidant sensitive thiol proteins by fluorescence labeling and two-dimensional electrophoresis. Proteomics 2: 1261–1266.

Becher, D., J. Bernhardt, S. Fuchs and K. Riedel. 2007. Metaproteomics to unravel major microbial players in leaf litter and soil environments: challenges and perspectives. Proteomics 13: 2895–2909.

Benndorf, D., G.U. Balcke, H. Harms and M. von Bergen. 2007. Functional metaproteome analysis of protein extracts from contaminated soil and groundwater. ISME J. 1: 224–234.

Bonetto, V. and P. Ghezzi. 2006. Thiol-disulfide oxidoreduction of protein cysteines: old methods revisited for proteomics. pp. 101–122. *In*: I. Dalle-Donne, A. Scaloni and D.A. Butterfield [eds.]. Redox Proteomics: From Protein Modifications to Cellular Dysfunction and Diseases. John Wiley & Sons, Inc., Hoboken, New Jersey, USA.

Bonilla-Valverde, D., J. Ruiz-Laguna, A. Muñoz, J. Ballesteros, F. Lorenzo, J.L. Gómez-Ariza and J. López-Barea. 2004. Evolution of biological effects of Aznalcollar mining spill in the Algerian mouse (*Mus spretus*) using biochemical biomarkers. Toxicology 197: 123–138.

Braconi, D., G. Bernardini and A. Santucci. 2011. Linking protein oxidation to environmental pollutants: Redox proteomic approaches. J. Proteomics 74: 2324–2337.

Bradley, B.P., E.A. Schrader, D.G. Kimmel and J.C. Meiller. 2002. Protein expression signatures: an application of proteomics. Marine Environ. Res. 45: 373–377.

Butterfield, D.A. and I. Dalle-Donne. 2012. Redox proteomics. Antioxid. Redox Signal 17: 1487–1489.

Cabiscol, E. and J. Ros. 2006. Oxidative damage to proteins: structural modifications and consequences in cell function. pp. 399–471. *In*: I. Dalle-Donne, A. Scaloni and D.A. Butterfield [eds.]. Redox Proteomics: From Protein Modifications to Cellular Dysfunction and Diseases. John Wiley & Sons, Inc., Hoboken, New Jersey, USA.

Cabrera, F., C.G. Toca and P. Arambarri. 1984. Acid mine-water and agricultural pollution in a river skirting the Doñana National Park (Guadiamar river, South West Spain). Water Res. 18: 1469–1482.

Charles, R., T. Jayawardhana and P. Eaton. 2013. Gel-based methods in redox proteomics. Biochim. Biophys. Acta 1840: 830–837.

Chiappetta, G., S. Ndiaye, A. Igbaria, C. Kumar, J. Vinh and M.B. Toledano. 2010. Proteome screens for Cys residues oxidation: the redoxome. Methods Enzymol. 473: 199–216.

Chora, S., B. McDonagh, D. Sheehan, M. Starita-Geribaldi, M. Romeo and M.J. Bebianno. 2008. Ubiquitination and carbonylation as markers of oxidative-stress in *Ruditapes decussatus*. Mar. Environ. Res. 66: 95–97.

Chora, S., B. McDonagh, D. Sheehan, M. Starita-Geribaldi, M. Romeo and M.J. Bebianno. 2010. Ubiquitination and carbonylation of proteins in the clam *Ruditapes decussatus*, exposed to nonylphenol using redox proteomics. Chemosphere 81: 1212–1217.

Chouchani, E.T., A.M. James, I.M. Fearnley, K.S. Lilley and M.P. Murphy. 2011. Proteomic approaches to the characterization of protein thiol modification. Curr. Opin. Chem. Biol. 15: 120–128.

Company, R., A. Torreblanca, M. Cajaraville, M.J. Bebianno and D. Sheehan. 2012. Comparison of thiol subproteome of the vent mussel *Bathymodiolus azoricus* from different Mid-Atlantic Ridge vent sites. Sci. Total Environ. 437: 413–421.

Costa, P.M., E. Chicano-Gálvez, J. López Barea, T.A. DelValls and M.H. Costa. 2010. Alterations to proteome and tissue recovery responses in fish liver caused by a short-term combination treatment with cadmium and benzo[a]pyrene. Environ. Pollut. 158: 3338–3346.

Costa, P.M., E. Chicano-Gálvez, S. Caeiro, J. Lobo, M. Martins, A.M. Ferreira, M. Caetano, C. Vale, J. Alhama-Carmona, J. López-Barea, T.A. DelValls and M.H. Costa. 2012. Hepatic proteome changes in *Solea senegalensis* exposed to contaminated estuarine sediments: a laboratory and *in situ* survey. Ecotoxicology 21: 1194–1207.

Dalle-Donne, I., A. Scaloni and D.A. Butterfield. 2006. Redox Proteomics: From Protein Modifications to Cellular Dysfunction and Diseases. John Wiley & Sons, Inc., Hoboken, New Jersey, USA.

David, R.M., V. Dakic, T.D. Williams, M.J. Winter and J.K. Chipman. 2011. Transcriptional responses in neonate and adult *Daphnia magna* in relation to relative susceptibility to genotoxicants. Aquat. Toxicol. 104: 192–204.

Davies, M.J. 2005. The oxidative environment and protein damage. Biochim. Biophys. Acta 1703: 93–109.

Diniz, M., D. Madeira, H. Santos, E. Araújo and J.L. Capelo. 2017. Effects of climate change in marine organisms: a proteomic approach. *In*: T. García-Barrera and J.L. Gómez-Ariza [eds.]. Environmental Problems in Marine Biology: Methodological Aspects and Applications. Science Publishers Inc., Enfield, New Hampshire, USA.

Eaton, P. 2006. Protein thiol oxidation in health and disease: techniques for measuring disulfides and related modifications in complex protein mixtures. Free Radic. Biol. Med. 40: 1889–1899.

Farnleitner, A.H., L. Hocke, C. Beiwl, G.G. Kavka and R.L. Mach. 2002. Hydrolysis of 4-methylumbelliferyl-beta-D-glucuronide in differing sample fractions of river waters and its implication for the detection of fecal pollution. Water Res. 36: 975–981.

Feng, S., S.J. Cokus, X. Zhang, P.Y. Chen, M. Bostick, M.G. Goll, J. Hetzel, J. Jain, S.H. Strauss, M.E. Halpern, C. Ukomadu, K.C. Sadler, S. Pradhan, M. Pellegrini and S.E. Jacobsen. 2010. Conservation and divergence of methylation patterning in plants and animals. Proc. Natl. Acad. Sci. USA 107: 8689–8694.

Fernández-Cisnal, R. 2008. Proteómica ambiental en "*Carcinus maenas*": Biomonitorización de la costa de Túnez. Master Thesis, University of Cordoba, Cordoba, Spain.

Fernández-Cisnal, R., J. Alhama, N. Abril, C. Pueyo and J. López-Barea. 2014. Redox proteomics as biomarker for assessing the biological effects of contaminants in crayfish from Doñana National Park. Sci. Total Environ. 490: 121–133.

Fiksdal, L., M. Pommepuy, M.P. Caprais and I. Midttun. 1994. Monitoring of fecal pollution in coastal waters by use of rapid enzymatic techniques. Appl. Environ. Microbiol. 60: 1581–1584.

Fiksdal, L. and I. Tryland. 2008. Application of rapid enzyme assay techniques for monitoring of microbial water quality. Curr. Opin. Biotechnol. 19: 289–294.

Fuentes-Almagro, C.A., M.J. Prieto-Alamo, C. Pueyo and J. Jurado. 2012. Identification of proteins containing redox-sensitive thiols after PRDX1, PRDX3 and GCLC silencing and/or glucose oxidase treatment in Hepa 1–6 cells. J. Proteomics 77: 262–279.

García-Sevillano, M.A., N. Abril, R. Fernández-Cisnal, T. García-Barrera, C. Pueyo, J. López-Barea and J.L. Gómez-Ariza. 2015. Functional genomics and metabolomics reveal the toxicological effects of cadmium in *Mus musculus* mice. Metabolomics 11: 1432–1450.

Ge, Y., J.F. Wang, S. Cristobal, D. Sheehan, F. Silvestre, X. Peng, H. Li, Z. Gong, S.H. Lam, H. Wentao, H. Iwahashi, J. Liu, N. Mei, L. Shi, M. Bruno, H. Foth and K. Teichman. 2013. Environmental OMICS: current status and future directions. J. Integr. OMICS 3: 75–87.

Georges, A.A., H. El-Swais, S.E. Craig, W.K. Li and D.A. Walsh. 2014. Metaproteomic analysis of a winter to spring succession in coastal northwest Atlantic Ocean microbial plankton. ISME J. 8: 1301–1313.

Ghedira, J. 2011. Etude des markeurs biochimiques de pollution chez le crustacé marin "*Carcinus maenas*" et characterization biochimique et toxicologique des cholinestrases. Ph.D. Thesis. University of Monastir, Sousse, Tunisia.

Ghedira, J., E. Chicano-Gálvez, R. Fernández-Cisnal, J. Jebali, M. Banni, L. Chouba, H. Boussetta, J. López-Barea and J. Alhama. 2015. Using environmental proteomics to assess pollutant response of *Carcinus maenas* along the Tunisian coast. Sci. Total Environ. 541: 109–118.

Gómez-Ariza, J.L., M. González-Fernández, T. García-Barrera, J. López-Barea and C. Pueyo. 2008. Integration of metallomics, proteomics and transcriptomics in environmental issues. Chem. Listy 102: s303–s308.

Grimalt, J.O., M. Ferrer and E. Macpherson. 1999. The mine tailing accident in Aznalcollar. Sci. Total Environ. 242: 3–11.

Groh, K.J., R.N. Carvalho, J.K. Chipman, N.D. Denslow, M. Halder, C.A. Murphy, D. Roelofs, A. Rolaki, K. Schirmer and K.H. Watanabe. 2015a. Development and application of the adverse outcome pathway framework for understanding and predicting chronic toxicity: II. A focus on growth impairment in fish. Chemosphere 120: 778–792.

Groh, K.J., R.N. Carvalho, J.K. Chipman, N.D. Denslow, M. Halder, C.A. Murphy, D. Roelofs, A. Rolaki, K. Schirmer and K.H. Watanabe. 2015b. Development and application of the adverse outcome pathway framework for understanding and predicting chronic toxicity: I. Challenges and research needs in ecotoxicology. Chemosphere 120: 764–777.

Hansen, R.E., D. Roth and J.R. Winther. 2009. Quantifying the global cellular thiol-disulfide status. Proc. Natl. Acad. Sci. USA 106: 422–427.

Heidelberg, K.B., J.A. Gilbert and I. Joint. 2010. Marine genomics: at the interface of marine microbial ecology and biodiscovery. Microb. Biotechnol. 3: 531–543.

Hill, B.G., C. Reily, J.Y. Oh, M.S. Johnson and A. Landar. 2009. Methods for the determination and quantification of the reactive thiol proteome. Free Radic. Biol. Med. 47: 675–683.

Ho, S.M. 2010. Environmental epigenetics of asthma: an update. J. Allergy Clin. Immunol. 126: 453–465.

Hogstrand, C., S. Balesaria and C.N. Glover. 2002. Application of genomics and proteomics for study of the integrated response to zinc exposure in a non-model fish species, the rainbow trout. Comp. Biochem. Physiol. B 133 B: 523–535.

Hu, W., S. Tedesco, R. Faedda, G. Petrone, S.O. Cacciola, A. O'Keefe and D. Sheehan. 2010. Covalent selection of the thiol proteome on activated thiol sepharose: a robust tool for redox proteomics. Talanta 80: 1569–1575.

Hu, W., S. Culloty, G. Darmody, S. Lynch, J. Davenport, S. Ramirez-Garcia, K.A. Dawson, I. Lynch, J. Blasco and D. Sheehan. 2014. Toxicity of copper oxide nanoparticles in the blue mussel, *Mytilus edulis*: a redox proteomic investigation. Chemosphere 108: 289–299.

Hwang, N.R., S.H. Yim, Y.M. Kim, J. Jeong, E.J. Song, Y. Lee, J.H. Lee, S. Choi and K.J. Lee. 2009. Oxidative modifications of glyceraldehyde-3-phosphate dehydrogenase play a key role in its multiple cellular functions. Biochem. J. 423: 253–264.

Jaafar, S.N., A.V. Coelho and D. Sheehan. 2015. Redox proteomic analysis of *Mytilus edulis* gills: effects of the pharmaceutical diclofenac on a non-target organism. Drug Test Anal. 7: 957–966.

James, P. 1997. Protein identification in the post-genome era: the rapid rise of proteomics. Q. Rev. Biophys. 30: 279–331.

Janssen-Heininger, Y.M., B.T. Mossman, N.H. Heintz, H.J. Forman, B. Kalyanaraman, T. Finkel, J.S. Stamler, S.G. Rhee and A. van der Vliet. 2008. Redox-based regulation of signal transduction: principles, pitfalls, and promises. Free Radic. Biol. Med. 45: 1–17.

Jebali, J., E. Chicano-Gálvez, M. Banni, H. Guerbej, H. Boussetta, J. López-Barea and J. Alhama. 2013. Biochemical responses in seabream (*Sparus aurata*) caged in-field or exposed to benzo(a)pyrene and paraquat. Characterization of glutathione *S*-transferases. Ecotoxicol. Environ. Saf. 88: 169–177.

Jebali, J., E. Chicano-Gálvez, R. Fernández-Cisnal, M. Banni, L. Chouba, H. Boussetta, J. López-Barea and J. Alhama. 2014. Proteomic analysis in caged Mediterranean crab (*Carcinus maenas*) and chemical contaminant exposure in Teboulba Harbour, Tunisia. Ecotoxicol. Environ. Saf. 100: 15–26.

Karlsen, O.A., D. Sheehan and A. Goksoyr. 2014. Alterations in the Atlantic cod (*Gadus morhua*) hepatic thiol-proteome after methylmercury exposure. J. Toxicol. Environ. Health A 77: 650–662.

Katsiadaki, I., T.D. Williams, J.S. Ball, T.P. Bean, M.B. Sanders, H. Wu, E.M. Santos, M.M. Brown, P. Baker, F. Ortega, F. Falciani, J.A. Craft, C.R. Tyler, M.R. Viant and J.K. Chipman. 2010. Hepatic transcriptomic and metabolomic responses in the Stickleback (*Gasterosteus aculeatus*) exposed to ethinyl-estradiol. Aquat. Toxicol. 97: 174–187.

Keiblinger, K.M., I.C. Wilhartitz, T. Schneider, B. Roschitzki, E. Schmid, L. Eberl, K. Riedel and S. Zechmeister-Boltenstern. 2012. Soil metaproteomics—Comparative evaluation of protein extraction protocols. Soil. Biol. Biochem. 54: 14–24.

Knigge, T., T. Monsinjon and O.K. Andersen. 2004. Surface-enhanced laser desorption/ionization-time of flight-mass spectrometry approach to biomarker discovery in blue mussels (*Mytilus edulis*) exposed to polyaromatic hydrocarbons and heavy metals under field conditions. Proteomics 4: 2722–2727.

Knigge, T., J. Letendre and T. Monsinjon. 2013. Sample preparation for two-dimensional gel electrophoresis: considering the composition of biological material. Proteomics 13: 3106–2108.

Larsson, D.G., C. de Pedro and N. Paxeus. 2007. Effluent from drug manufactures contains extremely high levels of pharmaceuticals. J. Hazard. Mater. 148: 751–755.

Leaver, M.J., A. Diab, E. Boukouvala, T.D. Williams, J.K. Chipman, C.F. Moffat, C.D. Robinson and S.G. George. 2010. Hepatic gene expression in flounder chronically exposed to multiply polluted estuarine sediment: Absence of classical exposure 'biomarker' signals and induction of inflammatory, innate immune and apoptotic pathways. Aquat. Toxicol. 96: 234–245.

Lemos, M.F., A.M. Soares, A.C. Correia and A.C. Esteves. 2010. Proteins in ecotoxicology—how, why and why not? Proteomics 10: 873–887.

Levine, R.L. 2002. Carbonyl modified proteins in cellular regulation, aging, and disease. Free Radic. Biol. Med. 32: 790–796.

López-Barea, J. 1995. Biomarkers in ecotoxicology: an overview. pp. 57–79. *In*: G.H. Degen, J.P. Seiler and P. Bentely [eds.]. Toxicology in Transition. Springer, Berlin, Germany.

López-Barea, J. and J.L. Gómez-Ariza. 2006. Environmental proteomics and metallomics. Proteomics 6 Suppl. 1: S51–62.

López-Pamo, E., D. Barettino, C. Antón-Pacheco, G. Ortiz, J.C. Arránz, J.C. Gumiel, B. Martínez-Pledel, M. Aparicio and O. Montouto. 1999. The extent of the Aznalcollar pyritic sludge spill and its effects on soils. Sci. Total Environ. 242: 57–88.

López, J.L., E. Mosquera, J. Fuentes, A. Marina, J. Vázquez and G. Alvarez. 2001. Two-dimensional electrophoresis of *Mytilus galloprovincialis*. Differences in protein expression between intertidal and cultured mussels. Marine Ecol. Progress Ser. 224: 149–156.

López, J.L., A. Marina, J. Vázquez and G. Alvarez. 2002. A proteomic approach to study the marine mussels *Mytilus edulis* and *M. galloprovincialis*. Marine Biology 141: 217–223.

Manzano, M., C. Ayora, C. Domenech, P. Navarrete, A. Garralon and M.J. Turrero. 1999. The impact of the Aznalcollar mine tailing spill on groundwater. Sci. Total Environ. 242: 189–209.

McDonagh, B., R. Tyther and D. Sheehan. 2005. Carbonylation and glutathionylation of proteins in the blue mussel *Mytilus edulis* detected by proteomic analysis and Western blotting: Actin as a target for oxidative stress. Aquat. Toxicol. 73: 315–326.

Menzorova, N.I., A.V. Seitkalieva and V.A. Rasskazov. 2014. Enzymatic methods for the determination of pollution in seawater using salt resistant alkaline phosphatase from eggs of the sea urchin *Strongylocentrotus intermedius*. Mar. Pollut. Bull. 79: 188–195.

Mirbahai, L., T.D. Williams, H. Zhan, Z. Gong and J.K. Chipman. 2011. Comprehensive profiling of zebrafish hepatic proximal promoter CpG island methylation and its modification during chemical carcinogenesis. BMC Genomics 12: 3.

Mirbahai, L. and J.K. Chipman. 2014. Epigenetic memory of environmental organisms: a reflection of lifetime stressor exposures. Mutat. Res. Genet. Toxicol. Environ. Mutagen. 764-765: 10–17.

Mirbahai, L., G. Yin, J.P. Bignell, N. Li, T.D. Williams and J.K. Chipman. 2015. DNA methylation in liver tumorigenesis in fish from the environment. Epigenetics 6: 1319–1333.

Monsinjon, T., O.K. Andersen, F. Leboulenger and T. Knigge. 2006. Data processing and classification analysis of proteomic changes: a case study of oil pollution in the mussel, *Mytilus edulis*. Proteome Sci. 4: 17.

Monsinjon, T. and T. Knigge. 2007. Proteomic applications in ecotoxicology. Proteomics 7: 2997–3009.

Montes-Nieto, R., C.A. Fuentes-Almagro, D. Bonilla-Valverde, M.J. Prieto-Alamo, J. Jurado, M. Carrascal, J.L. Gómez-Ariza, J. López-Barea and C. Pueyo. 2007. Proteomics in free-living *Mus spretus* to monitor terrestrial ecosystems. Proteomics 7: 4376–4387.

Montes Nieto, R., T. García-Barrera, J.L. Gómez-Ariza and J. López-Barea. 2010. Environmental monitoring of Domingo Rubio stream (Huelva Estuary, SW Spain) by combining conventional biomarkers and proteomic analysis in *Carcinus maenas*. Environ. Pollut. 158: 401–408.

O'Farrell, P.H. 1975. High resolution two-dimensional electrophoresis of proteins. J. Biol. Chem. 250: 4007–4021.

Oktyabrsky, O.N. and G.V. Smirnova. 2007. Redox regulation of cellular functions. Biochemistry (Mosc) 72: 132–145.

Olsson, B., B.P. Bradley, M. Gilek, O. Reimer, J.L. Shepard and M. Tedengren. 2004. Physiological and proteomic responses in *Mytitus edulis* exposed to PCBs and PAHs extracted from Baltic Sea sediments. Hydrobiologia 514: 15–27.

Panchout, F., J. Letendre, F. Bultelle, X. Denier, B. Rocher, P. Chan, D. Vaudry and F. Durand. 2013. Comparison of protein-extraction methods for gills of the shore crab, *Carcinus maenas* (L.), and application to 2DE. J. Biomol. Tech. 24: 218–223.

Pierre-Alain, M., M. Christophe, S. Severine, A. Houria, L. Philippe and R. Lionel. 2007. Protein extraction and fingerprinting optimization of bacterial communities in natural environment. Microb. Ecol. 53: 426–434.

Pueyo, C., J.L. Gómez-Ariza, M.A. Bello-López, R. Fernández-Torres, N. Abril, J. Alhama, T. García-Barrera and J. López-Barea. 2011. New methodologies for assessing the presence and ecological effects of pesticides in Doñana National Park (SW Spain). pp. 165–196. *In*: M. Sotytcheva [ed.]. Pesticides in the Modern World—Trends in Pesticides Analysis. INTECH open, Rijeka, Croatia.

Rainville, L.C., D. Carolan, A.C. Varela, H. Doyle and D. Sheehan. 2014. Proteomic evaluation of citrate-coated silver nanoparticles toxicity in *Daphnia magna*. Analyst 139: 1678–1686.

Rainville, L.C., A.V. Coelho and D. Sheehan. 2015. Application of a redox-proteomics toolbox to *Daphnia magna* challenged with model pro-oxidants copper and paraquat. Environ. Toxicol. Chem. 34: 84–91.

Roberts, R.A., D.L. Laskin, C.V. Smith, F.M. Robertson, E.M. Allen, J.A. Doorn and W. Slikker. 2009. Nitrative and oxidative stress in toxicology and disease. Toxicol. Sci. 112: 4–16.

Rocher, B., F. Bultelle, P. Chan, F. Le Foll, J. Letendre, T. Monsinjon, S. Olivier, R. Péden, A. Poret, D. Vaudry and T. Knigge. 2015. 2-DE mapping of the blue mussel gill proteome: The usual suspects revisited. Proteomes 3: 3–41.

Rodríguez-Carmona, F. 2015. Optimización de métodos de extracción de proteínas de microorganismos del suelo: aplicación de la metaproteómica para evaluar la contaminación de Doñana. BSc. Thesis. University of Cordoba. Cordoba, Spain.

Rodríguez-Ortega, M.J., B.E. Grosvik, A. Rodríguez-Ariza, A. Goksoyr and J. López-Barea. 2003. Changes in protein expression profiles in bivalve molluscs (*Chamaelea gallina*) exposed to four model environmental pollutants. Proteomics 3: 1535–1543.

Romero-Ruiz, A., M. Carrascal, J. Alhama, J.L. Gómez-Ariza, J. Abian and J. López-Barea. 2006. Utility of proteomics to assess pollutant response of clams from the Doñana bank of Guadalquivir Estuary (SW Spain). Proteomics 6 Suppl. 1: S245–255.

Ruhoy, I.S. and C.G. Daughton. 2007. Types and quantities of leftover drugs entering the environment via disposal to sewage—revealed by coroner records. Sci. Total Environ. 388: 137–148.

Ruíz-Laguna, J., C. García-Alfonso, J. Peinado, S. Moreno, L.A. Ieradi, M. Cristaldi and J. López-Barea. 2001. Biochemical biomarkers of pollution in Algerian mouse (*Mus spretus*) to assess the effects of Aznalcóllar disaster on Doñana Park (Spain). Biomarkers 6: 146–160.

Ruiz-Laguna, J., N. Abril, T. García-Barrera, J.L. Gómez-Ariza, J. López-Barea and C. Pueyo. 2006. Absolute transcript expression signatures of *Cyp* and *Gst* genes in *Mus spretus* to detect environmental contamination. Environ. Sci. Technol. 40: 3646–3652.

Ryzinska-Paier, G., T. Lendenfeld, K. Correa, P. Stadler, A.P. Blaschke, R.L. Mach, H. Stadler, A.K. Kirschner and A.H. Farnleitner. 2014. A sensitive and robust method for automated on-line monitoring of enzymatic activities in water and water resources. Water Sci. Technol. 69: 1349–1358.

Salas-Leiton, E., B. Cánovas-Conesa, R. Zerolo, J. López-Barea, J.P. Cañavate and J. Alhama. 2009. Proteomics of juvenile senegal sole (*Solea senegalensis*) affected by gas bubble disease in hyperoxygenated ponds. Mar. Biotechnol. (NY) 11: 473–487.

Santos, E.M., J.S. Ball, T.D. Williams, H. Wu, F. Ortega, R. van Aerle, I. Katsiadaki, F. Falciani, M.R. Viant, J.K. Chipman and C.R. Tyler. 2010. Identifying health impacts of exposure to copper using transcriptomics and metabolomics in a fish model. Environ. Sci. Technol. 44: 820–826.

Sellami, B., H. Louati, M. Dellali, P. Aissa, E. Mahmoudi, A.V. Coelho and D. Sheehan. 2014a. Effects of permethrin exposure on antioxidant enzymes and protein status in Mediterranean clams *Ruditapes decussatus*. Environ. Sci. Pollut. Res. Int. 21: 4461–4472.

Sellami, B., A. Khazri, H. Louati, F. Boufahja, M. Dellali, D. Sheehan, P. Aissa, M.R. Driss, E. Mahmoudi and H. Beyrem. 2014b. Effects of permethrin on biomarkers in Mediterranean clams (*Ruditapes decussatus*). Bull. Environ. Contam. Toxicol. 92: 574–578.

Sellami, B., A. Khazri, A. Mezni, H. Louati, M. Dellali, P. Aissa, E. Mahmoudi, H. Beyrem and D. Sheehan. 2015a. Effect of permethrin, anthracene and mixture exposure on shell components, enzymatic activities and proteins status in the Mediterranean clam *Venerupis decussata*. Aquat. Toxicol. 158: 22–32.

Sellami, B., A. Khazri, H. Louati, M. Dellali, M.R. Driss, P. Aissa, E. Mahmoudi, B. Hamouda, A.V. Coelho and D. Sheehan. 2015b. Effects of anthracene on filtration rates, antioxidant defense system, and redox proteomics in the Mediterranean clam *Ruditapes decussatus* (Mollusca: Bivalvia). Environ. Sci. Pollut. Res. Int. 22: 10956–68.

Sheehan, D., B. McDonagh and J.A. Bárcena. 2010. Redox proteomics. Expert Rev. Proteomics 7: 1–4.

Shepard, J.L. and B.P. Bradley. 2000. Protein expression signatures and lysosomal stability in *Mytilus edulis* exposed to graded copper concentrations. Mar. Environ. Res. 50: 457–463.

Shepard, J.L., B. Olsson, M. Tedengren and B.P. Bradley. 2000. Protein expression signatures identified in *Mytilus edulis* exposed to PCBs, copper and salinity stress. Mar. Environ. Res. 50: 337–340.

Shukor, M.Y., N. Masdor, N.A. Baharom, J.A. Jamal, M.P. Abdullah, N.A. Shamaan and M.A. Syed. 2008. An inhibitive determination method for heavy metals using bromelain, a cysteine protease. Appl. Biochem. Biotechnol. 144: 283–291.

Shukor, M.Y., N.A. Baharom, N.A. Masdor, M.P. Abdullah, N.A. Shamaan, J.A. Jamal and M.A. Syed. 2009. The development of an inhibitive determination method for zinc using a serine protease. J. Environ. Biol. 30: 17–22.

Siggins, A., E. Gunnigle and F. Abram. 2012. Exploring mixed microbial community functioning: recent advances in metaproteomics. FEMS Microbiol. Ecol. 80: 265–280.

Simpson, R.J. 2003. Proteins and Proteomics. A Laboratory Manual. Cold Spring Harbor Laboratory Press.

Taylor, E.B. and M.A. Williams. 2010. Microbial protein in soil: influence of extraction method and C amendment on extraction and recovery. Microb. Ecol. 59: 390–399.

Taylor, N.S., R. Merrifield, T.D. Williams, J.K. Chipman, J.R. Lead and M.R. Viant. 2015. Molecular toxicity of cerium oxide nanoparticles to the freshwater alga *Chlamydomonas reinhardtii* is associated with supra-environmental exposure concentrations. Nanotoxicology 4: 1–10.

Tedesco, S., H. Doyle, G. Redmond and D. Sheehan. 2008. Gold nanoparticles and oxidative stress in *Mytilus edulis*. Mar. Environ. Res. 66: 131–133.

Tedesco, S., S.N.T. Jaafar, A.V. Coelho and D. Sheehan. 2012. Protein thiols as novel biomarkers in ecotoxicology: A case study of oxidative stress in *Mytilus edulis* sampled near a former industrial site in Cork Harbour, Ireland. J. Integr. OMICS 2: 39–47.

Tell, G. 2006. Early molecular events during response to oxidative stress in human cells by differential proteomics. pp. 369–397. *In*: I. Dalle-Donne, A. Scaloni and D.A. Butterfield [eds.]. Redox Proteomics: From Protein Modifications to Cellular Dysfunction and Diseases. John Wiley & Sons, Inc., Hoboken, New Jersey, USA.

Tomanek, L. 2011. Environmental proteomics: changes in the proteome of marine organisms in response to environmental stress, pollutants, infection, symbiosis, and development. Ann. Rev. Mar. Sci. 3: 373–399.

Tomanek, L., M.J. Zuzow, A.V. Ivanina, E. Beniash and I.M. Sokolova. 2011. Proteomic response to elevated PCO_2 level in eastern oysters, *Crassostrea virginica*: evidence for oxidative stress. J. Exp. Biol. 214: 1836–1844.

Tomanek, L. 2012. Environmental proteomics of the mussel *Mytilus*: implications for tolerance to stress and change in limits of biogeographic ranges in response to climate change. Integr. Comp. Biol. 52: 648–664.

Tomanek, L. 2014. Proteomics to study adaptations in marine organisms to environmental stress. J. Proteomics 105: 92–106.

Ünlü, M., M.E. Morgan and J.S. Minden. 1997. Difference gel electrophoresis: a single gel method for detecting changes in protein extracts. Electrophoresis 18: 2071–2077.

Van Aggelen, G., G.T. Ankley, W.S. Baldwin, D.W. Bearden, W.H. Benson, J.K. Chipman, T.W. Collette, J.A. Craft, N.D. Denslow, M.R. Embry, F. Falciani, S.G. George, C.C. Helbing, P.F. Hoekstra, T. Iguchi, Y. Kagami, I. Katsiadaki, P. Kille, L. Liu, P.G. Lord, T. McIntyre, A. O'Neill, H. Osachoff, E.J. Perkins, E.M. Santos, R.C. Skirrow, J.R. Snape, C.R. Tyler, D. Versteeg, M.R. Viant, D.C. Volz, T.D. Williams and L. Yu. 2010. Integrating omic technologies into aquatic ecological risk assessment and environmental monitoring: hurdles, achievements, and future outlook. Environ. Health Perspect. 118: 1–5.

Vioque-Fernández, A., E. Alves de Almeida and J. López-Barea. 2009a. Assessment of Doñana National Park contamination in *Procambarus clarkii*: integration of conventional biomarkers and proteomic approaches. Sci. Total Environ. 407: 1784–1797.

Vioque-Fernández, A., E.A. de Almeida and J. López-Barea. 2009b. Biochemical and proteomic effects in *Procambarus clarkii* after chlorpyrifos or carbaryl exposure under sublethal conditions. Biomarkers 14: 299–310.

Wang, D.Z., Z.X. Xie and S.F. Zhang. 2014. Marine metaproteomics: current status and future directions. J. Proteomics 97: 27–35.

Washburn, M.P., D. Wolters and J.R. 3rd Yates. 2001. Large-scale analysis of the yeast proteome by multidimensional protein identification technology. Nat. Biotechnol. 19: 242–247.

Williams, T.D., A.M. Diab, S.G. George, R.E. Godfrey, V. Sabine, A. Conesa, S.D. Minchin, P.C. Watts and J.K. Chipman. 2006. Development of the GENIPOL European flounder (*Platichthys flesus*) microarray and determination of temporal transcriptional responses to cadmium at low dose. Environ. Sci. Technol. 40: 6479–6488.

Williams, T.D., A.M. Diab, M. Gubbins, C. Collins, I. Matejusova, R. Kerr, J.K. Chipman, R. Kuiper, A.D. Vethaak and S.G. George. 2013. Transcriptomic responses of European flounder (*Platichthys flesus*) liver to a brominated flame retardant mixture. Aquat. Toxicol. 142-143: 45–52.

Williams, T.D., L. Mirbahai and J.K. Chipman. 2014a. The toxicological application of transcriptomics and epigenomics in zebrafish and other teleosts. Brief Funct. Genomics 13: 157–171.

Williams, T.D., I.M. Davies, H. Wu, A.M. Diab, L. Webster, M.R. Viant, J.K. Chipman, M.J. Leaver, S.G. George, C.F. Moffat and C.D. Robinson. 2014b. Molecular responses of European flounder (*Platichthys flesus*) chronically exposed to contaminated estuarine sediments. Chemosphere 108: 152–158.

Winterbourn, C.C. and M.B. Hampton. 2008. Thiol chemistry and specificity in redox signaling. Free Radic. Biol. Med. 45: 549–561.

Wu, H., C. Ji, L. Wei and J. Zhao. 2013. Evaluation of protein extraction protocols for 2DE in marine ecotoxicoproteomics. Proteomics 13: 3205–3210.

Yan, L.J. and M.J. Forster. 2011. Chemical probes for analysis of carbonylated proteins: a review. J. Chromatogr. B 879(17-18): 1308–1315.

Ying, J., N. Clavreul, M. Sethuraman, T. Adachi and R.A. Cohen. 2007. Thiol oxidation in signaling and response to stress: detection and quantification of physiological and pathophysiological thiol modifications. Free Radic. Biol. Med. 43: 1099–1108.

9

Protein Expression Profiles in Marine Organisms Exposed to Nanoparticles

Maria João Bebianno,[*,1] *Tânia Gomes*[1,2] and *Sheehan David*[3]

ABSTRACT

Nanoparticles either occurring naturally or resulting from anthropogenic activities or designed for specific purposes have attracted increasing attention due to their capacity to alter physical and chemical properties of conventional materials and their wide range of technological applications. The development and increase in the production and use of nanomaterials make it likely that human and environmental exposure to these materials will inevitably occur. However, there is a need to identify clear end points to access the stressor effects of these nanomaterials to marine organisms. Therefore, this chapter discusses the main types of nanomaterials produced nowadays, their applications, and the main inputs to the environment and to the marine environment in particular. In addition, levels of NPs in biological systems are discussed. This chapter then focuses on the main advantages of using new technologies, proteomics in particular, to detect the mode of action related to the impact of engineered nanoparticles on marine organisms, namely bivalves and fish species.

Introduction

Naturally-occurring nanoparticles have existed on earth for millions of years (e.g., colloids, volcanic eruptions, forest fires) and organisms have found many ways to adapt to their presence. Anthropogenic activities have also contributed to the generation of nano-scale materials inadvertently formed as by-products

[1] CIMA—Centre of Marine and Environmental Research, University of Algarve, Campus de Gambelas, 8000-139 Faro, Portugal.
Email: mbebian@ualg.pt

[2] Present address: Norwegian Institute for Water Research (NIVA), Gaustadalléen 21, NO-0349 Oslo, Norway.
Email: tania.gomes@niva.no

[3] Proteomics Research Group, School of Biochemistry and Cell Biology, Environmental Research Institute, University College Cork, Cork Ireland.
Email: d.sheehan@ucc.ie

* Corresponding author

of industrial processes, such as fumes generated during welding, metal smelting, and from automobile exhausts. Anthropogenic nano-sized particles were also designed and produced for specific purposes due to their particular characteristics and these are referred to as engineered nanoparticles (ENPs) or manufactured nanoparticles (Nowack and Bucheli 2007; Oberdörster et al. 2005). In recent years, ENPs have attracted increasing attention due to their capacity to alter physical and chemical properties of conventional materials and to their technological applications (nanotechnology).

Several definitions have been attributed to nanotechnology, most of which are generally in agreement. Nevertheless, no internationally accepted definition of nanotechnology has yet been agreed on. The Royal Academy of Engineering of the United Kingdom defines nanotechnology as the design, characterization, production, and application of structures, devices, and systems by controlling shape and size at the nanometer scale (Royal Society and Royal Academy of Engineering 2004). Similar definitions have been proposed by The National Nanotechnology Initiative that defines nanotechnology as: (i) research and technology development involving structures with at least one dimension on the 0–100 nm range; (ii) as structures, devices, and systems that have novel properties and functions because of their nanometer scale dimensions, and (iii) the ability to control or manipulate particles on the atomic scale (www.nano.gov). Other definitions are more specific, such as the one most commonly used by several authors, stating that nanotechnology is defined as the understanding and control of matter at dimensions of roughly 0–100 nm, where unique physical properties make novel applications possible (e.g., Borm et al. 2006; Handy et al. 2008a,b; Moore 2006; Nowack and Bucheli 2007). Therefore, any materials that have structures or components on their structure that are 100 nm or less in at least one dimension are examples of nanomaterials (NMs) (Dowling et al. 2004; Nowack and Bucheli 2007). Figure 1 puts in perspective the definition of a nanoscale object compared with cellular structures of the human body and a flea.

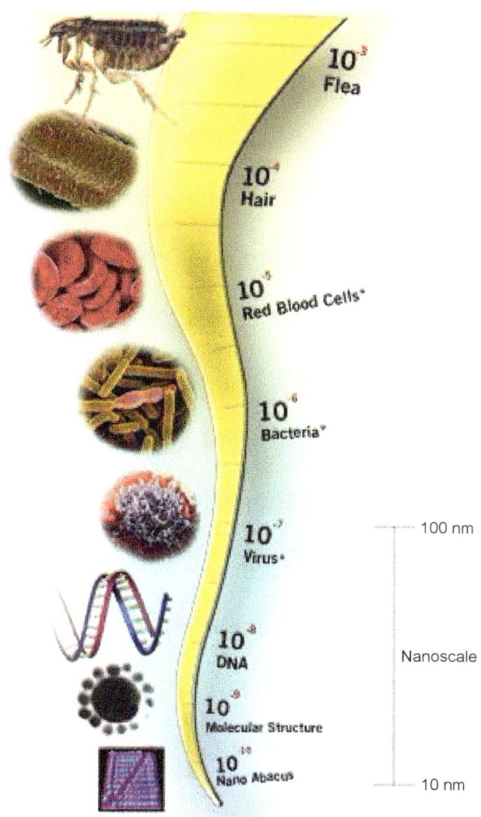

Fig. 1: Comparison of the size of biological molecules and structures on a nanometre scale (adapted from http://www.discovernano.northwestern.edu).

Among these novel NMs, nanoparticles (NPs) with 3 dimensions between 1 and 100 nm play a central role in the advance of nanotechnology (Ju-Nam and Lead 2008). NPs production massively increased in the last decade and, nowadays, these materials are used in a wide range of applications such as electronics, biomedical, pharmaceutical, cosmetics, energy, environmental, catalytic, and material applications (Tiede et al. 2009; Tedesco and Sheehan 2010). The importance and potential of NMs have catapulted nanotechnology to become one of the most rentable and expanding technologies of the 21th century, with a worldwide increase in investment, research, and development and with projections of nano-containing products to achieve sales in the order of trillions of dollars (Guzman et al. 2006). For example, levels of funding in nanotechnology research and development in 2008 reached $18.2 billion worldwide, with the United States and Japan leading investment and it is estimated that the annual revenue for all nanotechnology-related products will be more than $1 trillion by 2015 (Roco 2005). Accordingly, it is expected that the quantity of ENPs in use will increase rapidly over the next few years, with an estimated production rate of 58,000 metric tons per year between 2011–2020 (Maynard 2006; Royal Society and Royal Academy of Engineering 2004).

The development and increase in the production and use of NMs predicted for the coming years make it likely that human and environmental exposure to these materials will inevitably occur. As a result, NPs' potentially adverse effects are beginning to come into the public arena and discussion about their safety in terms of human health and the environment has become an increasingly important priority for governments, private sector, and the public all over the world (Nowack and Bucheli 2007; Roco 2005).

Types of NPs and their Applications

The increasing expansion of nanotechnology is directly related to the ability to exploit and manipulate molecules of exact specifications and properties allowing the creation of novel materials or the improvement of existing ones by making them more efficient. Materials can behave quite differently in their nano form than in bulk form because of their various physical and chemical characteristics. This different behavior makes them especially attractive for novel applications. Two important factors determine the different behavior between nanomaterials and their bulk counterparts: surface and quantum effects. Due to their especially small size, NPs have higher surface area/volume ratio, allowing a higher proportion of atoms to be present at their surface rather than in the particle's interior. This drastic increase in surface area partly determines the reactivity of NPs. At the nanoscale, quantum effects are also significant, namely electron localization, binding energy shift, surface collective charge excitation, thus affecting mechanical, optical, electric, and magnetic properties of NPs. Accordingly, these properties make NPs more appealing for new potential applications (Buzea et al. 2007; Castro et al. 2006; Dowling 2004; Klaine et al. 2012; Nel 2006). These differences in structure also contribute to other characteristics, for example reduction of melting point (Roduner 2006). However, it also means that their behavior/toxicity may be different from that of the micrometer or bulk form (Dowling 2004).

The nanotechnology sector is extremely diverse given its potential to synthesize, manipulate, and create a wide range of products/materials that have a range of technological uses. Given the necessity for innovation, a wide range of ENPs with differing composition, shape, and size are currently being created and commercialized. They are produced using bottom-up (e.g., organic synthesis, self-assembly, colloidal aggregation) or top-down strategies (e.g., photolithography, laser-beam processing, mechanical techniques) (Borm et al. 2006; Dowling 2004; Ju-Nam and Lead 2008). Accordingly, the task of classifying these materials into groups or categories is difficult, as ENPs should be broken down into a series of classes and not considered as a single homogenous group (Christian et al. 2008; Ju-Nam and Lead 2008). There are several ways of classifying ENPs with their chemical composition and properties being the most commonly used. Other classifications and terminologies are also employed in the literature to refer to specific groups of NPs, based on their dimensions, morphology, composition, uniformity, and agglomeration (Fig. 2) (Buzea et al. 2007; Ju-Nam and Lead 2008).

Regardless of how these materials are classified, the extensive variety of NPs, even within a single chemical (e.g., size, specific surface area, shape), will result in different chemical reactivity, bioavailability, and ecotoxicity (Handy et al. 2008b). Five main groups form the basis of chemical composition of ENPs

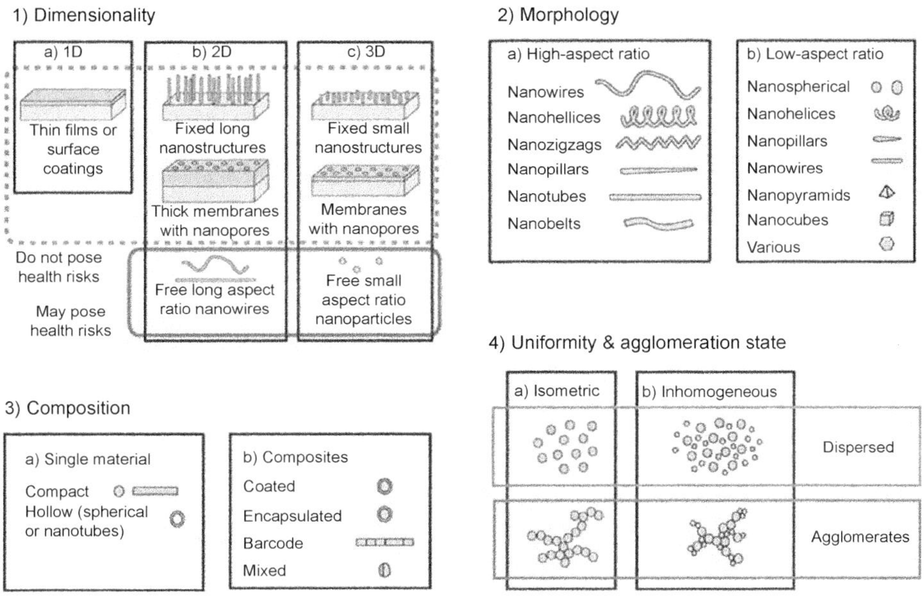

Fig. 2: Classification of nanostructured materials from the point of view of nanostructure dimension, morphology, composition, uniformity, and agglomeration state, adapted from (Buzea et al. 2007).

namely: carbon based NPs, metal-containing NPs (including metal oxides), quantum dots, zero-valent metals, and dendrimers (Bhatt and Tripathi 2011; Ju-Nam and Lead 2008; Klaine et al. 2008). Nowadays, toxicological research has mainly focused on the effects of three of the five classes of NPs based on their composition: carbon-based NPs (e.g., carbon nanotubes (CNTs) and fullerenes) (e.g., Oberdörster 2004; Smith et al. 2007) and metal or metal-oxide NPs (e.g., Ag NPs, CuO NPs, TiO_2NPs) (e.g., Buffet et al. 2011; Heinlaan et al. 2008; Ringwood et al. 2010).

Carbon based NPs

Carbon-based materials include CNTs, fullerene compounds, nanowires, etc. Fullerenes are perhaps the best-known of the carbon-based NPs that possess a spherical molecular structure formed from 60 atoms of carbon (C60). This buckminster fullerene C60 (a.k.a. buckyball) was first discovered in 1985 by Heath et al. (1985) and marked the origin of this type of NP. Subsequently, a cylindrical fullerene derivative, the CNT was produced, formed from sheets of carbon atoms covalently bound to form a one-dimensional hollow cylindrical shape. There are two main forms of CNTs, single (SWCNT) and multi-walled carbon nanotubes (MWCNT) that differ in structure by the presence of one or more layers of graphene sheets (Klaine et al. 2008; Bhatt and Tripathi 2011; Ju-Nam and Lead 2008; Nowack and Bucheli 2007).

Fullerenes and CNTs possess good thermal, electrical, photochemical, optical, mechanical, and elastic properties and therefore have potential use in batteries and fuel cell production, plastics, orthopedic implants, catalysts, water purification systems, and in components used in the electronics, aircraft, aerospace, and automotive industries (Klaine et al. 2008; Bhatt and Tripathi 2010; Ju-Nam and Lead 2008). Furthermore, the ability to chemically modify fullerenes by addition of functional groups to overcome their limited solubility in water has conferred on these carbon-based NPs many prospective uses in biological and medical applications as well as in drug delivery and gene therapy (Ju-Nam and Lead 2008; Klumpp et al. 2006; Bosi et al. 2003).

Metal-containing NPs

Metal-containing NPs comprise the largest number of NPs, which includes oxides such as zinc oxide (ZnO), titanium dioxide (TiO_2), cerium dioxide (CeO_2), copper oxide (CuO), chromium dioxide (CrO_2), molybdenum trioxide (MoO_3), bismuth trioxide (Bi_2O_3) and binary oxides such as barium titanate ($BaTiO_3$), lithium cobalt dioxide ($LiCoO_2$), or indium tin oxide (InSnO) (Bhatt and Tripatti 2011; Klaine et al. 2008). The synthesis of these NPs is very common and is achieved by hydrolysis of the transition metal ions (e.g., TiO_2 and ZnO) (Masala and Seshadri 2004). Metal oxide NPs have received considerable attention and have been massively produced over the last few years due to their extensive use in food, material science, chemical, and biological areas (Aitken et al. 2006). Amongst the metal oxides, CuO NPs are amongst the most commonly used NPs due to elevated thermal and electrical conductivity. Cu NPs are intensively used as heat transfer fluid in machine tools (Chang et al. 2005), as well as in polymers and plastics, gas sensors (Li et al. 2007), wood preservation, conductive inks for printing electronic components (Lee et al. 2008), and coatings on integrated circuits and batteries (Dhas et al. 1998). Additionally, these Cu NPs are applied in several materials as skin products and textiles mainly due to their antimicrobial properties (Cioffi et al. 2005; Ren et al. 2008).

Zero-valent metals NPs

The third class of NPs is zero-valent metals such as iron (Fe), gold (Au), and silver (Ag) NPs. These NPs are typically made by reduction or co-reduction of metal salts (Li et al. 2006; Masala and Seshadri 2004), whose physical properties are controlled by varying the reductant type and reduction conditions. These metal NPs have unique optical properties being largely implemented in electronics (Wang et al. 2008a,b). Nevertheless other potential applications have emerged such as water, sediment, and soil remediation (zero-valent iron) (Zhang 2003), industrial catalysis (Au) (Kim et al. 2009), and pharmaceuticals and drug delivery (Au and Ag) (Choi et al. 2007). From the variety of NPs that are currently being developed in nanotechnology, silver NPs (Ag NPs) have the highest degree of commercialization mainly due to their antibacterial properties. These particles have unique physic-chemical properties, including a high electrical and thermal conductivity, catalytic activity, and chemical stability (e.g., Fabrega et al. 2011; Farkas et al. 2011; Luoma 2008). Nevertheless, it is their antibacterial activity that makes them of special interest in textiles, eating utensils, food storage, cosmetics and personal hygiene, and household appliances (e.g., washing machines, vacuum cleaners) (www.nanoproject.org). Recently, Ag NPs have been used in medical equipment, such as catheters, infusion systems, and medical textiles (Markarian 2006; Simpson 2003).

Quantum dots

This class of NMs includes semi-conductor nanocrystals, also known as quantum dots, made of metals or semiconductors such as cadmium selenide (CdSe), cadmium telluride (CdTe), indium phosphide (InP), or zinc selenide (ZnSe) (Klaine et al. 2008; Bhatt and Tripatti 2010). Potential applications of quantum dots include medical applications such as medical imaging (Scown et al. 2010; Rocha et al. 2014), detection of biomolecules (Scown et al. 2010) and targeted therapeutics (Scown et al. 2010; Bhatt and Tripatti 2010) and most recently solar cells and photovoltaics (Scown et al. 2010), light emitting diodes in displays (Scown et al. 2010), and security inks and telecommunications (Bhatt and Tripatti 2010).

Dendrimers

Dendrimers are spherical polymeric molecules that consist of an inner core surrounded by a series of branches, formed by a nanoscale hierarchical process of self-assembly (Klaine et al. 2008; Bhatt and Tripatti 2010). The properties of dendrimers are dominated by the core molecules of functional groups and branches, making it possible to chemically alter their molecular surface (e.g., more hydrophilic, hydrophobic, or charged). This capacity to change size, topology, flexibility, and molecular weight during dendrimer synthesis offers a range of interesting properties within the pharmaceutical field, namely as nanoscale carrier molecules in targeted drug delivery, controlled drug release, and DNA chips (Klaine et al. 2008). Dendrimers are also used in conventional applications such as coatings and inks, as well as in environmental remediation (through metal ions trap) (Klaine et al. 2008; Bhatt and Tripatti 2010).

Inputs of NPs to the Environment

It is estimated that 50,000 kg/year of nano-sized materials are being produced through non-industrial and unintended processes, i.e., atmospheric emissions (e.g., waste combustion, diesel-exhaust, and other combustion processes), leaching from NP-containing products in landfills or soil-applied sewage sludge (Dowling 2004; Klaine et al. 2008). Some will also enter the environment during the product's life cycle (e.g., by erosion of the materials, accidental spills), from production facilities (i.e., initial and downstream manufacturers) and wastewater treatment plants (Nowack and Bucheli 2007; Oberdörster et al. 2005). Other types of sources resulting from the use of products that contain ENPs (e.g., paints, fabrics, and personal health care products) may be released into the environment in a manner proportional to their use from both direct (e.g., bathing) and indirect (e.g., sewer) sources (Biswas and Wu 2005; Zhang 2003). Diffuse release associated with wear and erosion from general use is also possible (e.g., wear of car tires, urban air pollution), as well as intentional use in remediation of contaminated environmental media (soil and water) (e.g., use of iron NPs to remediate groundwater) (Bhatt and Thripathi 2011; Zhang and Elliot 2006). Independent of the source (water, soil, or atmosphere), emitted particles will ultimately end up on land or in waterways (e.g., drainage ditches, rivers, lakes, estuaries, and coastal waters) where they have the potential to contaminate soil, migrate into surface and ground waters, and interact with biota (Klaine et al. 2008; Moore 2006; Nowack and Bucheli 2007). All these routes will allow ENPs to disperse through the environment (Fig. 3); but the fate of each of these NPs will vary (Klaine et al. 2012).

The amount of NMs that may be found in the environment is difficult to predict, so to ensure sustainable development of nanotechnology, proper tools and methodologies for the detection and characterization of NPs in atmospheric, aquatic, and terrestrial environments have to be applied to facilitate accurate ecological risk assessment of these materials (Hasselöv et al. 2008; Klaine et al. 2008; Maynard 2006). Nowadays the development of analytical methods to detect and quantify ENPs is still in its infancy and, consequently, little is known about their concentrations in different environmental

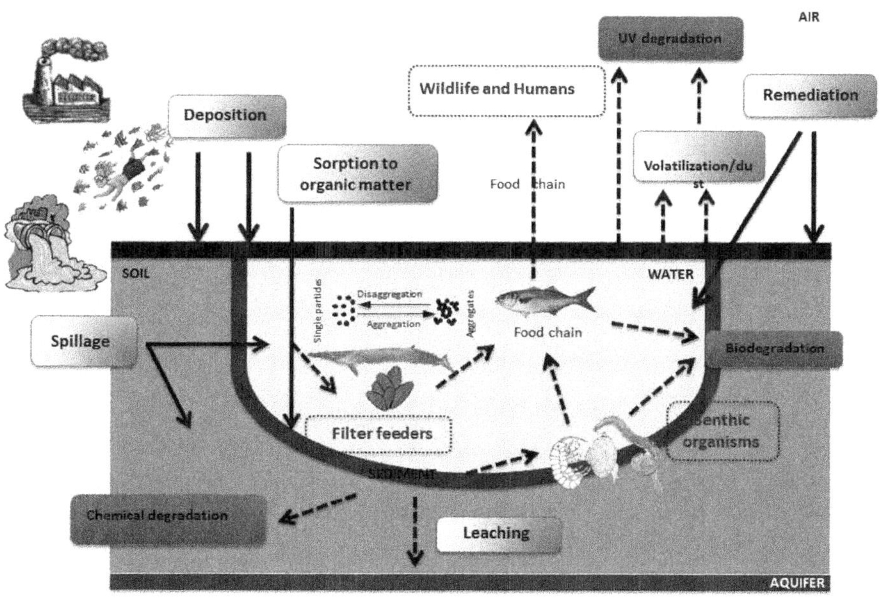

Fig. 3: Routes of exposure, uptake, distribution, and degradation of NPs in the environment.

compartments (air, soils, and water) or on their transport, fate, and behavior (Nowack and Bucheli 2007; Scown et al. 2010a). Several challenges have emerged not only deriving from their novel properties (e.g., size, shape, surface charge, composition, degree of dispersion) but also from possible alterations (e.g., ageing or weathering) and interactions during their environmental life cycle (Klaine et al. 2012). Classical exposure assessments are dependent on the soluble fraction of a given contaminant and, when dealing with NPs, the chemical speciation concept used needs to include their physical forms (dissolved, colloidal, and particulate). Knowledge already available on colloidal behavior is also very useful for the assessment of ENMs. Nevertheless, many of the well-established techniques for pure systems may not necessarily be appropriate to accurately detect NPs in complex environmental systems (Klaine et al. 2008, 2012).

Currently there are no standardized measures or methods to determine actual concentrations of ENPs in the environment, although some modelling approaches were undertaken to estimate the likely emission and load of ENPs in various environmental compartments including the aquatic environment (e.g., Boxall et al. 2008; Mueller and Nowack 2008). These modelling approaches highlighted uncertainties in the prediction of environmental concentrations and levels of ENPs based only on emission scenarios (from production volumes and life cycle assessments) and partitioning parameters (fate and behavior) where most of them require validation through measurements of actual environmental concentrations (Klaine et al. 2008).

Presence of NPs in the Aquatic Environment

As nanotechnology industries start to emerge with larger-scale production, it is inevitable that their products and by-products will end up in the environment in quantities likely to dramatically increase in the near future (Bhatt and Tripathi 2011; Moore 2006; Scown et al. 2010a). Although nano-sized particles have always occurred in nature, the latest developments in their use and production have raised concern over their potential release and side effects, not only on human health, but also on the natural environment. In order to determine the environmental fate and behavior of NPs it is necessary to understand their potential risks: (i) do NPs retain some of their nominal properties (e.g., size, structure, and reactivity) when they reach aquatic systems?; (ii) will NPs interact and associate with other colloidal and particulate constituents?; (iii) what will be the effects of dissolution and physical (e.g., flow) conditions of NPs behavior?; (iv) in what way will NPs affect aquatic and sedimentary biota and will this differ from their bulk counterparts; and (v) will biota (e.g., biofilms and invertebrates) interact and modify the behavior of these particles? Accordingly, it is necessary to further improve the knowledge of NPs' chemical behavior, transport within and between environmental and biological compartments, ultimate environmental fate, mechanisms of biological uptake, and toxic implications for living systems (Dowling et al. 2004; Ju-Nam and Lead 2008; Klaine et al. 2008).

Levels of NPs in Biological Systems

Although clear benefits are expected from the expansion of nanotechnology products, concern is growing about the potential toxicity and ecotoxicology associated with their unusual properties. The altered quantum effects and enhanced chemical reactivity that makes NPs so attractive and useful in numerous applications are also fundamental for their complex interactions and effects in the environment and their unpredictable/unknown effects on living organisms and on marine organisms in particular. Additionally, small particle size and high specific surface area will also benefit passage across biological barriers of some organisms and facilitate entry into cells through membranes and junctions between cells (Handy et al. 2008a; Scown et al. 2010a). Collectively, these properties give NPs the potential to induce adverse cellular effects and damage to living organisms. Most research on ecotoxicity of NPs has focused on mammals, and various effects have been reported and debated (e.g., Colvin 2003; Handy and Shaw 2007; Handy et al. 2008a), while ecotoxicity of NPs in aquatic organisms is still an emerging research area. Interaction of NPs with aquatic biota is a function of the physic-chemical properties of the particles (e.g., size, aggregation state, chemical composition), as well as of the biology of the target organisms. At least three primary biological targets can interact with NPs in the aquatic environment: (i) filter feeders, which can be exposed to high

NP concentrations present in surface waters; (ii) pelagic species (from phytoplankton to fish and mammals) exposed during vertical migration of the particles; and (iii) benthic species that are likely to interact to aggregated or adsorbed NPs or NPs deposited in sediment biofilms. A wide variety of parameters influence the fate and effects of NPs in cells and living organisms, and a given NP can have differing toxic effects depending on the particle properties (e.g., presence and type of coating), method of preparation (e.g., use of dispersants), or species sensitivity (Barber et al. 2009; Bhatt and Tripathi 2011; Handy et al. 2008a; Matranga and Corsi 2012). Most of the ecotoxicity of NPs directed towards aquatic organisms is centered on acute toxicity in species used for regulatory toxicology (i.e., short-term acute toxicity tests, chronic, and life-cycle effects), namely in freshwater species such as *Daphnia magna* and model fish species such as *Danio rerio* (zebrafish), *Pimephales promelas* (fathead minnows), and *Oncorhynchus mykiss* (rainbow trout) (Baun et al. 2008; Cattaneo et al. 2009; Handy et al. 2008a). Despite the information provided by acute experiments, they do not provide complete information about interactions of NPs with other species that cover a wide range of trophic levels and there is a need to direct research towards invertebrates using long-term exposure to better understand NPs' toxicity mechanisms (Baun et al. 2008; Griffitt et al. 2008; Heinlaan et al. 2008, 2011). NP exposure induces a wide range of biological effects towards aquatic organisms. For example, carbon-based and metallic NPs are toxic to embryonic fish models, in terms of developmental abnormalities and mortality, as well as in invertebrate species (such as *D. magna*), in terms of mortality, fertilization rates, and reduced molting success (e.g., Gaiser et al. 2011; Heinlaan et al. 2008; Oberdörster 2004; Scown et al. 2010b). There are clearly still significant gaps in the knowledge about NP toxicity, and current knowledge of the fate, concentrations, behavior, and toxicity of these particles in the marine environment is mainly based on hypothesis rather than experimental information, in addition to a lack of an ecological and physiological endpoint in most laboratory studies. This is especially true for species living in estuarine and coastal areas that are major depositional reservoirs for NPs. For example, in commercially harvested aquatic species, such as edible bivalves, such information is essential for the understanding of the potential transfer of NPs to humans (Canesi et al. 2012; Matranga and Corsi 2012; Ward and Kach 2009).

New Technologies for the Detection of NPs Mode of Action

Different types of NPs induce a wide range of biological effects in aquatic organisms both *in vivo* and *in vitro* at different levels of cellular organization. They are known to cause oxidative stress, lipid peroxidation, and DNA damage, inflammatory processes and apoptosis in a wide variety of cell types (e.g., Ahamed et al. 2010; Canesi et al. 2012; Ivask et al. 2010; Yeo and Pak 2008). Oxidative stress induces a wide range of reversible and irreversible modifications to proteins and their side chains (Davies 2005; McDonagh and Sheehan 2006) and therefore they are one of the key targets to assess the adverse effects of NPs due to their abundance in biological systems (Klaine et al. 2008). Conversely, conventional biomarkers have been extensively used in nanotoxicology impact assessment; although many of the toxic responses (e.g., oxidative stress, LPO, enzymatic activation/inhibition, genotoxicity) are common to several contaminants, including NPs ionic/bulk form (Handy et al. 2012; Gomes et al. 2012, 2014). They also present some disadvantages, as they are influenced by confounding factors (e.g., abiotic factors), highly-dependent on the route of exposure, bioaccumulation tendency, and detoxification mechanisms of stressors. This require a deep knowledge of the toxic mechanisms of stressors, and prevent a more comprehensive view of toxicity by focusing on only a few proteins (Vioque-Fernández et al. 2009). Accordingly, there is a need to develop nano-specific biological end points to differentiate nano-specific responses and mechanisms of action from their similar ionic/bulk counterparts. The OMICs approach and, in particular, proteomics-based methods provide a more insightful view on the global changes of protein expression indicative of NP exposure or effect by looking for specific molecular signatures (Amelina et al. 2007; López-Barea and Gómez-Ariza 2006). This approach enables us to identify new biomarkers in response to conventional stressors particularly in marine organisms (e.g., Apraiz et al. 2006; López-Barea and Gómez-Ariza 2006; Shepard and Bradley 2000). With the progress of sequencing technologies and bioinformatics, omics are becoming ubiquitous in all science fields. So the application of proteomics to nanotoxicology may help to identify protein pathways affected by these ENPs, providing a deeper understanding of the molecular mechanisms of NP-induced stress syndrome in organisms, and help to clarify and differentiate the toxic mode of action between macro- and nano-particles of similar substances.

OMICs

Nanomaterials can readily enter organisms and cross important barriers such as skin, intestine, lung mucosa, and even the blood-brain barrier (Grassian et al. 2007). They are also capable of penetrating cells by a wide variety of routes such as phagocytosis and receptor-mediated uptake (Hemmerich and von Mikecz 2013). As with all aspects of their toxicology, the propensity of any individual nanomaterial taking a particular route of uptake seems to be a consequence of a complicated set of variables such as geometry, size, chemical composition, capping groups, dispersants, and composition of a protein corona coating the particle. Omics approaches exploit high-throughput methodologies capable of profiling complex biochemical consequences of the interactions between nanomaterials and target cells or organisms. These consist of studies of the transcriptome (transcriptomics), proteome (proteomics) metabolome (metabolomics), or metal profiles (metallomics; especially significant for metal and metal oxide nanomaterials). Of these, proteomics approaches based on two-dimensional electrophoresis combined with mass spectrometry technologies have been most extensively applied especially in studies with aquatic sentinel species which reveal protein expression changes in response to nanomaterials (Gomes et al. 2014; Trevisan et al. 2014; Hu et al. 2014). As oxidative stress is often triggered in nanotoxicology, redox proteomics which profiles effects both on protein abundance and also on protein oxidative status, such as carbonylation and protein thiol oxidation, is also an important tool and are also especially informative (Hu et al. 2014; Rainville et al. 2014). While gel-based techniques display some limitations related to proteome coverage, gel-free techniques such as LC-MS are considered fast and low-cost high-throughput approaches. Furthermore, the development of shotgun proteomics will likely improve the use of the proteomic approach. An elegant example of transcriptomics is a study of effects on gene expression changes in the kidney as a result of intratracheal exposure to cadmium-based nanomaterials (Coccini et al. 2012). Such approaches have also been adopted in other aquatic models such as zebrafish (Griffit et al. 2011, 2013; Oliviera et al. 2014), daphnids (Scanlan et al. 2013), and epibenthic crustaceans (Poynton et al. 2013). Metabolomics approaches are much less developed than proteomics or transcriptomics in sentinel species although a proton NMR study of earthworms exposed to titanium oxide nanoparticles has been performed (Aslund et al. 2012). As so many environmentally released nanomaterials are metal or metal oxide in nature, there is particular potential for studying the metallome giving insights to metal speciation and distribution within test organisms (reviewed by Li et al. 2014).

Protein Expression Profiles in Marine Organisms

Proteomics describes the study of the proteome, the total complement of proteins expressed by a genome within a cell, tissue or organism, under specific conditions. The proteome is very dynamic, where the protein content of cells varies in response to alterations in the environment, physiological state of the cell, drug administration, health, and disease. Proteomics analysis is characterized by high-throughput methodologies that enable high-resolution separations and display of the proteins in the tissue in a form that allows subsequent analysis and comparison. From the hundreds to thousands of proteins that can be obtained in a single experiment, those that are either expressed under a given condition or suppressed can subsequently be identified (Alban et al. 2003; Knigge et al. 2004; Liebler 2002; Rabilloud 2000; Snape et al. 2004).

Environmental proteomics (or ecotoxicoproteomics) aims to analyze the proteome of organisms and to identify variations in proteins induced by contaminants without the need for detailed knowledge of toxicity mechanisms. A comparison of proteomes from stress conditions (versus controls) has the potential to identify not only single protein markers, but also to generate protein patterns that react to a specific type and degree of stress and, consequently, differentiate exposure and/or effect to contaminants. Differential expression of proteins can be compared among chemicals, concentrations, or complex mixtures in different natural environments, and the obtained up- or down-regulated proteins combined within protein patterns are specific to the stressor and the level of environmental stress. These alterations are identified as protein expression signatures (PESs), sets of proteins that potentially offer greater understanding of underlying toxic mechanisms of stress response (Knigge et al. 2004; Nesatyy and Sutter 2007; Shepard and Bradley 2000). In this context, proteomics has been extensively used in ecotoxicological research in the past few years, where field and laboratory studies showed that different PESs respond significantly to different kinds of contaminants, thus identifying candidate proteins for further study (e.g., Apraiz et al. 2006; López-Barea and Gómez-Ariza 2006; Vioque-Fernández et al. 2009; Trevisan et al. 2014). Since PES itself constitutes a biomarker, the need to explicitly identify altered proteins is not indispensable to diagnose adverse environmental effects (Shepard and Bradley 2000). This is especially useful when using bioindicator species whose genomes or proteomes have not yet been fully sequenced, as is the case of most bivalve species. A change in PES pattern alone may be sufficient to demonstrate contamination effects (López-Barea and Gómez-Ariza 2006; Monsinjon et al. 2006; Dowling and Sheehan 2006). These PESs can be quantified, identified, and used as novel and unbiased biomarkers of NP exposure and effect. Nevertheless, as sequencing information becomes ever-more available from approaches such as next-generation sequencing (Sheehan 2013), identification of proteins from non-model marine organisms (as bivalves) will provide greater understanding of the modes of action of toxic compounds and of how they affect organisms and ecosystem quality (Monsinjon and Knigge 2007; Nesatyy and Suter 2008). However, proteomics approaches to marine organisms are still very scarce.

Omics Approach of NPs in Bivalves

Bivalves, due to their fundamental role in marine ecosystems, their relevant economic importance, and application as sentinel organisms in marine pollution studies, are a relevant group of bioindicators to investigate the effects of NPs (Moore 2006). They filter contaminants in the dissolved and particulate form directly from water through their gills or indirectly through their digestive system. So, whether ENPs reach the aquatic environment in suspension or in an aggregated form, they will be taken up by these organisms and accumulate in their tissues. However, the general application of proteomics to assess effects induced by ENPs has been hampered by the lack of genome sequences available for these species. The recent publication of the draft and complete genome sequences for two oyster species, the pearl oyster *Pinctada fucata* (Takeuchi et al. 2012) and the Pacific oyster *Crassostrea gigas* (Zhang et al. 2012) will greatly improve proteomic applications. Table 1 shows the current transcriptomic databases already available for bivalve species. Meanwhile, to overcome the caveat of the lack of genome databases, alternative approaches, namely *de novo* 454-pyrosequencing of transcriptomes using next generation sequencing platforms, have been implemented (Suárez-Ulloa et al. 2013). Therefore, Suárez-Ulloa et al. (2013) published a recent review

Table 1: Marine bivalve transcriptomic databases.

Database	Species	N° Sequence	Tissues	References
Mytbase	*M. galloprovincialis*	7112	Digestive gland, gill, haemocytes	http://mussel.cribi.unipd.it
GigasDatabase	*C. gigas*	29745	Digestive gland, gills, gonad, hemocytes, mantle-edge, muscle	http://public-contigbrowser.sigenae.org:9090/ Crassostrea_gigas/index.html
RuphiBase	*Ruditapes phillipinarum*	32606	Several tissues	http://compgen.bio.unipd.it/ruphibase
ChameleaBase	*Chamelea gallina*	36422	Muscle	http://compgen.bio.unipd.it/chameleabase
DeepSeaVent	*Bathymodiolus azoricus*	35903	Gills	http://transcriptomics.biocant.pt:8080/deepSeaVent

on currently-available web-accessible molecular data to contribute to increased proteomic applications to monitor the effects of harmful compounds such as NPs in bivalve species.

In most nanotoxicology studies with bivalve species, the use of conventional biomarkers have been extensive, nevertheless, many of these toxic responses (e.g., oxidative stress, LPO, enzymatic activation/inhibition, genotoxicity) are common to conventional contaminants, including NP ionic and/or bulk forms (Handy et al. 2012; Gome et al. 2012, 2013; Buffet et al. 2011). Consequently, there is a pressing need to develop novel nano-specific tools to distinguish nano-specific responses and mechanisms of action from their ionic/bulk counterparts, as well as other stressors. With this in mind, proteomics-based approaches have been used in nanotoxicology in recent years to help identify protein pathways affected by NPs and possibly to yield greater insights to their toxic mode of action.

The first study using a proteomic approach was the application of redox proteomics to *M. edulis* tissues exposed to Au NPs-citrate (13 nm; 750 $\mu g.L^{-1}$; 24 h). Protein separation by one dimensional electrophoresis (1DE) showed higher protein carbonylation in the gills in comparison with the digestive gland, where higher protein ubiquitination occurred. Further studies using 1DE and 2D-SDS PAGE in the digestive gland of *M. edulis* exposed to the same Au NPs (~15 nm), showed a reduction in protein thiol oxidation as a response to ROS formation (Tedesco et al. 2010a). Changes in protein patterns in the sub-proteome of thiol-containing proteins were also reported by the same authors when using a smaller Au NPs size (5.3 ± 1 nm), consistent with higher protein thiol oxidation in response to smaller size. However, in this case, no proteins were identified (Tedesco et al. 2010a,b). Changes in carbonyl and protein thiols were also reported for *M. edulis* in response to CuO NPs (50 nm, 400, 700, 1000 $mg.L^{-1}$, 1 h) using redox-proteomics, where an increase in protein carbonyl and a decrease in reduced protein thiols were detected in gill extracts (Hu et al. 2014). Peptide mass fingerprinting (PMF) in combination with Mass Spectrometry (MS) allowed identification of six proteins: alpha- and beta-tubulin, actin, tropomyosin, triosephosphate isomerase, and Cu–Zn superoxide dismutase, indicative of significant protein oxidation of cytoskeleton and key enzymes in response to CuO NPs (Table 2) (Hu et al. 2014). These data highlighted the fact that redox proteomics approaches reveal more specific effects than traditional biomarkers of oxidative stress in sentinel species. Proteomes obtained by 2DE were also investigated in *M. galloprovincialis* gill and digestive gland in response to CuO NPs (31 ± 10 nm, 10 $\mu g.L^{-1}$, 15 days). Alterations in the proteomes of exposed mussels show the specificity of PES to CuO NPs. Protein identification further indicated that the biochemical pathways of cytoskeleton and cell structure oxidative stress as observed in gills of *M. edulis* (Hu et al. 2014) and stress response, transcription regulation, energy metabolism, apoptosis and proteolysis are altered during CuO NPs exposure. Thus, these processes may play putative roles in cellular toxicity and consequent cell death in mussels. Apart from the traditional molecular targets of CuO NP exposure in mussel tissues (e.g., HSPs, GST, ATP synthase), potential new targets were also identified (caspase 3/7-1, cathepsin L, Zn-finger protein, and precollagen-D) and these were considered to be putative new biomarkers for CuO NP exposure (Gomes et al. 2014).

A similar approach was also used to compare the effects of Ag NPs toxicity (42 ± 10 nm, 10 $\mu g.L^{-1}$, 15 days) in the gills and digestive gland of *M. galloprovincialis*. Tissue-specific PESs reflect differences in uptake, tissue-specific functions, redox requirements, and modes of action from their ionic counterpart. Protein analysis by PMF in combination with MS led to the identification of 15 proteins: catchin, myosin heavy chain, HSP70, GST, nuclear receptor subfamily 1G, precollagen-P, ATP synthase F0 subunit 6, NADH dehydrogenase subunit 2, putative C1q domain containing protein, actin, α-tubulin, major vault protein, paramyosin and ras, partial (Table 2). The exclusive identification of the major vault protein, paramyosin and ras, partial to Ag NP exposure suggests their use as new putative candidate molecular biomarker to assess Ag NPs toxicity (Gomes et al. 2013). Although the use of proteomics has already been able to suggest novel biomarkers of NPs exposure in marine organisms (Table 2), due to the exploratory nature of the proteomic approaches applied in these studies, further work is necessary to confirm and validate the usefulness of the identified proteins as putative biomarkers of NPs exposure and effect in a more realistic environmental exposure and risk perspective/scenario (Gomes et al. 2014).

Table 2: Proteins identified by proteomics as a result of NPs exposure in marine bivalves and fish.

NPs	Size (nm)	Concentration	Species	Tissue	Time of exposure	Protein identification Method	Identified Proteins	Function	Reference
Ag	42 ± 10	10 μg/l	*Mytilus galloprovincialis*	Gills	15 d	2DE MS/MS	Precollagen-P	Adhesion and mobility	Gomes et al. 2013
							Paramyosin	Cytoskeleton and cell structure	
							Catchin protein	Cytoskeleton and cell structure	
							Catchin protein	Cytoskeleton and cell structure	
							Glutahione s-transferase	Oxidative stress	
							ATP synthase F0 subunit 6	Stress response	
							HSP70	Stress response	
							Nuclear receptor subfamily 1G	Transcription regulation	
				Digestive gland			Ras, partial	Stress response	
							NADH dehydrogenase subunit 2	Stress response	
CuO	31 ± 10	10 μg/l	*Mytilus galloprovincialis*	Gills	15 d	2DE MS/MS	Actin	Cytoskeleton and cell structure	Gomes et al. 2014
							ATP synthase F0 subunit 6	Energy metabolism	
							Glutahione S-transferase	Oxidative stress	
							Heat shock cognate 71	Stress response	
							Putative C1q domain containing protein	Stress response	
							Nuclear receptor subfamily 1G	Transcription regulation	
							Zinc-finger BED domain-containing protein 1	Transcription regulation	

Table 2 contd....

Table 2 contd....

NPs	Size (nm)	Concentration	Species	Tissue	Time of exposure	Protein identification Method	Identified Proteins	Function	Reference
				Digestive gland			Caspase 3/7–1	Apoptosis	
							Paramyosin	Cytoskeleton and cell structure	
							Cathepsin L	Proteolysis	
							Heat shock cognate 71	Stress response	
							Actin	Cytoskeleton and cell structure	
CuO	50	400, 700, 1000 ppb	*Mytilus edulis*	Gills	1 hr	Redox proteomics	Putative beta-tubulin (Fragment)	Cytoskeleton	Hu et al. 2014
							Alpha-tubulin (Fragment)	Cytoskeleton	
							Alpha-tubulin (Fragment)	Cytoskeleton	
							Tropomyosin	Cytoskeleton	
							Tropomyosin	Cytoskeleton	
							Actin (Fragment)	Cytoskeleton	
							Actin (Fragment)	Cytoskeleton	
							Actin (Fragment)	Cytoskeleton	
							Actin (Fragment)	Cytoskeleton	
							Triosephosphate isomerase (Fragment)	Oxidative stress	
							Superoxide dismutase [Cu-Zn]	Oxidative stress	
							Superoxide dismutase [Cu-Zn]	Oxidative stress	
							Superoxide dismutase [Cu-Zn]	Oxidative stress	
C60/PhTS	39.1		*Danio rerio*	Embryos	48 h	SDS-PAGE & MALDI-TOF	Cytochrome P450 2K6	Oxidoreductase activity	
							ATPase 2A2	Energy metabolism	Kuznetsova et al. 2014
							Transferrin A	Ferric ion binding	
							Vitellogenin	Reproduction	

Proteomic Approach of NPs in Fish

The application of a proteomic approach of ENPs to fish is very scarce. However, thin SDS-PAGE sections with subsequent identification of embryonic proteins by MALDI-TOF indicated that embryonic exposure of *Danio rerio* for 48 h to phosphatidylcholine-based phospholipid nanoparticles containing C60 fullerenes (39.1 nm) has not led to direct lesions in the embryos. Instead, oxidative stress induced by the fullerenes caused changes in cytochrome P450 2K6 (57,345 Da) and in the content of vitellogenin along with the development of cardiomyopathy (changes in ATPase 2A2-113,508 Da) in the embryos and in transferrin A (45,868 Da) indicating a reproductive toxicity of these NPs to these embryos (Kuznetsova et al. 2014).

These endpoints obtained for these marine species constitute an important starting point for the creation of a protein database to compare nanotoxicological responses and decipher the modes of action of NPs in marine organisms.

Conclusions

The identification of proteins altered by NPs summarized in this chapter allowed a sneak peak of their action at the molecular level in bivalve and fish species and even provided novel and unbiased biomarkers of exposure and effect. However, there is still a long way to go to fully understand the molecular mechanisms of NP-induced stress syndrome in marine organisms and unravel other relevant protein pathways than those affected by oxidative stress. The information gathered with these first proteomic studies will constitute an important starting point for a protein database to compare nanotoxicological responses and decipher the modes of action of NPs in marine organisms, which also needs to be expanded to other relevant marine species.

The development of shotgun proteomics as well as progress in developing sequencing technologies and bioinformatics for nanotoxicology research will greatly improve our understanding of protein pathways affected by NPs.

Acknowledgements

M.J. Bebianno and T. Gomes would like to acknowledge the funding of the Portuguese Foundation for Science and Technology (FCT) through the project (NANOECOTOX; PTDC/AAC-AMB/121650/2010).

Keywords: Nanoparticles, omics, proteomics, 2DE eletrophoresis, protein expression profiles, aquatic environment, aquatic species

References

Ahamed, M., R. Posgai, T.J. Gorey, M. Nielsen, M.H. Saber and J.J. Rowe. 2010. Silver nanoparticles induce heat shock protein 70, oxidative stress and apoptosis in Drosophila melanogaster. Toxicology and Applied Pharmacology 242: 263–269.

Aitken, R.J., M.Q. Chaudhry, A.B.A. Boxall and M. Hull. 2006. Manufacture and use of nanomaterials: current status in the UK and global trends. Occupational Medicine 56: 300–306.

Alban, A., S.O. David, L. Jorkesten, C. Andersson, E. Sloge, S. Lewis and I. Currie. 2003. A novel experimental design for comparative difference gel electrophoresis standard. Proteomics 3: 36–44.

Amelina, H., I. Apraiz, W. Sun and S. Cristobal. 2007. Proteomics–based method for the assessment of marine pollution using liquid chromatography coupled with two–dimensional electrophoresis. Journal of Proteome Research 6: 2094–2104.

Apraiz, I., J. Mi and S. Cristobal. 2006. Identification of proteomic signatures of exposure to marine pollutants in mussels (*Mytilus edulis*). Molecular and Cellular Proteomics 5: 1274–1285.

Aslund, M.L.W., H. McShane, M.J. Simpson, A.J. Simpson, J.K. Whalen, W.H. Hendershot and G.I. Sunahara. 2012. Earthworm sublethal responses to titanium dioxide nanomaterial in soil detected by H-1 NMR metabolomics. Environ. Scien. Technol. 46: 1111–1118.

Barber, D.S., N.D. Denslow, R.J. Griffitt and C.J. Martyniuk. 2009. Sources, fate and effects of engineered nanomaterials in the aquatic environment. pp. 227–246. *In*: S.C. Sahu and D.A. Casciano [eds.]. Nanotoxicity. John Wiley and Sons, Ltd., Chichester, UK.

Baun, A., N.B. Hartmann, K. Grieger and K.O. Kusk. 2008. Ecotoxicity of engineered nanoparticles to aquatic invertebrates: a brief review and recommendations for future toxicity testing. Ecotoxicol. 17: 387–395.

Bhatt, I. and B.N. Tripathi. 2011. Interaction of engineered nanoparticles with various components of the environment and possible strategies for their risk assessment. Chemosphere 82: 308–317.

Biswas, P. and P. Wu. 2005. Nanoparticles and the environment. J. Air Waste Manag. Assoc. 55: 708–746.

Borm, P.J.A., D. Robbins, S. Haubold, T. Kuhlbusch, H. Fissan, K. Donaldson, R. Schins, V. Stone, W. Kreyling, J. Lademann, J. Krutmann, D. Warheit and E. Oberdorster. 2006. The potential risks of nanomaterials: a review carried out for ECETOC. Particle Fibre Toxicol. 3: 11.

Bosi, S., T.D. Ros, G. Spalluto and M. Prato. 2003. Fullerene derivatives: an attractive tool for biological applications. Eur. J. Med. Chem. 38: 913–923.

Boxall, A.B.A., Q. Chaudhry, C. Sinclair, A. Jones, R. Aitken, B. Jefferson and C. Watts. 2008. Current and future predicted environmental exposure to engineered nanoparticles. Report by Central Science Laboratory for Department for Environment, Food and Rural Affairs, Her Majesty's Government, UK.

Buffet, P.E., O.F. Tankoua, J.F. Pan, D. Berhanu, C. Herrenknecht, L. Poirier, C. Amiard–Triquet, J.C. Amiard, J.B. Bérard, C. Risso, M. Guibbolini, M. Roméo, P. Reip, E. Valsami–Jones and C. Mouneyrac. 2011. Behavioural and biochemical responses of two marine invertebrates *Scrobicularia plana* and *Hediste diversicolor* to copper oxide nanoparticles. Chemosphere 84: 166–174.

Buzea, C., I.P. Blandino and K. Robbie. 2007. Nanomaterials and nanoparticles: Sources and toxicity. Biointerph. 2: MR17–MR172.

Canesi, L., C. Ciacci, R. Fabbri, A. Marcomini, G. Pojana and G. Gallo. 2012. Bivalve molluscs as a unique target group for nanoparticle toxicity. Mar. Environ. Res. 76: 16–21.

Castro, A.T., E.L. Cuéllar, U.O. Méndez and M.J. Yacamán. 2006. Advances in developing TiNi nanoparticles. Mater. Sci. Engin. A 438–440: 411–413.

Cattaneo, A.G., R. Gornati, M. Chiriva–Internati and G. Bernardini. 2009. Ecotoxicology of nanomaterials: the role of invertebrate testing. Invert. Surv. J. 6: 78–97.

Chang, H., C. Jwo, C. Lo, T. Tsung, M. Kao and H. Lin. 2005. Rheology of CuO nanoparticle suspension prepared by ASNSS. Rev. Advan. Mat. Sci. 10: 128–132.

Choi, M.-R., K.J. Stanton–Maxey, J.K. Stanley, C.S. Levin, R. Bardhan, D. Akin, S. Badve, J. Sturgis, J.P. Robinson and R. Bashir. 2007. A cellular trojan horse for delivery of therapeutic nanoparticles into tumors. Nano Letters 7: 3759–3765.

Christian, P., F. Von der Kammer, M. Baalousha and Th. Hofmann. 2008. Nanoparticles: structure, properties, preparation and behavior in environmental media. Ecotoxicol. 17: 326–343.

Cioffi, N., N. Ditaranto, L. Torsi, R.A. Picca, E. De Giglio, L. Sabbatini, L. Novello, G. Tantillo, T. Bleve–Zacheo and P.G. Zambonin. 2005. Synthesis, analytical characterization and bioactivity of Ag and Cu nanoparticles embedded in poly–vinyl–methyl–ketone films. Anal. Bioanal. Chem. 382: 1912–1918.

Colvin, V.L. 2003. The potential environmental impact of engineered nanomaterials. Nature Biotechnol. 21: 1166–1170.

Coccini, T., E. Roda, M. Fabbri, M.G. Sacco, L. Gribaldo and L. Manzo. 2012. Gene expression profiling in rat kidney after intratracheal exposure to cadmium-doped nanoparticles. J. Nanopart. Res. 14: article 925.

Davies, M.J. 2005. The oxidative environment and protein damage. Biochimica et Biophysica Acta. 1703: 93–109.

Dhas, N.A., C.P. Raj and A. Gedanken. 1998. Synthesis, characterization, and properties of metallic copper nanoparticles. Chem. Mater. 10: 1446–1452.

Dowling, V.A. and D. Sheehan. 2006. Proteomics as a route to identification of toxicity targets in ecotoxicology. Proteomics 6: 5597–5604.

Fabrega, J., S.N. Luoma, C.R. Tyler, T.M. Galloway and J.R. Lead. 2011. Silver nanoparticles: Behaviour and effects in the aquatic environment. Environ. Int. 37: 517–531.

Farkas, J., P. Christian, J.A.G. Gallego–Urrea, N. Roos, M. Hassellöv, K.E. Tollefsen and K.V. Thomas. 2011. Uptake and effects of manufactured silver nanoparticles in rainbow trout (*Oncorhynchus mykiss*) gill cells. Aquat. Toxicol. 101: 117–125.

Gaiser, B.K., A. Biswas, P. Rosenkranz, M.A. Jepson, J.R. Lead, V. Stone, C.R. Tyler and T.F. Fernandes. 2011. Effects of silver and cerium dioxide micro- and nano-sized particles on *Daphnia magna*. J. Environ. Monitor. 13: 1227–1235.

Gomes, T., S. Chora, C.G. Pereira, C. Cardoso and M.J. Bebianno. 2014. Proteomic response of mussels *Mytilus galloprovincialis* exposed to CuO NPs and Cu2+: An exploratory biomarker discovery. Aquat. Toxicol. 155: 327–336.

Gomes, T., C.G. Pereira, C. Cardoso and M.J. Bebianno. 2013. Differential protein expression in mussels *Mytilus galloprovincialis* exposed to nano and ionic Ag. Aquat. Toxicol. 136: 79–90.

Gomes, T., O. Araújo, R. Pereira, A.C. Almeida, A. Cravo and M.J. Bebianno. 2013. Genotoxicity of copper oxide and silver nanoparticles in the mussel *Mytilus galloprovincialis*. Marine. Environ. Res. 84: 51–59.

Gomes, T., C.G. Pereira, C. Cardoso, V.S. Sousa, M.R. Teixeira, J.P. Pinheiro and M.J. Bebianno. 2014. Effects of silver nanoparticles exposure in the mussel *Mytilus galloprovincialis*. Mar. Environ. Res. 101: 208–214.

Gomes, T., C.G. Pereira, C. Cardoso, J.P. Pinheiro, I. Cancio and M.J. Bebianno. 2012. Accumulation and toxicity of copper oxide nanoparticles in the digestive gland of *Mytilus galloprovincialis*. Aquat. Toxicol. 118-119: 72–79.

Grassian, V.H., P.T. O'Shaughnessy, A. Adamcakova-Dodd, J.M. Pettibone and P.S. Thorne. 2007. Inhalation exposure study of titanium dioxinde nanoparticles with a primary particle size of 2 to 5 nm. Environ. Health Perspt. 115: 397–402.

Griffitt, R.J., A. Feswick, R. Weil, K. Hyndman, P. Carpinone, K. Powers, N.D. Denslow and D.S. Barber. 2013a. Chronic nanoparticulate silver exposure results in tissue accumulation and transcriptomic changes in zebrafish. Aquat. Toxicol. 130: 192–200.

Griffitt, R.J., C.M. Lavelle, A.S. Kane, N.D. Denslow and D.S. Barber. 2013b. Investigation of acute nanoparticulate aluminium toxicity in zebrafish. Environ. Toxicol. 26: 541–551.

Griffitt, R.J., J. Luo, J. Gao, J.C. Bonzongo and D.S. Barber. 2008. Effects of particle composition and species on toxicity of metallic nanomaterials in aquatic organisms. Environ. Toxicol. Chem. 27: 1972–1978.

Guzman, K.A.D., M.R. Taylor and J.F. Banfield. 2006. Environmental risks of nanotechnology: national nanotechnology initiative funding, 2000–2004. Environ. Sci. Technol. 40: 1401–1407.

Handy, R.D., R. Owen and E. Valsami–Jones. 2008a. The ecotoxicology of nanoparticles and nanomaterials: current status, knowledge gaps, challenges, and future needs. Ecotoxicol. 17: 315–325.

Handy, R.D., F. Kammer, J.R. Lead, M. Hassellöv, R. Owen and M. Crane. 2008b. The ecotoxicology and chemistry of manufactured nanoparticles. Ecotoxicol. 17: 287–314.

Handy, R.D. and B.J. Shaw. 2007. Toxic effects of nanoparticles and nanomaterials: Implications for public health, risk assessment and the public perception of nanotechnology. Health Risk Soc. 9: 125–144.

Hassellöv, M., J.W. Readman, J.F. Ranville and K. Tiede. 2008. Nanoparticle analysis and characterization methodologies in environmental risk assessment of engineered nanoparticles. Ecotoxicol. 17: 344–361.

Heath, J.R., S.C. O'Brien, Q. Zhang, Y. Liu, R.F. Curl, F.K. Tittel and R.E. Smalley. 1985. Lanthanum complexes of spheroidal carbon shells. J. Am. Chem. Soc. 107: 7779–7780.

Heinlaan, M., A. Ivask, I. Blinova, H.C. Dubourguier and A. Kahru. 2008. Toxicity of nanosized and bulk ZnO, CuO and TiO_2 to bacteria *Vibrio fischeri* and crustaceans *Daphnia magna* and *Thamnocephalus platyurus*. Chemosphere 71: 1308–1316.

Heinlaan, M., A. Kahru, K. Kasemets, B. Arbeille, G. Prensier and H.-C. Dubourgier. 2011. Changes in the *Daphnia magna* midgut upon ingestion of copper oxide nanoparticles: A transmission electron microscopy study. Wat. Res. 45: 179–190.

Hemmerich, P.H. and A.H. von Mikecz. 2013. Defining the subcellular interface of nanoparticles by live-cell imaging. PLoS One 8: e62018.

Hu, W., S. Culloty, G. Darmody, S. Lynch, J. Davenport, S. Ramirez-Garcia, K.A. Dawson, I. Lynch, J. Blasco and D. Sheehan. 2014. Toxicity of copper oxide nanoparticles in the blue mussel, *Mytilus edulis*. Chemosphere 108: 289–299.

Ju–Nam, Y. and J.R. Lead. 2008. Manufactured nanoparticles: An overview of their chemistry, interactions and potential environmental implications. STOTEN 400: 396–414.

Kim, S., S.W. Bae, J.S. Lee and J. Park. 2009. Recyclable gold nanoparticle catalyst for the aerobic alcohol oxidation and C–C bond forming reaction between primary alcohols and ketones under ambient conditions. Tetrahedron 65: 1461–1466.

Klaine, S.J., P.J.J. Alvarez, G.E. Batley, T.F. Fernandes, R.D. Handy, D.Y. Lyon, S. Mahendra, M.J. McLaughlin and J.R. Lead. 2008. Nanomaterials in the environment: behavior, fate, bioavailability and effects. Environ. Toxicol. Chem. 27(9): 1825–1851.

Klaine, S.J., A.A. Koelmans, N. Horne, S. Carley, R.D. Handy, L. Kapustka, B. Kowack and F. von der Kammer. 2012. Paradigms to assess the environmental impact of manufactured nanomaterials. Environ. Toxicol. Chemi. 31: 3–14.

Klumpp, C., K. Kostarelosc, M. Prato and A. Bianco. 2006. Functionalized carbon nanotubes as emerging nanovectors for the delivery of therapeutics. Biochimica et Biophysica Acta 758: 404–412.

Knigge, T., T. Monsinjon and O.K. Andersen. 2004. Surface–enhanced laser desorption/ionization–time of flight–mass spectrometry approach to biomarker discovery in blue mussels (*Mytilus edulis*) exposed to polyaromatic hydrocarbons and heavy metals under field conditions. Proteomics 4: 2722–2727.

Kuznetsova, G.P., O.V. Larina, N.A. Petushkova, Yu. S. Kisrieva, N.F. Samenkova, O.P. Trifonova, I.I. Karuzina, O.M. Ipatova, K.V. Zolotaryov, Yu A. Romashova and A.V. Lisitsa. 2014. Effects of Fullerene C60 on Proteomic Profile of *Danio Rerio* Fish Embryos. Bull. Experim. Biol. Med. 156: 694–698.

Li, Y.F., Y.X. Gao, Z.F. Chai and C.Y. Chen. 2014. Nanometallomics: An emerging field studying the biological effects of metal-related nanomaterials. Metallomics 6: 220–232.

Li, Y., J. Liang, Z. Tao and J. Chen. 2007. CuO particles and plates: Synthesis and gas–sensor applications. Mater. Res. Bull. 43: 2380–2385.

Liebler, D.C. 2002. Proteomic approaches to characterize protein modifications: new tools to study the effects of environmental exposures. Environ. Health Perspectives 110: 3–9.

Luoma, S.N. 2008. Silver Nanotechnologies and the Environment: Old Problems or New Challenges. Project on Emerging Nanotechnologies. Publication 15. Woodrow Wilson International Centre for Scholars and PEW Charitable Trusts, Washington, DC.

Markarian, J. 2006. Steady growth predicted for biocides. Plastics Addit. Compound 8: 30–3.

Masala, O. and R. Seshadri. 2004. Synthesis routes for large volumes of nanoparticles. Ann. Rev. Mater. Res. 34: 41–81.

Matranga, V. and I. Corsi. 2012. Toxic effects of engineered nanoparticles in the marine environment: Model organisms and molecular approaches. Mar. Environ. Res. 76: 32–40.

Maynard, A. 2006. Nanotechnology: a research strategy for addressing risk. Project on Emerging Nanotechnologies. Publication 3. Woodrow Wilson International Centre for Scholars and PEW Charitable Trusts, Washington, DC.

McDonagh, B. and D. Sheehan. 2006. Redox proteomics in the blue mussel Mytilus edulis: Carbonylation is not a pre–requisite for ubiquitination in acute free radical–mediated oxidative stress. Aquatic Toxicology 79(4): 325–333.

Monsinjon, T., O.K. Andersen, F. Leboulenger and T. Knigge. 2006. Data processing and classification analysis of proteomic changes: a case study of oil pollution in the mussel, *Mytilus edulis*. Proteome Sci. 4: 1–13.

Monsinjon, T. and T. Knigge. 2007. Proteomic applications in ecotoxicology. Proteomics 7: 2997–3009.

Moore, M.N. 2006. Do nanoparticles present ecotoxicological risks for the health of the aquatic environment? Environ. Intern. 32: 967–976.

Mueller, N.C. and B. Nowack. 2008. Exposure modelling of engineered nanoparticles in the environment. Environmental Science and Technology 42(12): 4447–4453.

Nel, A. 2006. Toxic potential of materials at the nanolevel. Science 311: 622–627.

Nesatyy, V.J. and M.J.-F. Suter. 2007. Proteomics for the analysis of environmental stress responses in organisms. Environ. Sci. Technol. 41: 6891–6900.

Nowack, B. and T.D. Bucheli. 2007. Occurrence, behavior and effects of nanoparticles in the environment. Environ. Pollut. 150: 5–22.

Oberdörster, E. 2004. Manufactured nanomaterials (Fullerenes, C60) induce oxidative stress in the brain of juvenile largemouth bass. Environ. Health Perspect. 112: 1058–1062.

Oberdorster, E. 2006. The potential risks of nanomaterials: a review carried out for ECETOC. Particle Fibre Toxicol. 3: 11.

Oberdörster, G., E. Oberdörster and J. Oberdörster. 2005. Nanotoxicology: An emerging discipline evolving from studies of ultrafine particles. Environ. Health Perspect. 113: 823–839.

Oliviera, E., M. Casado, M. Faria, A.M.V.M. Soares, J.M. Navas, C. Barata and B. Pina. 2014. Transcriptomic response of zebrafish embryos to polyaminoamine (PAMAM) dendrimers. Nanotoxicol. 8: 92–99.

Poynton, H.C., J.M. Lazorchak, C.A. Impellitteri, B. Blalock, M.E. Smith, K. Struewing, J. Unrine and D. Roose. 2013. Toxicity and transcriptomic analysis in *Hyalella Azteca* suggests increased exposure and susceptibility of epibenthic organisms to zinc oxide nanoparticles. Environ. Sci. Technol. 47: 9453–9460.

Rabilloud, T. 2000. Proteome research: Two–Dimensional Gel Electrophoresis and Identification Methods. Springer Verlag, Heidelberg. 244 pp.

Rainville, L.-C., D. Carolan, A.C. Varela, H. Doyle and D. Sheehan. 2014. Proteomic evaluation of citrate-coated silver nanoparticles toxicity in *Daphnia magna*. Analyst 139: 1678–1686.

Ranville, J.F., C.D. Vulpe and B. Gilbert. 2013. Silver nanowire exposure results in internalization and toxicity to *Daphnia magna*. ACS Nano. 7: 10681–10694.

Ren, G., D. Hu, E.W.C. Cheng, M.A. Vargas–Reus, P. Reip and R.P. Allaker. 2008. Characterisation of copper oxide nanoparticles for antimicrobial applications. Internat. J. Antimicrob. Agents 33: 587–590.

Ringwood, A.H., M. McCarthy, T.C. Bates and D.L. Carroll. 2010. The effects of silver nanoparticles on oyster embryos. Mar. Environ. Res. 69: 549–551.

Rocha, T.L., T. Gomes, C. Cardoso, J. Letendre, J.P. Pinheiro, V.S. Sousa, M.R. Teixeira and M.J. Bebianno. 2014. Immunocytotoxicity, cytogenotoxicity and genotoxicity of cadmium-based quantum dots in the marine mussel *Mytilus galloprovincialis*. Mar. Environ. Res. 101: 29–37.

Roco, M.C. 2005. Environmentally responsible development of nanotechnology. Environ. Sci. Technol. 39: 106A–112A.

Roduner, E. 2006. Size matters: why nanomaterials are different? Chem. Soc. Rev. 35: 583–592.

Royal Society and Royal Academy of Engineering. 2004. Nanoscience and nanotechnologies: opportunities and uncertainties. RS policy document 19/04. London. 113 pp.

Scanlan, L.D., R.B. Reed, A.V. Loguinov, P. Antczak, A. Tagmount, S. Aloni, D.T. Nowinski, P. Luong, C. Tran, N. Karunaratne, D. Pham, X.X. Lin, F. Falciani, C.P. Higgins, Scown, T.M., R.V. Aerle and C.R. Tyler. 2010a. Review: do engineered nanoparticles pose a significant threat to the aquatic environment? Critic. Rev. Toxicol. 40: 653–670.

Scown, T.M., R.V. Aerle and C.R. Tyler. 2010. Review: do engineered nanoparticles pose a significant threat to the aquatic environment? Critic. Rev. Toxicol. 40: 653–670.

Sheehan, D. 2013. Next-generation genome sequencing makes non-model organisms increasingly accessible for proteomic studies. J. Proteomics Bioinform. 6: e21. DoI: 10.4172/jpb.10000e21.

Shepard, J.L. and B.P. Bradley. 2000. Protein expression signatures and lysosomal stability in *Mytilus edulis* exposed to graded copper concentrations. Mar. Environ. Res. 50: 457–463.

Simpson, K. 2003. Using silver to fight microbial attack. Plastics, Addit. Compound 5: 32–35.

Smith, C.J., B.J. Shaw and R.D. Handy. 2007. Toxicity of single walled carbon nanotubes to rainbow trout (*Oncorhynchus mykiss*): Respiratory toxicity, organ pathologies, and other physiological effects. Aquat. Toxicol. 82: 94–109.

Snape, J.R., S.J. Maund, D.B. Pickford and T.H. Hutchinson. 2004. Ecotoxicogenomics: the challenge of integrating genomics into aquatic and terrestrial ecotoxicology. Aquat. Toxicol. 67: 143–154.

Suárez-Ulloa, V., J. Fernández-Tajes, C. Manfrin, M. Gerdol, P. Venier and J.-M. Eirín-López. 2013. Bivalve omics: state of the art and potential applications for the biomonitoring of harmful marine compounds. Mar. Drugs 11: 4370–4389.

Takeuchi, T., T. Kawashima, R. Koyanagi, F. Gyoja, M. Tanaka, T. Ikuta, E. Shoguchi, M. Fujiwara, C. Shinzato, K. Hisata, M. Fujie, T. Usami, K. Nagai, K. Maeyama, K. Okamoto, H. Aoki, T. Ishikawa, T. Masaoka, A. Fujiwara, K. Endo, H. Endo, H. Nagasawa, S. Kinoshita, S. Asakawa, S. Watabe and N. Satoh. 2012. Draft genome of the pearl oyster *Pinctada fucata*: A platform for understanding bivalve biology. DNA Res. 19: 117–130.

Tedesco, S., H. Doyle, J. Blasco, G. Redmond and D. Sheehan. 2010a. Exposure of the blue mussel, *Mytilus edulis*, to gold nanoparticles and the pro–oxidant menadione. Comp. Biochem. Physiol. C 151: 167–174.

Tedesco, S., H. Doyle, J. Blasco, G. Redmond and D. Sheehan. 2010b. Oxidative stress and toxicity of gold nanoparticles in *Mytilus edulis*. Aquat. Toxicol. 100: 178–186.

Tedesco, S. and D. Sheehan. 2010. Nanomaterials as emerging environmental threats. Current Chem. Biol. 4: 151–160.

Tiede, K., M. Hassellöv, E. Breitbarth, Q. Chaudhry and A.B.A. Boxall. 2009. Considerations for environmental fate and ecotoxicity testing to support environmental risk assessments for engineered nanoparticles. J. Chromatogr. A 1216: 503–509.

Trevisan, R., G. Delapedra, D.F. Mello, M. Arl, E.C. Schmidt, Z.L. Bouzon, A. Fisher, D. Sheehan and A.L. Dafre. 2014. Gills are an initial target of zinc oxide nanoparticles in oysters *Crassostrea gigas*, leading to mitochondrial disruption and oxidative stress. Aquat. Toxicol. 153: 27–38.

Valsami-Jones, E. and C. Mouneyrac. 2012. Fate of isotopically labeled zinc oxide nanoparticles in sediment and effects on two endobenthic species, the clam *Scrobicularia plana* and the ragworm *Hediste diversicolor*. Ecotoxicol. Environ. Saf. 84: 191–198.

Vioque-Fernández, A., E.A. de Almeida and J. López-Barea. 2009. Assessment of Doñana National Park contamination in *Procambarus clarkii*: Integration of conventional biomarkers and proteomic approaches. STOTEN 407: 1784–1797.

Wang, C., Y.J. Hu, C.M. Lieber and S.H. Sun. 2008a. Ultrathin Au nanowires and their transport properties. J. Amer. Chem. Soc. 130: 8902–8903.

Wang, H.-T., O.A. Nafday, J.R. Haamheim, E. Tevaarwerk, N.A. Amro, R.G. Sanedrin, C.-Y. Chang, F. Ren and S.J. Pearton. 2008b. Toward conductive traces: Dip Pen Nanolithography® of silver nanoparticle–based inks. Appl. Phys. Lett. 93: 143105.

Ward, J.E. and D.J. Kach. 2009. Marine aggregates facilitate ingestion of nanoparticles by suspension–feeding bivalves. Mar. Environ. Res. 68: 137–142.

Zhang, W. 2003. Nanoscale iron particles for environmental remediation: an overview. J. Nanoparticle Res. 5: 323–332.

Zhang, G., X. Fang, X. Guo, L. Li, R. Luo, F. Xu, P. Yang, L. Zhang, X. Wang, H. Qi et al. 2012. The oyster genome reveals stress adaptation and complexity of shell formation. Nature 490: 49–54.

10

Effects of Climate Change on Marine Organisms

A Proteomic Approach

Mário Diniz,[a,*] *Diana Madeira*[b] and *Eduardo Araújo*[c]

ABSTRACT

The effects of climate change, namely sea warming and ocean acidification, cause changes in the physiology, phenology, and biogeographical distribution of organisms. The stress induced in marine organisms at the proteome level, altering the expression of proteins and their modifications, is still largely unknown. The most common techniques used in marine proteomics regarding climate change are 2D electrophoresis and protein identification by mass spectrometry. Typically, the workflow includes several sequential steps: (1) sample preparation, (2) protein denaturation and reduction, (3) protein (peptide) separation, enzymatic digestion, and mass spectrometry analysis, and (4) bioinformatics and protein identification. As such the "omics", and particularly proteomics, have the potential to provide new insights into the integrative functional responses of organisms to environmental stresses. Studies developed to date on proteome changes suggest a convergence toward a common set of stress-induced proteins related to metabolism, the cytoskeleton, chaperones, and diverse protective proteins. Therefore, this chapter discusses the main aspects related to climate change that affect marine organisms, with a special emphasis on the typical proteomic workflow used for marine organisms, the most relevant studies, the main problems, and future challenges.

Introduction

Some of the major consequences of climate change, namely sea warming and ocean acidification, are caused by anthropogenic activities. In fact, currently there is global agreement in the scientific community that the anthropogenic impact on the oceans is weakening marine biodiversity. The warming of the sea and ocean acidification seems to increase the risks from global warming for marine biota.

Department of Chemistry-UCIBIO, Faculty of Sciences and Technology, University, NOVA de Lisboa, Campus de Caparica, 2829-516 Caparica, Portugal.

[a] Email: mesd@fct.unl.pt

[b] Email: dianabmar@gmail.com

[c] Email: jeduardoaraujo88@gmail.com

* Corresponding author

The increase of atmospheric CO_2 from anthropic sources is not only responsible for global warming but also for shifts in ocean chemistry. Indeed, the ocean takes up part of the atmospheric CO_2, which reacts with seawater to form carbonic acid, decreasing the pH of seawater and reducing oxygen levels (Feely et al. 2009; Pineiro et al. 2010). Meanwhile, our comprehension of the effects of climate change and its potential consequences on a large number of marine organisms is still rudimentary, leading to increased concerns in society about the risks for marine ecosystems. Nonetheless, several studies have already showed that sea warming and ocean acidification greatly impact the physiology, phenology, and biochemistry of the marine biota (Visser and Both 2005; Somero 2010; Padilla-Gamino et al. 2013; Crozier et al. 2014). However, there have been very few studies on the effects of environmental stress at the proteome level in relation to climate change (see Tomanek 2011 for an extensive review). Environmental proteomics is defined as the study of the expression profile of proteins from the cells of living organisms. The most frequent techniques used in marine proteomics regarding climate change are 2D electrophoresis and protein identification by mass spectrometry using MALDI-TOF-TOF followed by LC-MS-MS (ESI). The typical workflow includes: (1) sample preparation, (2) protein denaturation, (3) protein (peptide) separation, (4) enzymatic digestion, (5) mass spectrometry analyses, and (6) bioinformatic approaches and protein identification. However, developing research on the effects of environmental stress on proteomics is of great importance as it enables the study of the entire set of proteins within a cell and the changes between cell types over time. Thus, the proteome can be viewed as a measure of the cell phenotype and can be directly related to fitness (Feder and Walser 2005; Dupont et al. 2007). In addition, according to Tomanek (2011), the proteomic approach for environmental stress studies is an excellent tool to provide some understanding on proteome changes reflecting alterations in protein synthesis, post-translational modifications, or degradation.

Regarding thermal stress, studies using the proteomic approach have shown that several pathways are affected by this environmental variable (see Tomanek 2011, 2014 for an extensive review). For instance, a wide range of studies have reported changes in protein functions related to chaperoning, metabolism, proteolysis, cytoskeleton composition, oxidative damage, electron transport chain, and life span (e.g., López et al. 2002; Gardeström et al. 2007; Tomanek and Zuzow 2010; Tomanek 2011; Fields et al. 2012). In general, proteomic studies concerning thermal stress in marine organisms have been performed on fish (McLean et al. 2007; Ibarz et al. 2010), sea squirts (Serafini et al. 2011), mussels (Tomanek 2012), clams (Truebano et al. 2013), algae (Choi and Lee 2012), Daphnia (Schwerin et al. 2009), shrimps (Fan et al. 2013), crabs (Wang et al. 2007), gastropods (Gardeström et al. 2007; Martínez-Fernández et al. 2008) and polychaetes (Dilly et al. 2012).

Protein identification can be very useful in studies of environmental stress, such as ones related to the effects of climate change on the marine biota. However, many proteomic studies have encountered several difficulties in matching peptide masses or peptide fragmentation patterns to known sequences, as databases are frequently incomplete for many species. Nonetheless, there are several reasons to focus on proteomics, such as the molecular phenotype of cells, which are represented by proteins that have a direct effect on the physiology and fitness of the organisms. Moreover, proteomic analyses complements transcriptomic analyses, as variation in transcript levels in general explains less than half of the variation in protein abundance (Feder and Walser 2005).

While the primary goal of proteomics is to characterise all forms of proteins, including post-translational modifications (PTMs), current technical limitations restrict access to the entire proteome. However, despite the limitations, proteomic studies based on mass spectrometry have been successful in discovering new insights into the complexities of biological systems (Cravatt et al. 2007). Therefore, this chapter will focus on the most relevant aspects of proteomic techniques and cellular responses when applied to samples from marine organisms subjected to either experimental sea warming or ocean acidification. There are very few proteomic studies on the effects of climate change on the marine biota when compared to proteomic studies in biomedicine or even on the effects of aquatic pollutants on the marine proteome. The complexity and diversity of marine taxa makes marine proteomics a huge challenge for the decades to come, but it could greatly benefit from advances expected in the near future. Therefore, the most relevant studies on this subject will be discussed and the typical proteomic workflow used for environmental proteomics will be briefly described.

Climate Change, Sea Warming, and Ocean Acidification

The ocean is the largest habitat on Earth, covering approximately 70 percent of the Earth's surface and containing 97 percent of the planet's water. In addition, the marine ecosystem is the major aquatic system on Earth, playing an essential role in regulating climate and weather systems (Wang et al. 2014). It presents huge richness in terms of biodiversity, with about 50 to 80 percent of all the life on earth being found under the surface of the sea. However, climate change may compromise marine ecosystems, with some species potentially at risk of extinction if no measures are taken (IOC/UNESCO et al. 2011).

Human activities are considered the major drivers of climate change, altering the earth's surface and atmospheric composition (Fabry et al. 2008). The scientific community is now convinced that the global mean surface temperature has increased since the late 19th century, becoming successively warmer with each decade. Over the next two decades, a warming of about 0.2°C is expected. It is clear that the upper ocean has also warmed since the 1870s. If greenhouse gas emissions continue at or above the same rates further warming will occur and many changes will occur in the global climate system during the 21st century. However, even if greenhouse gas concentrations were stabilised, anthropogenic warming and sea level would still continue to rise for centuries due to the time scales linked to climate processes and feedback mechanisms (IPPC 2013).

The concentrations of atmospheric greenhouse gases, carbon dioxide (CO_2), methane (CH_4), and nitrous oxide (N_2O), have increased mostly due to anthropic activities, leading to global warming resulting in environmental and ecological alterations in marine ecosystems (Caldeira and Wicket 2003; Orr et al. 2005). Current atmospheric CO_2 levels have reached an unparalleled 400 ppm and are expected to rise up to 730–1020 ppm by 2100 (Meehl et al. 2007; IPPC 2013).

However, absorption of CO_2 by the oceans has decreased the ocean's capacity to absorb additional atmospheric CO_2; therefore, future emissions will result in inevitable and rapid global warming (Doney et al. 2009; Caldeira and Wicket 2003; Feeley et al. 2004). It is expected that the global average surface and vertical average ocean temperature will increase in the short-term and in the long-term. An increase of up to 2°C is expected in the top 100 meters by 2100 (IPPC 2013), promoting an increase in hypoxia in coastal areas, while the oceanic uptake of anthropic CO_2 will result in a progressive acidification of the ocean (Rosa et al. 2012). This shift in ocean chemistry is caused by a reaction between CO_2 and seawater resulting in a net increase in oceanic inorganic carbon concentrations and causing a decrease in pH and carbonate ion concentrations (Wei et al. 2015).

Since the beginning of the industrial revolution the pH of ocean surface water has decreased by about 0.1 units and models predict a decrease of up to 0.5 units by 2100 (Rosa et al. 2014; Meehl et al. 2007). Such a noticeable reduction in pH is unparalleled for modern marine organisms, which have evolved in a relatively constant marine environment (Pearson and Palmer 2000). However, there are some areas, such as coastal zones or estuaries that have been impacted more by human activities than others, where the decreasing pH of sea water is even more pronounced (Madeira et al. 2014a; Rosa et al. 2014). Additionally, the ocean's oxygen concentrations have decreased in many ocean regions, mainly in the open ocean thermocline but also in coastal areas (Doney et al. 2012; Fabry et al. 2008). These future changes in the ocean's chemistry are expected to cause difficulties for the marine biota that must adapt to a new environment (IPPC 2013; Hönisch et al. 2012).

According to the IPPC report (2013), the length and number of heat waves has increased worldwide and more extreme events are expected to occur by the end of the 21st century, increasing in intensity, duration, and frequency, as a consequence of global warming. Coastal and estuarine living organisms may be particularly at risk since they can experience events within or above their physiological limits of tolerance and the ecological and biological effects are poorly understood. Therefore, one of the major questions we need to address is how marine organisms will adapt to a changing environment, in terms of both gradual warming and heat waves. Diverse field and laboratory studies have contributed to our understanding of the adaptive mechanisms and stress responses involved when marine animals experience heat and acidification stress (e.g., Portner et al. 2004; Somero 2010; Tomanek 2010; Hendriks et al. 2015). Knowledge of how temperature affects physiological and cellular processes enables us to understand how vulnerable a species is to sea warming and predict how it will react to ongoing changes. Filling this gap is important for understanding how species and communities will be affected by climate change and by what means the organisms are able to tolerate and resist environmental stress. Studies of environmental plasticity have focused on the effects of some abiotic factors within an individual's lifetime. However, little is known about the effects at the molecular level, particularly on marine species proteome and metabolome.

The effects of sea warming and ocean acidification on marine animals

The effects of climate change on the marine biota are already occurring, affecting species physiology, biodiversity, and marine phenology (e.g., the timing of seasonal activities such as breeding (Trübenbach et al. 2013; Visser and Both 2005)). Moreover, global warming can threaten the geographical distribution of marine animals, which may lead to the extinction of some species previously common in given locations (Portner 2004). Although species may have or develop limited physiological/biochemical adaptability to environmental stressors (OA and hypoxia), their synergism narrows their thermal tolerance limits (Rosa et al. 2012). Therefore, seawater CO_2 chemistry, O_2, and temperature shifts will directly affect their development, physiology, survival, and distribution (Fabry et al. 2008; Rosa et al. 2012; Madeira et al. 2014), especially among poikilotherms, such as copepods and amphipods, which are often selected for biological studies due to their ecological relevance, sensitivity to environmental stress, and the ease with which they can be cultured and tested (Cruz et al. 2013).

Indeed, future changes expected in the chemistry of the oceans will lead to increasing difficulties for calcifying organisms, especially key species (IPCC 2007). Along with sea warming and ocean acidification, hypoxia will grow in coastal areas threatening organisms' survival (Rosa et al. 2012). Yet temperature is one of the most important factors affecting organisms as it impacts the kinetic energy of molecules, influencing biochemical reactions, and thus the animal's physiology and behavior (Gracey 2007; Wang et al. 2009; Rosa et al. 2014). There are several studies regarding the effects of ocean acidification and sea warming on the marine biota (e.g., Rosa and Seibel 2008; Hofmann et al. 2010; Poloczanska et al. 2013; Trübenbach et al. 2013; Rosa et al. 2016), but little is known about the effects at the molecular level, particularly in the proteome of marine species adapting to environmental changes. With regards to OA, the problem is more complex as most studies have focused on more vulnerable target key species, with little research devoted to organisms or taxa that might be less vulnerable to the predicted changes. Yet, the real problems with respect to hypercapnia and temperature occur very early in ontogeny, with early life stages being particularly sensitive to environmental stresses (Melzner et al. 2009). Thus, in the near future several marine organisms will deal with pCO_2 and temperature increases and many of the biological consequences are still unknown. As many of these organisms form the primary products of oceans, any change in their life cycle may potentially impact all marine ecosystems (Doney et al. 2009; Feeley et al. 2004). Nonetheless, some effects on the marine biota are predictable given that many marine organisms make shells or supporting plates out of calcium carbonate ($CaCO_3$) in a process called calcification.

Recently, there have been studies discussing and investigating the heat shock response at the molecular level of different marine species from distinct habitats (Madeira et al. 2014a,b, 2015, 2016a; Vinagre et al. 2012, 2014; Rosa et al. 2014). These authors showed the different responses of marine species to stress oxidative enzymes, indicators of tissue injury (e.g., lipid peroxides), or Hsp70 production patterns in a global warming scenario. Other studies provided a comprehensive and integrated view of biological responses to future warming in invertebrate keystone species, showing that heat shock proteins (Hsp70/Hsc70) and antioxidant enzyme activities constituted an integrated stress response to ocean warming (Madeira et al. 2012; Rosa et al. 2012; Madeira et al. 2014a). Alongside oxidative stress enzymes, heat shock proteins and other biomarkers of proteomics can provide new insights and produce additional information in biomarker and physiology studies, allowing an integrated approach and promoting the understanding of the complex mechanisms involved in the response to environmental stressors.

The "Omics" versus *proteomics*

The term "omics" is used to refer to large-scale data analysis in biology, such as genomics, proteomics, or metabolomics, that have a high potential for changing how we comprehend the function of cells and organisms at the systems-level (Dowd 2012). Over the last decades "Omics" technologies, mainly applied in biomedical sciences, have been widely implemented in many different fields of biological sciences, including environmental stress studies. While metabolomics allows for the characterisation, identification, and quantification of metabolites resulting from a wide range of biochemical processes in living systems (Go 2010), proteome profiling can provide greater insights into the mechanistic nature of a response to environmental stressors (Veldhoen et al. 2012). Indeed, by studying protein abundance and modification patterns, proteomics has great potential to clarify the functional responses of organisms to environmental stressors (Tomanek 2011). Genetic constitution and physiological tolerance are major biological factors that determine a species' ability to respond to climate change scenarios. Functional genomics has increasingly been applied in order to investigate fundamental biological processes directly associated to species physiological responses and evolutionary adaptation to environmental stressors (Gracey 2007).

Several authors argue that identification and expression pattern analysis of candidate/target genes involved in the adaptive response to ocean warming, ocean acidification, and hypoxia makes transcriptomics (Wang et al. 2009), together with proteomics (Tomanek 2011), metabolomics (Lankadurai et al. 2013), physiological and morphological-based tools, a valuable resource for assessing the vulnerability of species physiological responses to environmental change. However, although transcriptomic and proteomic data do not tell the whole story of what might be occurring in a cell, metabolic profiling can give a snapshot of the cell's physiology. Thus, one of the challenges of systems biology is to integrate proteomic, transcriptomic, and metabolomic information to give a more complete picture of a living organism's biological processes (Go 2010; Lankadurai et al. 2013).

Nonetheless, the proteome reflects the biological context of a particular biological system, at a particular time and also the structural and functional understanding of proteins, their modifications, interactions between them, and the intracellular localization and quantification of their expression. From this perspective, proteomics is the large-scale study of a proteome from various aspects (Pineiro et al. 2010):

a. Expression proteomics (focused on large-scale characterization and quantification of all components of a particular proteome)
b. Differential expression proteomics (compares dynamic changes in the proteome of two or more different samples
c. Cellular map and interaction proteomics (studies the relationships among proteins in a certain biological system)
d. Post-translational modification studies-PTMs (proteins suffer PTMs in response to a wide range of extra- and intra-cellular signals.

Thus, environmental studies of proteomics are increasing in quantity and quality as a consequence of improvements and advances in protein separation methods and peptide ionization and identification tools, leading to a more extensive use of proteomics in different species (Veldohen 2012). Additionally, proteomic tools, in particular those based on mass spectrometry, have proved to be useful in identifying proteins and uncovering new insights concerning responses to environmental stressors (Tomanek 2011). However "omics" techniques still have several constraints, such as high costs, limited access to expensive instruments, incomplete databases, a shortage of genomic sequence information, a need for better analytical methods for large data sets, and high throughput bioinformatics tools that must be solved to move forward.

Proteomic basic strategies

The basic idea of proteomics is that peptides match protein identities through the use of mass spectrometry (Mann et al. 2001). Therefore, protein profiling is the most common type of proteome study. Briefly, the common approaches include the following steps: (1) Bottom-up, two samples are labeled with different stable isotope labels, the two unfractionated protein extracts are mixed, digested with trypsin, and then separated by one or more separation methods prior to being introduced to a mass spectrometer. The MS/MS spectra and the relative abundances of each protein are determined using ratios of the signals for the light and heavy forms of identified labeled peptides (Yates et al. 2009). (2) Top-down protein analysis, during which separate intact proteins are measured using one or more separation modes and quantitative changes in levels of individual proteins. The proteins of interest are then typically identified by fragmenting the protein with trypsin, followed by mass spectrometry analysis (Yates et al. 2009).

The most common workflow in environmental proteomics involves protein isolation by gel electrophoresis (1D and 2D), followed by gel image analysis for protein (spot) detection and quantification (Gulcicek et al. 2005). The gel bands or spots are then excised and in-gel digested, normally using trypsin. This approach has been increasingly used to study proteome responses in many non-model species (Chora et al. 2008). According to Knigge et al. (2004), even when genomic data is lacking, we can use the concept of protein expression signature (PES), consisting of the set of proteins expressed at any given time, some of which may represent a particular stress response or be associated with a transient developmental stage (Bradley et al. 2002). Thus, establishing the PES for animals exposed to various stressors, proteomics can potentially be used to study the stress response of a non-model species at the molecular level. Whereas, the most common strategies used in proteomic studies relating to the effects of sea warming and ocean acidification in marine organisms always comprise of sample collection, protein extraction techniques, purification and eventually sample fractionation, followed by mass spectrometry analysis for protein identification using bioinformatics tools (Wang et al. 2014).

Proteomics Approach in Climate Change Studies

Experimental design and results

When performing proteomic studies the design is fundamental, with a clear and full description of the experiment, including the methods used that are required to allow other researchers to repeat the experiment. Proper planning of the assays ensures that all variables are taken into consideration. For climate change assays there is a published "Guide to best practices for ocean acidification research and data reporting" from Riebesell et al. (2010) that helps with the planning and designing of the experiments, while for proteomic research scientific journals have several indications that may guide authors. For instance, in some journals as the Journal of Proteomics, Proteomics or JIOMICS "authors are encouraged to use proteomics data in the context of interdisciplinary projects based on biological hypotheses and, by doing so, validate the proteomics data through hypothesis-driven experiments. They advise authors that, when performing data analysis in 2DE and MS-based experiments, they should provide details of the number of biological and analytical replicates used, not just fold differences, and that in comparative proteomics multivariate analysis of the variance should be used. Moreover, identification of proteins based solely on mass fingerprinting should be considered only if the sequence of the genome from the organism under study is known. In addition, for peptide mass fingerprinting, the number of peptides that match the sequence, the number of unmatched peaks and the total percent of sequence coverage must be stated. Identification of proteins from organisms with an unknown genome sequence will be accepted only if MS/MS-derived peptide sequence data has been used during database searches or BLAST analysis. The score of the highest ranked hit for a homologous, orthologous or paralogous protein should be indicated. Some journals also ask for proteomic findings to be confirmed through independent and complementary validation experiments. Moreover, when using animals authors must have the approval of national authorities, state if the procedures used were in accordance with animal welfare standards and be certified to perform the experiments at a national and/ or international level".

Finally, the method used to generate the mass spectrometry data must be described, "including the methods used to create peak lists from raw MS or MS/MS data. The name and version of the program(s) used for database searches, the values for the critical search parameters (e.g., parent ion and fragment mass tolerance, cleavage rules used and allowance for number of missed cleavages) and the name and version of the database(s) searched must be provided. For each protein identified, measures of certainty must be provided (e.g., *p*-values). For MS/MS, the number of peptides used to identify a protein must be given, as well as the sequence and charge state of each peptide. For peptide mass fingerprinting, the number of peptides that match the sequence and the total percent of sequence coverage must be mentioned. Full details of any software, bioinformatics tools or image capture processes used for data processing or analysis should be provided. In comparative or shotgun-like discovery studies, confirmatory data are required data (e.g., from validated immunoassays, immunohistochemistry, alternative MS-based methods, Western blotting, etc.) using independent replication sets for at least a subset of proteins".

Typical workflow for proteomics

The typical workflow used in proteomics combines IEF and SDS-PAGE providing a powerful tool for separating complex mixtures of proteins (Gorg et al. 2000, 2004). Figure 1 represents the main steps for a typical workflow for proteomics. It is well established that protein extraction and clean up procedures are decisive for ensuring optimal resolution, reduced variability in the 2D gels, and obtaining the best results in proteomics. Thus, treating samples to prevent interference from substances such as salts, detergents, denaturants, or organic solvents is a fundamental step in successful analysis (Gorg et al. 2004). The most common proteomic experiments can be simplified into a series of fundamental steps that include the extraction, fractionation, and solubilisation of the sample proteins from the cells, tissues, or organisms. They may also include a processing step for protein depletion of high-abundance proteins and/or enrichment of less abundant proteins in the sample to increase the sensitivity of downstream analyses. The treated samples are then electrophoretically separated in one dimension (SDS-PAGE) or two sequential dimensions for more complex mixtures (IEF followed by SDS-PAGE). Next, the gels are stained, followed by imaging analysis for visualisation and quantification of proteins. Isolated spots/bands of interest are excised from the gel, digested with a protease (e.g., trypsin), and finally identified by mass spectrometry with the assistance of bioinformatic software (Capelo et al. 2009). It is also very important to always wear gloves throughout sample processing to prevent contamination (Shevchenko et al. 2006).

Thus, this section provides a simplified overview of the most common procedures employed in environmental proteomics, especially the techniques that have been used in climate change proteomics.

Sample preparation

Appropriate sample preparation prior to MS-based analysis is essential in the analytical workflow, as it influences the quality and reproducibility of the MS analysis. The main aim of this step is thorough solubilisation of all (or a specific subset) of the proteins within a given sample. Although there is not a

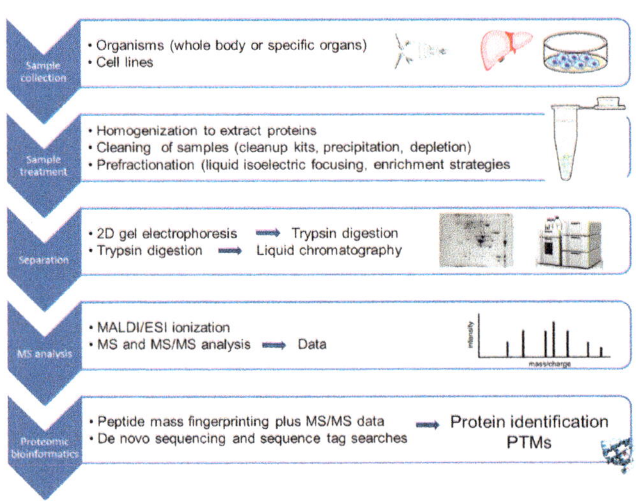

Fig. 1: Scheme of a typical proteomic workflow.

universal standard method for protein extraction, and each protocol must be first optimised for the type and nature of the sample, a typical extraction procedure starts by obtaining whole-cell lysates (Bodzon-Kulakowska et al. 2007; Martínez-Maqueda et al. 2013). Therefore, the principal stages considered in sample treatment are:

- Cell disruption
- Proteases inhibition
- Clean up: removal of interfering contaminants
- Solubilisation of the proteins

Cell disruption can be achieved by chemical or physical methods, such as osmotic lysis, freeze-thawing, detergent lysis, enzymatic lysis of the cell wall, sonication, grinding with nitrogen, high pressure (e.g., French press), homogenization with glass beads, high-speed rotor blade homogenization, or grinding with a glass/teflon Wheaton Potter Elvejhem tissue grinder (Ahmed 2009; Bodzon-Kulakowska et al. 2007) as shown in Fig. 2. All of these methods have some constraints and benefits depending on the type and nature of the sample to be processed. Usually, lysis is carried out using a proper buffer that can vary in its composition depending on the type of sample to be analysed.

Compounds that are present in the samples and that may interfere with the analysis must then be removed and/or inactivated. For instance, proteolytic enzymes must be inhibited, most often by using *Protease Inhibitor Cocktails*, which are available commercially from several companies (e.g., Protease Inhibitor Cocktail—Sigma, Halt Protease Inhibitor Cocktail—Thermo Scientific), to prevent protein degradation that may cause charge artefacts. Samples containing high concentrations of salts may also interfere with electrophoretic separation and can be removed either by (spin)dialysis, precipitation of the proteins (e.g., DOC/TCA/acetone), or by using a 2DE clean-up kit (e.g., GE Healthcare 2-D clean up kit, Ready Prep—BioRad) to remove the interfering substances, such as salts, lipids, phenols, nucleic acids, and detergents (Bodzon-Kulakowska et al. 2007). Another alternative is to dilute the sample below a critical salt concentration and apply a larger volume onto the IPG gel, preferably by sample in-gel rehydration, thereby performing a sort of 'in-gel desalting' (Gorg et al. 2004, 2007). Nonetheless, some protein loss may occur when using clean up strategies, such as protein precipitation with TCA/acetone, due to insufficient re-solubilisation of the proteins (Gorg et al. 2007). It is also important to follow the manufacturer's instructions regarding the loading volumes per type of strip.

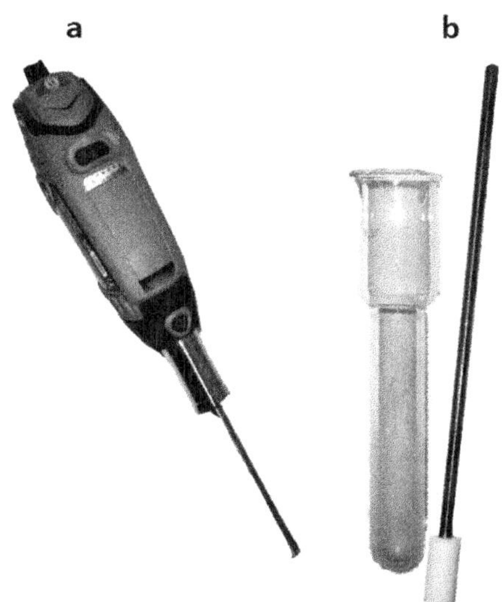

Fig. 2: Scheme of (a) high-speed rotor blade –Omni tissue homogenizer and (b) Wheaton Potter Elvejhem tissue grinder.

Protein solubilisation

After cell disruption, proteins must be solubilised. The main objective of sample solubilisation is to break all non-covalent protein bonds to prevent protein aggregation or precipitation and obtain a soluble sample for 2DE separation (Gorg et al. 2004; Ahmed 2009). Typically solubilisation solutions contain chaotropes, non-ionic, and/or zwitterionic detergents (e.g., CHAPS, Triton X-100), reducing agents, buffers, salts, and carrier ampholytes (Rabilloud 2009). They must be chosen taking into consideration the electrophoretic technique to be used and their compatibility with the samples. The most common buffers used for sample solubilisation contain urea and/or thiourea, which are efficient in disrupting hydrogen bonds when combined with adequate detergents resulting in protein unfolding and denaturation. Reducing agents, such as DTT, are also used for reduction and to prevent reoxidation of disulphide bonds during sample preparation (Gorg et al. 2004; Rabilloud 1996, 1998).

It is important to note that solubilising agents must not alter the original pI of the proteins prior to the IEF, which excludes the use of strong ionic detergents (e.g., SDS). However, low amounts up to (0.03% w/v) of ionic detergents can be used, provided that conditions favouring the exchange of SDS (sodium dodecyl sulfate) for other non-ionic detergents are used in IEF (Bodzon-Kulakowska et al. 2007). This ensures the removal of bound SDS from the proteins, but this also means that the benefits of SDS are lost for the IEF dimension. However, the use of SDS has often been recommended as a way to ensure a complete initial solubilisation before IEF (ReadyStrip IPG Strip Manual, Biorad).

Extracts containing proteins should be centrifuged (60 min, 40,000 × g, 15°C) to remove any insoluble material, as solid particles may obstruct the pores of the gel. The resulting protein extracts can be used immediately or stored in aliquots at –70°C for up to several months until further analysis. However, samples must not be exposed to repeated thawing and freezing (ReadyStrip IPG™ Strip Manual, Biorad).

Fractionation and protein depletion

Eukaryotic cells show a high dynamic range and diversity of expressed proteins. As such it is desirable to perform a pre-fractionation step to reduce the complexity of a protein sample and/or to promote enrichment of certain proteins (Bodzon-Kulakowska et al. 2007). To carry out the pre-fractionation step, samples can be high-speed centrifuged in a sucrose gradient to separate cell organelles and/or compartments. Alternative methods are based on the precipitation of proteins (e.g., TCA/acetone), electrophoresis in the liquid phase, IEF in Sephadex gels, chromatography and/or affinity purification of protein complexes, sequential extraction procedures with increasingly powerful solubilising buffers (usually aqueous buffers), organic solvents such as ethanol or chloroform/methanol, and detergent-based extraction solutions. However, cross-contamination between the individual fractions may be a problem, especially with the latter procedure (Bodzon-Kulakowska et al. 2007). Depletion removes abundant proteins, reducing sample complexity and consequently increasing the representation of less abundant ones (Bodzon-Kulakowska et al. 2007). The depletion method intends to remove albumin/IgG from samples, usually employing commercial kits (e.g., Aurum serum mini-kit-BioRad; Proteoprep®-Sigma; Pierce™ Albumin/IgG Removal Kit-Themo Fisher or ACN method). Depletion can also be used for reducing dynamic ranges by decreasing the amount of high-abundance proteins and the enrichment of less abundant proteins, which are not detected through traditional methods (e.g., ReadyPrepTM or ProteoMiner-BioRad; Enrichment kits-Genotech or by DTT method).

Protein quantification

The most common method used for protein quantification is colorimetric assays, in which the presence of protein causes a shift in colour that is measured spectrophotometrically. In all these assays, standard curves are constructed with a series of diluted concentrations of a known protein (e.g., BSA). Protein quantification is a common procedure when processing samples prior to electrophoretic separation are used to estimate the amount of protein needed to load a gel. Proteins are commonly quantified by classical methods such as BCA and Lowry and Bradford assays (Noble and Bailey 2009). This quantification is needed to ensure that the amount of protein is adequate for the gel size and staining method and facilitates comparison among similar samples when performing image-based analysis.

The Bradford assay (range 1–50 μg) was first described by Bradford (1976) and is used to determine the concentration of proteins in a solution. The method is based on the binding of an acidic solution of Coomassie dye (Brilliant Blue G-250) to the proteins in the solution. The protein-dye complex binds to aromatic amino acids, causing a metachromatic shift from 465 to 595 nm; the amount of absorption is proportional to the amount of protein present. The method is suitable for microplates and standard assays. Moreover, Bradford Reagent is compatible with reducing agents, which are often used to stabilise proteins in solutions. However, Bradford Reagent is only compatible with low concentrations of detergents. If the protein sample has detergents present in the buffer the BCA protein determination procedure should be used (Noble and Bailey 2009).

The Lowry assay (range 5–100 μg) is used for soluble protein quantification. It is based on a two-step chemical reaction; the first step involves the reduction of copper (Cu^{2+} to Cu^{+}) by proteins in alkaline solutions; this is followed by the enhancement stage where the reduction of the *Folin–Ciocalteu* reagent (phosphomolybdate and phosphotungstate) produces a purple colour with an absorbance maximum of 650–750 nm (Lowry et al. 1951).

The BCA (bicinchinonic acid) assay (range 0.2–50 μg), first described by Smith et al. (1985), is similar to the Lowry procedure, as both are based on the alkaline reduction of the cupric ion to the cuprous ion (Cu^{2+} to Cu^{+}) using aromatic amino acids. The amount of reduction is proportional to the amount of protein present. The BCA reaction produces an intense purple colour that can be read at 562 nm. The BCA assay is more sensitive than the Lowry assay and shows less variability than the Bradford assay.

Isoelectric focusing and second dimension electrophoresis

In general, the first dimension electrophoresis of 2DE is based on a protein's isoelectric point using IGP strips that can be selected according to pH gradient. The second dimension of 2DE is performed in a standard SDS-PAGE and proteins are separated based on molecular weight (Gorg et al. 2009). A brief description of a typical protocol used in our laboratory and based on several sources such as Gorg et al. (2004, 2009), Li and Franz (2014), Link (1999), and ReadyStrip IPG Strip Manual (Biorad, USA) follows:

1. Rehydration overnight (7 M urea, 2M thiourea, 0.5% 2/v CHAPS, 0.2% v/v IPG buffer, 0.002% bromophenol blue, 10 mM DTT.
2. IEF run.
3. Equilibration buffer for protein reduction (6 M urea, 75 mMTris-HCl, 20% v/v glycerol, 2% w/v SDS—sodium dodecyl sulfate, 0.002% bromophenol blue, 2% w/v DTT) for 15 min with continuous shaking and then equilibration buffer II (6 M urea, 75 mMTris-HCl, 20% v/v glycerol, 2% w/v SDS, 0.002% bromophenol blue, 2,5% w/v IAA—iodoacetamide) for 15 min with continuous shaking.
4. Afterwards, IPG strips are placed on top of 12.5% polyacrylamide gels and were covered with an agarose sealing solution (0.5% w/v agarose and 0.002% bromophenol blue in running buffer—25 mMTris base, 192 mM glycine, 0.1% SDS).
5. Gels are run in a IEF instrumentation (e.g., Mini-Protean® 3 Cell, Bio-Rad) at 200 V for 45 min and are then stained for 48 h with a solution of colloidal Coomassie Blue G-250 (0.12% w/v Coomassie G-250, 10% w/v ammonium sulphate, 10% w/v orthophosphoric acid, 20% methanol). Following, gels are destained with milli-Q water in several washes.

Prior to 2D electrophoresis, proteins from the samples must be denatured, disaggregated, reduced, and solubilised to achieve disruption of the molecular interactions (Capelo et al. 2009). These procedures are designed to prepare samples for 2D electrophoresis, to improve the quality of the 2DE gels and to reduce streaking, background staining, and other artefacts caused by contaminants (Gorg et al. 2009).

The IGP strip (Fig. 3) must be selected according to the pH range required for optimum separation of the proteins. The strips must then be rehydrated before the IEF run using one of two basic methods: rehydration without the sample or application of the sample during rehydration (Fig. 4). Rehydration with protein sample minimises solubility problems and allows for larger sample loads (Rabilloud et al. 1997). Note that the rehydration buffer can be prepared fresh before use or alternatively stored in aliquots at –70°C until further use. However DTT should be added to the buffer just at the moment of using.

For rehydration, a common protocol performed in our laboratory is as follows:

1. In rehydration with sample—strips are rehydrated in buffer containing sample (sample in-gel rehydration). This method allows loading larger amounts of sample (up to 1 mg), while preventing sample precipitation. The sample is normally prepared in a buffer that contains: 8–9.8 M Urea, 1–4% CHAPS+15–100 mM DTT (or 2 mM Tributylphosphine), 0–0.2% (w/v) carrier ampholytes; 0.001% Orange G or Bromophenol Blue. The IPG strips are placed gel side down onto the sample and care is taken to not trap air bubbles beneath the IPG strip. Then, strips are covered with mineral oil to prevent evaporation through rehydration process and the rehydration tray is covered with a plastic lid. Strips are incubated overnight for rehydration and the protein sample is loaded (ReadyStrip IPG Strip Manual, Biorad).
2. In rehydration without the sample, the sample is applied to the IGP strips after rehydration just prior to focusing. The IGP strips are rehydrated in a rehydration buffer (8–9.8 M Urea, 0,5% CHAPS+10 mM DTT (or 2 mM Tributylphosphine), 0–0.2% (w/v) Bio-lytes; 0.001% Orange G or Bromophenol Blue) without the sample. The rehydrated strips are then placed in the focusing tray and covered

Fig. 3: Scheme of IPG strips.

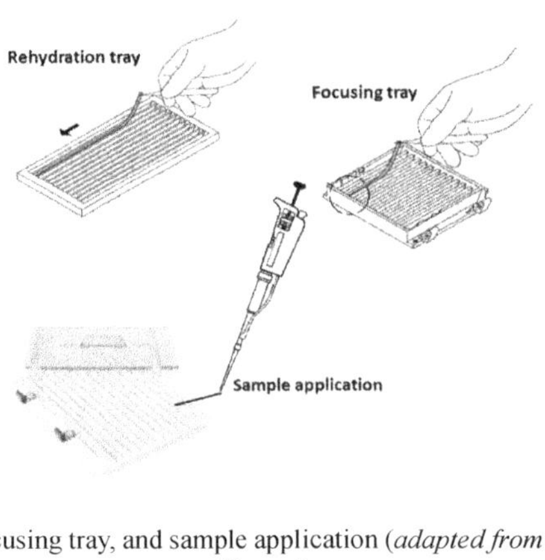

Fig. 4: Scheme of the rehydration, focusing tray, and sample application (*adapted from* Protean IEF Cell Instruction Manual, Bio-rad).

with mineral oil. Then samples are added to the sample loading wells on the channel in the focusing tray. The ideal buffer composition and protein load depends on the sample and can be optimized. Afterwards, the IGP strips are then laid gel side-down, into the channels avoiding trapping air bubbles and covered with mineral oil, to prevent drying, and rehydrated overnight at approximately 20°C, lower temperatures > 10°C should be avoided to prevent urea crystallization on the IPG gel (ReadyStrip IPG Strip Manual, Biorad).

Following rehydration, but prior to IEF, the rehydrated IPG trips are rinsed with distilled water for a few seconds and then blotted between two sheets of moist filter paper to remove excess rehydration buffer to prevent urea crystalization on the gel surface. It is also important that the acidic ends of the IPG gel strips face towards the anode. Electrode paper strips (cut from 1 mm thick filter paper) are then soaked with deionized water, blotted with filter paper to remove excess liquid, and placed on top of the aligned IPG gel strips at the cathodic and anodic ends (ReadyStrip IPG Strip Manual, Biorad). IPG strips rehydrated with sample solution are directly placed onto the apparatus cooling plate. The IGP strips can be stored at –70°C after the run by using a rehydration tray and lid can be used to store the IGP strips (Gorg et al. 2009).

When performing the IEF run, the initial voltage should be limited to 150 V for 60 min to allow maximal sample entry and then it can be progressively increased until 3500 V is attained. Current and power settings should be limited to 0.05 mA and 0.2 W per IPG strip, respectively. After sample entry, the filter papers beneath the anode and cathode should be replaced with new ones. This is because salt contaminants have moved through the gel and have now collected in the electrode papers. For higher salt concentrations, it is recommended to change the filter paper strips more frequently (ReadyStrip IPG Strip Manual, Biorad).

As an example, some typical running conditions are given in Table 1 for Protean IEF (Bio-Rad).

Equilibration and second dimension

After running the first dimension (IEF) and before performing the second dimension, the first equilibration step is achieved by equilibrating IPG strips in a equilibration buffer (6 M urea, 0.375 M Tris-HCl, pH 8.8, 2% SDS, 2% (w/v) DTT), for reduction of sulfhydryl. The solution is discarded and a second equilibration buffer is added (6 M urea, 0.375 M Tris-HCl, pH 8.8, 2% SDS, 20% glycerol, 2.5% (w/v) iodoacetamide) for alkylation of the reduced sulfhydryl groups (Gorg et al. 2009). After equilibration, IPG strips are blotted along the edge using filter paper for 1 min to remove excess liquid before application to horizontal or vertical second dimension SDS-PAGE gels.

Then for loading the equilibrated IPG strip, the gel cassettes are placed in a vertical position and filled with 2–3 ml of hot (75°C) agarose solution (0.5% agarose in electrode buffer (25 mMTris, 192 mM glycine, 0.1% w/v SDS, and 0.03% w/v bromophenol blue)). Then, using forceps, the IPG strip is carefully inserted between the glass plates with a spatula and brought in close contact with the upper edge of the SDS gel. A filter paper (cut square), soaked with molecular marker protein solution (5–7 μl), can also be inserted. Then allow the agarose to solidify and gel run through the second dimension (usual settings: 200 V, 150 mA) as shown in Fig. 5.

Afterwards, several methods can be used for protein imaging and analysis involving stain or stain-free techniques in combination with specialized imaging devices and respective software (López 2007).

Table 1: Focusing conditions for Strips (pH 3–10, 4–7, 5–8).

	Start Voltage	End Voltage	Volt-Hours	Ramp	Temperature
7 cm	0 V	4,000 V	8–10,000 V-hr	Rapid	20°C
11 cm	0 V	8,000 V	20–35,000 V-hr	Rapid	20°C
17 and 18 cm	0 V	10,000 V	40–60,000 V-hr	Rapid	20°C
24 cm	0 V	10,000 V	60–80,000 V-hr	Rapid	20°C

(*In* ReadyStrip IPG Strip Manual, Bio-rad)

Staining, imaging and protein quantitation

Following 2DE, proteins are revealed through in-gel staining (Fig. 6). The most common staining methods used for 2DE gels are Coomassie staining, silver staining, negative staining with metal cations, fluorescence and radioactive isotope staining using autoradiography, fluorography or phosphorimaging (Granvogl et al. 2007; Shevchenko et al. 2006). Of all the stains available, Coomassie shows the broadest spectrum of proteins, while silver staining is very sensitive but has a limited dynamic range and does not react with all proteins evenly (see www.bio-rad.com and www.thermofisher.com for more staining details).

After running a SDS-PAGE the gels are fixed in ethanol/acetic acid/water for several hours. The gels can then be stained with Coomassie (0.12% w/v Coomassie Brilliant blue G-250 in 20% methanol, 10% w/v ammonium sulphate, 10% w/v orthophosphoric acid), followed by destaining with subsequent washes in water (López 2007; Westermeyer and Marouga 2005). However, there are also stain-free technologies available which enable the visualisation of gels without a staining step, allowing visualisation to be done within five minutes with the same sensitivity of Coomassie.

After staining, 2DE gels are digitised using a documentation imaging system for evaluating the complex 2D gels (Westermeyer and Marouga 2005). The image analysis software obtains quantitative and qualitative data, followed by image comparisons using computer-aided image analysis programs (e.g., PDQuest-BioRad; Samespots-Progenesis) as shown in Fig. 6.

Spot excision

The analysis of protein spots by mass spectrometry requires careful excision and removal of 2D gel either using an automated spot cutter or manually (Shevchenko et al. 2006). Differentially expressed proteins identified by image analysis software are marked in a printed image of the 2D gel and the spots can then be

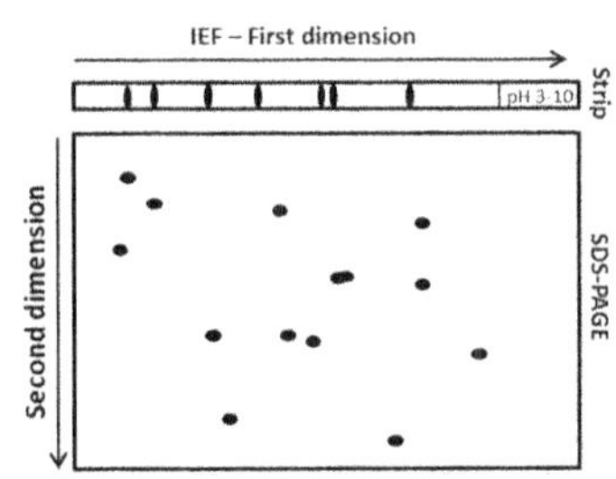

Fig. 5: Scheme representing the first dimension (IEF) followed by the second dimension.

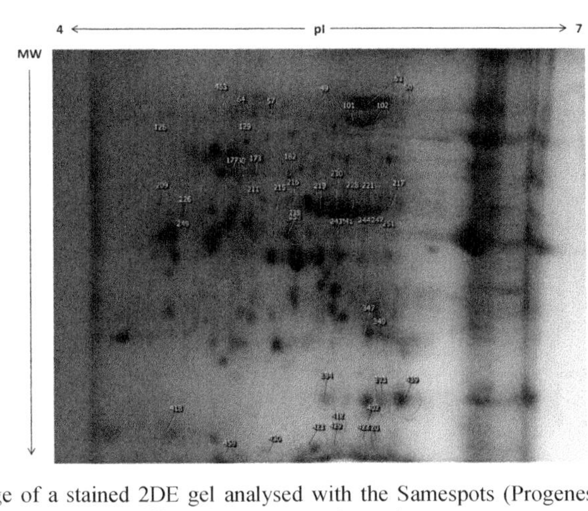

Fig. 6: Representative image of a stained 2DE gel analysed with the Samespots (Progenesis) program, highlighting the differentially expressed protein spots.

manually cut from the gel or excised using an automated spot cutter (e.g., ExQuest spot cutter—BioRad). Manual excision can be carried out using a scalpel, a razor blade or by cutting pipette tips to the desired diameter. The gel pieces are then transferred to clean microcentrifuge tubes and immediately processed or stored (–80°C).

Protein digestion

The identification of proteins is usually achieved by in-gel digestion of protein spots followed by MS analysis of peptides extracted from gel (Capelo et al. 2009; Li and Franz 2014). Thus, according to these authors while 2D gels are previously reduced and alkylated, for 1DE gels, proteins in excised gel pieces are first reduced and alkylated to minimize disulphide cross-linking and cysteine oxidation. The reduction can be carried out by adding 0.2 ml DTT (10 mM dithiothreitol) in 100 mM NH_4HCO_3, pH 8 (make stock solutions fresh) to the gel piece, followed by incubation at 60°C with agitation for 30 min. The reducing solution is removed and replaced with 0.2 ml of 55 mM iodoacetamide in 100 mM NH_4HCO_3 (pH 8), and incubated at room temperature in the dark for 30 min. The alkylating solution is removed and the gel piece rinsed for 20 min in 0.5 ml 100 mM NH_4HCO_3.

Then gel is dehydrated with increasing volumes of organic solvent. The NH_4HCO_3 rinse solution is removed and replaced with 0.5 ml 50% acetonitrile, 100 mM NH_4HCO_3 (pH 8), and the gel piece is incubated for 20 min at room temperature. After removing the 50% acetonitrile buffer solution, gel pieces are transferred to a clean 0.65 ml tube. If the gel pieces are long and skinny, they are cut into 2–4 pieces using a clean razor blade, so that they fit tightly into the bottom of the digestion tube. Dehydration is carried out by adding 50 ml of 100% acetonitrile to the gel piece followed by incubation for 15 min at room temperature. The gel piece should shrink noticeably and become opaque. Excess acetonitrile is removed and the sample then dried completely in a Speedvac centrifuge (10 min). The transfer tubes should be low peptide binding, particularly if low amounts of protein are analyzed. Many brands of polypropylene tubes work well (e.g., Titertubes (BioRad), nonsilanized PCR tubes (Midwest Scientific, Sarstedt), and Optimum Bestubes (Life Science Products)). Alternatively, 96-well plates that are low peptide binding can be used for high throughput protocols. Another step incorporates protease into the gel, normally trypsin, since it is small and relatively easy to diffuse into the gel pieces, and it produces positively charged peptides favourable for mass spectrometry analysis. Then, concentrated formic acid or trifluoroacetic acid (TFA) is added to a final concentration of 5% (v/v) to acidify the digestion mixture prior to extraction. Subsequently the gel supernatant must be sonicated (30 min) using an ultra-sonic bath, and transferred to a new tube. For highly abundant proteins, the supernatant can be analysed directly by MALDI- or LC-MS. While for low abundance proteins a concentration step is required by evaporation in a Speedvac system; for optimal recovery, the sample should not be reduced to dryness. Additionally, some protocols suggest further extraction of the gel piece by adding 50–100 μl of 1% TFA/60% acetonitrile/40% water. Depending on the ionization technique to be used (e.g., MALDI, LC-ESI, or nanospray) the protocol steps should be optimized and tested before using research samples.

Mass Spectrometry analysis

For mass spectrometry analysis, peptide digests are mixed with matrix, and spotted onto gold-plated or stainless steel target plates. Typically α-cyano 4-hydroxy cinnamic acid (α-CHCA) or 2,5-dihydroxybenzoic acid (DHB) are efficient matrices for peptide analysis using MALDI-TOF or MALDI-quadrupole-TOF instruments. The sample is mixed with the matrix (1:1), placed on target plates and let the drop dry (Capelo et al 2009; Santos et al. 2010).

Protein identification can be accomplished using a variety of approaches including, peptide mass fingerprinting (PMF) using matrix-assisted laser desorption ionization with time of flight mass spectrometry (MALDI—TOF MS), MALDI tandem MS (MALDI—TOF—TOF MS) or liquid chromatography (LC) tandem MS using reversed-phase chromatography coupled online to a mass spectrometer via an electrospray ionization source (ESI). In the first approach, proteins are identified by comparing the peptide masses against a protein sequence database. In the latter two approaches, proteins are identified using peptide masses and their MS/MS fragmentation patterns to search the protein database (Reinders et al. 2004). Once candidate proteins are identified, the spectra are analysed again for peaks corresponding to the predicted peptides. Statistical scoring of the top candidate proteins, and the difference between the scores of the top and next best candidates, provides an important indicator of accuracy in protein identification (Mann et al. 2001).

Recently, MALDI interfaces for ion trap, hybrid quadrupole-TOF or TOF-TOF mass analyzers have made MALDI an accessible and easy method for analysing in-gel digests. However, ESI-MS provides a greater coverage than MALDI and may be advantageous for mass fingerprinting, when performed on instruments with high mass accuracy (e.g., quadrupole TOF) (Yates et al. 2009).

Database search

The existence of bioinformatic tools for protein analysis freely available on the web and accessible to the public has been essential for the successful analysis of MS and MS/MS data derived from in-gel digests (Kumar and Mann 2009). Some examples of the search engines widely used to assist in the analysis of peptide mass fingerprints, peptide sequencing and to account for post-translational modifications are Protein Prospector (http://prospector.ucsf.edu), Profound (http://prowl.rockefeller.edu), and Peptide Search (http://www.narrador.embl-heidelberg.de). These programs are accessible and intuitive to use and frequently have tutorial pages available. There are other commercial programs such as Sequest (Thermo Finnigan) and Sonar (Genomic Solutions). However, the most popular search engine for protein identification is Mascot (http://www.matrixscience.com), which is available either under licensing or on a public website with a limited number of data files (Resing and Ahn 2004).

Proteomics in Marine Animals Under Climate Changes: What do We Know?

Most studies on environmental proteomics follow the typical proteomic workflow (see previous sections). However, the majority are focused on the effects of pollutants in the organism's proteome (López-Barea and Gómez-Ariza 2006), while the effects of ocean acidification and sea warming are less well studied. Some of the most representative reviews on proteomics and climate change effects were carried out by Tomanek (2010, 2012). We can see from the studies currently available on the effects of sea warming and ocean acidification that the results on organisms' responses seem to share some common changes at the proteome level. Most of the proteins identified are related to energy metabolism, cytoskeletal function, and oxidative stress. However, sensitivity to the shift of these abiotic variables varies among species, life stages, and type of tissue analyzed. As such, interpretations of proteomic studies can be very complex and we must be cautious regarding the conclusions.

To date proteomic studies show that molecular stress responses and differentially expressed proteins seem to converge toward a common set of stress-induced proteins common to a wide range of taxa. These include molecular chaperones that stabilise denaturing proteins during cellular stress, proteases that regulate protein turnover, and proteins that repair DNA and RNA damage and that are involved in fatty acid metabolism. Moreover, proteins involved in energy metabolism are also represented, as are proteins involved in redox regulation, i.e., emphasising the detoxification of reactive oxygen species and the production of reducing equivalents such as NADPH (Kultz 2005; Petrak et al. 2008). For instance, Tomanek et al. (2011) studied the proteomic response to elevated pCO_2 in the mantle of oysters (*Crassostrea virginica*) and found that two main functional categories of proteins were upregulated in response to hypercapnia: those associated with the cytoskeleton (e.g., diverse actin isoforms) and those associated with oxidative stress (e.g., superoxide dismutase, peroxiredoxins and thioredoxin-related nucleoredoxin), suggesting that reactive oxygen species (ROS) induced oxidative stress and caused alterations in the cytoskeleton. Moreover, elements from the G-protein signalling processes were changed in response to stress. In addition, other proteins such as mitochondrial malate dehydrogenase (Krebs cycle enzyme) and proteins related to protein synthesis (40S ribosomal) and degradation (proteasome type-3) were also found to be upregulated (Fig. 7a).

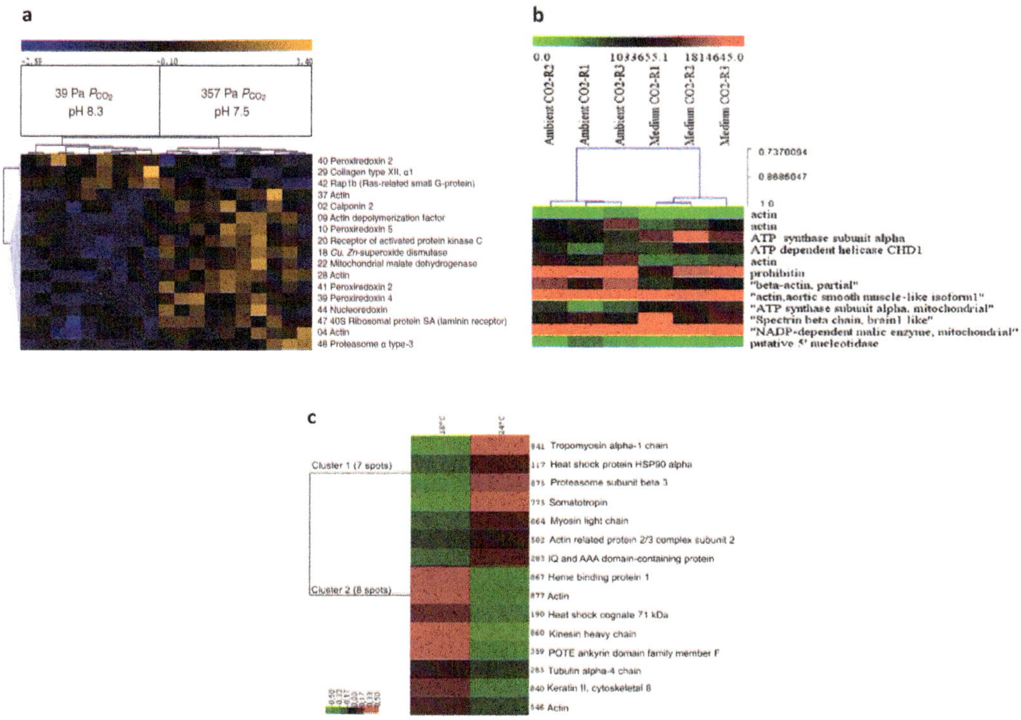

Fig. 7: Representative cluster analysis of differentially expressed proteins in (a) *C. virginica* (Tomanek et al. 2011); (b) Atlantic herring (*Clupeaharengus*) larvae (Maneja et al. 2014) both exposed to elevated pCO_2; (c) Sea bream (*Sparusaurata*) larvae exposed to heat stress (Madeira et al. 2016b).

In another interesting study from Tomanek and Zuzow (2010), proteomic changes in the Mediterranean blue mussel, *Mytilus galloprovincialis* (an invasive species in California), and in the native congener, *Mytilus trossulus*, were investigated by comparing proteome responses to heat stress. The authors found that levels of Hsp70 (Hsp 70) isoforms and Hsp families were higher in response to heat stress in both congeners. Although the general alteration pattern was similar in both congeners, interspecific differences were also found in the Hsp isoforms. For example, three Hsp70 isoforms were found to be upregulated at 32°C in *M. trossulus* but not in *M. galloprovincialis*. Thus, *M. trossulus* changed at a lower temperature than the warm-adapted *M. galloprovincialis*. In effect, according to the authors, the principal role that thermal and oxidative damage to proteins plays in heat stress is further displayed by the changes detected in the levels of proteins involved in proteolysis, with major differences seen between the species. Four of the five proteasome isoforms identified in *M. galloprovincialis* showed significantly lower levels during heat stress, while proteasomes showed higher levels during heat stress in *M. trossulus*. Additionally, the authors presented several possible explanations for a strategy that temporarily reduces ATP consumption by the proteasomes, thus providing energy for the higher ATP-consuming chaperoning activity by Hsp70 and other molecular chaperones. Moreover, several proteins related to energy metabolism (e.g., pyrophosphatase, nucleoside diphosphate kinase) were lower in *M. galloprovincialis* at 28°C or 32°C and arginine kinase (suggested to increase invertebrate ability to cope with environmental stress such as hypoxia and acidosis) showed lower levels in *M. trossulus* at 28°C and 32°C compared to 24°C. The authors hypothesised that the decrease of some proteins in *M. galloprovinicialis* was related to the mussel's metabolism in response to heat stress, while other proteins identified suggest a role of PTMs (Post Translational Modifications), possibly phosphorylation.

Acute heat stress also induced changes in the quantities of cytoskeletal elements (e.g., tubulin, actin) in *M. galloprovincialis*, while only one cytoskeletal protein was identified in *M. trossulus*. However, chaperones, signalling (e.g., Ras-related small G-proteins) energy metabolism and oxidative stress proteins were also identified in these studies. The same authors (Tomanek et al. 2012), using the same congener species but exposed to hyposaline stress, went on to show changes in the abundance of proteins involved in protein chaperoning, vesicle transport, cytoskeletal adjustments by actin-regulatory proteins, energy metabolism, and oxidative stress. Similar results were also obtained by Fields et al. (2012), where *M. galloprovincialis* and *M. trossulus* exposed to cold and heat stress showed changes in cytoskeletal and energy metabolism proteins, which were differentially expressed in both species. The authors proposed that the Mediterranean blue mussel showed a stronger response to cold stress possibly because it is more adapted to a warm environment.

In a different study, Fields et al. (2012) examined a hierarchical clustering of proteins from *Mytilus* sp., showing altered protein expression profiles (Fig. 7b) that revealed that *M. trossulus* acclimated to 7 and 13°C have roughly the same protein expression patterns (i.e., only two major protein clusters differentiate 7 and 13°C together from 19°C). Similar results were found by Dineshram et al. (2012) in oyster larvae (*Crassostreagigas*) exposed to increases in water CO_2 (OA), revealing changes in proteins belonging to the energy, calcium, and cellular metabolism, as well as in cytoskeleton and respiration related proteins. Moreover, a stress related protein (annexin), similar to Hsps, was upregulated in response to OA treated larvae. The authors argued that the lack of other Hsp identified proteins was the result of an insufficient abundance to permit identification; however, they may also have been differentially expressed. Furthermore, energy metabolism (cytochrome c oxidase), calcium metabolism (calmodulin and troponin C), cell metabolism (nucleoside diphosphate kinase) and cytoskeleton proteins (beta-tubilin, tekin) were found to be downregulated. In fact, the authors suggest that the global protein expression pattern of oyster larvae exposed to high CO_2 shows a reduction of protein expression associated with this massive downregulation and larval metabolic depression, possibly due to the organism's strategies for tolerating OA stress. Similarly, in larval sea urchin the expression of many genes in the biomineralisation, metabolic, and apoptotic pathways were downregulated in response to OA (Todgham and Hofmann 2009; O'Donnell et al. 2010). In addition, sea squirts (Ciona sp.) exposed to acute heat stress showed differentially expressed proteins, such as molecular chaperones, extracellular matrix proteins, calcium-binding proteins, cytoskeletal proteins, and energy metabolism proteins, despite some differences found between genus responses to heat stress (Serafini et al. 2011).

The proteome response to heat stress in the Antarctic clam (*Laternulaelliptica*) was studied by Truebano et al. (2013). Effects were found in the global protein expression patterns, identifying cytoskeletal (tubulin), molecular chaperone (TCP-1) and enzymes, enolase and aldehyde dehydrogenase, proteins related to redox regulation. Another invertebrate species studied was the tubeworm *Hydroideselegans*; however, no changes in the proteome were found when these creatures were exposed to heat stress (Mukherjee et al. 2013). Yet, under OA stress this invertebrate showed a high degree of plasticity. For example, metabolic and calcification-related proteins, as well as several stress proteins, were downregulated. As a consequence metamorphosis was affected, due to a disruption in ATP production (affecting the energy for cells). Proteins related to the cytoskeleton were also downregulated (Mukherjee et al. 2013). Moreover, barnacle larvae (*Balanusamphitrite*) showed a low percentage of changed proteins as a result of exposure to elevated pCO_2 and appear to be resistant to stress caused by OA (Wong et al. 2011). However, some energy metabolism, respiration, and molecular chaperone proteins were differentially regulated. In fact, following exposure to elevated pCO_2 some changes were detected in the larvae proteome, for example, hemoglobin beta chain-like proteins, associated with respiration processes, and protein elongation factor 2 (EF2), associated with energy metabolism and essential in polypeptide elongation during protein production, were upregulated. In contrast, the molecular chaperone (Hsp83, Hsp90) and cathepsin L-like protein (related with protein degradation) were downregulated. This may represent a strategy used by barnacle larvae to tolerate OA, as they can adjust their protein expression patterns in the short-term (Wong et al. 2011). According to Mukherjee et al. (2013), calcifying marine invertebrates with complex life cycles are particularly vulnerable to climate change stresses, as they suffer an abrupt ontogenetic shift during larval metamorphosis. The same author states that although the comprehension of larval responses to climate change stress is evolving, the plasticity shown by the proteome associated to compensatory responses to these stresses is largely unknown. Moreover, it has been argued that proteome plasticity to stress is related to PTMs of proteins. Despite the fact that the majority of proteomic studies have focused on the most abundant proteins, recent studies have generated new hypotheses regarding organisms' responses to environmental stresses, for instance the mode as PTMs may regulate a cell's response to oxidative stress (see Sheehan 2007; Tomanek 2011).

In research carried out by Fan et al. (2013), white shrimp (*Litopenaeusvannamei*) exposed to cold stress showed changes in proteins related to energy metabolism. For instance, enzymes (cystathionase, glyceraldehyde 3-phosphate dehydrogenase, and glyoxalase 1), actin depolymerizing factor, and an immune related protein (oncoprotein nm23) were upregulated and hemocyanin, hemocytetransglutaminase, and an enzyme from the pentose phosphate pathway (transketolase) were downregulated. In addition, according to Doney et al. (2009), marine organisms respond differently (as some species are more robust than others) to OA stress due to their use of diverse strategies to compensate for OA stress at the level of protein expression.

There are few studies on the effects of climate change on fish proteome. Nonetheless, Ibarz et al. (2010) exposed sea bream (*Sparusaurata*) to cold stress, showing differentially expressed proteins. For example, actin, enzymes for amino acid metabolism, and oxidative stress enzymes (e.g., glutathione-S-transferase and catalase) were identified. In addition, proteins with protective roles were upregulated (peroxiredoxin, thioredoxin, and lysozyme), but the upregulation of proteases, proteasome activator protein, and trypsinogen-like protein indicates an increase in proteolysis as well. For the authors of this study, this was indicative of oxidative damage in hepatocytes caused by exposure to cold stress. Recent proteomic studies from Madeira et al. (2016b) in *S. aurata* larvae exposed to heat stress, revealed differences in the expression of proteins related to molecular chaperones (Hsc70 and Hsp90), the cytoskeleton, the proteasome system and porphyrin metabolism (Fig. 7c), suggesting that oxygen transport is compromised. These findings are supported by the down-regulation of heme-binding protein1 protein (Fig. 7c), which binds porphyrin. However, no alterations were found regarding energetic metabolism related proteins.

In a different study, it was shown that the proteome of Atlantic herring (*Clupeaharengus*) larvae is resistant to elevated pCO_2, with just a small number of proteins being altered between the control and treatment, suggesting that larvae can cope with levels of CO_2 predicted for the near future without significant proteome-wide changes (Maneja et al. 2014). On the other hand, Atlantic cod exposed to OA (elevated pCO_2 levels) showed a two-fold increase in the expression of ATPase (Melzner et al. 2009). A similar resistance to increased pCO_2 was also found in a marine coccolithophore (*Emilianiahuxleyi*), with no significant changes detected in the proteome.

Although several studies have been carried out there is still a huge lack of information on the proteomic responses of the marine biota to sea warming and OA. This is due to the large biodiversity found in marine ecosystems and also related to the scarce information available in databases. However, as databases become more complete and are combined with advances in proteomic techniques and instrumentation, some developments are expected regarding our understanding of proteome changes in response to environmental stresses.

Conclusions

The increase in the emission of gases to the atmosphere mainly due to anthropogenic activities, are provoking a global warming with still unpredictable consequences for living organisms. Major consequences to the marine ecosystems caused by sea warming and ocean acidification are already occurring and may affect the entire trophic web. Moreover, physiology, geographical distribution, biodiversity, and marine phenology will also be affected. The scientific community also agrees that biodiversity is becoming weak. Even the long-term implications for humans are largely unknown, but it seems certain that tropical diseases can emerge in temperate countries.

Concerning the effects at the proteome level driven by climate changes there is little information and more data must be produced to understand all the variables involved in an organism's response to climate change stresses. Major tools that can be used to elucidate the mechanisms of response at cellular and molecular level are the so called "omics", namely proteomics. In fact, proteomic methodologies can provide useful information on differentially expressed proteins in organisms exposed to climate change stresses. In this sense, the techniques such as 2DE and protein identification by mass spectrometry are the most common techniques used in this type of studies. Furthermore, from the results currently available there seems to be convergence toward a common set of stress-induced proteins in widely diverse species, which includes metabolic, cytoskeletal, molecular chaperones, antioxidant, repairing, and protective proteins suggesting a response to fight stress and minimize injury to cells.

Acknowledgements

Authors would like to thank the following organisation for its support of this work: the "Unidade de Ciências Biomoleculares Aplicadas-UCIBIO" which is financed by national funds from FCT/MEC (UID/Multi/04378/2013) and co-financed by the ERDF under the PT2020 Partnership Agreement (POCI-01-0145-FEDER-007728) and project grants PTDC/MAR/119068/2010 from FCT/MEC. The authors also acknowledge all staff of the Bioscope group.

Keywords: climate change, sea warming, ocean acidification, proteomic approach, effects on marine biota, marine proteomics, differentially expressed proteins

References

Ahmed, F.E. 2009. Sample preparation and fractionation for proteome analysis and cancer biomarker discovery by mass spectrometry. J. Sep. Sci. 32: 771–798.

Bodzon-Kulatowska, A., A. Brerczynska-Krzysik, T. Dylag, A. Drabik, P. Suder, M. Noga, J. Jarzebaska and J. Siberring. 2007. Methods for samples preparation in proteomic research. J. Chromatogr. 849: 1–31.

Bradford, M. 1976. A rapid and sensitive method for the quantification of microgram quantities of protein utilizing the principle of protein-dye binding. Anal. Biochem. 72: 248–254.

Bradley, B.P., E.A. Shrader, D.G. Kimmel and J.C. Meiller. 2002. Protein expression signatures: an application of proteomics. Mar. Environ. Res. 54: 373–377.

Caldeira, K. and M.E. Wickett. 2003. Anthropogenic carbon and ocean pH. Nature 425: 365–365.

Caldeira, K. and M.E. Wickett. 2005. Ocean model predictions of chemistry changes from carbon dioxide emissions to the atmosphere and ocean. J. Geophys. Res. 110: C09S04.

Capelo, J.L., R. Carreira, M. Diniz, L. Fernandes, M. Galesio, C. Lodeiro, H.M. Santos and G. Vale. 2009. Overview on modern approaches to speed up protein identification workflows relying on enzymatic cleavage and mass spectrometry-based techniques. Anal. Chim. Acta 650: 151–159.

Choi, K.M. and M.Y. Lee. 2012. Differential protein expression associated with heat stress in Antarctic microalga. Biochip J. 6: 271–279.

Chora, S., B. McDonagh, D. Sheehan, M. Starita-Geribaldi, M. Roméo and M.J. Bebianno. 2008. Ubiquitination and carbonylation as markers of oxidative-stress in *Ruditapesdecussatus*. Mar. Environ. Res. 66(1): 95–97.

Cravatt, B.F., G.M. Simon and J.R. Yates. 2007. The biological impact of mass-spectrometry–based proteomics. Nature 450: 991–1000.

Crozier, L.G. and J.A. Hutchings. 2014. Plastic and evolutionary responses to climate change in fish. Evol. Appl. 7: 68–87.

Cruz, J., S. Garrido, M.S. Pimentel, R. Rosa, A. Santos and P. Ré. 2013. Reproduction and respiration of a climate change indicator species: effect of temperature and variable food in the copepod *Centropageschierchiae*. J. Plankton Res. 35(5): 1046–1058.

Dilly, G.F., C. Robert Young, W.S. Lane, J. Pangilinan and P.R. Girguis. 2012. Exploring the limit of metazoan thermal tolerance via comparative proteomics: thermally induced changes in protein abundance by two hydrothermal vent polychaetes. Proc. R. Soc. B 279: 3347–56.

Dineshram, R., A. Lane, K. Wang, S. Xiao, Z. Yu, P.Y. Qian and V. Thiyagarajan. 2012. Analysis of Pacific oyster larval proteome and its response to high-CO_2. Mar. Poll. Bull. 64: 2160–2167.

Doney, S.C., V.J. Fabry, R.A. Feely and J.A. Kleypas. 2009. Ocean acidification: the other CO_2 problem. Annu. Rev. Mar. Sci. 1: 169–92.

Doney, S.C., M. Ruckelshaus, J.E. Duffy, J.P. Barry, F. Chan, C.A. English, H.M. Galindo, J.M. Grebmeier, H.B. Hollowed, N. Knowlton, J. Polovina, N.N. Rabalais, W.J. Sydeman and L.D. Talley. 2012. Climate change impacts on marine ecosystems. Annu. Rev. Mar. Sci. 4: 11–37.

Dowd, W.W. 2012. Challenges for biological interpretation of environmental proteomics data in non-model organisms. Integr. Comp. Biol. 52(5): 705–20.

Dupont, S., K. Wilson, M. Obst, H. Sköld, H. Nakano and M.C. Thorndyke. 2007. Marine ecological genomics: when genomics meets marine ecology. Mar. Ecol. Prog. Ser. 332: 257–273.

Fabry, V.J., B.A. Seibel, R.A. Feely and J.C. Orr. 2008. Impacts of ocean acidification on marine fauna and ecosystem processes. ICES J. Mar. Sci. 65: 414–32.

Fan, L., A. Wang and Y. Wu. 2013. Comparative proteomic identification of the hemocyte response to cold stress in white shrimp, *Litopenaeusvannamei*. J. Proteomics 80C: 196–206.

Feder, M.E. and J.C. Walser. 2005. The biological limitations of transcriptomics in elucidating stress and stress responses. J. Evol. Biol. 18: 901–10.

Feely, R.A., C.L. Sabine, K. Lee, W. Berelson, J. Kleypas, V.J. Fabry and F.J. Millero. 2004. Impact of anthropogenic CO_2 on the $CaCO_3$ system in the oceans. Science 305(5682): 362–366.

Feely, R.A., J. Orr, V.J. Fabry, J.A. Kleypas, C.L. Sabine and C. Langdon. 2009. Present and future changes in seawater chemistry due to ocean acidification. *In*: B.J. Mcpherson and E.T. Sundquist [eds.]. Carbon Sequestration and Its Role in the Global Carbon Cycle, American Geophysical Union, Washington, D.C., USA. doi: 10.1029/2005GM000337.

Fields, P.A., M.J. Zuzow and L. Tomanek. 2012. Comparative proteomics of blue mussel (*Mytilus*) congeners to temperature acclimation. J. Exp. Biol. 215: 1106–16.

Gardestrom, J., T. Elfwing, M. Lof, M. Tedengren, J.L. Davenport and J. Davenport. 2007. The effect of thermal stress on protein composition in dogwhelks (*Nucella lapillus*) under normoxic and hyperoxic conditions. Comp. Biochem. Physiol. A Comp. Physiol. 148: 869–75.

Go, Y.M., H. Park, M. Koval, M. Orr, M. Reed, Y. Liang, D. Smith, J. Pohl and D.P. Jones. 2010. A key role for mitochondria in endothelial signaling by plasma cysteine/cystine redox potential. Free Radic. Biol. Med. 48: 275–283.

Gorg, A., C. Obermaier, G. Boguth, A. Harder, B. Scheibe, R. Wildgruber and V. Weiss. 2000. The current state of two-dimensional electrophoresis with immobilized pH gradients. Electrophoresis 21: 1037–1053.

Gorg, A., W. Weiss and M.J. Dunn. 2004. Current two dimensional electrophoresis technology for proteomics. Proteomics 4: 3665–3685.

Gorg, A., A. Klaus, C. Luck, F. Weiland and W. Weiss. 2007. Two-Dimensional Electrophoresis with Immobilized pH Gradients for Proteome Analysis. A Laboratory Manual. TechnischeUniversitätMünchen. http://www.wzw.tum.de/proteomik.

Gorg, A., O. Drews, C. Luck, F. Weiland and W. Weiss. 2009. 2-DE with IPGs. Electrophoresis 30: S122–S132.

Gracey, A.Y. 2007. Interpreting physiological responses to environmental change through gene expression profiling. J. Exp. Biol. 210(9): 1584–1592.

Granvogl, B., M. Plöscher and L.A. Eichacker. 2007. Sample preparation by in-gel digestion for mass spectrometry-based proteomics. Anal. Bioanal. Chem. 389: 991–1002.

Gulcicek, E.E., C.M. Colangelo, W. McMurray, K. Stone, K. Williams, T. Wu, H. Zhao, H. Spratt, A. Kurosky and B. Wu. 2005. Proteomics and the analysis of proteomic data: an overview of current protein-profiling technologies. Curr. Protoc. Bioinformatics 1-40, doi:10.1002/0471250953.bi1301s10.

Hendriks, I.E., C.M. Duarte, Y.S. Olsen, A. Steckbauer, L. Ramajo, T.S. Moore, J.A. Trotter and M. McCulloch. 2015. Biological mechanisms supporting adaptation to ocean acidification in coastal ecosystems. Estuar. Coast. Shelf Sci. 152: A1–A8.

Hofmann, G.E., J.P. Barry, P.J. Edmunds, R.D. Gates, D.A. Hutchins, T. Klinger and M.A. Sewell. 2010. The effect of ocean acidification on calcifying organisms in marine ecosystems: an organism-to-ecosystem perspective. Annu. Rev. Ecol. Evol. Syst. 41: 127–47.

Hönisch, B., A. Ridgwell, D.N. Schmidt, E. Thomas, S.J. Gibbs, A. Sluijs, R. Zeebe, L. Kump, R.C. Martindale, S.E. Greene, W. Kiessling, J. Ries, J.C. Zachos, D.L. Royer, S. Barker, T.M. Marchitto, R. Moyer, C. Pelejero, P. Ziveri, G.L. Foster and B. Williams. 2012. The geological record of ocean acidification. Science 335: 1058–1063.

Ibarz, A., M. Martin-Perez, J. Blasco, D. Bellido, E. de Oliveira and J. Fernandez-Borras. 2010. Gilthead sea bream liver proteome altered at low temperatures by oxidative stress. Proteomics 10: 963–75.

IOC/UNESCO, IMO, FAO, UNDP. 2011. A Blueprint for Ocean and Coastal Sustainability. Paris: IOC/UNESCO, 42 pp.

IPCC. 2007. Climate Change. 2007: The Physical Science Basis. *In*: S. Solomon, D. Qin, M. Manning, Z. Chen, M. Marquis, K.B. Averyt, M. Tignor and H.L. Miller [eds.]. Contribution of Working Group I to the Fourth Assessment Report of the Intergovernmental Panel on Climate Change. Cambridge University Press, Cambridge, United Kingdom and New York, NY, USA, 996 pp.

IPCC. 2013. Climate Change. 2013: The Physical Science Basis. *In*: T.F. Stocker, D. Qin, G.-K. Plattner, M. Tignor, S.K. Allen, J. Boschung, A. Nauels, Y. Xia, V. Bex and P.M. Midgley [eds.]. Contribution of Working Group I to the Fifth Assessment Report of the Intergovernmental Panel on Climate Change. Cambridge University Press, Cambridge, United Kingdom and New York, NY, USA, 1535 pp.

Knigge, T., T. Monsinjon and O.K. Andersen. 2004. Surface-enhanced laser desorption/ionization-time of flight mass spectrometry approach to biomarker discovery in blue mussels (*Mytilus edulis*) exposed to polyaromatic hydrocarbons and heavy metals under field conditions. Proteomics 4: 2722–27.

Kültz, D. 2005. Molecular and evolutionary basis of the cellular stress response. Annu. Rev. Physiol. 67: 225–57.

Kumar, C. and M. Mann. 2009. Bioinformatics analysis of mass spectrometry-based proteomics data sets. FEBS Lett. 583(11): 1703–1712.

Lankadurai, B., E. Nagato and M. Simpson. 2013. Environmental metabolomics: an emerging approach to study organism responses to environmental stressors. Environ. Rev. 21: 180–205.

Li, X. and T. Franz. 2014. Up to date sample preparation of proteins for mass spectrometric Analysis. Arch. Physiol. Biochem. 120(5): 188–91.

Link, A.J. [ed.]. 1999. 2-D Proteome Analysis Protocols. Methods in Molecular Biology 112. Humana Press, Totowa, NJ, USA.

López, J.L., A. Marina, J. Vazquez and G. Alvarez. 2002. A proteomic approach to the study of marine mussels *Mytilus edulis* and *M. galloprovincialis*. Mar. Biol. 141: 217–23.

López, J.L. 2007. Two-dimensional electrophoresis in proteome expression analysis. J. Chromatogr. B 849: 190–202.

López-Barea, J. and J.L. Gómez-Ariza. 2006. Environmental proteomics and metallomics. Proteomics 6: 51–62.

Lowry, O.H., N.J. Rosebrough, A.L. Farr and R.J. Randall. 1951. Protein measurement with the Folin phenol reagent. J. Biol. Chem. 193: 265–275.

Madeira, D., L. Narciso, H.N. Cabral, M.S. Diniz and C. Vinagre. 2012. Thermal tolerance of the crab *Pachygrapsusmarmoratus*: intraspecific differences at a physiological (CTMax) and molecular level (Hsp70). Cell Stress Chaperon. 17: 707–716.

Madeira, D., L. Narciso, H.N. Cabral, M.S. Diniz and C. Vinagre. 2014a. Role of thermal niche in the cellular response to thermal stress: Lipid peroxidation and HSP70 expression in coastal crabs. Ecol. Indic. 36: 601–606.

Madeira, D., C. Vinagre, P.M. Costa and M.S. Diniz. 2014b. Histopathological alterations, physiological limits and molecular responses of a juvenile eurythermal fish (*Sparusaurata*) to thermal stress. Mar. Ecol. Prog. Ser. 505: 253–266.

Madeira, D., V. Mendonça, M. Dias, J. Roma, P.M. Costa, C. Vinagre and M.S. Diniz. 2015. Physiological, cellular and biochemical thermal stress response of intertidal shrimps with different vertical distributions: *Palaemonelegans* and *Palaemon serratus*. Comp. Biochem. Physiol. 183: 107–15.

Madeira, D., C. Vinagre and M.S. Diniz. 2016a. Are fish in hot water? Effects of warming on oxidative stress metabolism in the commercial species *Sparusaurata*. Ecol. Indic. 63: 324–331.

Madeira, D., J.E. Araújo, R. Vitorino, J.L. Capelo, C. Vinagre and M.S. Diniz. 2016b. Ocean warming alters cellular metabolism and induces mortality in fish early life stages: A proteomic approach. Environ. Res. 148: 164–176.

Maneja, R.H., R. Dineshram, V. Thiyagarajan, A.B. Skiftesvik, A.Y. Frommel, C. Clemmesen, A.J. Geffen and H.I. Browman. 2014. The proteome of Atlantic herring (*Clupeaharengus* L.) larvae is resistant to elevated pCO_2. Mar. Poll. Bull. 86: 154–160.

Mann, M., R.C. Hendrickson and A. Pandey. 2001. Analysis of proteins and proteomes by mass spectrometry. Annu. Rev. Biochem. 70: 437–73.

Martínez-Fernandez, M., A.M. Rodrıguez-Pineiro, E. Oliveira, M. Paez de la Cadena and E. Rolan-Alvarez. 2008. Proteomic comparison between two marine snail ecotypes reveals details about the biochemistry of adaptation. J. Proteome Res. 7: 4926–34.

Martínez-Maqueda, D., B. Hernández-Ledesma, L. Amigo, B. Miralles and J.Á. Gómez-Ruiz. 2013. Extraction/Fractionation techniques for proteins and peptides and protein digestion. Chapter 2, pp. 21–50. *In*: F. Toldrá and L.M.L. Nollet [eds.]. Proteomics in Foods: Principles and Applications, Food Microbiology and Food Safety 2, Springer Science, New York, USA.

McLean, L., I.S. Young, M.K. Doherty, D.H. Robertson, A.R. Cossins, A.Y. Gracey, R.J. Beynon and P.D. Whitfield. 2007. Global cooling: cold acclimation and the expression of soluble proteins in carp skeletal muscle. Proteomics 7: 2667–81.

Meehl, G.A., T.F. Stocker, W.D. Collins, P. Friedlingstein, A.T. Gaye, J.M. Gregory, A. Kitoh, R. Knutti, J.M. Murphy, A. Noda, S.C.B. Raper, I.G. Watterson, A.J. Weaver and Z.-C. Zhao. 2007. Global Climate Projections. *In*: S. Solomon, D. Qin, M. Manning, Z. Chen, M. Marquis, K.B. Averyt, M. Tignor and H.L. Miller [eds.]. Climate Change 2007: The Physical Science Basis. Contribution of Working Group I to the Fourth Assessment Report of the Intergovernmental Panel on Climate Change. Cambridge University Press, Cambridge, United Kingdom and New York, NY, USA.

Melzner, F., M.A. Gutowska, M. Hu and M. Stumpp. 2009. Acid-base regulatory capacity and associated proton extrusion mechanisms in marine invertebrates: an overview. Comp. Biochem. Physiol. A 153A: S80–S80.

Mukherjee, J., K. Wong, K. Chandramouli, P. Qian, P. Leung, R. Wu and V. Thiyagarajan. 2013. Proteomic response of the marine invertebrate larvae to acidification and hypoxia at the time of metamorphosis and calcification. J. Exp. Biol. 216: 4580–4589.

Noble, J.E. and M.J.A. Bailey. 2009. Quantitation of protein. Methods Enzymol. 463: 73–95.

O'Donnell, M.J., A.E. Todgham, M.A. Sewell, L.T.M Hammond, K. Ruggiero, N.A. Fangue, M.L. Zippay and G.E. Hofmann. 2010. Ocean acidification alters skeletogenesis and gene expression in larval sea urchins. Mar. Ecol. Prog. Ser. 398: 157–171.

Orr, J.C., V.J. Fabry, O. Aumont, L. Bopp, S. Doney, R. Feely, A. Gnanadesikan, N. Gruber, A. Ishida, F. Joos, R.M. Key, K. Lindsay, E. Maier-Reimer, R. Matear, P. Monfray, A. Mouchet, R.G. Najjar, G.-K. Plattner, K.B. Rodgers, C.L. Sabine, J.L. Sarmiento, R. Schlitzer, R.D. Slater, I.J. Totterdell, M.-F. Weirig, Y. Yamanaka and A. Yool. 2005. Anthropogenic ocean acidification over the twenty-first century and its impact on calcifying organisms. Nature 437: 681–686.

Padilla-Gamino, J.L., M.W. Kelly, T.G. Evans and G.E. Hofmann. 2013. Temperature and CO_2 additively regulate physiology, morphology and genomic responses of larval sea urchins, *Strongylocentrotuspurpuratus*. Proc. R. Soc. B: Biol. Sci. 280(1759): 20130155.

Pearson, P.N. and M.R. Palmer. 2000. Atmospheric carbon dioxide concentrations over the past 60 million years. Nature 406: 695–699.

Petrak, J., R. Ivanek, O. Toman, R. Cmejla, J. Cmejlova, D. Vyoral, J. Zivny and C.D. Vulpe. 2008. Déejà vu in proteomics. A hit parade of repeatedly identified differentially expressed proteins. Proteomics 8(9): 1744–49.

Pineiro, C., B. Canas and M. Carrera. 2010. The role of proteomics in the study of the influence of climate change on seafood products. Food Res. Int. 43: 1791–1802.

Poloczanska, E.S., C.J. Brown, W.J. Sydeman, W. Kiessling, D.S. Schoeman, P.J. Moore, K. Brander, J.F. Bruno, L.B. Buckley, M.T. Burrows, C.M. Duarte, B.S. Halpern, J. Holding, C.V. Kappel, M.I. O'Connor, J.M. Pandolfi, C. Parmesan, F. Schwing, S.A. Thompson and A.J. Richardson. 2013. Global imprint of climate change on marine life. Nature Clim. Change 3(10): 919–925.

Pörtner, H.O., M. Langenbuch and A. Reipschläger. 2004. Biological impact of elevated ocean CO_2 concentrations: lessons from animal physiology and earth history. J. Oceanogr. 60: 705–718.

Protean IEF® Cell Instruction Manual. Catalog Number 165-4000. Bio-Rad laboratories, 47 p. (available at: http://www.bio-rad.com/webroot/web/pdf/lsr/literature/4006164B.pdf).

Rabilloud, T. 1996. Solubilisation of proteins for electrophoretic analyses. Electrophoresis 17: 813–829.

Rabilloud, T., C. Adessi, A. Giraudel and J. Lunardi. 1997. Improvement of the solubilisation of proteins in two-dimensional electrophoresis with immobilised pH gradients. Electrophoresis 18: 307–316.

Rabilloud, T. 1998. Use of thiourea to increase the solubility of membrane proteins in two dimensional electrophoresis. Electrophoresis 19: 758–60.

Rabilloud, T. 2009. Detergents and chaotropes for protein solubilization before two-dimensional electrophoresis. Methods Mol. Biol. 528: 259–267.

ReadyStrip IPG™ Strip Manual. Catalog #163-2099. Bio-Rad, 38 p. (available at: http://www.bio-rad.com/webroot/web/pdf/lsr/literature/4006166G.pdf).

Reinders, J., U. Lewandrowski, J. Moebius, Y. Wagner and A. Sickmann. 2004. Challenges in mass spectrometry-based proteomics. Proteomics 4: 3686–3703.

Resing, K.A. and N. Ahn. 2004. Protein identification by in-gel digestion and mass spectrometry. pp. 163–182. *In*: D.W. Speicher [ed.]. Proteome Analysis, Interpreting the Genome. Elsevier B.V., Amsterdam, The Netherlands.

Riebesell, U., V.J. Fabry, L. Hansson and J.-P. Gattuso [eds.]. 2010. Guide to best practices for ocean acidification research and data reporting. Publications Office of the European Union, Luxembourg. Available at: http://www.epoca-project.eu/index.php/guide-to-best-practices-for-ocean-acidification-research-and-data-reporting.html.

Rosa, R. and B.A. Seibel. 2008. Synergistic effects of climate-related variables suggest future physiological impairment in a top oceanic predator. PNAS 105: 20776–20780.

Rosa, R., M.S. Pimentel, J. Boavida-Portugal, T. Teixeira, K. Trübenbach and M.S. Diniz. 2012. Ocean warming enhances malformations, premature hatching, metabolic suppression and oxidative stress in the early stages of a keystone squid. PLoS One 7(6): 1–11 e38282.

Rosa, R., A.R. Lopes, M. Neves, M. Pimentel, F. Faleiro, M. Baptista, L. Narciso, T. Repolho, R. Calado and M. Diniz. 2014. Ocean's cleaning stations under a changing climate: biological responses of tropical and temperate fish-cleaner shrimps to global warming. Global Change Biol. 20: 3068–3079.

Rosa, R., M. Pimentel, J. Paula, A.R. Lopes, M. Baptista, V.M. Lopes, C. Santos, D. Campos, V.M.F. Almeida-Val, M. Diniz, R. Calado and T. Repolho. 2016. Neuro-oxidative damage and aerobic potential loss of sharks under climate change. Mar. Biol. 163: 119 (in press). doi: 10.1007/s00227-016-2898-7.

Santos, H., D. Glez-Peña, M. Reboiro-Jato, F. Fdez-Riverola, M.S. Diniz, C. Lodeiro and J.L. Capelo. 2010. A novel 18-O inverse labeling-based workflow for accurate bottom-up mass spectrometry quantification of proteins separated by gel electrophoresis. Electrophoresis 31(20): 3363–3507.

Schwerin, S., B. Zeiss, T. Lamkemeyer, R.P. Paul, M. Koch, J. Madlung, C. Fladerer and R. Pirow. 2009. Acclimatory responses of the *Daphnia pulex* proteome to environmental changes. II. Chronic exposure to different temperatures (10 and 20°C) mainly affects protein metabolism. BMC Physiol. 20099: 8.

Serafini, L., J.B. Hann, D. Kultz and L. Tomanek. 2011. The proteomic response of sea squirts (genus Ciona) to acute heat stress: a global perspective on the thermal stability of proteins. Comp. Biochem. Phys. D 6: 322–34.

Sheehan, D. 2007. The potential of proteomics for providing new insights into environmental impacts on human health. Rev. Environ. Health 22(3): 175–194.

Shevchenko, A., A. Tomas, J. Havlis, J.V. Olsen and M. Mann. 2006. In-gel digestion for mass spectrometric characterization of proteins and proteomes. Nat. Protoc. 1(6): 2856–2860.

Smith, P.K., R.I. Krohn, G.T. Hermanson, A.K. Mallia, F.H. Gartner, M.D. Provenzano, E.K. Fujimoto, N.M. Goeke, B.J. Olson and D.C. Klenk. 1985. Measurement of protein using bicinchoninic acid. Anal. Biochem. 150: 76–85.

Somero, G.N. 2010. The physiology of climate change: how potentials for acclimatization and genetic adaptation will determine 'winners' and 'losers'. J. Exp. Biol. 213: 912–920.

Todgham, A.E. and G.E. Hofmann. 2009. Transcriptomic response of sea urchin larvae *Strongylocentrotuspurpuratus* to CO_2-driven seawater acidification. J. Exp. Biol. 212: 2579–2594.

Tomanek, L. 2010. Variation in the heat shock response and its implication for predicting the effect of global climate change on species' biogeographical distribution ranges and metabolic costs. J. Exp. Biol. 213(6): 971–9.

Tomanek, L. and M.J. Zuzow. 2010. The proteomic response of the mussel congeners *Mytilus galloprovincialis* and *M. trossulus* to acute heat stress: implications for thermal tolerance and metabolic costs of thermal stress. J. Exp. Biol. 213: 3559–74.

Tomanek, L. 2011. Environmental proteomics: changes in the proteome of marine organisms in response to environmental stress, pollutants, infection, symbiosis, and development. Annu. Rev. Mar. Sci. 3: 373–399.

Tomanek, L. 2012. Environmental Proteomics of the *Mussel Mytilus*: Implications for tolerance to stress and change in limits of biogeographic ranges in response to climate change. Integr. Comp. Biol. 52(5): 648–664.

Tomanek, L. 2014. Proteomics to study adaptations in marine organisms to environmental stress. J. Proteomics 105: 92–106.

Tomanek, L., M.J. Zuzow, A.V. Ivanina, E. Beniash and I.M. Sokolova. 2011. Proteomic response to elevated PCO2 level in eastern oysters, *Crassostrea virginica*: evidence for oxidative stress. J. Exp. Biol. 214: 1836–1844. doi:10.1242/jeb.055475.

Truebano, M., A.P. Diz, M.A.S. Thorne, M.S. Clark and D.O.F. Skibinski. 2013. Proteome response to heat stress in the Antarctic clam *Laternulaelliptica*. J. Integr. OMICS 3: 34–43.

Trübenbach, K., T. Teixeira, M. Diniz and R. Rosa. 2013. Hypoxia tolerance and antioxidant defense system of juvenile jumbo squids in oxygen minimum zones. Deep-Sea Res. Pt. II 95: 209–217.

Veldhoen, N., M.G. Ikonomou and C.C. Helbing. 2012. Molecular profiling of marine fauna: Integration of omics with environmental assessment of the world's oceans. Ecotox. Environ. Safe. 76: 23–38.

Vinagre, C., D. Madeira, L. Narciso, H.N. Cabral and M.S. Diniz. 2012. Impact of climate change on coastal versus estuarine nursery areas: cellular and whole-animal indicators in juvenile seabass, *Dicentrarchuslabrax*. Mar. Ecol. Progr. Ser. 464: 237–243.

Vinagre, C., D. Madeira, V. Mendonça, M. Dias, J. Roma and M.S. Diniz. 2014. Effect of temperature in multiple biomarkers of oxidative stress in coastal shrimp. J. Therm. Biol. 41: 38–42.

Visser, M. and C. Both. 2005. Shifts in phenology due to global climate change: the need for a yardstick. Proc. R. Soc. B 272: 2561–2569.

Wang, D.-Z., Z.-X. Xie and S.-F. Zhang. 2014. Marine metaproteomics: current status and future directions. J. Proteomics 91: 27–35.

Wang, N., L. Mackenzie, A.G. DeSouza, H. Zhong, G. Goss and L. Li. 2007. Proteome profile of cytosolic component of zebra fish liver generated by LC-ESIMS/MS combined with trypsin digestion and microwave-assisted acid hydrolysis. J. Proteome Res. 6: 263–272.

Wang, P., F.G. Bouwman and E.C. Mariman. 2009. Generally detected proteins in comparative proteomics-a matter of cellular stress response? Proteomics 9: 2955–2966.

Wei, L., Q. Wang, H. Wu, C. Ji and J. Zhao. 2015. Proteomic and metabolomic responses of Pacific oyster *Crassostreagigas* to elevated pCO_2 exposure. J. Proteomics 112: 83–94.

Westermeyer, R. and R. Marouga. 2005. Protein detection methods in proteomics research. Biosci. Rep. 25: 19–32.

Wong, K.K.W., A.C. Lane, P.T.Y. Leung and V. Thiyagarajan. 2011. Response of larval barnacle proteome to CO_2-driven seawater acidification. Comp. Biochem. Phys. D 6(3): 310–21.

Yates, J.R., C.I. Ruse and A. Nakorchevsky. 2009. Proteomics by mass spectrometry: approaches, advances, and applications. Annu. Rev. Biomed. Eng. 11: 49–79.

11

Environmental Metabolomics and Toxicometabolomics in Marine Pollution Assessment

José Luis Gómez-Ariza,[a,*] *Tamara García-Barrera,*[b] *Belén Callejón-Leblic*[c] and *Gema Rodríguez Moro*[d]

ABSTRACT

Environmental metabolomics is a developing topic that provides global information about free-living organisms since it is reflecting the changes in phenotype and therefore accounts for the influence of external factors. This methodology has an especial interest in the marine environment where many variables affect the organisms, both extreme natural factors such as salinity, temperature, or food restriction and anthropogenic factors, mainly metal contamination, organic compounds, pesticides, drugs, and nanoparticles. Most conventional metabolomic methods are based on the use of 1 HNMR, but it slow sensitivity has triggered the development and implementation of procedures based on mass spectrometry, which allows fingerprinting or profiling of metabolites altered by the exogenous factors. Instrumental couplings based on HPLC- and GC-MS have been proposed in the metabolomics approach, although direct infusion of the sample in high resolution mass spectrometers also provides valuable feasibility and high throughput analysis. Other critical aspects in the metabolomics workflow are suitable extraction of metabolites to get a complete coverage of them and statistic treatment of a result. One important point is related to the overriding role of metals in the metabolism and wellbeing of living organisms whose fluxes and homeostasis have to be maintaining, which generates a host of metal-containing molecules and opens the door to possible interactions and competitions between metals, particularly toxic metals, whose characterization is fundamental to understanding the changes and evolution of metabolic cycles. Therefore, metabolomic techniques for metal-containing metabolites (Metallomics) are also included in most of the recent approaches in this field, incorporating atomic mass spectrometry to the work scheme. In addition, the complementary application of environmental metabolomics

Department of Chemistry, Faculty of Experimental Sciences, Campus El Carmen, University of Huelva, Fuerzas Armadas Ave., 21007 Huelva, Spain.

[a] Email: ariza@uhu.es

[b] Email: tamara@dqcm.uhu.es

[c] Email: belen.callejon@dqcm.uhu.es

[d] Email: gema.moro@dqcm.uhu.es

* Corresponding author

with other omics, such as transcriptomics and proteomics has also been proposed. Metabolomics has been applied to marine organisms, ranging from microbes to marine mammals, although most of the studies have been focused on bivalves and crustaceans, since application of metabolomics to big organisms such as Pinnipeds and cetaceans is still very limited.

Introduction

Metabolomics is today a well-known set of analytical approaches related to the presence of low molecular weight organic and organometallic metabolites naturally occurring in the different organs of living organisms, as well as cells, tissues, and biofluids, which can be perturbed by external factors (e.g., temperature), the presence of pollutants, or the action of pathological mechanisms associated with disease. In connection to this, environmental metabolomics plays an important role in deciphering the consequences of temperature, dessication, salinity or food availability on free-living organisms in certain areas with respect to others, including seasonal changes (Viant et al. 2009). On the other hand, global warming at a world worldwide scale (Mayor et al. 2015) besides the presence of contaminants, such as metals, organic pollutants, pharmaceuticals and nanoparticles have important consequences on animal and plant metabolisms that allows the comparison and evaluation of contaminant and non-contaminant scenaries (Gómez Ariza et al. 2013; Gago-Tinoco et al. 2014; García-Sevillano et al. 2014a, 2015). Generally, metabolomics studies in the field (García-Sevillano et al. 2012, 2014a; Gago-Tinoco et al. 2014) can be complemented with the results from laboratory exposure experiments (García-Sevillano et al. 2013; García-Sevillano et al. 2014b,c), considering not only metabolic changes associated to serious alterations of parameters from natural or anthropogenic sources, but also metabolic alterations related to biotic-biotic interactions, as is the case of solitary- and crowd-reared forms of locust *Schistocerca gregaria*, which generates different levels of metabolites depending on the form considered, such as putrescine, trehalose (both increased in the solitary phase), acetate, and ethanol (both decreased), which allows researchers speculate that polyamine metabolism might be linked to hormone production that could affect swarming behavior (Miller et al. 2008). Metabolomics provides several benefits for studying organism–environment interactions and for assessing organism function and health at the molecular level. Metabolomic results can be easily related with the organism phenotype providing the actual picture of its functional situation and the influence of external factors. In addition, metabolomics is based more on questions than on hypothesis, which can provide unexpected relationships and metabolite responses, which in itself can lead to hypothesis generation (Kell and Oliver 2004; Glass and Hall 2008). As a consequence, metabolomics is having increasing applications in environmental studies, including the study of biological response of organisms to conventional and emergent contaminants, climate change, as well as natural factors (temperature, salinity), biotic-biotic interactions, and presence of microorganisms, among others.

One important point in environmental metabolomics is the study of biological response to toxic elements, such as As, Cd, Cu, Hg, Ni, Pb, and Zn, and their consequences on metabolic pathways. This metal-based metabolomics (metallomics) constitutes a subdiscipline of metabolomics, and requires analytical techniques with high sensitivity for metals, such as ICP-MS. Furthermore, more conventional metabolomics needs instrumentation able to perform molecule identification, such as nuclear magnetic resonance (NMR) (Goldsmith et al. 2010) and mass spectrometry (Viant and Sommer 2013). NMR has been the analytical approach more conventionally used in environmental metabolomics, especially marine pollution metabolomics (see Tables 2 to 5), although it is occurring its gradual replacement by mass spectrometry because of their greater sensitivity, by using direct introduction of the sample by infusion in the mass spectrometer (DIMS) (Viant and Sommer 2013; González-Domínguez et al. 2014; García-Sevillano et al. 2015), which allows high throughput metabolomics analysis, or by instrumental coupling of MS with gas or liquid chromatography (Czech et al. 2012; Ibáñez et al. 2013). In addition, due to the high complexity of living-organisms metabolism, the complementary use of several techniques has been proposed (García-Sevillano et al. 2014a,b,c), as well as the combined use of metabolomics with others omics (Veldhoen et al. 2012).

Finally, the application of metabolomics in marine environment assessment is transforming the existing vision of this important ecological system. Although, a number of studies have been performed on marine aquatic species most of them are focused on chemical, physic-chemical, or biological stress on microbe, bivalves, crustaceans, and teleost. However, the study of marine mammals such as pinnipeds and cetaceans is still in its infancy and the application of omics-based assessment is underway (Veldhoen et al. 2012).

The approaches for these latter specimens are rather limited due to the difficulty of select gene transcripts and metabolites from the current limitations in available molecular information. While species-specific database development continues, cross-species omics tools can be used complementarily to marine mammal sentinels, on the basis that evolutionary conservation in protein function is often reflected in substantial similarity in the protein encoding regions of gene transcripts between species within a general animal group (Bininda-Emonds et al. 2007; Arnason et al. 2008).

The purpose of this chapter is to consider the possibilities of metabolomics for the study of marine aquatic species, since the number of publications in this field is still low, mainly considering the assessment of environmental stress, changes induced by extreme natural conditions, influence of others animals and microbes, harmful conditions during embryo and alevin stage, and others. Metabolomics is a very versatile methodology that can tackle any situation that represents a change in the usual living conditions of organisms under study and provides a powerful approach to deep insight into environmental stress, consequences of extreme natural factors, episodes of anthropogenic pollution, overpopulation, food limitation, and any disruption of ecological balance, setting with other omics a holistic approach for the study of marine living organisms and their environment.

Environmental Metabolomics

Metabolomics is one of the most recent omics that provides information about the external factors affecting biological systems, which is one of its main advantages over other omics, such as transcriptomic or proteomics, since it reflects well the phenotype of living-organisms and cells, and, therefore, the response to the environment. Thus, metabolomics is a good methodological support for environmental studies, not only to assess the influence on living-organisms of changes in conventional variables such as temperature, water and food availability, salinity, light/circadian rhythm, atmospheric gases and other seasonal factors, but also by the presence of different contaminants (Viant et al. 2009). Although most of metabolomics approaches consider only the changes in organic metabolites, more recently, alterations in metal-containing metabolites (metal-metabolomics) in tissues and fluids of free-living organisms under environmental stress, mainly caused by metals, is being considered, since metals play an important role in the function of more than 30% of biomolecules involved in these organisms functions, as proteins, enzymes, and metabolites (Gómez-Ariza et al. 2011). It has to be considered that a continuous flux of essential elements has to be maintained in all organisms to assure the homeostasis of these elements and consequently their wellbeing (Fig. 1). Environmental xenobiotics as toxic metals and organic contaminants can adversely affect the aforementioned homeostatic processes, which produce changes in the metabolism that can lead to pathological processes and even death. This transport processes is especially needed in invertebrates with circulatory systems and particularly in mollusks that has been considered in detail by Robinson (Robinson 2013), who discusses an integrative model of metal regulation and transport, including transmembrane transport of metal ions via channels and carriers (Simkiss and Taylor 1995), intracellular binding of metals to metallothioneins and Ca concretions (Del Castillo and Robinson 2008; Wang and Rainbow 2010) and ion regulation within extracellular fluids (Na^+, K^+, Ca^{2+}) (Zanotto and Wheatly 2006). These critical process(es), whereby metals are transported in invertebrate circulatory systems and selectively transferred to tissues, imply selective and coordinated mechanisms involving metal recognition (likely in the form of metal specie), protein-to-protein communication, transmembrane metal transfer, and intracellular processing. Robinson et al. 2013 studies metal transport in bivalves (the blue mussel *Mytilus edulis* and the quahog *Mercenaria mercenaria*) considering exposure to Cd, but also Ca and Pb. The major plasma metal binding protein in bivalve was the histidine-rich glycoprotein (HRG) (Nair and Robinson 2001a,b), which bound the majority of Cd (and Ca) in bivalve blood (Nair and Robinson 1998, 1999) and latterly transported to specific tissues of marine bivalve mollusks.

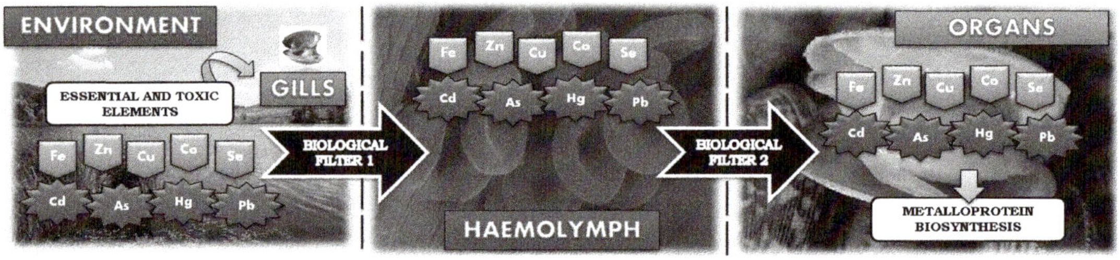

Fig. 1: Essential elements homeostatic cycles and metabolic changes caused by contaminants (e.g., toxic metals).

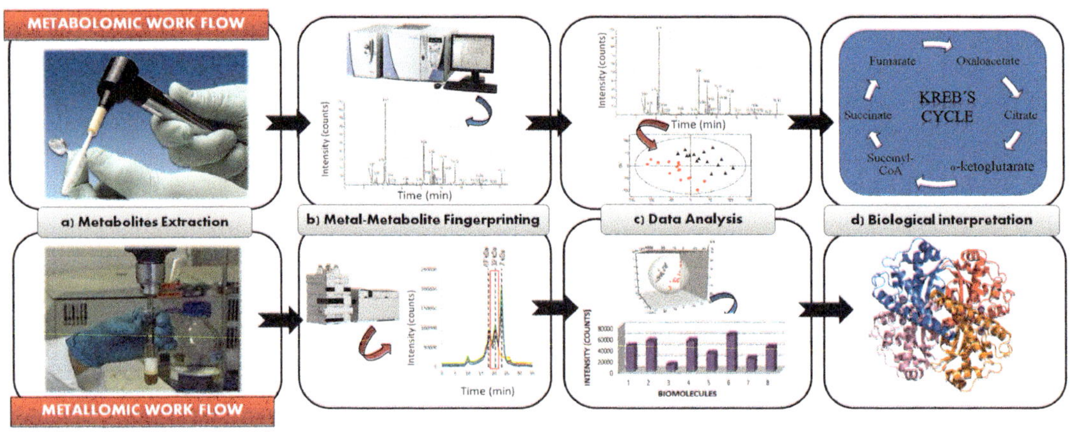

Fig. 2: Workflows used in environmental metabolomics and metallomics.

Analytical approaches in environmental metabolomics

The basic workflow used in environmental metabolomics is outlined in Fig. 2, which comprises several well-established steps: (a) metabolites extraction; (b) metabolite fingerprinting or profiling, depending on the use of direct shotgun techniques, such as direct infusion mass spectrometry (DIMS) or NMR, or the coupling of a chromatographic separation (GC or HPLC) with mass spectrometry. Alternatively, for metal-metabolomics elemental mass spectrometry (ICP-MS) is used for detection of metal-tagged molecules, generally coupled to liquid chromatography (HPLC-ICP-MS); (c) data analysis by means of complex bioinformatics tools; and (d) biological interpretation of the results.

Sample extraction

It is crucial in the use of suitable sample extraction procedures to assure the recovery of as many metabolites as possible from any biological fluid or tissue. Therefore, the choice of an appropriate sample treatment approach is very critical to guarantee representative metabolites isolation and appropriate biological interpretation of data (Duportet et al. 2011). Generally, a multi-step extraction approach is advisable to increase metabolite coverage, in order to obtain comprehensive and reproducible metabolomics profiles. These procedures must be nonselective, simple and fast for non-targeted metabolomics using liquid chromatography–mass spectrometry (Vuckovic 2012).

The analysis of biological fluids is normally carried out after a protein precipitation step (Zhang et al. 2012), while tissues extraction requires a previous lysis of cells using a specific homogenizer in combination with a simple extraction protocol based on the use of solvents mixtures (Römisch-Margl et al. 2011). These multiplexed extractions procedures can be designed for selective recovery of lipophilic

(Watson 2006) or water-soluble metabolites (Cheng et al. 2010), using solvents with different polarity. On the other hand, low abundance metabolites can be concentrated by condensation, solvent extraction, cut-off centrifugal filters, column extraction, or more recently using the vacuum concentrator. However, the low detection limits of many metabolites can also be related to the insufficient performance of the chromatographic device or inadequate ionization of MS (Zhou et al. 2012). For this reason, the use of GC-MS with a preceding step of chemical derivatization has been proposed as a good alternative to improve separation, detection, and sensitivity of metabolite analysis (García-Sevillano et al. 2014d). The application of chemical derivatization in LC-MS analysis has been increasing in recent years (Halket et al. 2005), and has been applied to different functional groups in metabolite moiety, such as amino, carbonyl, carboxyl, and hydroxyl, using esterification reactions of carboxyl group with amines, hydrazines, or alcohol and amino acids with dansyl chloride (Jia et al. 2011) while detection of aldehydes and ketones is assisted by the formation of Schiff bases (Gao et al. 2005; Xu et al. 2011).

Most recently, Pawliszyn et al. (Vuckovic et al. 2011) have drawn attention to the consequences of sampling and sample preparation in the composition of the measured metabolome, so the analytical results may not adequately reflect the true metabolome at the time of sampling (Teahan et al. 2006; Bolten et al. 2007; Moco et al. 2007). This is due to insufficient metabolism quenching during handling procedures, which contribute to unsuitable metabolite loss and/or degradation. Therefore, a sample preparation method based on solid-phase microextraction (SPME) has been proposed for *in-vivo* metabolomics analysis of living systems using liquid chromatography mass spectrometry (LC-MS). *In vivo* SPME allows accurate extraction of the metabolome directly from the tissue or fluid of freely moving animals, which eliminates the operation of sampling where the amount of analyte extracted by SPME is independent of the sample volume (Lord et al. 2003; Musteata et al. 2008).

High throughput metabolomic techniques for overall metabolites appraisal

Metabolomic analysis is based on the use of high-throughput techniques for comprehensive metabolic appraisal of changes associated to exogenous or endogenous factors. Two main spectroscopic techniques have been currently used for this purpose, NMR and MS. NMR achieved extensive use due to compatibility with a wide range of environmental matrices capability of quantitative analysis (Moing et al. 2004), and reproducibility in interlaboratory exercises (Viant et al. 2009). However, the low sensitivity associated to this technique has driven the researchers to others more sensitive techniques, mainly mass spectrometry, which is used with direct introduction (infusion) of the sample into the mass device (DIMS) (Viant and Sommer 2013; González-Domínguez et al. 2014; García-Sevillano et al. 2015) by coupling of any separation unit to the mass spectrometer, typically by liquid chromatography (LC-MS) (Ibáñez et al. 2013) gas chromatography (GC-MS) (Czech et al. 2012) or capillary electrophoresis (CE-MS) (Ibáñez et al. 2013). Optimum ionization of molecules before the introduction into the mass analyzer is also a very critical step for suitable identification and characterization of metabolites. Electrospray ionization (ESI) is the most common ionization source employed in metabolomic studies because it is able to ionize compounds of a wide range of mass and polarity (Boernsen et al. 2005; Cubbon et al. 2007; Monton and Soga 2007; Evans et al. 2009; Lapainis et al. 2009). For less polar compounds, the alternative use of atmospheric pressure chemical ionization (APCI) is more suitable (Williams et al. 2006; Sato et al. 2012). Finally, for little polar or non-polar compounds the use of atmospheric pressure photoionization (APPI) has been recommended, although it has scarcely been reported in metabolomics. The APPI source is based on application of a photoionization lamp and a dopant flow to form dopant radical ions that can directly ionize non-polar analytes through charge exchange reactions. On the other hand, for polar compounds dopant photoions produce intermediate reactive species by reactions with solvent or oxygen molecules, in positive and negative ionization modes respectively, followed by a proton transfer reaction with the analyte (Robb and Blades 2008). Therefore, simultaneous analysis of both polar and non-polar compounds is possible if a careful selection of solvent, dopant, and theirs flows is performed. In this way, APPI has a considerable potential in metabolomics due to its character of universal source but also by the low susceptibility to matrix effects and high linear range, usually higher than that for ESI (Cavaliere et al. 2006; Theron et al. 2007). In addition, APPI requires less heat for desolvation than APCI that makes the technique more suitable for the analysis of thermally labile compounds (Greig et al. 2003). It can be mentioned that the application of metabolomics based on liquid chromatography mass spectrometry integrates different ionization modes: ESI, APCI, and APPI, to the study of potential biomarkers of lung cancer (An et al. 2010), comprehensive lipidomics analysis (Cai and Syage 2006) and Alzheimer disease study (González-Domínguez et al. 2015). Although, APPI can be directly connected to the MS device, setting a high-throughput approach with several advantages in metabolomics, such as fast and reproducible analysis and more comprehensive non-target metabolite coverage. In addition, the use of high resolution systems such as time-of-flight (TOF) or Fourier transform ion cyclotron resonance (FTICR) allows overcome problems associated to isobaric interferences, avoiding the use of chromatographic separation prior to detection (Han et al. 2008).

A large number of mass analyzers have been used in metabolomics, mainly high resolution and accuracy instruments, such as quadrupole time-of-flight (Q-TOF-MS), Fourier transform ion cyclotron resonance (FT-ICR) and Orbitrap detectors (Dunn and Ellis 2005; Fiehn 2008; Han et al. 2009), which are especially recommended in DIMS. This latter approach has undoubted advantages in high throughput metabolomics analysis because it avoids time-resolved introduction of metabolites into the MS as consequence of chromatographic separation, increasing analysis rapidity and reproducibility, as well as metabolite coverage (Draper et al. 2012). Furthermore, the lack of a time-consuming separation step before MS detection allows faster analysis of sample (less than 50 s for DIMS against several minutes for other methods). Although DIMS also has important drawbacks associated to the lack of differentiation for isobaric compounds and difficulty of quantification due to ion suppression (González-Domínguez et al. 2014). In order to overcome isobaric problems associated to isobaric interferences, high resolution instruments have been proposed, while matrix effects and ion suppression are attenuated by optimum dilution of the sample (González-Domínguez et al. 2014).

Nevertheless, electrospray ionization mass spectrometry coupled to LC (LC/ESI-MS) is the most common analytical coupling used in untargeted metabolomics (Lin et al. 2010; Want et al. 2010), in which the main variable for analysis optimization are both the mobile phase and the column used in the LC system that have to be complemented with a suitable selection of the ion source and mass detector. The mobile phase plays a critical role in metabolite ionization; therefore a correct election of this variable is essential. Normally, acetonitrile, methanol and water are used for this purpose, but finally, the solvent has to be selected according to chemical properties of metabolites. In addition, mobile phase properties are modified with additives, such as formic acid, ammonium acetate or alkylamines (tributyl- or trimethylamine), that modify the pH, promote the formation of adducts or form ion pairs, respectively, in order to increase metabolite sensitivity (Coulier et al. 2006). New columns based on hydrophilic interaction liquid chromatography (HILIC) provide complementary information to the more usual reverse phase column, usually recommended in metabolomics, increasing the coverage of polar metabolites (Tang et al. 2014).

Metabolomic techniques for metal-containing metabolites (Metallomics)

As previously discussed contaminants and particularly toxic metals have important consequences in relation to metabolites and essential metals homeostasis, proteins and metalloproteins expression, alteration of metabolic cycles, toxic metal species and metal-metabolites generation, as well as other processes associated to metals that trigger oxidative stress and other biological and health concerns (Fig. 1). Consequently, the identification and quantification of metal-containing metabolites and metalloproteins, related to organisms' defense mechanisms against heavy metals stress, metals mode of action (MOA) in the field, or exposure experiments under controlled laboratory conditions, in toxicological studies, have a great interest.

In this context, it is necessary to define clearly the problem under consideration in order to design suitable analytical methods. We can consider two approaches: (a) targeted analytical methods focused on well-known low molecular organometallic chemical species from anthropogenic origin, which can suffer transformation by living organisms after intake to reduce their toxicity—Speciation methods, and (b) untargeted analytical approaches for metal-biomolecules present in biological systems, normally with unknown structure, size and composition—Metallomics approaches. In Fig. 3 are compared the analytical techniques involved in chemical speciation and metallomics.

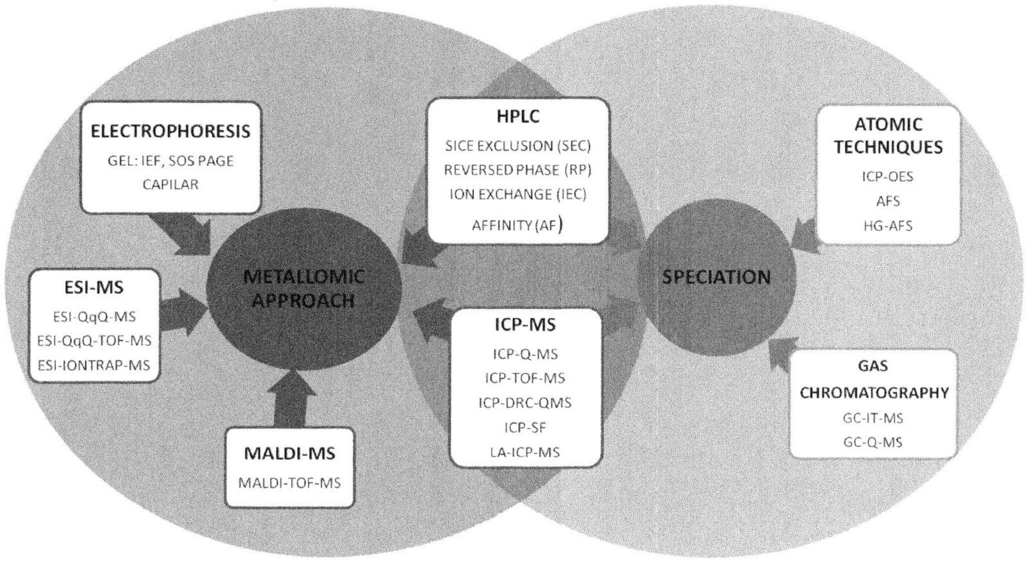

Fig. 3: Comparison of analytical techniques used in speciation and metallomics.

Speciation uses two-dimensional instrumental couplings combining the species-discriminant potential of chromatographic techniques, such as high-performance liquid chromatography (HPLC), capillary electrophoresis (CE), or gas chromatography (GC), and any sensitive atomic detector (mainly, atomic fluorescence spectrometry—AFS and inductively coupled mass spectrometry—ICP-MS) to trace the metals present in the molecule. The Table 1 shows the most common metal species addressed in speciation studies in relation to marine environment. In speciation commercial standards are available for the analyst or they can be synthetized, which allows validation of the methods and intercomparison exercises that finally provide certified reference materials, such as tuna fish (Quevauviller et al. 1996)and seafood (Ibáñez-Palomino et al. 2012) containing methylmercury, arsenic in algae (van Elteren et al. 2007), butyl and phenyltin species in marine sediments (Inagaki et al. 2007), and selenium in serum with indicative levels of selenoproteins (Jitaru et al. 2010).

The metallomics approaches consider that metal-binding biomolecules represent a substantial part of metabolism-involved molecules, in which metals play usually an important role in the biomolecule function (cofactor), and they can be used as a tag to trace the traffic, interactions, homeostasis and other bioprocesses related to both overproduction and inhibition of metallometabolites and metalloproteins triggered by exogenous factors (Garcia-Sevillano et al. 2014a). One important point in metallomics is that metal-biomolecules studied are "a priori" structurally unknown, which oblige to use at least three-dimensional instrumental arrangements: (a) a separation unit (HPLC, GC, CE or 2D-PAGE) to isolate target biomolecules from the matrix and their time-resolved introduction into the detector; (b) an atomic sensitive detector to trace metal-containing biomolecules (normally ICP-MS), and (c) an unit for structural molecule characterization (generally, a high resolution tandem mass spectrometer) in order to identify the unknown molecules (Gómez-Ariza et al. 2004; Bettmer et al. 2009; Mounicou et al. 2009). As a consequence no standards and reference materials exist in metallomics and metal-biomolecules nature has to be characterized by molecular mass spectrometry.

Combination of analytical metabolomics platforms

The complexity of metabolome from biological samples makes advisable the combination of several analytical platforms to get more global information about changes, overexpression and inhibition of metabolites. Usually, these instrumental designs combine couplings for polar and non-polar metabolites. In this sense, the combined use of DIMS for the study of more polar metabolites and GCMS for

Table 1: Metal specie in marine organisms.

Element	Species	Marine organism	References
Arsenic	❖ **Frequent species** 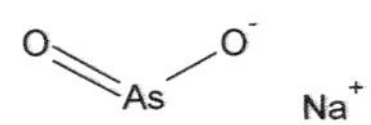 **Sodium arsenite (As (III))** 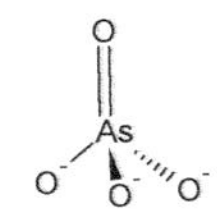 **ion Arsenato (As(V))** 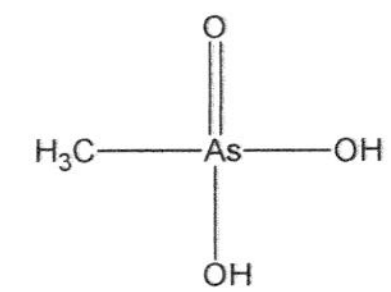 **Monomethylarsonic acid (MMA)** 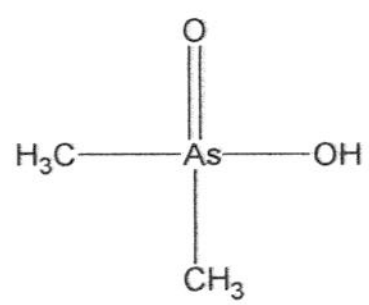 **methylarsonic acid (DMA)** 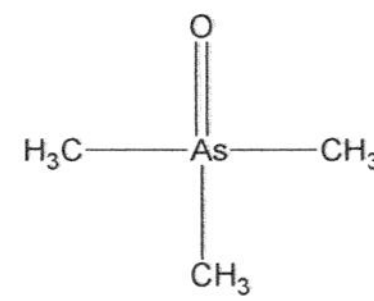 **Trimethylarsine oxide (TMAO)** 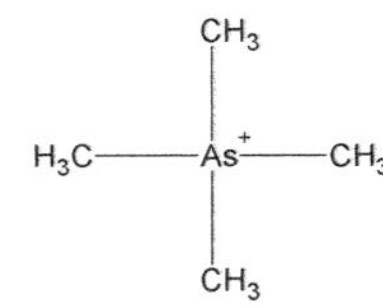 **Tetramethylarsonium ion (TMAs)** 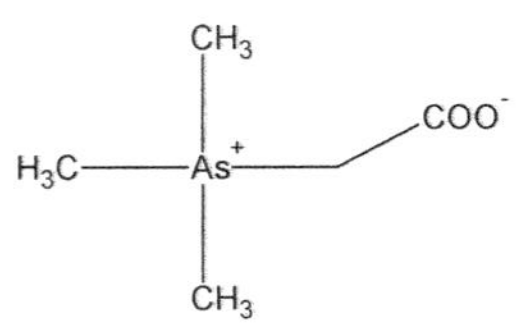 **Arsenobetaine (AsBet)** 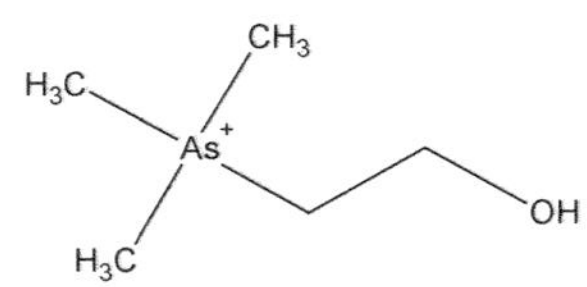 **Arsenocholine (AsC)** 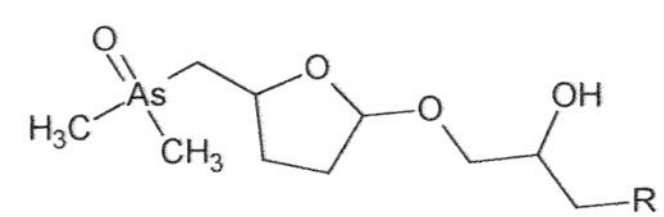 **Arsenosugars**	Anemone Bivalves Algae	(Contreras-Acuña et al. 2013) (McSheehy et al. 2000)

❖ **Arsenolipids**	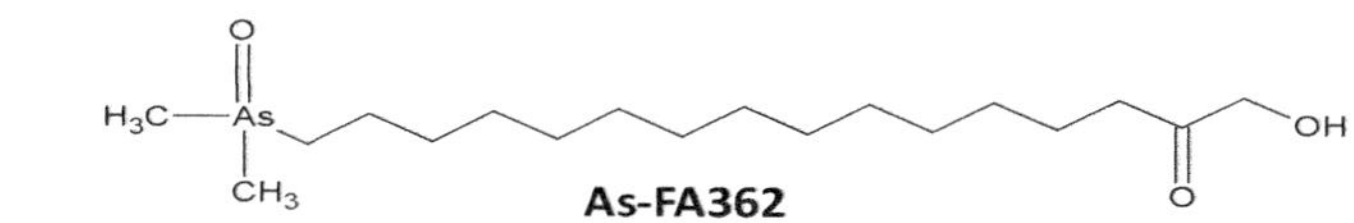 As-FA362 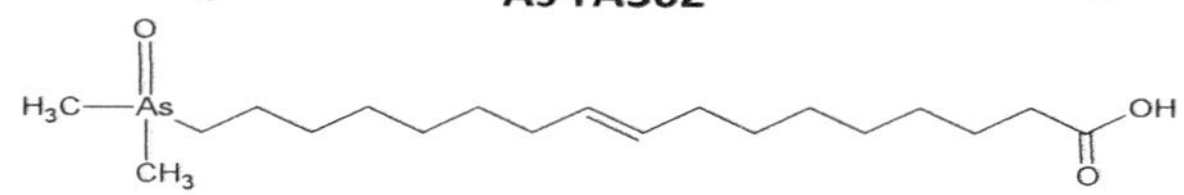 As-FA388 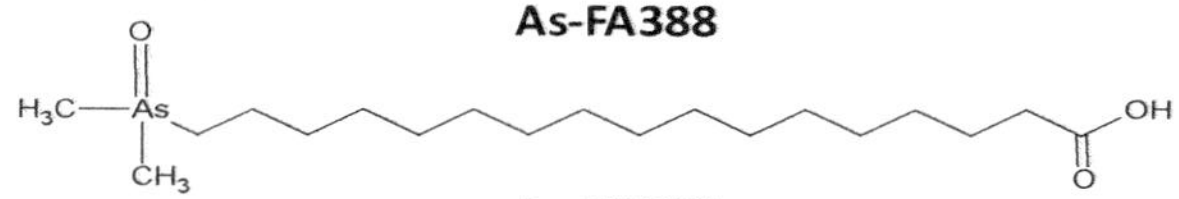 As-FA390 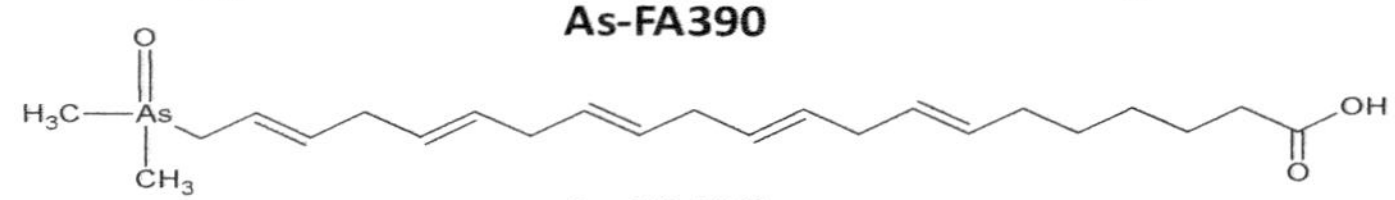 As-FA436 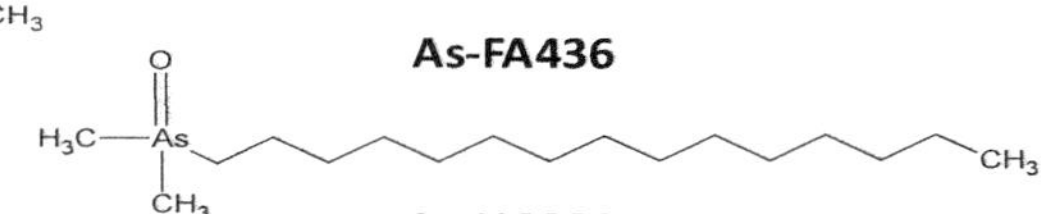 As-HC332 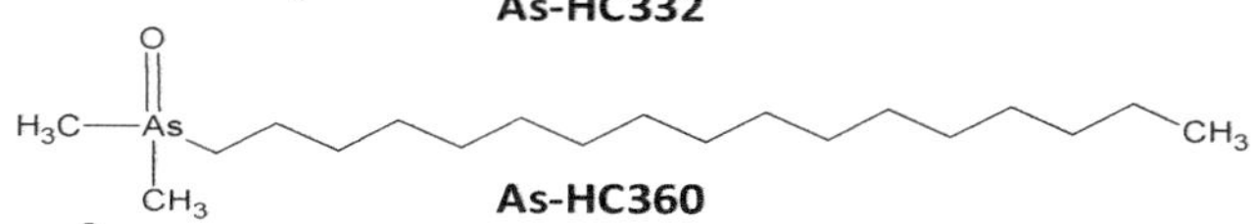 As-HC360 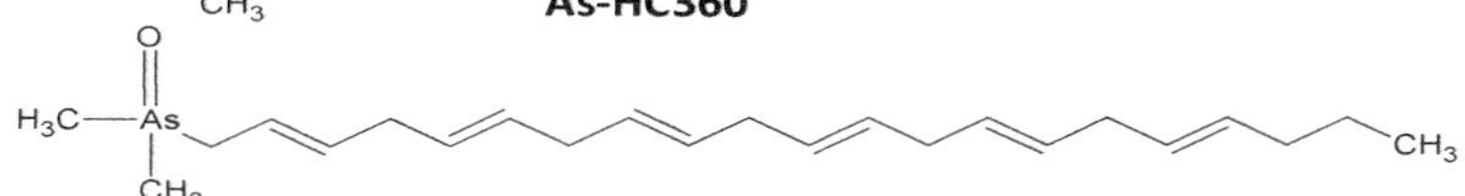 As-HC404	Fish Algae	(Schmeisser et al. 2005)

Table 1 contd....

Table 1 contd...

Element	Species	Marine organism	References
Arsenic	❖ **Arsenohydrocarbons** 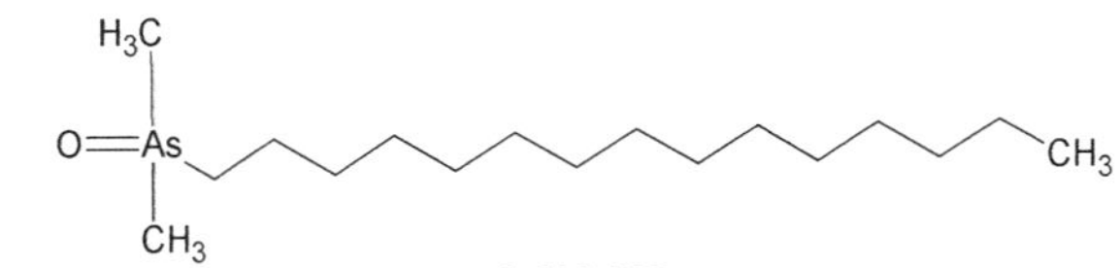 **AsHC 332** 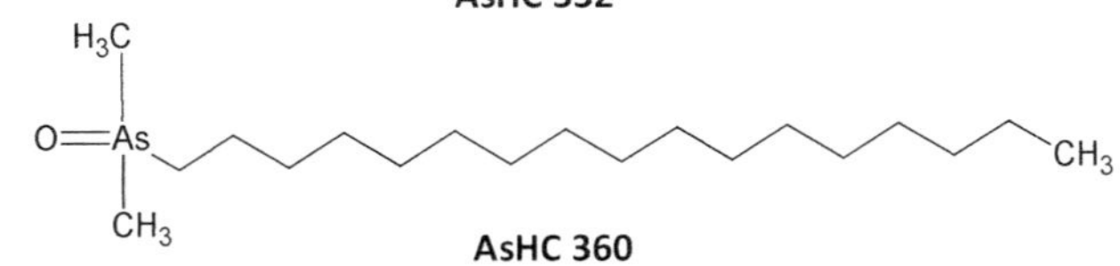 **AsHC 360** 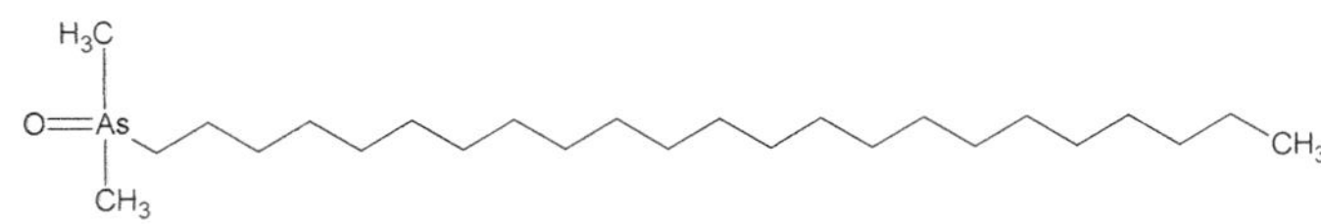 **AsHC444**	Fish oil Tuna Cod Herring Brown algae	(Meyer et al. 2014)
	❖ **Rare species** 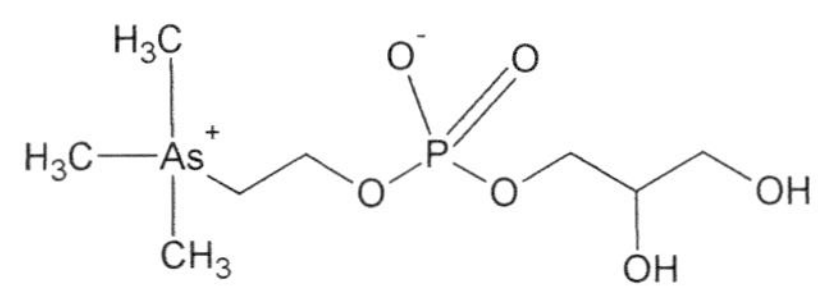 **Glycerophosphoryl-arsenocholine (GPAsC)** 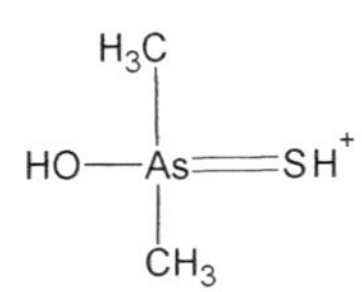 **Dimethylarsenothioic acid (DMAS)**	Anemone (*Anemoniasulcata*)	(Contreras-Acuña et al. 2014)

Mercury	❖	Frequent species	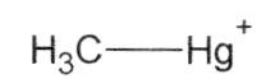 **Monomethylmercury (MeHg)**	Fish Tuna Swordfish Seafood	(Ebdon et al. 2002) (Bosch et al. 2016) (Cano-Sancho et al. 2015) (Sevillano-Morales et al. 2015)
	❖	Less abundant	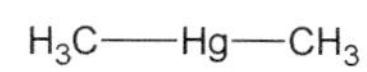 **Dimethylmercury (Me_2Hg)**		
			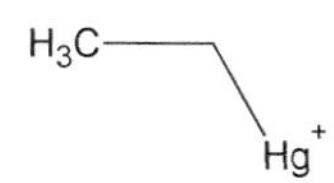 **Ethylmercury (EtHg)**	Seafood	(Bowman et al. 2015) (Yin et al. 2008)
			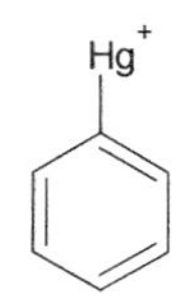 **Phenylmercury (PhHg)**		

Table 1 contd....

Table 1 contd...

Element	Species	Marine organism	References
Tin	 Monobutyltin (MBT) 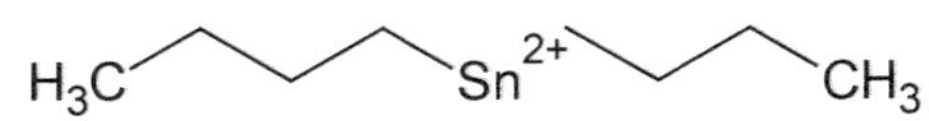 Dibutyltin (DBT) 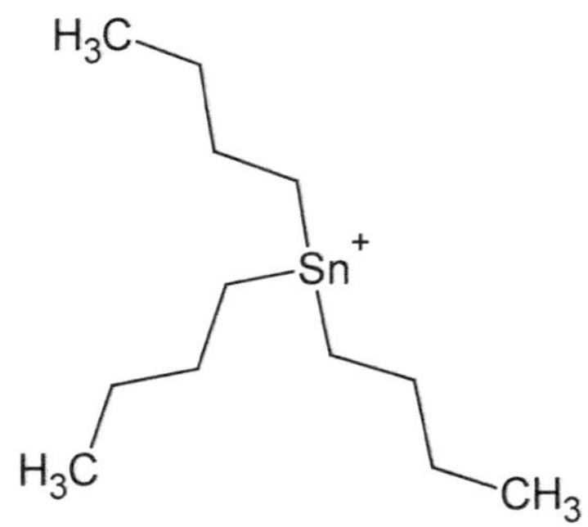 Tributyltin (TBT) 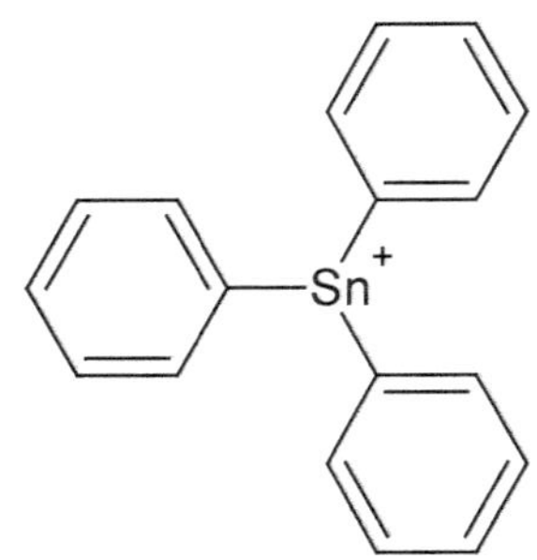 Triphenyltin (TPT)	Bivalves Gastropods Flatfishes	(Gomez-Ariza et al. 1999) (Gomez-Ariza et al. 2001) (Gomez-Ariza et al. 1998)dibutyltin (DBT (Gómez-Ariza et al. 2006) (Ten Hallers-Tjabbes et al. 2003)

Selenium	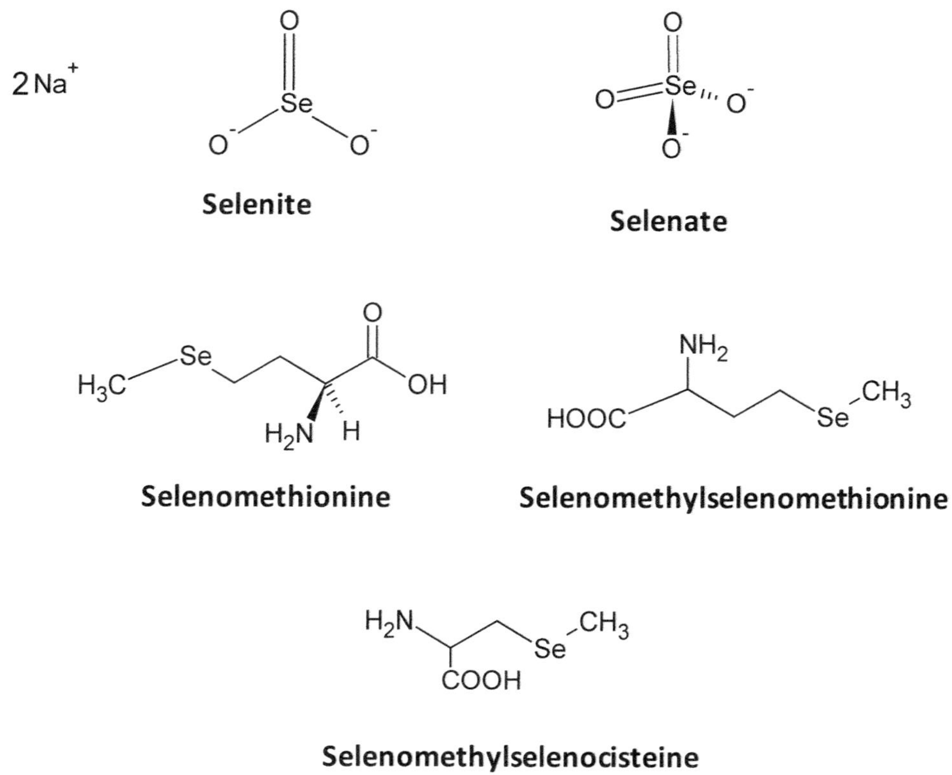	Atlantic cod Atlantic salmon Greenland halibut Atlantic herring Blue mussel Common crab, scallop Copepod (*calanus*) Bivalve (*Corbiculafluminea*) Clam Fish Crustacean	(Bryszewska and Måge 2015) (Adam-Guillermin et al. 2009) (Fournier et al. 2006) (Fournier et al. 2005) (Gómez-Ariza et al. 2002) (Cabañero et al. 2004) (Cabañero et al. 2005) (Bjerregaard and Christensen 2012)

low polarity ones, together to HPLC-ICP-MS to tracemetal-containing biomolecules have provided very promising results in deciphering toxicological issues, such as action of toxic metals in the field (Garcia-Sevillano et al. 2014a) or metal exposition exposure experiments in the laboratory, as example can be cited exposure experiments to mercury (García-Sevillano et al. 2014b), arsenic (García-Sevillano et al. 2013) arsenic/cadmium (García-Sevillano et al. 2014c), and arsenic/cadmiun/mercury (García-Sevillano et al. 2014e).

Environmental metabolomics studies in marine organisms

The application of metabolomics to environmental monitoring is a very complex task due to many factors involved in the response of marine free-living organisms. There are many studies focused on free-living organisms in the field, considering the action of natural factor reaching high levels, such as temperature, salinity and pCO_2. Others account the consequences of anthropogenic contamination, although frequently complementary studies based on controlled exposure are performed to complement and clarify the results from field survey (Kammenga et al. 2000). Complementary analytical techniques are used for this purpose and conventional approaches are extended by others that consider the influence in the metabolomic response of less frequent factors such as the size of the specimen under study or the occurrence of crowded or isolated animal colonies. There are many metabolomic studies on marine organisms but the research on the numerous existing species is uneven, with more focus on bivalves and crustaceans, and still insufficient approach on higher size organisms, and especially marine mammals. The most outstanding works in this regards are summarized below.

Marine Microbes

Marine microbes play an important role in global biogeochemical cycles due to their metabolic interactions with dissolved organic matter (DOM) in the ocean. The suite of compounds produced by a microbe colony reflects its metabolic state, which is a consequence of both genetic and environment (Kido Soule et al. 2015). The interactions with neighboring organisms, availability of nutrients and the effect of the media also have a remarkable influence on the phenotypic metabolic profile and hence on microbial contributions to the DOM pool. Heterotrophic and many autotrophic microorganisms use the low molecular weight fraction (< 600 Da) of DOM as base of their growth and metabolism (Kujawinski 2011). Therefore, it is important to identify the metabolites that pass through the pool of DOM and affect microbial species, in order to understand the impact of microbes on the global carbon cycle. Metabolomics methods can be applied to the study of microbial communities in their environment, including marine systems. In this way, untargeted metabolomics enables the discovery of new or unexpected metabolites, thereby expanding the metabolic pathways as consequence of microbial consortia; this is the case of microbial exuded metabolites that promote cooperative interactions.

With this purpose have been studied the species-level chemical diversity of two marine actinobacterial species *Salinispora arenicola* and *Salinispora pacifica*, isolated from sponges distributed across the Great Barrier Reef (GBR). Secondary metabolite profiles have been obtained using UHPLC-QToF-MS-based metabolomics. The results showed a high level of inter-species diversity of metabolites in strains from these two bacterias. It was also found rifamycins and saliniketals produced exclusively by *S. arenicola*, as the main secondary metabolites differentiating of the two species. In addition, it was discovered the presence of 57 candidate compounds that clearly increases the small number of secondary metabolites previously known to be produced by these species. Finally, it was found for the first time the production of rifamycin O and W, a type of ansamycin compounds, in *S. arenicola*.

One important point is related to the metabolomics changes caused by the infection of marine microbial community by viruses at any given time. This infection can reshape host metabolism at a global level and the effect of this alteration on the cellular material released after viral lysis is few understood. To deep insight into this knowledge, the growth dynamics, metabolism and extracellular lysate of roseophage-infected *Sulfitobacter* sp. *2047* was studied using a variety of techniques, including liquid chromatography–tandem mass spectrometry (LC-MS/MS)-based metabolomics (Ankrah et al. 2014). Quantitative analysis estimates that total amount of carbon and nitrogen sequestered into particulate biomass after phage infection redirects about 75% of nutrients into virions.

A number of 82 metabolites were measured along the time over the infection cycle, of which a 71% presented markedly elevated concentrations in infected populations raising the metabolic activity. The profile of metabolites from infected cultures showed that about 70% of 56 quantified compounds had decreased concentrations in the lysate relative to uninfected controls, suggesting that these small, labile, nutrients were being utilized by surviving cells (Table 2).

These results indicate that virus-infected cells are physiologically distinct from their uninfected counterparts, which has implications for microbial community ecology and biogeochemistry.

Heterotrophic bacteria *Pseudovibrio* sp. *FO-BEG1* has been studied in order to evaluate the influence of oceanic DOM composition in the preferential release of certain compounds. However, they also secrete other molecules depending on physiological state, environmental conditions and growth of the colony. The exo-metabolome of this marine bacterium has been include in this study for the first time using ultra-high resolution mass spectrometry (ESI-FT-ICR-MS), considering the influence of phosphate. The bacteria can be isolated from marine invertebrates and present an exo-metabolome large and diverse, consisting of hundreds of compounds drastically affected by the physiological state of the strain. In addition, it has been proved that phosphate limitation greatly influences both the amount and the composition of the secreted molecules. Correlatively, it has been observed that under phosphate surplus conditions the secreted molecules were mostly peptides and highly unsaturated compounds. In contrast, under phosphate limitation the composition of the exo-metabolome showed an increase in highly unsaturated, phenolic, and polyphenolic compounds (Romano et al. 2014).

Polychaetes

Infaunal deposit-feeding invertebrates such as *Nereis virens* can be extensively exposed to oil hydrocarbons, which can be accumulated on the basis of intakes and efficiency of biotransformation and excretion of this organism. Since *N. virens* is a food source for higher organisms, the well-known biotransformation of PAHs for this polychaete can provide information about the mechanism underlying oil hydrocarbons degradation in this species (McElroy 1985, 1990; Jørgensen et al. 2005). Metabolomics can help to identify the metabolites that discriminate between crude oil exposed and non-exposed marine polychaetes, clarifying the mechanisms related to tissue response. Experimental set up is very critical in this kind of studies and optimization of sample preparation, extraction and data preprocessing method for untargeted metabolomics is crucial (Fernández-Varela et al. 2015). These authors compare the extraction efficiency and metabolite coverage using two-steps liquid extraction and pressurized solvent extraction (PSE), this later assisted and non-assisted with C18 solid phase. In summary, the two-step extraction with 80:20 (v/v) methanol:water on freeze dried sample provides more suitable results since it gets a compromise between simplicity, speed of analysis, extraction efficiency and reproducibility. Although PLE provided higher extraction efficiency, the analytical (intermediate) reproducibility appears to be low compared to the two-step extraction. Although PLE is more suitable for the analysis of non-polar compounds (such as cholesterols) when C18 material were not added, these compounds are also detected with the two-extraction procedure although with lower sensitivity. In addition, PLE leads to more matrix interferences with consequent deterioration of GC column and liner. PLE is more time consuming and higher operating costs. The metabolomics study in polychaetes reveals that glycine, sucrose, and methylphosphate present higher levels in expose specimen. However, compounds as phosphoric acid, histidine, l-tyrosine, isoleucine and l-valine present higher concentrations in non-exposed samples (Table 2). Glycine has an important role in this organism, being precursor in the synthesis of porphrins, creatine and purine nucleotides. The increase in sucrose could be linked to the disruption of the energy metabolism. These results suggest that amino acid and carbohydrate metabolic pathways are affected by exposure to crude oil (Fernández-Varela et al. 2015). Liquid-liquid extraction has been exhaustively evaluated by Alvarez et al. (Alvarez et al. 2010) using the polychaete *Hediste diversicolor* as bioindicator, methanol:water extraction at a 1:1 ratio, drying the sample under nitrogen provides the best reproducibility and yield. For the extraction of non-polar fraction chloroform shows the best results.

Table 2: Metabolic changes caused by contamination in marine microorganisms and polychaetes.

Marine organism	Contaminant	Metabolite change	Target organ	Analytical technique	Reference
Salinispora arenicola *Salinispora pacifica*	Change of environmental parameters	Production of rifamycins and saliniketals 57 new secondary metabolites	Biomass from culture	UHPLC-QToF-MS	(Kido Soule et al. 2015)
Sulfitobacter sp. 2047	Alteration of metabolites after infection with roseophage	*Amino acids* ↑Asparagine, ↑methionine, ↓threonine, ↑homoserine, ↓N-acetylhornithine, ↑malonyl CoA, ↓ 2-oxoglutarate, ↑3-phosphoglycerate, ↓phosphoenolpyruvate	Cell filtrates	LC-MS/MS	(Ankrah et al. 2014)
		Lipids ↑Phosphatidylcholine, ↑phosphatidylethanolamine, ↑propionyl CoA, ↑ethanolamine, ↓farnesylpyrophosphate			
		Nucleic acids, nucleosides and nucleotides ↑Deoxyadenosine, ↑Thymine, ↑Cytosine, ↓TDP, ↓GMP, ↓5-Methyldeoxycytidine			
		Cofactors, vitamins and electron carriers ↑Acetyl CoA, ↑Pyridoxine ↓Thiamine, ↓FAD, ↓NAD			
		Otros ↑urea, ↑ácidopipecólico↓citaconate, ↓phenuldyruvate, ↓lactate			
Nereisvirens	Oil hydrocarbons	Exposed ↑glycine, ↑sucrose, ↑methylphosphate	Whole body	GC-MS	(Fernández-Varela et al. 2015)
		Non-exposed ↑phosphoric acid, ↑histidine, ↑l-tyrosine,↑isoleucine,↑l-valine			

Bivalves

Bivalves have frequently used as sentinel organisms to monitor metal and other contaminants effects on marine emplacements. These bioindicators are sessile and present a long life cycle, exhibiting an elevated tolerance to contaminants, so they accumulate high concentrations of these pollutants. In addition they are easily sampled and handled, providing suitable amount of sample for the analysis (Elder and Collins 1991), which represents undoubted advantages in environmental metabolomics.

Manila clam, *Ruditapes philippinarum*, has been much studied for its wide geographic distribution on the China coast and the current human consumption, although other bivalves have also been considered in toxicological metabolomics (Table 3). Primarily, it is important to consider the effect of salinity changes, the results showed that *V. philippinarum* presented high mortality at lower salinities (0 and 7 g/L) but tolerated high salinity levels (35 and 42 g/L). The quantification of ionic content revealed that clams had the capacity to maintain ionic homeostasis along the salinity gradient, mainly changing the concentration of Na, but also with the contribution of Mg and Ca, which maintain osmoregulation. The osmoregulation regulation process involves high energy cost that forces organisms to spend more energy resources, explaining the massive reduction of glycogen levels at low and high salinities. Osmoregulation may be achieved with inorganic cations and organic molecules such as taurine and betaine (osmolytes), as well as alpha-amino acids (Yancey 2005). However, results show that taurine and betaine decrease both at low and high salinity in comparison with control conditions. This contradiction can be explained considering that osmoregulation can be achieved not only by regulation of ionic species or osmolytes, but also by the closure of shell valves that induces hypoxia (Kim et al. 2001) with important consequences on metabolism. The resulting low O_2 concentrations in cells will decrease the oxidative phosphorylation of ATP that induce the accumulation of metabolites to feed the respiratory chain, activating alternative metabolic pathways of ATP production. The low efficiency of these routes is energetically expensive yielding the consumption of glycogen stores that are rapidly consumed, and cells have to resort to protein catabolism to provide energy. The increase of protein catabolism triggers, consequently, the increase of amino acids levels. The lower activity of the electron respiratory chain decreases the oxidation of amino acids obtained by protein catabolism, leading to their accumulation, or the accumulation of their degradation intermediate metabolites, such as succinic acid (in isoleucine, threonine and methionine metabolism) or formic acid (in serine metabolism). An exception is asparagine since this amino acid can be converted into glutamate and glutamine (which are increased) with ATP production. Other amino acids (leucine, lysine, phenylalanine, tryptophan and tyrosine) are degraded into ketone bodies increasing the levels of acetoacetic acid with formation of glutamate (Carregosa et al. 2014), Table 3. Metal contamination can also have a differential response in two dominant pedigree of the clam *R. philippinarum* (White and Zebra) that can be used for marine environmental toxicology (Ji et al. 2015). The results showed marked differences between White and Zebra clams on the basis of metabolomics profiles and antioxidant enzyme activities. In this way, the levels of amino acids and succinate are higher in the digestive gland of Zebra clam and levels of ATP are higher in digestive gland of Zebra clam, respectively. On the other hand, superoxide dismutase activities in control White and Zebra clam samples were markedly different. Finally, White clam was more sensitive to Cd considering the accumulation of this element, enzymatic alterations and metabolic changes. It can be remarked that metabolic alterations induced by Cd in White clam digestive glands included over-expression of valine, isoleucine, leucine, arginine, glutamate, acetoacetate and succinate, and inhibition of hypotaurine and ATP (Table 3). The exposure of White clams to Zn increased the presence of valine, isoleucine, leucine and histidine; while simultaneous exposure to Cd and Zn increased valine, isoleucine, leucine, acetoacetate and glucose, decreasing aspartate, taurine, betaine and ATP. In Zebra clam it was observed the increase of ATP and decrease of homarine and histidine in digestive gland under Cd exposure, and the increase of glucose by the action of Zn. For combined Cd+Zn exposure Zebra clam showed over-expression of alanine, arginine, glycine, glucose and ATP (Table 3). The major metabolic differences between White and Zebra clam samples are related to amino acids (valine, isoleucine, leucine and histidine) and energy metabolism-related metabolites (succinate and ATP). The differences among amino acids can be related to the balance of intracellular osmolarity associated to different strategies to regulate this parameter by the two pedigree clams (Zhang et al. 2011a). As final conclusion, White clam may be a more suitable bioindicator for marine environmental toxicology.

Tabla 3: Metabolic changes caused by contamination in bivalves.

Marine organism	Contaminant	Metabolite change	Target organ	Analytical technique	Reference
Mytilus Galloprovincialis	Hg and PAHs	↑Isoleucine, ↑leucine, ↑valine, ↑arginine, ↑valine, ↑alanine, ↑arginine, ↑glutamate, ↑glutamine, ↑aspartate, ↑glycine, ↓tyrosine, ↑acetate, ↑acetoacetate, ↑succinate, ↑malonate, ↑glucose, ↑glycogen, ↑ATP, ↑ADP, ↑hypotaurine, ↑taurine, ↑betaine, ↑homarine, ↓acetylcholine, ↓tyrosine	Gills	^{1}H-NMR	(Cappello et al. 2013)
	Ni	↑Adenosine, ↑ATP/ADP, ↑glutamine, ↑guanine, ↑hypotaurine, ↑taurine, ↓allantoin, ↓arginine, ↓betaine, ↓glucose, ↓glutamate, ↓glycine, ↓glycogen, ↓hippurate, ↓histidine, ↓homarine, ↓isoleucine, ↓leucine, ↓malate, ↓PAG, ↓phenylalanine, ↓proline, ↓succinate, ↓tryptophan, ↓tyrosine, ↓uracil, ↓valine, ↑adenosinenucleotides, ↑TP and ADP, ↑glutamine, ↑guaninenucleotides, ↑hypotaurine, ↑taurine	Digestive gland	^{1}H-NMR	(Jones et al. 2008)
	Chlorpyrifos	↓Arginine, ↓glutamate, ↓hippurate, ↓isoleucine, ↓lactate, ↓leucine, ↓phenylacetylglycine, ↓phenylalanine, ↓proline, ↓tyrosine, ↓valine			
Venerupis philippinarum	Salinity	↑leucine, ↑isoleucine, ↑valina, ↑threonine/lactate, ↑alanine, ↓arginine, ↑glutamate, ↑glutamine, ↓asparagine,↑glycine, ↑glycine, ↑tyrosine, ↑phenylalanine, ↑aceroacetate, ↑succinate, ↓formate ↑hypotaurine, ↓taurine, ↓betaine, ↑homarine, ↓glucose, ↓glycogen	Clam soft tissue	^{1}H-NMR	(Carregosa et al. 2014)
	Cd	White clam:↑valine, ↑isoleucine, ↑leucine, ↑arginine, ↑glutamate, ↑acetoacetate, ↑succinate ↓hypotaurine, ↓ATP Zebra clams:↑ATP, ↓homarine, ↓histidine			
	Zn	White clam:↑valine, ↑isoleucine, ↑leucine, ↑histidine Zebra clams:↑glucose	Digestive gland	^{1}H-NMR	(Ji et al. 2015)
	Cd+Zn	White clam:↑valine, ↑isoleucine, ↑leucine, ↑acetoacetate, ↑glucose↓aspartate, ↓taurine, ↓betaine, ↓ATP Zebra clams: ↑alanine, ↑arginine, ↑glycine, ↑glucose, ↑ATP			

	As	↑Succinate, ↑fumarate, ↑betaine, ↓amino acids (leucine, isoleucine, threonine, valine and tyrosine), ↑ATP	Gills	^{1}H-NMR	(Wu et al. 2013)
	Cd	↑Branched chain amino acids, ↑aspartate, ↑betaine, ↓alanine, ↓succinate, ↓hypotaurine, ↓homarine, ↑glutamate, ↓acetate	Gills	^{1}H-NMR	(Zhang et al. 2011)
	Cu	Low levels of Cu: ↓aspartate, ↓dimethylamine, ↑hypotaurine, ↓alanine, ↓acetate, ↓succinate, ↑citrate, ↓acetylcholine, ↑branched chain amino acids, ↑betaine, ↑homarine, ↓ATP/ADP, ↓taurine High levels of Cu:↑aspartate, ↓dimethylamine, ↑hypotaurine, ↓alanine, ↓acetate, ↓succinate, ↑citrate, ↓acetylcholine, ↑branched chain amino acids, ↑betaine, ↑homarine, ↓ATP/ADP, ↓taurine, ↑glycogen, ↑glycine	Gills	^{1}H-NMR	(Zhang et al. 2011a)
Ruditapes philippinarum	Hg	White clams: ↑lactate, ↑succinate, ↑taurine, ↑acetylcholine, ↑betaine, ↑homarine, ↓alanine, ↓arginine, ↓glutamine, ↓glutamate, ↓acetoacetate, ↓glycine, ↓ATP/ADP Red clams:↑succinate, taurine, ↑acetylcholine, ↓glutamine, ↓glycine, ↓aspartate Zebra clams: ↑branched-chain amino acids, ↑lactate, ↑succinate, ↑acetylcholine, ↑homarine, ↓alanine, ↓acetoacetate, ↓glycine, ↓taurine	Muscle	^{1}H-NMR	(Liu et al. 2011a)
		White clams:↑3-hydroxybutyrate, ↑alanine, ↑glutamate, ↑succinate, ↑taurine, ↓acetoacetate, ↓hypotaurine, ↓glycine Red clams:↑branched-chain amino acids (valine, leucine, and isoleucine), ↑alanine, ↑glutamate, ↑succinate, ↑aspartate, ↑betaine, ↑homarine, ↓arginine, ↓glucose, ↓ATP/ADP Zebra clams: ↑citrate, ↑taurine, ↑homarine, ↓alanine, ↓arginine, ↓glutamate, ↓acetylcholine, ↓ATP/ADP, ↓glycine	Digestive gland	^{1}H-NMR	(Liu et al. 2011b)

Table 3 contd....

Table 3 contd...

Marine organism	Contaminant	Metabolite change	Target organ	Analytical technique	Reference
Cassostrea gigas	pCO_2	↓glutamate, ↓glycine, ↓homarine, ↓ATP, ↑threonine, ↑taurine, ↑dimethylglycine, ↑succinate	Gills	^{1}H-NMR	(Wei et al. 2015)
		↑Branched chain amino acids (aspartate, threonine, alanine), ↑osmolytes (homarine), ↓hypotaurine, ↑ATP, glucose and glycogen	Digestive gland		
	Metals	↑Branched chain amino acids (valine, leucine, isoleucine), ↑arginine, ↑aspartate, ↑lysine, ↑phosphocholine, ↑proline, ↑AMP, ↑tyrosine↓alanine, ↓acetoacetate, ↓β-alanine, ↓betaine, ↓malonate, ↓taurine, ↓homarine	Gills		
Corbiculaflumilea	Cd + Zn	Small clams: ↑malonate, ↑adenosine, ↑glucose, ↑glycogen, ↑turanose, ↑betaine, ↓alanine, ↓glutamine, ↓isoleucine, ↓lysine, ↓threonine, ↓valine, ↓leucine, ↓arginine, ↓glutamate	Whole Soft tissue	1H-NMR + GCMS	(Spann et al. 2011)

Other studies related to exposure of *R. philipinarum* to metals reveal that metabolic changes observed are associated to disturbance in energy metabolism, osmotic regulation and neurological effects caused by different toxic metals, such as As (Ji et al. 2013; Wu et al. 2013a), Cd (Zhang et al. 2011a), Cu (Zhang et al. 2011b) and Hg (Liu et al. 2011a,b). In these studies gills were normally used as target organs, since they can accumulate considerable amounts of contaminants, mainly involved in nutrients uptake, digestion and gas exchange, being the first organ to be affected by pollutants (Table 3).

Mussels have also been considered in metal environmental metabolomics (Table 3), having checked the severe morphological damage caused by PAHs and Hg on *Mytilus galloprovincialis* gills caged during 30 days in a petrochemical area, and the corresponding metabolic alterations (Cappello et al. 2013). Gills were selected as target organ directly affected by pollutants by their involvement in nutrient uptake, digestion and gas exchange. Using ^{1}HNMR spectroscopy and multivariate statistics were observed specific changes in metabolites related to osmotic regulation, energy metabolism and neurotransmission, which suggests that the mussels transplanted to the contaminated field site were suffering from adverse environmental conditions. Metabolite changes were calculated via the ratio between the averages of the stressed and control peak areas, $P < 0.05$. Elevated levels of amino acids, energy metabolites such as acetate, acetoacetate, succinate, malonate, glucose, ATP and ADP, osmolytes such as hypotaurine, taurine, betaine and homarine, together with the decreased concentration of tyrosine, glycogen and acetylcholine were tested (Table 3). Additional studies in petrochemical areas reveal de neurotoxicological effects of these activities, perturbation of serotonergic (with alteration of levels of serotonin, 5-HT, and its receptor, 5-HT3R), cholinergic (acetylcholine, acetylcholinesterase, AChE, and choline acetyltransferase, ChAT), and dopaminergic systems (i.e., tyrosine and tyrosine hydroxylase, TH) (Cappello et al. 2015). In addition, exposure to Ni, the organophosphorus insecticide chlorpyrifos (Chlp) or a combination of both chemicals produces clear metabolic changes using 1H-NMR-based metabolomics (Jones et al. 2008). Certain metabolic alterations were common across all groups, especially in the concentrations of amino acids. This fact was particularly remarkable in the Ni exposed group where concentrations of many more compounds were affected than in Chlp and Ni+Chlp groups. Metabolic changes in Table 3 seem to be related to reduce feeding rates in response to the stress of pollutant exposure (calorie restriction). During food restriction, gluconeogenesis is increased to produce glucose from a range of sources, including lactate, amino acids and ketone bodies. This catabolic period is also usually characterized by degradation of lipids, glycogen and proteins. However, there is some evidence that marine mollusks are less involved on fatty acid and ketone body metabolism compared with other organisms (Rosenblum et al. 2003). The rationale for this fact may be that some marine mollusks use high intracellular concentrations of free amino acids to balance the intracellular osmolarity with the environment, which can also be used in cellular energy metabolism.

The effects of exposure to elevated pCO_2 have been studied in gills and hepatopancreas of *Crassostrea gigas* by integration of proteomic and metabolomic approaches (Wei et al. 2015). Metabolic responses showed that high CO_2 levels caused disturbances in energy metabolism and osmotic regulation marked by differentially altered ATP, glucose, glycogen, amino acids and organic osmolytes in oysters, and the reduction of ATP in gills. In addition, accumulation of ATP, glucose and glycogen in hepatopancreas accounted for the difference in energy distribution between these two tissues. The summarized results obtained in the metabolomics study of these samples follows:

Gills: (a) the levels of amino acids (glutamate and glycine), and organic osmolyte (homarine) were significantly decreased; (b) the amino acid threonine and organic osmolytes (taurine and dimethylglycine) were significantly increased in oyster gills under elevated pCO_2 exposure, which denotes osmotic stress in oyster gills induced by elevated pCO_2 exposure; (c) the level of ATP (energy storage compound) was significantly decreased, and succinate (an intermediate in citrate cycle) was significantly increased. These two later altered metabolites are related to disturbance in energy metabolism (Table 3).

Hepatopancreas: (a) The elevated amino acids (branched chain amino acids-BCAA, aspartate, threonine and alanine) and osmolytes (homarine), and the depletion of hypotaurine suggested a severe osmotic stress caused by elevated pCO_2 exposure; (b) the abundance of aspartate, threonine, alanine and BCAA indicated that protein degradation was elevated in oyster hepatopancreas; (c) the high levels of the energy storage metabolites (ATP, glucose and glycogen) in oyster hepatopancreas can be related to the reduction of energy demand that disturbed the energy metabolism (Table 3).

Under these results is unclear the metabolic behavior of both organs. The increased taurine and threonine in oyster hepatopancreas were altered similarly to those in gills under elevated pCO_2 exposure, which could suggest similar biological effects in both tissues. However, other metabolites such as ATP and homarine reflect important differences between gills and hepatopancreas. In addition, the reduction of ATP in gills and the accumulation of ATP, glucose and glycogen in hepatopancreas point to the difference in energy distribution between these two both tissues. Gills overcome the stress caused by high pCO_2 by consuming more ATP, while hepatopancreas reduced energy demand and synthesized more glucose and glycogen in order to accumulate energy for cellular physiological regulations. Therefore, the clam redistributes the energy in gill and hepatopancreas to defend against seawater acidification maintaining the homeostasis. However, it is not yet a complete understanding of the effects of high pCO_2 levels on physiological responses of oysters. As a consequence, the decreased seawater pH caused by the increasing input of CO_2 to the atmosphere (Ocean acidification—OA) is affecting the physiology of many marine calcifying organisms, and metabolomics approaches can contribute to deep insight into physiological responses of marine mollusk to the OA stress.

The oyster *Crassostrea hongkongensis* is able to accumulate different metals, such as Cd, Zn, and Cu, at significant levels in the tissues (Liu and Wang 2012). This bivalve is consumed in Southern China, so that is interesting to use this bioindicator to characterize the biological stress caused by metal pollution. In relation to this fact omics approaches and in particular metabolomics are providing a global overview about the biological response induced by metal pollution on the basis of non-targeted or unsupervised analysis, which allows discover new biomarkers related to this kind of contamination (Ji et al. 2015b). This approach improves more conventional methods based on targeted biomarkers for characterization of biological stress, such as lysosomal membrane stability related to the activation of catabolic processes, and anti-oxidative biochemical indices, in connection to oxidative stress (Regoli 2000; de Almeida et al. 2004; Rank et al. 2007).

The concentration of eight metals (Cr, Ni, Mn, Cu, Zn, Ag, Cd and Pb) and one metalloid (As) in the gills of *C. hongkongensis* oyster collected from clean and metal-contaminated sites has been estimated, revealing differences among them on the basis of principal component analysis (PCA) of the results. The metabolomics study conducted on the ^{1}HNMR spectral datasets from gills tissue reveals the predominant presence of organic osmolytes, particularly betaine and taurine, and the Krebs cycle intermediate succinate. Although other classes of metabolites were also identified, including amino acids (e.g., valine, leucine, isoleucine, alanine, arginine, glutamate, glutamine, aspartate, and glycine), osmolytes (e.g., betaine, taurine, homarine, hypotaurine), energy storage compounds (e.g., glucose, glycogen) and Krebs cycle intermediates (succinate). Principal component analysis treatment of metabolic results confirms the similar response of metal-exposed groups and even being stated differences between them (Ji et al. 2015b).

In addition, O-PLS-DA treatment of data on control and metal pollution-exposed groups shows good classifications between control and metal pollution exposed groups. The loading plots showed overexpression of branched chain amino acids-BCAAs (valine, leucine and isoleucine), arginine, aspartate, lysine, phosphocholine, proline, AMP and tyrosine and depletion of alanine, acetoacetate, β-alanine, betaine, malonate, taurine and homarine in gills. Although, changes in metabolic expression is depending of the differential presence of toxic metals in the area of study, as demonstrated by Wu et al. (Wu et al. 2013b) by the increased level of glycogen in the gill of *Ruditapes philippinarum* exposed to Cd against other ones with a lower presence of this element.

The potential of metabolomics analytical procedures can even distinguish differences in the response of bivalve to pollution in sediment (low levels of Cd and Zn), in relation to the size of specimens (Spann et al. 2011). Two different size classes of the Asian freshwater clam *Corbicula fluminea* were exposed to low levels of Cd and Zn in sediments, tracing the metabolic changes caused by these elements by NMR and gas chromatography–mass spectrometry (GC–MS). The condition index (ratio of tissue dry weight to volume inside the shell valves) was also measured. Small and large clams presented different metabolic responses to these contaminants, the main perturbations affecting amino acid and energy metabolism. In the small clams, levels of amino acids decreased under Cd+Zn exposure stress, whereas large clams suffered the opposite effect. The results for this freshwater bivalve show that, in contrast to marine bivalves, free amino acids concentration increases in response to stress, presenting a negative relationship between size (measured as shell length) and concentration of free amino. Cadmium lowered the adenylate energy charge (AECh = (ATP + 1/2 ADP)/(ATP + ADP + AMP)), which can be used as a measure of instant energy availability (Giesy et al. 2008), which can be considered as biomarker of stress response, considering that pollution induces an increase in the energy requirements. It has been observed that small clams respond to Cd and Zn exposure with lower levels of ADP and ATP, but not larger specimens. Conversely, in large clams of the levels of carbohydrates like glycogen, glucose and turanose decreased in response to Cd and Zn treatment. The change in these energy molecules can also be considered a general stress response (Spann et al. 2011). We can conclude that small clams used amino acids as an energy resource in response to Cd and Zn exposure, and larger ones primarily drawing on their larger storage reserves of carbohydrates. In Table 3 are collected the most important changes of metabolites in *C. fluminea* under Cd and Zn exposure.

Crustaceans

Daphnia magna planktonic crustaceans have been used in assessment of metabolic changes under metal contamination in marine ecosystems. *D. magna* exposed to As, Cu, or Li (Nagato et al. 2013) showed remarked metabolites variations related to a number of amino acids, as well as uracil and glycerophosphocholine, which contributes to the separation observed in the principal components analysis (PCA) score plots. For copper and lithium exposures, metabolites that contributed significantly to the PCA separation from the control include: alanine, leucine, isoleucine, valine, methionine, glutamate, glycine, and glycerophosphocholine. For arsenic exposure, alanine, isoleucine, glycine, methionine, glutamate, threonine and arginine contributed to the variation from the control. Specific metabolites that suffer statistically significant changes with exposure are alanine, depleted under the action of arsenic, and lysine, overexpressed in this experiment. Under both copper and lithium exposures amino acids such as alanine, lysine, leucine, methionine, valine, glutamine, glycine, phenylalanine and glutamate decreased as well as uracil, and increases glycerophosphocholine (Table 4). Both lithium and copper is suspected to disrupt the Na^+/K^+ ATPase function that affect the Na^+ levels (Bianchini and Wood 2008), which explains the similarities of metabolomes of *D. magna* under copper and lithium exposure. In addition, different amino acids (alanine, leucine, glutamate, valine and methionine), intermediaries of the tricarboxylic acid (TCAr) cycle, decreased to alleviate metal toxicity. Particularly depletion is suffered by alanine involved in the conversion to pyruvate in the TCAr cycle (Ekman et al. 2007; Southam et al. 2008), which suggests an increase in energy metabolism as survival mechanism against metal exposure (Spann et al. 2011). This fact is related to the oxidative stress caused by copper (Barata et al. 2005; Lushchak 2011), which can affect the basal energetic metabolism of these living organisms (Connon et al. 2008; Booth et al. 2011). On the other hand, the depletion of amino acids suggests the increased of metabolic burden (De Schamphelaere and Janssen 2004) or impaired respiration (Khangarot and Rathore 2003; Giarratano et al. 2007). As a consequence, the lack of available amino acids can compromise downstream protein synthesis, thereby inhibiting organismal growth. The marked decrease of uracil under the exposure to copper can be related to alterations of nucleotide metabolism that may reduce RNA production, protein synthesis and organismal growth (Boer et al. 2010), which is consistent with the observed amino acids decrease.

Although there are many metabolomics studies of crustaceans based on ^{1}H-NMR, Taylor et al. proposed the use of direct infusion high resolution mass spectrometry for high throughput metabolomics analysis of this bioindicator of contamination based on FT-ICR-MS (Taylor et al. 2008). Due to the reduced size of *D. magna* optimization of biomass used for analysis have been checked, establishing that one whole adult could be suitable for the analysis or alternatively thirty neonates. Copper exposure induced cellular toxicity due to the participation of this element in the formation of reactive oxygen species (ROS) (Stohs 1995; Gaetke 2003), which is the consequence of significant reduction of glutathione (GSH) that can be related to the protective effect of this molecule in the stabilization of Cu^{2+} oxidizing state, preventing redox cycling and generation of free radicals. This fact can also explain the important reduction of GSH-analogue ophthalmic acid that can substitute GSH along depletion of this metabolite as a consequence of oxidative stress (Soga et al. 2006). In addition, ROS can also oxidize amino acids, which explain the decreased concentration of most susceptible to oxidation amino acids

Table 4: Metabolic changes caused by contaminationin crustaceans.

Marine organism	Contaminant	Metabolite change	Target organ	Analytical technique	Reference
Daphnia magna Planktonic crustaceans	Cu	↓Threonine, ↓proline, ↓glutamic acid, ↓methionine, ↓histidine, ↓leucine, ↓isoleucine, ↓lysine, ↓arginine, ↓phenylalanine, ↓tyrosine, ↓tryptophan, ↓glutathione ↓opthalmic acid	Whole body-adult 30-neonates	FT-ICR-MS	(Taylor et al. 2008)
	Cd	↑Glutathione, ↑glutamylcysteine, ↑fructose, ↑phosphate, ↑chitobiose, ↑arginine, ↓N-acetylglucosamine, ↓arachidonic acid	Whole body Hemolymph	DIMS	(Taylor et al. 2010)
		↓Lauric acid, ↓myristic acid, ↓palmitic acid, ↓stearic acid, ↓proline, ↓methionine, ↓histidine, ↓arginine, ↓phenylalanine, ↓tryptophan, ↓lysine	Hemolymph	DIMS and 1H-NMR	(Poynton et al. 2011)
Procambarus clarkia crab	Metal contamination	↓Phospholipids, ↓carnosine, ↓ascorbate, ↓niacinamide, ↓pantothenic acid, ↓ acetoacetate, ↑acetyl-carnitine, ↓alanine, ↓TGs	Digestive gland	DIMS	(Gago-Tinoco et al. 2014)
		↓Phospholipids, ↑choline, ↑uric acid, ↓TGs	Gills		
		↓Phospholipids, ↑phosphocholine, ↓glucose	Antennal gland		
		↓Phospholipids, ↓glucose, ↓glucose-6-phosphate, ↓alanine, ↓arginine	Muscle		

(Table 4). Further, amino acids can be used in metabolic repairing mechanisms, such as synthesis of stress proteins and DNA repair enzymes (Knops et al. 2001; Smolders et al. 2005).

Direct infusion Fourier transform ion cyclotron resonance mass spectrometry (FTICR-MS) metabolomics was also applied to evaluate the toxicological effects of Cd in whole-organism homogenate in comparison with hemolymph of single adult *D. magna*. It can be concluded that metabolic profiles of whole *D. magna* homogenates are more discriminatory for evaluation of toxicant action than hemolymph (Taylor et al. 2010). Additionally, similar studies were performed by the complementary use of 1H-NMR showing different altered metabolites depending on the analytical technique (Taylor et al. 2010; Poynton et al. 2011), Table 4.

Other study considers the metabolic response of the crab *Procambarus clarkii* for aquatic environmental issues assessment. The metabolic approach is based in the application of DIMS to monitor the contamination of Doñana Natural Park (southwest Spain). Different organs of the crab were used in the study such as digestive gland, gill and muscle. Several metabolites have been proposed as biomarkers of metal pollution: (a) decreased levels of carnosine, alanine, niacinamide, acetoacetate, pantothenic acid, ascorbate, glucose-6-phosphate, arginine, glucose, lactate, phospholipids, and tryglicerides; and (b) overexpression of acetyl carnitine, phosphocholine, choline, and uric acid. In connection to this, metal-induced toxicity could be related to metabolic impairments, principally oxidative stress, metabolic dysfunction, and dyslipidemia (Gago-Tinoco et al. 2014), Table 4.

Marine Teleost

Salmonids are among the more studied marine teleost from toxicological point of view and they have been the focus of the different analytical profiling tools, such as transcriptomics, proteomics and metabolomics. Most of these studies have dealt with the freshwater rainbow trout (*Onchorhynchus mykiss*) as model salmonid. However, the impact on muscle and liver metabolomes of Chinook salmon pre-smolts and smolts exposed to crude oil treated with and without chemical dispersant was recently investigated (Tjeerdema 2008; Lin et al. 2009; Van Scoy et al. 2010). Although the water-accommodated fraction of oil alone (WAF) showed an altered metabolite profile for both tissues, exposure to chemically-dispersed crude oil (CEWAF) resulted in more significant changes, although this form of contaminant oil is less toxic by lethality measures. The hydrocarbon-related metabolite signature included altered abundance of amino acids, formate, lactate, ATP, glycerophosphorylcholine, and phosphocreatine (Tjeerdema 2008; Lin et al. 2009; Van Scoy et al. 2010). In muscle, amino acids including valine, alanine, arginine/phosphoarginine, glutamate, and glutamine increased after either WAF or CEWAF exposure, but each metabolite's pattern was unique. Significant increases were observed at low (glutamine), intermediate (alanine), and high WAF concentrations (arginine/phosphoarginine). Significant increases were also observed at both low and high—but not intermediate—CEWAF concentrations (i.e., valine). The increase in taurine was significant at a high WAF concentration, and at the lowest CEWAF concentrations, while glycerophosphorylcholine decreased significantly at all concentrations of CEWAF (Table 5). In liver, hepatic valine concentrations were elevated significantly in smolts exposed to high WAF or CEWAF concentrations, glutamine and glycine were found to be significantly elevated only at intermediate and high CEWAF concentrations, respectively. Glycerophosphorylcholine decreased significantly at low to intermediate WAF concentrations. Hepatic β-glucose increased with exposure to intermediate CEWAF concentration. Formate increased dramatically after exposure to intermediate and high concentrations of WAF, while AMP decreased after exposure to high concentrations of either WAF or CEWAF (Lin et al. 2009) (Table 5).

Together with observations on metabolomic responses of early life stage Chinook (eyed eggs and alevins) to the agricultural pesticides, dinoseb, diazinon, and esfenvalerate, it is apparent that metabolome profile signatures can be obtained following sublethal exposures that aid in identification of the nature of the chemical contaminant (Viant et al. 2006a; Tjeerdema 2008; Lin et al. 2009). A similar study was performed by Viant et al. (Viant et al. 2006b) considering the toxic actions of dinoseb in medaka (*Oryziaslatipes*) embryos. The metabolite changes caused by this herbicide are related to increased levels of lactate and decreased levels of ATP, phosphocreatine, alanine and tyrosine in exposed fish, Table 5.

Table 5: Metabolic changes caused by contamination in teleost.

Marine organism	Contaminant	Metabolite change	Target organ	Analytical technique	Reference
Chinook salmon	Crude oil (WAF)	↑Valine, ↑alanine, ↑arginine/phosphoarginine, ↑glutamate, ↑glutamine	Muscle	^{1}H-NMR	(Lin et al. 2009)
	WAF-low	↑Glutamine			
	WAF-intermediate	↑Alanine			
	WAF-high	↑Arginine/phosphoarginine, ↑taurine			
	Chemically-dispersed crude oil (CEWAF)	↑Valine, ↑alanine, ↑arginine/phosphoarginine, ↑glutamate, ↑glutamine			
	CEWAF-low	↑Valine, ↑taurine, ↓glycerophosphocholine			
	CEWAF-intermediate	↓Glycerophosphocholine			
	CEWAF-high	↑Valine, ↓glycerophosphocholine			
	WAF-low	↓Glycerophosphocholine	Liver		
	WAF-intermediate	↑Valine, ↓glycerophosphocholine, ↑formato			
	WAF-high	↑Formato, ↓AMP			
	CEWAF-low	↑Glutamine			
	CEWAF-intermediate	↑β-glucose			
	CEWAF-high	↑Valine, ↑glycine, ↓AMP			
Medaka (*Oryziaslatipes*)	Dinoseb (herbicide)	↑Lactate↓ATP, ↓phosphocreatine, ↓ alanine, ↓ tyrosine	Embryos	^{1}H-NMR	(Viant et al. 2006a)
European flounder (*Platichthys flesus*)	PAHs, PCBS, heavy metals, TBT (In mesocosm)	6 metabolites (non-identified) significantly increased in contaminated sediment 12 significantly decreased : choline, glucose, glutamine, glycine, lactate, malonate, maltose, taurine, N,N-dimethylglycine, o-phosphocholine, taurine, tyrosine and valine are similarly expressed in contaminated and non-contaminated sediments	Muscle	^{1}H-NMR	(Williams et al. 2014)

Flounder fish (*Platichthys flesus*) were exposed in mesocosms during seven months to contaminated estuarine sediment containing high organics and high metals and tributyltin, and the results compared with non-contaminated control sediment. The contaminated sediment contained higher concentrations of key environmental pollutants, including polycyclic aromatic hydrocarbons (PAHs), chlorinated biphenyls and heavy metals (Williams et al. 2014). The metabolomics analysis reveals that six metabolites significantly increased in concentration in contaminated sediment and 12 significantly decreased, but none of these could be identified. A number of metabolites similarly expressed in both groups were identified choline, glucose, glutamine, glycine, lactate, malonate, maltose, taurine, N,N-dimethylglycine, o-phosphocholine, taurine, tyrosine and valine (Table 5).

Conclusions

Marine environmental metabolomics is a very powerful approach to research the changes affecting marine organisms associated to contamination and extreme natural conditions, such as temperature, pCO_2, salinity, and climate global change. Other factors linked to the habitat of these living organisms, mainly food limitation or occurrence in crowded or scarce colonies are also manifested in the metabolome, since it mirror the organism phenotype and the influence of exogenous factors.

Conventional metabolomics approaches are mainly based on ^{1}HNMR, and many marine organisms studies apply this technique, although it is being progressively substituted by MS due to the higher sensitivity. Generally, the couplings HPLC-MS and GC-MS are the choice for environmental metabolites profiling, although DIMS provides very good results in high throughput analysis, in which extraction procedure of metabolites is crucial and demand further contributions to assure metabolite coverage and metabolite quenching during handling procedures, which is leading to *in-vivo* sampling techniques. In addition, HPLC-ICP-MS, used in combination with the previous techniques, provides information about metal-containing metabolites (metallomics) that complete the general picture of metabolites changes caused by marine environment in the organisms that inhabit.

The application of metabolomics to marine organisms is a challenge task that has been mainly focused on bivalves and crustacean, in which important metabolic cycles affected by contamination or extreme marine conditions suffer important changes such as osmotic regulation, energy metabolism and neurotransmission. However, additional research is now guided to fishes and especially to marine mammals as pinnipeds and cetaceans.

Acknowledgements

This work has been supported by the projects CTM2012-38720-C03-01 and CTM2015-67902-C2-1-P from the Spanish Ministry of Economy and Competitiveness and P12-FQM-0442 from the Regional Ministry of Economy, Innovation, Science and Employment (Andalusian Government, Spain). Finally, authors are grateful to FEDER (European Community) for financial support, grants number UNHU13-1E-1611 and UNHU15-CE-3140.

Keywords: Environmental metabolomics, metallomics, speciation, mass spectrometry, ^{1}HNMR, GC-MS, HPLC-MS, HPLC-ICP-MS, omics integration, bivalves, gills, hepatopancreas, marine microbe, crustacean, teleost, marine mammals

References

Adam-Guillermin, C., E. Fournier, M. Floriani, V. Camilleri, J.-C. Massabuau and J. Garnier-Laplace. 2009. Biodynamics, subcellular partitioning, and ultrastructural effects of organic selenium in a freshwater bivalve. Environ. Sci. Technol. 43: 2112–2117.

Alvarez, M. del C., J.A. Donarski, M. Elliott and A.J. Charlton. 2010. Evaluation of extraction methods for use with NMR-based metabolomics in the marine polychaete ragworm, Hediste diversicolor. Metabolomics 6: 541–549.

An, Z., Y. Chen, R. Zhang, Y. Song, J. Sun, J. He, J. Bai, L. Dong, Q. Zhan and Z. Abliz. 2010. Integrated ionization approach for RRLC-MS/MS-based metabonomics: finding potential biomarkers for lung cancer. J. Proteome Res. 9: 4071–81.

Ankrah, N.Y.D., A.L. May, J.L. Middleton, D.R. Jones, M.K. Hadden, J.R. Gooding, G.R. LeCleir, S.W. Wilhelm, S.R. Campagna and A. Buchan. 2014. Phage infection of an environmentally relevant marine bacterium alters host metabolism and lysate composition. ISME J. 8: 1089–100.

Arnason, U., J.A. Adegoke, A. Gullberg, E.H. Harley, A. Janke and M. Kullberg. 2008. Mitogenomic relationships of placental mammals and molecular estimates of their divergences. Gene 421: 37–51.

Barata, C., I. Varo, J.C. Navarro, S. Arun and C. Porte. 2005. Antioxidant enzyme activities and lipid peroxidation in the freshwater cladoceran *Daphnia magna* exposed to redox cycling compounds. Comp. Biochem. Physiol. C. Toxicol. Pharmacol. 140: 175–86.

Bettmer, J., M. Montes Bayón, J.R. Encinar, M.L. Fernández Sánchez, M. del R. Fernández de la Campa and A. Sanz Medel. 2009. The emerging role of ICP-MS in proteomic analysis. J. Proteomics 72: 989–1005.

Bianchini, A. and C.M. Wood. 2008. Sodium uptake in different life stages of crustaceans: the water flea *Daphnia magna* Strauss. J. Exp. Biol. 211: 539–47.

Bininda-Emonds, O.R.P., M. Cardillo, K.E. Jones, R.D.E. MacPhee, R.M.D. Beck, R. Grenyer, S.A. Price, R.A. Vos, J.L. Gittleman and A. Purvis. 2007. The delayed rise of present-day mammals. Nature 446: 507–12.

Bjerregaard, P. and A. Christensen. 2012. Selenium reduces the retention of methyl mercury in the brown shrimp Crangon crangon. Environ. Sci. Technol. 46: 6324–9.

Boer, V.M., C.A. Crutchfield, P.H. Bradley, D. Botstein and J.D. Rabinowitz. 2010. Growth-limiting intracellular metabolites in yeast growing under diverse nutrient limitations. Mol. Biol. Cell 21: 198–211.

Boernsen, K.O., S. Gatzek and G. Imbert. 2005. Controlled protein precipitation in combination with chip-based nanospray infusion mass spectrometry. An approach for metabolomics profiling of plasma. Anal. Chem. 77: 7255–64.

Bolten, C.J., P. Kiefer, F. Letisse, J.C. Portais and C. Wittmann. 2007. Sampling for metabolome analysis of microorganisms. Anal. Chem. 79: 3843–9.

Booth, S.C., M.L. Workentine, J. Wen, R. Shaykhutdinov, H.J. Vogel, H. Ceri, R.J. Turner and A.M. Weljie. 2011. Differences in metabolism between the biofilm and planktonic response to metal stress. J. Proteome Res. 10: 3190–9.

Bosch, A.C., B. O'Neill, G.O. Sigge, S.E. Kerwath and L.C. Hoffman. 2016. Mercury accumulation in Yellowfin tuna (Thunnus albacares) with regards to muscle type, muscle position and fish size. Food Chem. 190: 351–6.

Bowman, K.L., C.R. Hammerschmidt, C.H. Lamborg and G. Swarr. 2015. Mercury in the North Atlantic Ocean: The U.S. GEOTRACES zonal and meridional sections. Deep Sea Res. Part II Top. Stud. Oceanogr. 116: 251–261.

Bryszewska, M.A. and A. Måge. 2015. Determination of selenium and its compounds in marine organisms. J. Trace Elem. Med. Biol. 29: 91–8.

Cabañero, A.I., Y. Madrid and C. Cámara. 2004. Selenium and mercury bioaccessibility in fish samples: an *in vitro* digestion method. Anal. Chim. Acta 526: 51–61.

Cabañero, A.I., C. Carvalho, Y. Madrid, C. Batoréu and C. Cámara. 2005. Quantification and speciation of mercury and selenium in fish samples of high consumption in Spain and Portugal. Biol. Trace Elem. Res. 103: 17–35.

Cai, S.-S. and J.A. Syage. 2006. Comparison of atmospheric pressure photoionization, atmospheric pressure chemical ionization, and electrospray ionization mass spectrometry for analysis of lipids. Anal. Chem. 78: 1191–9.

Cano-Sancho, G., G. Perelló, A.L. Maulvault, A. Marques, M. Nadal and J.L. Domingo. 2015. Oral bioaccessibility of arsenic, mercury and methylmercury in marine species commercialized in Catalonia (Spain) and health risks for the consumers. Food Chem. Toxicol. 86: 34–40.

Cappello, T., A. Mauceri, C. Corsaro, M. Maisano, V. Parrino, G. Lo Paro, G. Messina and S. Fasulo. 2013. Impact of environmental pollution on caged mussels *Mytilus galloprovincialis* using NMR-based metabolomics. Mar. Pollut. Bull. 77: 132–9.

Cappello, T., M. Maisano, A. Giannetto, V. Parrino, A. Mauceri and S. Fasulo. 2015. Neurotoxicological effects on marine mussel *Mytilus galloprovincialis* caged at petrochemical contaminated areas (eastern Sicily, Italy): ^{1}H NMR and immunohistochemical assays. Comp. Biochem. Physiol. C. Toxicol. Pharmacol. 169: 7–15.

Carregosa, V., E. Figueira, A.M. Gil, S. Pereira, J. Pinto, A.M.V.M. Soares and R. Freitas. 2014. Tolerance of Venerupis philippinarum to salinity: osmotic and metabolic aspects. Comp. Biochem. Physiol. A. Mol. Integr. Physiol. 171: 36–43.

Cavaliere, C., P. Foglia, E. Pastorini, R. Samperi and A. Laganà. 2006. Liquid chromatography/tandem mass spectrometric confirmatory method for determining aflatoxin M1 in cow milk: comparison between electrospray and atmospheric pressure photoionization sources. J. Chromatogr. A 1101: 69–78.

Cheng, H., G. Sun, K. Yang, R.W. Gross and X. Han. 2010. Selective desorption/ionization of sulfatides by MALDI-MS facilitated using 9-aminoacridine as matrix. J. Lipid Res. 51: 1599–609.

Connon, R., H.L. Hooper, R.M. Sibly, F.-L. Lim, L.-H. Heckmann, D.J. Moore, H. Watanabe, A. Soetaert, K. Cook, S.J. Maund, T.H. Hutchinson, J. Moggs, W. Coen, T. Iguchi and A. Callaghan. 2008. Linking molecular and population stress responses in *Daphnia magna* exposed to cadmium. Environ. Sci. Technol. 42: 2181–2188.

Contreras-Acuña, M., T. García-Barrera, M.A. García-Sevillano and J.L. Gómez-Ariza. 2013. Speciation of arsenic in marine food (Anemonia sulcata) by liquid chromatography coupled to inductively coupled plasma mass spectrometry and organic mass spectrometry. J. Chromatogr. A 1282: 133–41.

Contreras-Acuña, M., T. García-Barrera, M.A. García-Sevillano and J.L. Gómez-Ariza. 2014. Arsenic metabolites in human serum and urine after seafood (Anemonia sulcata) consumption and bioaccessibility assessment using liquid chromatography coupled to inorganic and organic mass spectrometry. Microchem. J. 112: 56–64.

Coulier, L., R. Bas, S. Jespersen, E. Verheij, M.J. van der Werf and T. Hankemeier. 2006. Simultaneous quantitative analysis of metabolites using ion-pair liquid chromatography-electrospray ionization mass spectrometry. Anal. Chem. 78: 6573–82.

Cubbon, S., T. Bradbury, J. Wilson and J. Thomas-Oates. 2007. Hydrophilic interaction chromatography for mass spectrometric metabonomic studies of urine. Anal. Chem. 79: 8911–8.

Czech, C., P. Berndt, K. Busch, O. Schmitz, J. Wiemer, V. Most, H. Hampel, J. Kastler and H. Senn. 2012. Metabolite profiling of Alzheimer's disease cerebrospinal fluid. PLoS One 7: e31501.

De Almeida, E.A., S. Miyamoto, A.C.D. Bainy, M.H.G. de Medeiros and P. Di Mascio. 2004. Protective effect of phospholipid hydroperoxide glutathione peroxidase (PHGPx) against lipid peroxidation in mussels Perna perna exposed to different metals. Mar. Pollut. Bull. 49: 386–92.

De Schamphelaere, K.A.C. and C.R. Janssen. 2004. Effects of chronic dietary copper exposure on growth and reproduction of daphnia magna. Environ. Toxicol. Chem. 23: 2038.

Del Castillo, E. and W.E. Robinson. 2008. Nuclear and cytosolic distribution of metallothionein in the blue mussel *Mytilus edulis* L. Comp. Biochem. Physiol. B. Biochem. Mol. Biol. 151: 46–51.

Draper, J., A.J. Lloyd, R. Goodacre and M. Beckmann. 2012. Flow infusion electrospray ionisation mass spectrometry for high throughput, non-targeted metabolite fingerprinting: a review. Metabolomics 9: 4–29.

Dunn, W.B. and D.I. Ellis. 2005. Metabolomics: Current analytical platforms and methodologies. TrAC Trends Anal. Chem. 24: 285–294.

Duportet, X., R.B.M. Aggio, S. Carneiro and S.G. Villas-Bôas. 2011. The biological interpretation of metabolomic data can be misled by the extraction method used. Metabolomics 8: 410–421.

Ebdon, L., M.E. Foulkes, S. Le Roux and R. Muñoz-Olivas. 2002. Cold vapour atomic fluorescence spectrometry and gas chromatography-pyrolysis-atomic fluorescence spectrometry for routine determination of total and organometallic mercury in food samples. Analyst 127: 1108–1114.

Ekman, D.R., Q. Teng, K.M. Jensen, D. Martinovic, D.L. Villeneuve, G.T. Ankley and T.W. Collette. 2007. NMR analysis of male fathead minnow urinary metabolites: a potential approach for studying impacts of chemical exposures. Aquat. Toxicol. 85: 104–12.

Elder, J.F. and J.J. Collins. 1991. Freshwater molluscs as indicators of bioavailability and toxicity of metals in surface-water systems. Rev. Environ. Contam. Toxicol. 122: 37–39.

Evans, A.M., C.D. DeHaven, T. Barrett, M. Mitchell and E. Milgram. 2009. Integrated, nontargeted ultrahigh performance liquid chromatography/electrospray ionization tandem mass spectrometry platform for the identification and relative quantification of the small-molecule complement of biological systems. Anal. Chem. 81: 6656–67.

Fernández-Varela, R., G. Tomasi and J.H. Christensen. 2015. An untargeted gas chromatography mass spectrometry metabolomics platform for marine polychaetes. J. Chromatogr. A 1384: 133–41.

Fiehn, O. 2008. Extending the breadth of metabolite profiling by gas chromatography coupled to mass spectrometry. Trends Analyt. Chem. 27: 261–269.

Fournier, E., C. Adam, J.-C. Massabuau and J. Garnier-Laplace. 2005. Bioaccumulation of waterborne selenium in the Asiatic clam Corbicula fluminea: influence of feeding-induced ventilatory activity and selenium species. Aquat. Toxicol. 72: 251–60.

Fournier, E., C. Adam, J.-C. Massabuau and J. Garnier-Laplace. 2006. Selenium bioaccumulation in chlamydomonas reinhardtii and subsequent transfer to Corbicula fluminea: role of selenium speciation and bivalve ventilation. Environ. Toxicol. Chem. 25: 2692.

Gaetke, L. 2003. Copper toxicity, oxidative stress, and antioxidant nutrients. Toxicology 189: 147–163.

Gago-Tinoco, A., R. González-Domínguez, T. García-Barrera, J. Blasco-Moreno, M.J. Bebianno and J.L. Gómez-Ariza. 2014. Metabolic signatures associated with environmental pollution by metals in Doñana National Park using *P. clarkii* as bioindicator. Environ. Sci. Pollut. Res. Int. 21: 13315–23.

Gao, S., Z.P. Zhang and H.T. Karnes. 2005. Sensitivity enhancement in liquid chromatography/atmospheric pressure ionization mass spectrometry using derivatization and mobile phase additives. J. Chromatogr. B. Analyt. Technol. Biomed. Life Sci. 825: 98–110.

García-Sevillano, M.A., M. González-Fernández, R. Jara-Biedma, T. García-Barrera, J. López-Barea, C. Pueyo and J.L. Gómez-Ariza. 2012. Biological response of free-living mouse Mus spretus from Doñana National Park under environmental stress based on assessment of metal-binding biomolecules by SEC-ICP-MS. Anal. Bioanal. Chem. 404: 1967–81.

García-Sevillano, M.A., T. García-Barrera, F. Navarro and J.L. Gómez-Ariza. 2013. Analysis of the biological response of mouse liver (Mus musculus) exposed to As2O3 based on integrated-omics approaches. Metallomics 5: 1644–55.

García-Sevillano, M.A., T. García-Barrera, F. Navarro, N. Abril, C. Pueyo, J. López-Barea and J.L. Gómez-Ariza. 2014a. Use of metallomics and metabolomics to assess metal pollution in Doñana National Park (SW Spain). Environ. Sci. Technol. 48: 7747–55.

García-Sevillano, M.A., T. García-Barrera, F. Navarro, J. Gailer and J.L. Gómez-Ariza. 2014b. Use of elemental and molecular-mass spectrometry to assess the toxicological effects of inorganic mercury in the mouse Mus musculus. Anal. Bioanal. Chem. 406: 5853–65.

García-Sevillano, M.A., T. García-Barrera, F. Navarro-Roldán, Z. Montero-Lobato and J.L. Gómez-Ariza. 2014c. A combination of metallomics and metabolomics studies to evaluate the effects of metal interactions in mammals. Application to Mus musculus mice under arsenic/cadmium exposure. J. Proteomics 104: 66–79.

García-Sevillano, M.A., M. Contreras-Acuña, T. García-Barrera, F. Navarro and J.L. Gómez-Ariza, J.L. 2014d. Metabolomic study in plasma, liver and kidney of mice exposed to inorganic arsenic based on mass spectrometry. Anal. Bioanal. Chem. 406: 1455–1469.

García-Sevillano, M.A., T. García-Barrera and J.L. Gómez-Ariza. 2014e. Application of metallomic and metabolomic approaches in exposure experiments on laboratory mice for environmental metal toxicity assessment. Metallomics 6: 237–48.

García-Sevillano, M.A., T. García-Barrera and J.L. Gómez-Ariza. 2015. Environmental metabolomics: Biological markers for metal toxicity. Electrophoresis 36: 2348–2365.

Giarratano, E., L. Comoglio and O. Amin. 2007. Heavy metal toxicity in Exosphaeroma gigas (Crustacea, Isopoda) from the coastal zone of Beagle Channel. Ecotoxicol. Environ. Saf. 68: 451–62.

Giesy, J.P., C.S. Duke, R.D. Bingham and G.W. Dickson. 2008. Changes in phosphoadenylate concentrations and adenylate energy charge as an integrated biochemical measure of stress in invertebrates: The effects of cadmium on the freshwater clam Corbicula fluminea. Toxicol. Environ. Chem. 6: 259–295.

Glass, D.J. and N. Hall. 2008. A brief history of the hypothesis. Cell 134: 378–81.

Goldsmith, P., H. Fenton, G. Morris-Stiff, N. Ahmad, J. Fisher and K.R. Prasad. 2010. Metabonomics: a useful tool for the future surgeon. J. Surg. Res. 160: 122–32.

Gomez-Ariza, J.L., E. Morales and I. Giraldez. 1998. Spatial distribution of butyltin and phenyltin compounds on the Huelva coast (Southwest Spain). Chemosphere 37: 937–950.

Gomez-Ariza, J.L., E. Morales and I. Giraldez. 1999. Uptake and elimination of tributyltin in clams, Venerupis decussata. Mar. Environ. Res. 47: 399–413.

Gomez-Ariza, J.L., I. Giraldez and E. Morales. 2001. Occurrence of organotin compounds in water, sediments and mollusca in estuarine systems in the Southwest of Spain. Water Air Soil Pollut. 126: 253–270.

Gómez-Ariza, J.L., M.A. Caro de la Torre, I. Giráldez, D. Sánchez-Rodas, A. Velasco and E. Morales. 2002. Pretreatment procedure for selenium speciation in shellfish using high-performance liquid chromatography-microwave-assisted digestion-hydride generation-atomic fluorescence spectrometry. Appl. Organomet. Chem. 16: 265–270.

Gómez-Ariza, J.L., T. García-Barrera, F. Lorenzo, V. Bernal, M.J. Villegas and V. Oliveira. 2004. Use of mass spectrometry techniques for the characterization of metal bound to proteins (metallomics) in biological systems. Anal. Chim. Acta 524: 15–22.

Gómez-Ariza, J.L., M.M. Santos, E. Morales, I. Giráldez, D. Sánchez-Rodas, N. Vieira, J.F. Kemp, J.P. Boon and C.C. Ten-Hallers-Tjabbes. 2006. Organotin contamination in the Atlantic Ocean off the Iberian Peninsula in relation to shipping. Chemosphere 64: 1100–8.

Gómez-Ariza, J.L., E.Z. Jahromi, M. González-Fernández, T. García-Barrera and J. Gailer. 2011. Liquid chromatography-inductively coupled plasma-based metallomic approaches to probe health-relevant interactions between xenobiotics and mammalian organisms. Metallomics 3: 566–77.

Gómez Ariza, J.L., T. García-Barrera, M.A. García-Sevillano, M. González-Fernández and V. Gómez-Jacinto. 2013. Heavy Metal Stress in Plants, Heavy Metal Stress in Plants. Springer Berlin Heidelberg, Berlin, Heidelberg.

González-Domínguez, R., T. García-Barrera and J.L. Gómez-Ariza. 2014. Using direct infusion mass spectrometry for serum metabolomics in Alzheimer's disease. Anal. Bioanal. Chem. 406: 7137–48.

González-Domínguez, R., T. García-Barrera and J.L. Gómez-Ariza. 2015. Application of a novel metabolomic approach based on atmospheric pressure photoionization mass spectrometry using flow injection analysis for the study of Alzheimer's disease. Talanta 131: 480–9.

Greig, M.J., B. Bolaños, T. Quenzer and J.M.R. Bylund. 2003. Fourier transform ion cyclotron resonance mass spectrometry using atmospheric pressure photoionization for high-resolution analyses of corticosteroids. Rapid Commun. Mass Spectrom. 17: 2763–2768.

Halket, J.M., D. Waterman, A.M. Przybarowska, R.K. Patel, P.D. Fraser and P.M. Bramley. 2005. Chemical derivatization and mass spectral libraries in metabolic profiling by GC/MS and LC/MS/MS. J. Exp. Bot. 56: 219–243.

Han, J., R.M. Danell, J.R. Patel, D.R. Gumerov, C.O. Scarlett, J.P. Speir, C.E. Parker, I. Rusyn, S. Zeisel and C.H. Borchers. 2008. Towards high-throughput metabolomics using ultrahigh-field Fourier transform ion cyclotron resonance mass spectrometry. Metabolomics 4: 128–140.

Han, J., R. Datla, S. Chan and C.H. Borchers. 2009. Mass spectrometry-based technologies for high-throughput metabolomics. Bioanalysis 1: 1665–84.

Ibáñez, C., C. Simó, D.K. Barupal, O. Fiehn, M. Kivipelto, A. Cedazo-Mínguez and A. Cifuentes. 2013. A new metabolomic workflow for early detection of Alzheimer's disease. J. Chromatogr. A 1302: 65–71.

Ibáñez-Palomino, C., J.F. López-Sánchez and A. Sahuquillo. 2012. Certified reference materials for analytical mercury speciation in biological and environmental matrices: do they meet user needs? A review. Anal. Chim. Acta 720: 9–15.

Inagaki, K., A. Takatsu, T. Watanabe, Y. Aoyagi, T. Yarita, K. Okamoto and K. Chiba. 2007. Certification of butyltins and phenyltins in marine sediment certified reference material by species-specific isotope-dilution mass spectrometric analysis using synthesized 118Sn-enriched organotin compounds. Anal. Bioanal. Chem. 387: 2325–34.

Ji, C., H. Wu, X. Liu, J. Zhao, J. Yu and X. Yin. 2013. The influence of salinity on toxicological effects of arsenic in digestive gland of clam Ruditapes philippinarum using metabolomics. Chinese J. Oceanol. Limnol. 31: 345–352.

Ji, C., L. Cao and F. Li. 2015a. Toxicological evaluation of two pedigrees of clam Ruditapes philippinarum as bioindicators of heavy metal contaminants using metabolomics. Environ. Toxicol. Pharmacol. 39: 545–54.

Ji, C., Q. Wang, H. Wu, Q. Tan and W.-X. Wang. 2015b. A metabolomic investigation of the effects of metal pollution in oysters Crassostrea hongkongensis. Mar. Pollut. Bull. 90: 317–22.

Jia, S., Y.P. Kang, J.H. Park, J. Lee and S.W. Kwon. 2011. Simultaneous determination of 23 amino acids and 7 biogenic amines in fermented food samples by liquid chromatography/quadrupole time-of-flight mass spectrometry. J. Chromatogr. A 1218: 9174–82.

Jitaru, P., M. Roman, C. Barbante, S. Vaslin-Reimann and P. Fisicaro. 2010. Challenges in the accurate speciation analysis of selenium in humans: first report on indicative levels of selenoproteins in a serum certified reference material for total selenium (BCR-637). Accredit. Qual. Assur. 15: 343–350.

Jones, O., F. Dondero, A. Viarengo and J. Griffin. 2008. Metabolic profiling of *Mytilus galloprovincialis* and its potential applications for pollution assessment. Mar. Ecol. Prog. Ser. 369: 169–179.

Jørgensen, A., A.M.B. Giessing, L.J. Rasmussen and O. Andersen. 2005. Biotransformation of the polycyclic aromatic hydrocarbon pyrene in the marine polychaete nereis virens. Environ. Toxicol. Chem. 24: 2796.

Kammenga, J.E., R. Dallinger, M.H. Donker, H.R. Köhler, V. Simonsen, R. Triebskorn and J.M. Weeks. 2000. Biomarkers in terrestrial invertebrates for ecotoxicological soil risk assessment. Rev. Environ. Contam. Toxicol. 164: 93–147.

Kell, D.B. and S.G. Oliver. 2004. Here is the evidence, now what is the hypothesis? The complementary roles of inductive and hypothesis-driven science in the post-genomic era. Bioessays 26: 99–105.

Khangarot, B.S. and R.S. Rathore. 2003. Effects of copper on respiration, reproduction, and some biochemical parameters of water flea *Daphnia magna* Straus. Bull. Environ. Contam. Toxicol. 70: 112–7.

Kido Soule, M.C., K. Longnecker, W.M. Johnson and E.B. Kujawinski. 2015. Environmental metabolomics: Analytical strategies. Mar. Chem. 177: 374–387.

Kim, W.S., H.T. Huh, S.-H. Huh and T.W. Lee. 2001. Effects of salinity on endogenous rhythm of the Manila clam, Ruditapes philippinarum (Bivalvia: Veneridae). Mar. Biol. 138: 157–162.

Knops, M., R. Altenburger and H. Segner. 2001. Alterations of physiological energetics, growth and reproduction of *Daphnia magna* under toxicant stress. Aquat. Toxicol. 53: 79–90.

Kujawinski, E.B. 2011. The impact of microbial metabolism on marine dissolved organic matter. Ann. Rev. Mar. Sci. 3: 567–99.

Lapainis, T., S.S. Rubakhin and J.V. Sweedler. 2009. Capillary electrophoresis with electrospray ionization mass spectrometric detection for single-cell metabolomics. Anal. Chem. 81: 5858–64.

Lin, C.Y., B.S. Anderson, B.M. Phillips, A.C. Peng, S. Clark, J. Voorhees, H.-D.I. Wu, M.J. Martin, J. McCall, C.R. Todd, F. Hsieh, D. Crane, M.R. Viant, M.L. Sowby and R.S. Tjeerdema. 2009. Characterization of the metabolic actions of crude versus dispersed oil in salmon smolts via NMR-based metabolomics. Aquat. Toxicol. 95: 230–8.

Lin, L., Q. Yu, X. Yan, W. Hang, J. Zheng, J. Xing and B. Huang. 2010. Direct infusion mass spectrometry or liquid chromatography mass spectrometry for human metabonomics? A serum metabonomic study of kidney cancer. Analyst 135: 2970–8.

Liu, F. and W.-X. Wang. 2012. Proteome pattern in oysters as a diagnostic tool for metal pollution. J. Hazard. Mater. 239-240: 241–8.

Liu, X., L. Zhang, L. You, M. Cong, J. Zhao, H. Wu, C. Li, D. Liu and J. Yu. 2011a. Toxicological responses to acute mercury exposure for three species of Manila clam Ruditapes philippinarum by NMR-based metabolomics. Environ. Toxicol. Pharmacol. 31: 323–32.

Liu, X., L. Zhang, L. You, J. Yu, M. Cong, Q. Wang, F. Li, L. Li, J. Zhao, C. Li and H. Wu. 2011b. Assessment of clam Ruditapes philippinarum as heavy metal bioindicators using NMR-based metabolomics. Clean - Soil Air Water 39: 759–766.

Lord, H.L., R.P. Grant, M. Walles, B. Incledon, B. Fahie and J.B. Pawliszyn. 2003. Development and evaluation of a solid-phase microextraction probe for *in vivo* pharmacokinetic studies. Anal. Chem. 75: 5103–5115.

Lushchak, V.I. 2011. Environmentally induced oxidative stress in aquatic animals. Aquat. Toxicol. 101: 13–30.

Mayor, D.J., U. Sommer, K.B. Cook and M.R. Viant. 2015. The metabolic response of marine copepods to environmental warming and ocean acidification in the absence of food. Sci. Rep. 5: 13690.

McElroy, A.E. 1985. *In vivo* metabolism of benz[a]anthracene by the polychaete Nereis virens. Mar. Environ. Res. 17: 133–136.

McElroy, A.E. 1990. Polycyclic aromatic hydrocarbon metabolism in the polychaete Nereis virens. Aquat. Toxicol. 18: 35–50.

McSheehy, S., M. Marcinek, H. Chassaigne and J. Szpunar. 2000. Identification of dimethylarsinoyl-riboside derivatives in seaweed by pneumatically assisted electrospray tandem mass spectrometry. Anal. Chim. Acta 410: 71–84.

Meyer, S., J. Schulz, A. Jeibmann, M.S. Taleshi, F. Ebert, K.A. Francesconi and T. Schwerdtle. 2014. Arsenic-containing hydrocarbons are toxic in the *in vivo* model Drosophila melanogaster. Metallomics 6: 2010–4.

Miller, G.A., M.S. Islam, T.D.W. Claridge, T. Dodgson and S.J. Simpson. 2008. Swarm formation in the desert locust Schistocerca gregaria: isolation and NMR analysis of the primary maternal gregarizing agent. J. Exp. Biol. 211: 370–6.

Moco, S., J. Vervoort, R.J. Bino, R.C.H. De Vos and R. Bino. 2007. Metabolomics technologies and metabolite identification. TrAC Trends Anal. Chem. 26: 855–866.

Moing, A., M. Maucourt, C. Renaud, M. Gaudillère, R. Brouquisse, B. Lebouteiller, A. Gousset-Dupont, J. Vidal, D. Granot, B. Denoyes-Rothan, E. Lerceteau-Köhler and D. Rolin. 2004. Quantitative metabolic profiling by 1-dimensional 1 H-NMR analyses: application to plant genetics and functional genomics. Funct. Plant Biol. 31: 889.

Monton, M.R.N. and T. Soga. 2007. Metabolome analysis by capillary electrophoresis-mass spectrometry. J. Chromatogr. A 1168: 237–46.

Mounicou, S., J. Szpunar and R. Lobinski. 2009. Metallomics: the concept and methodology. Chem. Soc. Rev. 38: 1119–38.

Musteata, F.M., I. de Lannoy, B. Gien and J. Pawliszyn. 2008. Blood sampling without blood draws for *in vivo* pharmacokinetic studies in rats. J. Pharm. Biomed. Anal. 47: 907–12.

Nagato, E.G., J.C. D'eon, B.P. Lankadurai, D.G. Poirier, E.J. Reiner, A.J. Simpson and M.J. Simpson. 2013. (1)HNMR-based metabolomics investigation of *Daphnia magna* responses to sub-lethal exposure to arsenic, copper and lithium. Chemosphere 93: 331–7.

Nair, P.S. and W.E. Robinson. 1998. Calcium speciation and exchange between blood and extrapallial fluid of the quahog Mercenaria mercenaria (L.). Biol. Bull. 195: 43–51.

Nair, P.S. and W.E. Robinson. 1999. Purification and characterization of a histidine-rich glycoprotein that binds cadmium from the blood plasma of the bivalve *Mytilus edulis*. Arch. Biochem. Biophys. 366: 8–14.

Poynton, H.C., N.S. Taylor, J. Hicks, K. Colson, S. Chan, C. Clark, L. Scanlan, A.V. Loguinov, C. Vulpe and M.R. Viant. 2011. Metabolomics of microliter hemolymph samples enables an improved understanding of the combined metabolic and transcriptional responses of *Daphnia magna* to cadmium. Environ. Sci. Technol. 45: 3710–7.

Quevauviller, P., I. Drabæk, H. Muntau, M. Bianchi, A. Bortoli and B. Griepink. 1996. Certified reference materials (CRMs 463 and 464) for the quality control of total and methyl mercury determination in tuna fish. TrAC Trends Anal. Chem. 15: 160–167.

Rank, J., K.K. Lehtonen, J. Strand and M. Laursen. 2007. DNA damage, acetylcholinesterase activity and lysosomal stability in native and transplanted mussels (*Mytilus edulis*) in areas close to coastal chemical dumping sites in Denmark. Aquat. Toxicol. 84: 50–61.

Regoli, F. 2000. Total oxyradical scavenging capacity (TOSC) in polluted and translocated mussels: a predictive biomarker of oxidative stress. Aquat. Toxicol. 50: 351–361.

Robb, D.B. and M.W. Blades. 2008. State-of-the-art in atmospheric pressure photoionization for LC/MS. Anal. Chim. Acta 627: 34–49.

Robinson, W.E. 2013. The need for integrative models of metal transport and transfer in marine invertebrates. JSM Environ. Sci. Ecol. 1: 5–6.

Romano, S., T. Dittmar, V. Bondarev, R.J.M. Weber, M.R. Viant and H.N. Schulz-Vogt. 2014. Exo-metabolome of Pseudovibrio sp. FO-BEG1 analyzed by ultra-high resolution mass spectrometry and the effect of phosphate limitation. PLoS One 9: e96038.

Römisch-Margl, W., C. Prehn, R. Bogumil, C. Röhring, K. Suhre and J. Adamski. 2011. Procedure for tissue sample preparation and metabolite extraction for high-throughput targeted metabolomics. Metabolomics 8: 133–142.

Rosenblum, E.S., R.S. Tjeerdema, M. Viant, E.S. Rosenblum and R.S. Tjeerdema. 2003. NMR-based metabolomics: a powerful approach for characterizing the effects of environmental stressors on organism health. Environ. Sci. Technol. 37: 4982–4989.

Satish Nair, P. and W.E. Robinson. 2001a. Histidine-rich glycoprotein in the blood of the bivalve *Mytilus edulis*: role in cadmium speciation and cadmium transfer to the kidney. Aquat. Toxicol. 52: 133–142.

Satish Nair, P. and W.E. Robinson. 2001b. Cadmium binding to a histidine-rich glycoprotein from marine mussel blood plasma: Potentiometric titration and equilibrium speciation modeling. Environ. Toxicol. Chem. 20: 1596–1604.

Sato, Y., I. Suzuki, T. Nakamura, F. Bernier, K. Aoshima and Y. Oda. 2012. Identification of a new plasma biomarker of Alzheimer's disease using metabolomics technology. J. Lipid Res. 53: 567–76.

Schmeisser, E., W. Goessler, N. Kienzl and K.A. Francesconi. 2005. Direct measurement of lipid-soluble arsenic species in biological samples with HPLC-ICPMS. Analyst 130: 948–55.

Sevillano-Morales, J.S., M. Cejudo-Gómez, A.M. Ramírez-Ojeda, F. Cámara Martos and R. Moreno-Rojas. 2015. Risk profile of methylmercury in seafood. Curr. Opin. Food Sci. 6: 53–60.

Simkiss, K. and M. Taylor. 1995. Transport of metals across membranes. pp. 1–43. *In*: A. Tessier and D. Turner [eds.]. Metal Speciation and Bioavailability in Aquatic Systems. Reading, UK.

Smolders, R., M. Baillieul and R. Blust. 2005. Relationship between the energy status of *Daphnia magna* and its sensitivity to environmental stress. Aquat. Toxicol. 73: 155–70.

Soga, T., R. Baran, M. Suematsu, Y. Ueno, S. Ikeda, T. Sakurakawa, Y. Kakazu, T. Ishikawa, M. Robert, T. Nishioka and M. Tomita. 2006. Differential metabolomics reveals ophthalmic acid as an oxidative stress biomarker indicating hepatic glutathione consumption. J. Biol. Chem. 281: 16768–76.

Southam, A.D., J.M. Easton, G.D. Stentiford, C. Ludwig, T.N. Arvanitis and M.R. Viant. 2008. Metabolic changes in flatfish hepatic tumours revealed by NMR-based metabolomics and metabolic correlation networks. J. Proteome Res. 7: 5277–85.

Spann, N., D.C. Aldridge, J.L. Griffin and O.A.H. Jones. 2011. Size-dependent effects of low level cadmium and zinc exposure on the metabolome of the Asian clam, Corbicula fluminea. Aquat. Toxicol. 105: 589–99.

Stohs, S. 1995. Oxidative mechanisms in the toxicity of metal ions. Free Radic. Biol. Med. 18: 321–336.

Tang, D.-Q., L. Zou, X.-X. Yin and C.N. Ong. 2014. HILIC-MS for metabolomics: An attractive and complementary approach to RPLC-MS. Mass Spectrom. Rev. 1–27.

Taylor, N.S., R.J.M. Weber, A.D. Southam, T.G. Payne, O. Hrydziuszko, T.N. Arvanitis and M.R. Viant. 2008. A new approach to toxicity testing in *Daphnia magna*: application of high throughput FT-ICR mass spectrometry metabolomics. Metabolomics 5: 44–58.

Taylor, N.S., R.J.M. Weber, T.A. White and M.R. Viant. 2010. Discriminating between different acute chemical toxicities via changes in the daphnid metabolome. Toxicol. Sci. 118: 307–17.

Teahan, O., S. Gamble, E. Holmes, J. Waxman, J.K. Nicholson, C. Bevan and H.C. Keun. 2006. Impact of analytical bias in metabonomic studies of human blood serum and plasma. Anal. Chem. 78: 4307–18.

Ten Hallers-Tjabbes, C.C., J.-W. Wegener, B. Van Hattum, J.F. Kemp, E. Ten Hallers, T.J. Reitsema and J.P. Boon. 2003. Imposex and organotin concentrations in Buccinum undatum and Neptunea antiqua from the North Sea: relationship to shipping density and hydrographical conditions. Mar. Environ. Res. 55: 203–233.

Theron, H.B., M.J. van der Merwe, K.J. Swart and J.H. van der Westhuizen. 2007. Employing atmospheric pressure photoionization in liquid chromatography/tandem mass spectrometry to minimize ion suppression and matrix effects for the quantification of venlafaxine and O-desmethylvenlafaxine. Rapid Commun. Mass Spectrom. 21: 1680–6.

Tjeerdema, R.S. 2008. Application of NMR-based techniques in aquatic toxicology: brief examples. Mar. Pollut. Bull. 57: 275–9.

Van Elteren, J.T., Z. Slejkovec, M. Kahn and W. Goessler. 2007. A systematic study on the extractability of arsenic species from algal certified reference material IAEA-140/TM (Fucus sp., Sea Plant Homogenate) using methanol/water extractant mixtures. Anal. Chim. Acta 585: 24–31.

Van Scoy, A.R., C. Yu Lin, B.S. Anderson, B.M. Philips, M.J. Martin, J. McCall, C.R. Todd, D. Crane, M.L. Sowby, M.R. Viant and R.S. Tjeerdema. 2010. Metabolic responses produced by crude versus dispersed oil in Chinook salmon pre-smolts via NMR-based metabolomics. Ecotoxicol. Environ. Saf. 73: 710–7.

Veldhoen, N., M.G. Ikonomou and C.C. Helbing. 2012. Molecular profiling of marine fauna: integration of omics with environmental assessment of the world's oceans. Ecotoxicol. Environ. Saf. 76: 23–38.

Viant, M.R., C.A. Pincetich, D.E. Hinton and R.S. Tjeerdema. 2006a. Toxic actions of dinoseb in medaka (*Oryzias latipes*) embryos as determined by *in vivo* 31P NMR, HPLC-UV and 1H NMR metabolomics. Aquat. Toxicol. 76: 329–42.

Viant, M.R., C.A. Pincetich and R.S. Tjeerdema. 2006b. Metabolic effects of dinoseb, diazinon and esfenvalerate in eyed eggs and alevins of Chinook salmon (Oncorhynchus tshawytscha) determined by 1HNMR metabolomics. Aquat. Toxicol. 77: 359–71.

Viant, M.R., D.W. Bearden, J.G. Bundy, I.W. Burton, T.W. Collette, D.R. Ekman, V. Ezernieks, T.K. Karakach, C.Y. Lin, S. Rochfort, J.S. Ropp, Q. Teng, R.S. Tjeerdema, J.A. Walter and H. Wu. 2009. International NMR-based environmental metabolomics intercomparison exercise. Environ. Sci. Technol. 43: 219–225.

Viant, M.R. and U. Sommer. 2013. Mass spectrometry based environmental metabolomics: A primer and review. Metabolomics 9: 144–158.

Vuckovic, D., I. de Lannoy, B. Gien, R.E. Shirey, L.M. Sidisky, S. Dutta and J. Pawliszyn. 2011. *In vivo* solid-phase microextraction: capturing the elusive portion of metabolome. Angew. Chem. Int. Ed. Engl. 50: 5344–8.

Vuckovic, D. 2012. Current trends and challenges in sample preparation for global metabolomics using liquid chromatography-mass spectrometry. Anal. Bioanal. Chem. 403: 1523–48.

Want, E.J., I.D. Wilson, H. Gika, G. Theodoridis, R.S. Plumb, J. Shockcor, E. Holmes and J.K. Nicholson. 2010. Global metabolic profiling procedures for urine using UPLC-MS. Nat. Protoc. 5: 1005–18.

Wang, W.-X. and P.S. Rainbow. 2010. Significance of metallothioneins in metal accumulation kinetics in marine animals. Comp. Biochem. Physiol. C. Toxicol. Pharmacol. 152: 1–8.

Watson, A.D. 2006. Thematic review series: Systems Biology Approaches to Metabolic and Cardiovascular Disorders. Lipidomics: a global approach to lipid analysis in biological systems. J. Lipid Res. 47: 2101–2111.

Wei, L., Q. Wang, H. Wu, C. Ji and J. Zhao. 2015. Proteomic and metabolomic responses of Pacific oyster Crassostrea gigas to elevated pCO_2 exposure. J. Proteomics 112: 83–94.

Williams, J., L. Pandarinathan, J. Wood, P. Vouros and A. Makriyannis. 2006. Endocannabinoid metabolomics: a novel liquid chromatography-mass spectrometry reagent for fatty acid analysis. AAPS J. 8: E655–60.

Williams, T.D., I.M. Davies, H. Wu, A.M. Diab, L. Webster, M.R. Viant, J.K. Chipman, M.J. Leaver, S.G. George, C.F. Moffat and C.D. Robinson. 2014. Molecular responses of European flounder (*Platichthys flesus*) chronically exposed to contaminated estuarine sediments. Chemosphere 108: 152–8.

Wu, H., X. Liu, X. Zhang, C. Ji, J. Zhao and J. Yu. 2013a. Proteomic and metabolomic responses of clam Ruditapes philippinarum to arsenic exposure under different salinities. Aquat. Toxicol. 136-137: 91–100.

Wu, H., C. Ji, Q. Wang, X. Liu, J. Zhao and J. Feng. 2013b. Manila clam Venerupis philippinarum as a biomonitor to metal pollution. Chinese J. Oceanol. Limnol. 31: 65–74.

Xu, F., L. Zou, Y. Liu, Z. Zhang and C.N. Ong. 2011. Enhancement of the capabilities of liquid chromatography-mass spectrometry with derivatization: general principles and applications. Mass Spectrom. Rev. 30: 1143–72.

Yancey, P.H. 2005. Organic osmolytes as compatible, metabolic and counteracting cytoprotectants in high osmolarity and other stresses. J. Exp. Biol. 208: 2819–30.

Yin, Y., J. Liu, B. He, J. Shi and G. Jiang. 2008. Simple interface of high-performance liquid chromatography-atomic fluorescence spectrometry hyphenated system for speciation of mercury based on photo-induced chemical vapour generation with formic acid in mobile phase as reaction reagent. J. Chromatogr. A 1181: 77–82.

Zanotto, F.P. and M.G. Wheatly. 2006. Ion regulation in invertebrates: molecular and integrative aspects. Physiol. Biochem. Zool. 79: 357–62.

Zhang, L., X. Liu, L. You, D. Zhou, J. Yu, J. Zhao, J. Feng and H. Wu. 2011a. Toxicological effects induced by cadmium in gills of Manila clam Ruditapes philippinarum using NMR-based metabolomics. CLEAN - Soil, Air, Water 39: 989–995.

Zhang, L., X. Liu, L. You, D. Zhou, H. Wu, L. Li, J. Zhao, J. Feng and J. Yu. 2011b. Metabolic responses in gills of Manila clam Ruditapes philippinarum exposed to copper using NMR-based metabolomics. Mar. Environ. Res. 72: 33–9.

Zhang, N., M. Venkateshwaran, M. Boersma, A. Harms, M. Howes-Podoll, D. den Os, J.-M. Ané and M.R. Sussman. 2012. Metabolomic profiling reveals suppression of oxylipin biosynthesis during the early stages of legume-rhizobia symbiosis. FEBS Lett. 586: 3150–8.

Zhou, B., J.F. Xiao, L. Tuli and H.W. Ressom. 2012. LC-MS-based metabolomics. Mol. Biosyst. 8: 470–81.

12

Fate and Toxicity of Inorganic Engineered Nanomaterials in the Marine Environment

Analytical Techniques and Methods

Francisco Laborda,[*,a] *Eduardo Bolea*[b] and *Javier Jiménez-Lamana*[c]

ABSTRACT

The extensive use of engineered nanomaterials in a large variety of emerging technologies and commercial products implies their release into the different ecosystems in the environment, including the marine environment. The study of the fate and transport of ENMs in marine environments, as well as their exposure and bioaccumulation in marine organisms requires obtaining analytical information about these ENMs in a wide range of marine samples. The strategies currently used are based on performing laboratory experiments on fate and transport of ENMs and testing their toxicity with different organisms, being limited to the use of a few analytical techniques, mostly atomic spectrometry, to determine total element contents or, in combination with fractionation techniques, to differentiate between dissolved and particulate forms of the element present in the ENM. In contrast, the detection, characterization, and quantification of ENMs at relevant environmental conditions and concentrations is a challenge in environmental nanosciences that requires multi-methodological approaches. In this chapter, fate and toxicity studies related to ENMs in the marine environment and the key issues related to the analytical tools currently used are reviewed, together with an overview about the state-of-the-art of ENMs analysis in complex samples.

Introduction

Although there is not a widely accepted definition for nanomaterials, all the available definitions are based on size, where NMs are defined as materials with external dimensions within 1–100 nm. NMs have always

Institute of Environmental Sciences (IUCA), Group of Analytical Spectroscopy and Sensors (GEAS), University of Zaragoza, Pedro Cerbuna, 12, 50009 Zaragoza, Spain.

[a] Email: flaborda@unizar.es

[b] Email: edbolea@unizar.es

[c] Email: jjlamana@unizar.es

* Corresponding author

been present in our environment, as natural NMs from weathering and volcanic emissions, or incidental NMs, inadvertently produced as a result of human activities. In contrast, engineered nanomaterials are developed and produced purposely to exploit their properties in a variety of applications, by incorporating them in industrial and consumer products. Due to the widespread use of ENMs, they are likely to be released into the environment during production, transport, use or disposal of the corresponding nano products, resulting in increased exposure of ENMs in the environment. Therefore, a deeper knowledge on potential impact of ENMs on both environment and human health is required, not only to minimize and control the risks, but also to ensure the sustainable development of the nanotechnology industry. In addition, a growing number of products containing ENMs are being developed specifically for marine applications, such as antifouling coatings and environmental remediation systems, increasing the need to address any potential risks for marine organisms and ecosystems (Corsi et al. 2014).

There are little empirical data concerning the release of ENMs and their concentration in the environment. Many estimations are based on models that may have little significance to actual released volumes, and few recognise the marine environment as an important sink for ENMs (Baker et al. 2014). ENMs can enter the marine environment via rainwater, wastewater, or river transport. ENMs can be bioaccumulated along the marine trophic chain, potentially affecting marine biological resources, and finally being transferred to humans through diet by consumption of contaminated seafood products. The estimations made by exposure modellings suggest that environmental concentrations of different types of ENMs are in the range of ng L^{-1} to μg L^{-1} in freshwaters (Gottschalk et al. 2011).

Among the large variety of nanomaterials currently developed, inorganic ENMs include metals (e.g., Ag, Au, Fe), oxides (e.g., CeO_2, CuO, Fe_xO_y, SiO_2, TiO_2, ZnO), and quantum dots (e.g., CdS, CdSe, ZnS). Oxides represent the largest worldwide production of ENMs, headed by TiO_2 with up to 10,000 t/year (Piccinno et al. 2012). These ENMs are generally produced as nanoparticles (NMs with all three external dimensions in the nanoscale). NPs can be available as free or aggregated entities, both as such or dispersed in liquids, or incorporated in bulk materials, either dispersed into the material or bound to the surface. Free NPs represent the higher risk, whereas NP incorporated to solids are likely to be released in the environment after an aging and weathering process.

The OECD has highlighted four metal and metal oxide NPs, namely cerium oxide (CeO_2), silver (Ag), zinc oxide (ZnO), and titanium dioxide (TiO_2) due to their inherent properties, widespread use, and commercial importance. TiO_2 is commonly employed in sunscreens blocking UV light (JRC-IRMM 2011), the same as ZnO, which is also found in paints, cosmetics, animal feeds, and fertilisers. Ag is used in numerous consumer products including textiles, personal care products, or food storage containers because of their bactericidal properties, whereas CeO_2 NPs are widely used as fuel additive in diesel engines and also as glass polishers and heat-resistant coatings (Baker et al. 2014).

Understanding the properties, behaviour, and effects of these ENMs in the environment requires the proper characterization and accurate quantification of ENMs at environmental conditions and concentrations. On the other hand, developing tools to measure the exposure and toxicity of ENMs in marine ecosystems presents many challenges. The inherent problems associated with the detection and characterization of ENMs in the variety of complex matrices involved in the different compartments under study (seawater, sediments, biofilms, tissues...), together with the transformations undergone by these ENMs and the resultant products, contribute to this challenge, as discussed in the following sections.

Fate of Nanomaterials in the Marine Environment

The production, use, and disposal of engineered nanomaterials (ENMs), result in the release of ENMs into the environment. The oceans have been found to be the sink for many classes of persistent environmental contaminants, and it is likely that ENMs will also find its way to the sea, water borne or via sediment transport (Klaine et al. 2008). ENMs may enter the marine environment via rainwater, wastewater, and river transportation, although not all marine discharges of ENMs would be derived from freshwater effluents (Baker et al. 2014). For instance, the use of TiO_2 and ZnO NPs in sunscreens (Wahie et al. 2007) represents a major source of these materials due to wash-off from individuals into the marine environment (Danovaro et al. 2008). In addition to accidental releases and exposures, a suite of new marine nanotechnologies, including antifouling paints and pollution remediation systems, are also being developed (Corsi et al. 2014). Besides, dumping wastes such as nano-waste, trash containing ENMs, harbour-dredging, mud containing ENMs into the sea, coastal runoff, atmospheric sedimentation, and others may all cause ENMs to enter the marine environment (Zhu and Cai 2012).

The fundamental behaviour and transport processes affecting engineered nanomaterials in seawater, such as dissolution and complexation, interactions with natural organic matter, stabilization and agglomeration, or sedimentation processes will be discussed next.

Transformations and Transport Processes of ENMs

To elucidate the environmental fate of ENMs, it is important to have a basic understanding of the type of transformations and transport processes ENMs of a certain type undergo. The time scales of these processes, and the effect of the environmental conditions, which in the case of marine environment has a strong influence on the properties and behaviour of ENMs should also be considered. In comparison with freshwater ecosystems, the marine environment has higher pH (6–9 in freshwaters, 8–9 in seawater) and ionic strength (10^{-4}–10^{-3}M in freshwater and relatively constant at 0.7 M in seawater) (Tiede et al. 2009), together with a wide variety of colloids and natural organic matter (Klaine et al. 2008). Dissolved organic matter, which represents the major fraction of natural organic matter in aquatic systems, may reach up to 8×10^{-4}M in freshwater, but usually do not exceed $(6–9) \times 10^{-5}$M in the ocean (Tiede et al. 2009). However, a large variability with respect to these average values can be expected when considering marine environments. For instance, organic matter will be present at higher concentrations in the coastal zone compared to pristine oceanic water. In addition, the oceans exhibit changes in physicochemical characteristics with depth that affect aggregation and colloid chemistry (Klaine et al. 2008). Seasonal variations also influence phytoplankton and microbial activity that produces exudates whose structure and abundance determine their potential roles in the fate of the suspended particulate matter, and thus of ENMs, and can also act as chelating agents for inorganic metal binding (Peralta-Videa et al. 2011).

Figure 1 summarizes the physicochemical mechanisms that appear determinant in the fate of ENMs in the seawater column and sediments. In the seawater column, both hydrodynamic flow and random diffusion favour the colloidal transport of particulate matter, competing with gravitational sedimentation (Corsi et al. 2014). The particle size and density determine the persistence of ENMs in suspension.

ENMs present in water do not form solutions, but dispersions, multiphase systems that are not thermodynamically stable (Handy et al. 2008; Praetorius et al. 2014). In such dispersion, ENMs will undergo

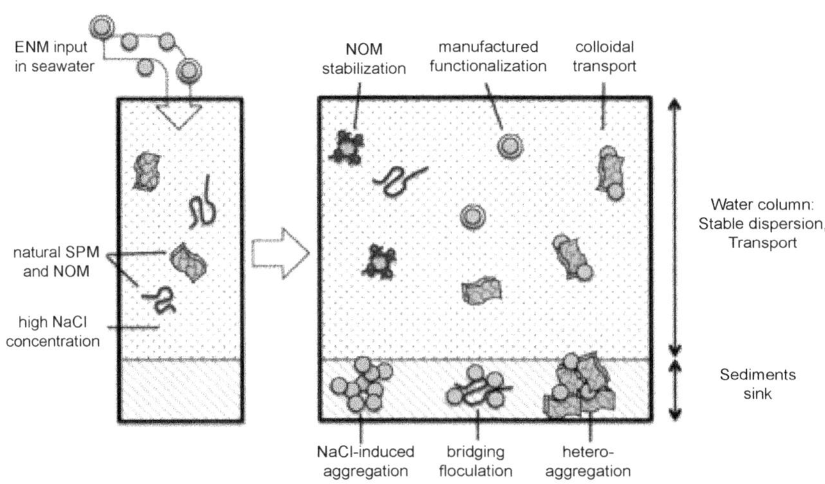

Fig. 1: Potential fate scenarios encountered by ENMs when released into seawater. Reprinted with permission from (Corsi et al. 2014). Copyright 2014 American Chemical Society.

homoaggregation with the same type of ENM particles, heteroaggregation with other types of suspended particles, and deposition to interfaces with other phases, such as surfaces of solid phases. For some ENMs (i.e., metallic) dissolution and chemical transformation should also be considered (Lead et al. 2014).

The aggregation of these ENMs results from the balance between repulsive and attractive interparticle forces. Surface charge, which strongly depends on pH and salinity, plays a determining role in the electrostatic dispersion and the stability of particles. Because of its ionic strength, seawater modifies the surface charge of ENMs generating more particle collisions, aggregation, and precipitation (Batley et al. 2012). In addition, the presence of organic matter in freshwaters also contributes to the stabilization of ENPs, especially for those with coatings that do not provide steric protection against aggregation (Casals et al. 2008; Labille and Brant 2010). This has been demonstrated for citrate (no steric protection) and polyvinylpyrrolidone (PVP) (with steric protection) coated Ag nanoparticles (Huynh and Chen 2011). The PVP coating assists in steric repulsion of particles, so slightly higher salt concentrations (0.1 M NaCl) were required to induce aggregation, compared to 0.03 M NaCl for the citrate-coated particles. Keller et al. (2010) showed that TiO_2 NPs rapidly agglomerates in seawater whereas it remained unaltered for 6 hours in the case of freshwater.

On the other hand, heteroaggregation accounts for a wide panel of colloidal dynamics, ranging from flocculation to colloidal stabilization. Natural minerals (clay, carbonates, etc.) and organic matter (algae, exopolymers, etc.) are suspended in the seawater column as geogenic and biogenic colloids (Wells et al. 1998), potentially interacting with the ENMs, leading to heteroaggregation phenomena. As an example, high aggregation and sedimentation rates of metal oxide and silver ENMs have been reported in low NOM seawaters, whereas they remained more stable in freshwater (low ionic strength, high NOM). Similarly, metal and oxide ENMs that aggregate at high salt concentrations are actually stabilized by interactions with proteins (Corsi et al. 2014).

Exposure of Marine Organisms to Engineered Nanomaterials

With regard to the ENMs exposure in the marine ecosystems, the stability of these ENMs is a key factor that determines their occurrence in the benthic or pelagic systems, depending on the residence time in the water column, as previously discussed. Those remaining in the colloidal size range, typically below 1 µm in size, promote further transport of ENMs and their exposure to pelagic organisms that feed at these zones (Batley et al. 2012), while the formation and sedimentation of larger flocs affects basically to the benthic ecosystem (Corsi et al. 2014), as shown in Fig. 2.

In general, the agglomeration, aggregation, and precipitation of ENPs in seawater would result in the NP deposition on sediment biofilms, with the subsequent accumulation in the sediment and exposure to sediment-dwelling organisms. In coastal marine environments, stable colloids brought by river encounter increasing salinity gradients that may cause their agglomeration and sedimentation. In these areas, ENMs bound to those colloids not only may bury and accumulate within sediments, but also they can persist in the water column (Stolpe and Hassellov 2007) (Fig. 2).

Finally, ENMs can also accumulate on the surface microlayer of the oceans, given the viscosity and surface tension shown by these microlayers (Klaine et al. 2008). In this case, ENMs would presumably present a route of aerosol exposure risk to marine birds and mammals, as well as the organisms living in the surface microlayer.

Toxicity of ENMs to Marine Organism

Toxicity of ENMs is related to the category, concentration, particle size, suspension preparation method, and solubility in seawater of the ENMs. Most of the currently available ecotoxicological data regarding ENMs are limited to species used in regulatory testing or freshwater species (Matranga and Corsi 2012). However, the different properties of ENMs in seawater may cause different toxic effects compared with freshwater and, hence, it is not clear yet how to extrapolate freshwater data to marine organisms (Baker et al. 2014). For example, the mechanism of action of ZnO NPs in freshwater green algae may be mainly

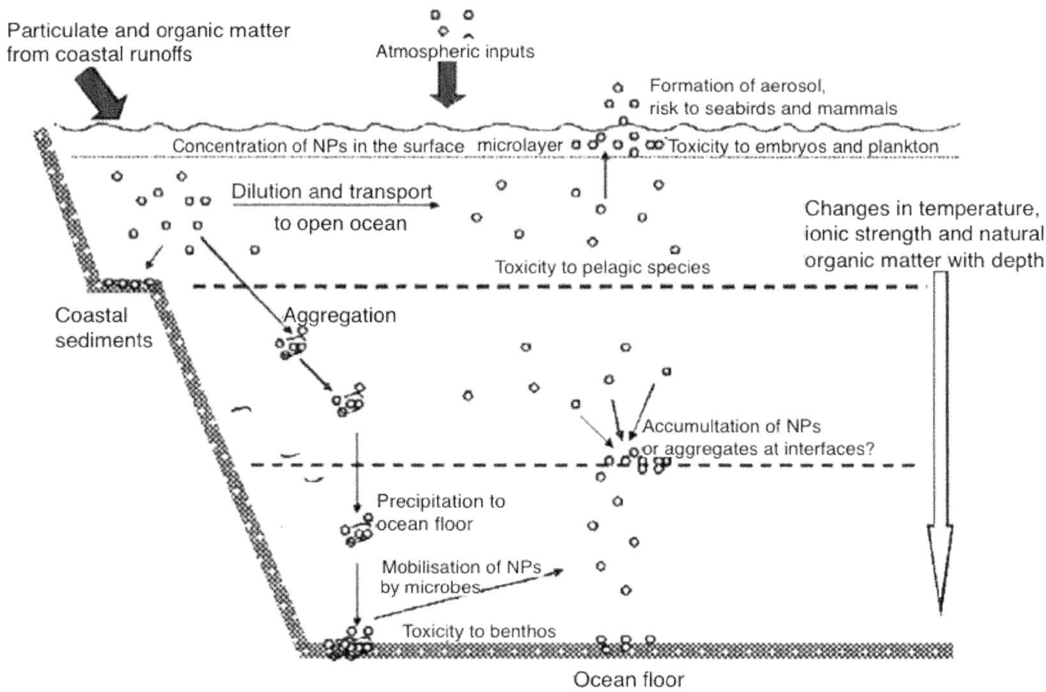

Fig. 2: Schematic diagram outlining the possible fate of ENMs in the marine environment and the organisms at risk of exposure. Reprinted from (Klaine et al. 2008) with permission. Copyright 2008 SETAC.

due to the release of Zn^{2+} (Franklin et al. 2007), whereas the toxic effects of ZnO NPs in marine algae may be the combined toxic effects of nanoparticles and metal ions, since the Zn^{2+} released from ZnO NPs in seawater is less bioavailable than in freshwater (Miller et al. 2010). Ag NPs are widely used as an antibacterial agent, but it fails to have a significant influence on the diversity of natural bacterial colonies in river mouth sediments (Bradford et al. 2009). TiO_2 NPs at concentrations up to 500 mg L^{-1} were not acutely toxic for zebrafish embryos in freshwater, but 10 mg L^{-1} significantly reduced embryo hatching and increased the malformation rate of the abalone *Haliotis diversicolor supertexta* in seawater (Zhu et al. 2011).

In view of the data available, the current opinion is that environmentally-realistic concentrations of inorganic ENMs are unlikely to cause significant adverse acute health problems to marine organisms, even though sub-lethal effects have been observed in many organisms which could result in chronic health impacts (Baker et al. 2014). Besides, the study of toxicity caused by ENMs on marine organisms requires the identification of the form responsible for the biological effects observed (i.e., the NP itself, the released element, or both). Table 1 summarizes toxicity studies performed with selected ENMs in marine organisms involving bioaccumulation tests. Special attention is paid to the toxic effects and the driver of such effects, namely the NPs themselves or the released element. For a comprehensive view of the toxicity of inorganic ENMs to marine organisms, the recent review by Baker et al. (2014) is recommended.

Silver nanoparticles

Ag^+ has been reported to determine the Ag NPs toxicity to the microalgae *Thalassiosira weissflogi*, since no toxicity was observed when the Ag^+ released from the NPs was removed or complexed (Miao et al. 2009). No phytotoxic effects were observed to the macroalga *Ulva lactuca* at concentrations up to 15 μg L^{-1} of Ag NPs, whereas $AgNO_3$ exposures showed negative effects at only 2.5 μg L^{-1}, suggesting the toxic role of the released Ag^+ (Turner et al. 2012). In the case of the clam *Scrobicularia plana* exposed to Ag NPs and ionic silver, similar effects were observed, which indicated that the Ag^+ released from Ag NPs was the main responsible for toxicity (Buffet et al. 2013b). Comparable genotoxic effects were induced in hemolymph cells of the mussel *Mytilus galloprovincialis* by Ag NPs and $AgNO_3$ although each species damaged different proteins, which suggest a nano-effect not explained just by the released of Ag^+ from Ag NPs (Gomes et al. 2013).

From these results it can be deduced that not always is possible to establish a unique responsible for the toxicity observed. A shared effect of Ag NPs and released silver was suggested as responsible for the growth inhibition observed in diatom *Thalassiosira pseudonana* (Burchardt et al. 2012). In a study conducted with oyster *Crassostrea virginica*, different results were obtained comparing Ag NPs and $AgNO_3$ exposures: hepatopancreas tissues proved to be more sensitive to Ag NPs while gill tissues were shown more prone to oxidative damage with dissolved Ag (McCarthy et al. 2013).

On the other hand, some studies have reported Ag NPs themselves as directly responsible for toxicity effects. In a study conducted with bivalve *Scrobicularia plana* and ragworm *Hediste diversicolor* exposed to Ag NPs and $AgNO_3$ at environmentally relevant concentrations (low $\mu g\ L^{-1}$), just NPs induced an immune response; moreover, genotoxicity was induced more severely by Ag NPs aggregates than by soluble Ag in the digestive gland of clams, suggesting a specific nano effect (Buffet et al. 2014a). Ag NPs have also been reported to induce genotoxic effects in the sediment-dwelling polychaete, *Nereis diversicolor*, at concentrations ≥ 25 mg g^{-1} (Cong et al. 2011). When compared to micron and ionic Ag exposures, Ag NPs had the greatest genotoxic effect, suggesting a nano-effect not related to released Ag^+. Surface coating may also have influence in Ag NPs' toxicity, as was shown in a study conducted with the marine polychaete *Platynereis dumerilii* exposed to citrate and humic acid-capped Ag NPs and $AgNO_3$ (García-Alonso et al. 2014). Humic acid capped NPs showed the highest acute toxicity in terms of mortality and abnormal development, which was associated with higher uptake rate.

Copper oxide nanoparticles

Most of the studies have reported CuO NPs as mainly responsible for toxicity to marine organisms. The toxic effects of CuO NPs at environmentally relevant concentrations (10 $\mu g\ L^{-1}$) in bivalve *Scrobicularia plana* and ragworm *Hediste diversicolor* have been studied in laboratory and outdoor mesocosms (Buffet et al. 2011 and 2013a). In the laboratory study, several defence biomarkers were activated, although no significant effects were observed. However, in the mesocoms study, CuO NPs caused DNA damage and affected burrowing kinetics in both organisms, whereas oxidative stress was only observed in ragworms. Since no detectable release of Cu^{2+} from CuO NPs was observed, the toxic effects were related to the NPs. Similarly, CuO NPs have also been reported to cause oxidative stress and induce DNA damage in hemolymph cells of mussel *Mytilus galloprovincialis* (Gomes et al. 2011, 2012, 2013). The observed oxidative stress was related to the NPs, with aggregation as a key factor, since only a small fraction of Cu^{2+} was released from the NPs. In the same sense, the CuO NPs toxicity to estuarine amphipod *Leptocheirus plumulosus* was also explained due to a nanoparticle effect (Hanna et al. 2013). In contrast, CuO NPs reduced growth population of diatom *Thalassiosira weissflogii* showing a similar toxicity that algae exposed to $CuCl_2$ revealed (Bielmyer-Fraser et al. 2014). On the other hand, no genotoxicity and no effects on mortality or burrowing behaviour were observed in deposit-feeding clam *Macoma balthica* exposed to sediments spiked with CuO NPs (Dai et al. 2013).

Zinc oxide nanoparticles

The toxicity of ZnO NPs is believed to be related to the release of Zn^{2+} from the NPs. For instance, released zinc justified the toxicity of ZnO NPs in oysters *Crassostrea gigas*, since similar toxicity was observed when oysters were exposed to dissolved zinc (Trevisan et al. 2014). In a study conducted using the marine diatoms *Skeletonema costatum* and *Thalassiosia pseudonana*, the crustaceans *Tigriopus japonicus* and *Elasmopus rapax*, and medaka fish *Oryzias melastigma*, ZnO NPs were found more toxic towards algae than bulk ZnO, but relatively less toxic towards crustaceans and fish, toxicity of ZnO NPs mainly attributed to dissolved Zn^{2+} ions (Wong et al. 2010). Changes found on growth rate, photosystem II quantum yield, and chlorophyll *a* content in diatom *Thalassiosira pseudonana* were also mainly explained by Zn^{2+} release,

Table 1: Summary of reported toxic effects of ENMs on marine organisms, suggested causes (nanoparticle/released element) and analytical methods used to study bioaccumulation.

Nanoparticle	Organism	Species	Toxic Effects	Toxicity Cause	Analytical Information	Sample Treatment	Analytical Technique	Reference
Ag	biofilm	-	biomass decrease	NP	total element	Digestion in a pressurized vial with HNO_3 for 7 h at 60°C	GF-AAS	Fabrega et al. 2011
Ag	annelid	*Nereis diversicolor*	genotoxicity	NP	total element in tissue	Digestion with HNO_3 in microwave oven	GF-AAS	Cong et al. 2011
Ag	annelid	*Platynereis dumerilii*	mortality abnormal development	NP	total element in tissue	Digestion with HNO_3 at 100°C	ICP-MS	García-Alonso et al. 2014
Ag	annelid	*Hediste diversicolor*	oxidative stress apoptosis genotoxicity immunomodulation	NP	total element in tissue	Digestion with HNO_3 at 90°C	GF-AAS	Buffet et al. 2014a
Ag	mollusc	*Mytilus galloprovincialis*	genotoxicity	released Ag^+ > NP	total element in tissue	Drying at 80°C. Digestion with HNO_3	AAS	Gomes et al. 2013
Ag	mollusc	*Scrobicularia plana*	oxidative stress	released Ag^+	total element in tissue	Digestion with HNO_3 at 90°C	GF-AAS	Buffet et al. 2013a
Ag	mollusc	*Scrobicularia plana*	oxidative stress apoptosis genotoxicity immunomodulation	NP	total element in tissue	Digestion with HNO_3 at 90°C	GF-AAS	Buffet et al. 2014a
Ag	mollusc	*Macoma balthica*	no effects	-	total element in tissue and shell	Lyophilization Microwave assisted digestion with HNO_3	GF-AAS F-AAS	Dai et al. 2013
Ag	macroalgae	*Ulva lactuca*	phytotoxicity	released Ag^+	total element in tissue	Digestion with HNO_3	ICP-MS	Turner et al. 2012

AgO	microalgae	*Thalassiosira weissflogii*	growth rate	-	distribution of element in cell components	Centrifugation at 3500 rpm for 15 min. Lysis of cells by sonication Separation of cell components by differential centrifugation	GF-AAS	Bielmyer-Fraser et al. 2014
Au	mollusc	*Mytilus edulis*	oxidative stress	NP	total element in gill, digestive gland, and mantle tissues	Digestion with HNO_3 and H_2O_2 at 95°C	ICP-OES	Tedesco et al. 2010a
Au	mollusc	*Ruditapes philippinarum*	oxidative stress	NP	total tissues subcellular location	Digestion with 2 mL HNO_3 at 95°C for 60 min	ICP-MS TEM EDX SEM-FEG	García-Negrete et al. 2013
CdS	annelid	*Hediste diversicolor*	oxidative stress	NP/released Cd^{2+}	total element in tissue	Digestion with HCl/ HNO_3 (3:1) at 90°C	ET-AAS	Buffet et al. 2014b
CeO_2	mollusc	*Mytilus galloprovincialis*	not studied	-	total element in tissue and faeces	Digestion with HNO_3 at 80°C for 1 h	ICP-OES	Montes et al. 2012
CuO	microalgae	*Thalassiosira weissflogii*	growth rate	NP	distribution of element in cell components	Centrifugation at 3500 rpm for 15 min. Lysis of cells by sonication Separation of cell components by differential centrifugation	GF-AAS	Bielmyer-Fraser et al. 2014
CuO	annelid	*Hediste diversicolor*	no effect	-	total element in tissue	Digestion with HCl/ HNO_3 (3:1) at 90°C	F-AAS	Buffet et al. 2011
CuO	annelid	*Hediste diversicolor*	genotoxicity oxidative stress	NP	total element in tissue	Digestion with HCl/ HNO_3 (3:1) at 90°C	F-AAS	Buffet et al. 2013a

Table 1 Cont....

Table 1 Cont...

Nanoparticle	Organism	Species	Toxic Effects	Toxicity Cause	Analytical Information	Sample Treatment	Analytical Technique	Reference
CuO	arthropods	*Leptocheirus plumulosus*	mortality increase	NP	total element in tissue	Drying at 60ºC for 72 h Digestion with HNO_3 at 60ºC for 2 h	ICP-OES	Hanna et al. 2013
CuO	mollusc	*Scrobicularia plana*	no effect		total element in tissue	Digestion with HCl/ HNO_3 (3:1) at 90ºC	F-AAS	Buffet et al. 2011
CuO	mollusc	*Scrobicularia plana*	genotoxicity oxidative stress burrowing kinetics	NP	total element in tissue	Digestion with HCl/ HNO_3 (3:1) at 90ºC	ET-AAS	Buffet et al. 2013a
CuO	mollusc	*Mytilus galloprovincialis*	oxidative stress genotoxicity	NP	total element in gills	Drying at 80ºC Digestion with HNO_3	ET-AAS	Gomes et al. 2011 Gomes et al. 2013
CuO	mollusc	*Macoma balthica*	no effect	-	total element in tissue and shell	Lyophilization Microwave assisted digestion with HNO_3	GF-AAS F-AAS	Dai et al. 2013
Fe	microalgae	*Pavlova lutheri* *Isochrysis galbana* *Tetraselmis suecica*	no effect	-	total element in test medium	Digestion with HCl	ICP-MS	Kadar et al. 2012
TiO_2	annelide	*Arenicola marina*	genotoxicity	NP	total element in tissue	Drying at 70ºC for 48 h. Digestion with 3 ml HNO_3 at room temperature for 3 weeks. Digestion with 1 ml H_2O_2 and 2 ml HNO_3 at 70ºC for 5 h	ICP-OES	Galloway et al. 2010

ZnO	microalgae	*Phaeodactylum tricornutum* *Alexandriumminutum* *Tetraselmis suecica*	growth rate reduction oxidative stress	released Zn^{2+}	total element	Filtration in fiberglass filters Drying of the filters at 60°C Digestion with H_2SO_4	ICP-MS	Castro-Bugallo et al. 2014
ZnO	microalgae	*Thalassiosira weissflogii*	growth rate	NP released Zn^{2+}	distribution of element in cell components	Centrifugation at 3500 rpm for 15 min. Lysis of cells by sonication Separation of cell components by differential centrifugation	GF-AAS	Bielmyer-Fraser et al. 2014
ZnO	annelid	*Hediste diversicolor*	burrowing behaviour feeding rate	-	total element in tissue	Digestion with HNO_3 and HCl at 75°C for 48 h	isotopic labelling ICP-MS	Buffet et al. 2012
ZnO	arthropods	*Leptocheirus plumulosus*	mortality	released Zn^{2+}	total element in tissue	Drying at 60°C for 72 h Digestion with HNO_3 at 60°C for 2 h	ICP-OES	Hanna et al. 2013
ZnO	mollusc	*Mytilus galloprovincialis*	not studied	-	total element in tissue and faeces	Digestion with HNO_3 at 80°C for 1 h	ICP-OES	Montes et al. 2012
ZnO	mollusc	*Scrobicularia plana*	burrowing behaviour feeding rate	-	total element in tissue	Digestion with HNO_3 and HCl at 75°C during 48 h	isotopic labelling ICP-MS	Buffet et al. 2012
ZnO	mollusc	*Crassostrea gigas*	mitochondrial damage oxidative stress	released Zn^{2+}	total element in gills and digestive gland	Acid digestion	ICP-OES	Trevisan et al. 2014

despite the presence of ZnO NPs (Miao et al. 2010). In the same sense, the ZnO NPs toxicity to estuarine amphipod *Leptocheirus plumulosus* in an exposure experiment through sediments resulted from released Zn^{2+} (Hanna et al. 2013).

In some other cases, it is not possible to stablish whether toxic effects are due to the release of the NPs or the metal ion. In an exposure experiment of ZnO NPs to four species of phytoplankton, *Thalassiosira pseudonana*, *Skeletonema marinoi*, *Dunaliella tertiolecta*, and *Isochrysis galbana*, a significantly depressed growth rate was found in all four species (Miller et al. 2010). The toxicity of ZnO NPs to phytoplankton was explained by the dissolution, release, and uptake of free zinc ions, although specific nanoparticulate effects could be difficult to differentiate from effects due to free zinc ions. On the other hand, the role of Zn^{2+} ions in ZnO NPs toxicity to the marine diatom *Skeletonema costatum* and the amphipod *Melta longidactyla* was not conclusive, since release of Zn^{2+} ion from ZnO NPs increased with decreasing temperature whereas toxicity increased with increasing temperature (Wong and Leung 2014). Similarly, in a study of the toxic effects of ZnO NPs in marine alga *Dunaliella tertiolecta* and by comparing the effects of $ZnCl_2$ and the dissolution properties of the ZnO, it was suggested that the behaviour of ZnO NPs could not be strictly related to the toxic action of zinc ions (Manzo et al. 2013). Environmental factors like salinity have been reported to affect ZnO NPs toxicity. Increasing salinity decreased the acute toxicity of ZnO NPs to copepod *Tigriopus japonicas* (Park et al. 2014). Results suggested that acute toxicity of ZnO NPs could not be explained by released metal ions alone.

The morphology of ZnO NPs can also influence its toxicity, as it has been shown in a study of three marine diatoms (*Thalassiosira pseudonana, Chaetoceros gracilis,* and *Phaeodactylum tricornutum*) exposed to spherical and rod-shaped NPs (Peng et al. 2011). Both forms stopped growth of *T. pseudonana* and *C. gracilis* at all concentrations tested, whereas *P. tricornutum* was the least sensitive, with its growth rate inversely proportional to nanoparticle concentration. Moreover, toxicity of rod-shaped particles to *P. triocornutum* was greater than that of the spherical, likely due to the increased surface area for dissolution.

Other nanoparticles

Titanium dioxide nanoparticles

No acute effect was found when abalone *Haliotis diversicolorsupertexta* was exposed to TiO_2 NPs in the range from 0.1 to 10 mg L^{-1} (Zhu et al. 2011). However, results shown that TiO_2 NPs might induce oxidative stress with an exposure concentration of ≥ 1.0 mg L^{-1}. Similarly, TiO_2 NPs showed no measurable effect on growth rates of fours species of phytoplankton: *Thalassiosira pseudonana*, *Skeletonema marinoi*, *Dunaliella tertiolecta,* and *Isochrysis galbana* (Miller et al. 2010). On the other hand, in the presence of levels of UV light similar to those found in nature, TiO_2 NPs can induce oxidative stress to the same species of phytoplankton (Miller et al. 2012). TiO_2 NPs caused sublethal impacts on the lugworm *Arenicola marina* exposed through sediments, although no evidence of nanoparticle dissolution was found (Galloway et al. 2010).

Gold nanoparticles

Like bulk gold, Au NPs are generally considered nontoxic. However, several studies have shown their toxicity to marine bivalves. *Mytilus edilus* exposed to 5 nm Au NPs showed significant increase in oxidative damage at concentrations of 750 μg L^{-1} (Tedesco et al. 2010a), whereas Au NPs of 15 nm, at identical concentration, only caused modest oxidative stress (Tedesco et al. 2010b). The different toxic effects were attributed principally to the size-related oxidative properties of Au NPs. Similar results were found in exposure experiments with bivalve *Ruditapes philippinarum*, where NPs caused modest oxidative stress (García-Negrete et al. 2013).

Iron nanoparticles

Zero-valent iron NPs are particularly interesting in remediation studies because of their high chemical reactivity. The effect of iron nanoparticles in three marine microalgae (*Pavlova lutheri, Isochrysis galbana,* and *Tetraselmis suecica*) was studied, showing normal algal growth in the presence of the NPs (Kadar et al. 2012). However, iron NPs have proved to cause fertilization impairments in mussel *Mytilus galloprovincialis*, urchin *Psammechinus milliaris,* and solitary tunicate *Ciona intestinalis* and disruption of embryo development in *P. milliaris* and *M. galloprovincialis* (Kadar et al. 2013).

Cadmium sulfide quantum dots

Oxidative stress was observed in ragworm *Hediste diversicolor* exposed to QD in seawater or via food and compared with soluble Cd^{2+} (Buffet et al. 2014b). Although QD partially dissolved, higher effect was induced by QD compared to soluble Cd, suggesting a specific nano-effect.

Techniques and methods for the analysis of ENMs in marine samples

Detection, characterization, and quantification of ENMs at environmentally relevant conditions and concentrations is a challenge in environmental nanosciences. This analytical information is decisive in studies on the environmental fate and transport of ENMs in marine environments, as well as to study the bioaccumulation of ENMs in marine organisms in the context of ecotoxicity tests. In any case, the inherent problems associated to pristine NPs are magnified by the variety of complex matrices involved in the different compartments studied (seawater, sediments, biofilms, tissues...) but also by the transformations undergone by the NPs in these compartments (dissolution, complexation, redox reactions, aggregation, surface modifications).

In spite of the final objective to detect, characterize, and quantify ENMs in naturally occurring marine samples (waters, sediments, organisms...), most of the works performed so far are oriented to the study of the fate and toxicity of ENMs added in the context of simulation experiments, as well as *in vitro* and *in vivo* assays. For such studies, the characterization of the original ENMs is also needed to establish correlations between the properties of the starting material and the ultimate effect of the ENM in the system being studied. The characterization of pristine ENMs is not so demanding in comparison with the analysis of marine samples containing ENMs; thus just techniques and methods oriented to the analysis of complex samples will be considered.

Current state of ENMs analysis in complex samples

There is currently no single analytical method able to detect, characterize, and quantify ENMs in complex systems. Thus a multimethod approach is often required to obtain analytical information (Montaño et al. 2014). On the other hand, although a wide range of analytical techniques is available to study ENMs, limitations become evident when they are applied to the analysis of ENMs in complex samples, like those from environmental or biological systems.

Analytical Techniques

Detection of a specific ENM within a complex matrix is probably the first challenge. Focusing on inorganic nanomaterials, the use of element-specific techniques is the most valuable tool to achieve this objective. Atomic spectrometry techniques like flame and graphite furnace AAS, ICP-OES, and ICP-MS can be used for detection, as well as for quantification, of the element/s present in the ENM and the sample. These techniques involve the introduction of the sample in liquid phase, what means that solid samples must undergo some sample treatment for the digestion of the matrix.

Currently, one of the most common techniques for the identification and quantification of inorganic ENMs involves the use of ICP-MS, due to the low detection limits attainable (down to ng L^{-1}). ICP-OES provide detection limits in the µg L^{-1} range, whereas graphite furnace AAS offers a half-way performance between ICP-MS and ICP-OES, and flame AAS the highest detection limits (mg L^{-1}). In any case, conventional atomic spectrometry techniques are sensitive to the element/s present in the sample that supposedly contains the ENM, but they are not capable of providing any information about the physicochemical form of the element (if present as dissolved species or as particulate), or any other information related to the ENM (size, aggregation...). In this respect, a new mode of ICP-MS, called single particle ICP-MS, is gaining significant interest (Laborda et al. 2014). Single particle ICP-MS is able to detect and quantify dissolved versus particulate forms of the element in the same sample, providing information about the size of the NPs, as well as about their number concentration at levels below ng L^{-1}.

Although microscopy techniques have been used for detection of ENMs in environmental release experiments (Kaegi et al. 2008) or *in vitro* and *in vivo* accumulation studies (García-Negrete et al. 2013), imaging techniques are mainly employed to obtain information about size and morphology of pristine ENMs at relatively high concentrations. TEM, due to its sub-nanometre resolution, is used more frequently than SEM and AFM. Electron microscopy techniques are combined with EDS to obtain bulk and mapped elemental composition at the percent concentration level of the imaged ENMs (Gelabert et al. 2014).

From the point of view of sizing, electron microscopy methods are counting methods, and size distributions are built up on a particle-by-particle basis. Although this approach makes EM a low-throughput method even with automated image processing, it does offer the potential to size ENMs in the presence of interfering particles. This is not possible with ensemble techniques, like DLS, where the instrument response arises from the measurement of large number of particles simultaneously. Dynamic light scattering is the most commonly employed technique to measure nanoparticle size in aqueous dispersions, but it is less useful for the analysis of polydisperse samples and mixtures of particles of different natures, because of the difficulties of interpretation of the scattering signal and its lack of selectivity. Nanoparticle tracking analysis is also a light scattering technique but it is based on tracking the Brownian motion of single particles to determine their diffusion coefficients and subsequently the size of the particles (Boyd et al. 2011). Additionally, particles of different composition might be identified by comparing scattered light intensity from particles of the same size.

Separation techniques such as field flow fractionation (asymmetric flow field flow fractionation, sedimentation-field flow fractionation) and hydrodynamic chromatography can provide size information about nanoparticles in aqueous matrices, as well as allow the collection of size fractions for further characterization. As with any other continuous separation techniques, FFF and HDC are coupled to one or more detectors. The most commonly used approaches are based on the coupling of one or more non-destructive detectors (e.g., UV-Vis, MALS) and a destructive element-specific detector (e.g., ICP-MS) at the end.

Apart from detection and size characterization, determination of the ENMs concentration is necessary to assess their occurrence and exposure. By using unspecific counting techniques, like NTA, number concentrations can be determined, although target nanoparticles cannot be differentiated from background ones. If atomic spectrometry techniques are used, element mass concentrations are obtained, although element-specific number concentrations can be measured by using ICP-MS in single particle mode. When FFF or HCD are coupled to an atomic spectrometry technique (e.g., ICP-MS), the mass concentration can be related to the size of the nanoparticles separated.

Sample preparation

Except when working with pristine ENMs, most samples require some kind of preparation prior to their analysis by most analytical techniques. Sample preparation methods related to remove the matrix or to separate the analyte from the matrix will be considered.

Sample digestion. Digestion of solid samples containing inorganic ENMs (e.g., tissues, microorganisms, sediments...) can involve the dissolution of the ENM, the degradation of the sample matrix or both. Concentrated oxidizing acids like nitric acid, alone or in combination with hydrogen peroxide or hydrochloric acid (aqua regia), are commonly used to digest organic matrices, by using conventional heating systems at atmospheric pressure or under pressure with microwave assisted techniques. Under acidic conditions some ENMs can be dissolved (e.g., silver, copper, zinc, and copper oxides), whereas others will require additional reagents (aqua regia for gold, hydrogen peroxide for CeO_2, or hydrofluoric acid for TiO_2). In any case, these acid-based digestions are oriented to get information of the total element content from the ENM in the sample.

Alternatives to acid digestions are those based on the use of alkaline reagents. Tetramethylammonium hydroxide is often employed for degradation of organic matrices without affecting the core of inorganic NPs, allowing the direct detection, quantification, and size characterization of the NPs themselves. In this line, silver and gold nanoparticles from *Daphnia magna* and *Lumbriculus variegatus* were detected, sized, and quantified by SP-ICP-MS after TMAH digestion (Gray et al. 2013). Alkaline digestion methods are also used in combination with enzymes (e.g., proteinase K) to improve the efficiency of the matrix degradation (Loeschner et al. 2014).

Centrifugation. By centrifugation, species are fractionated in liquid samples according to their density, or their size if having the same composition. Thus centrifugation is suitable for being used for separation of nanoparticles from dissolved species, although high centrifugal forces and long times are required. However, removal of nanoparticles from supernatants containing dissolved species is incomplete even at very harsh ultracentrifugation conditions (e.g., 150000 g for 60 min) (Xu et al. 2013) and centrifugation is not considered an efficient technique for the fractionation of nanoparticles (Tsao et al. 2011).

Ultrafiltration and dialysis. Ultrafiltration and dialysis can be considered more reliable than centrifugation for size-based fractionation in liquid samples. Dialysis is based on the diffusion of species through a semipermeable membrane. Membranes from different materials are available and they are characterized according to their nominal molecular-weight cutoff. Because the MWCO of a membrane is the result of the number and average size of its pores, MWCO is defined as the molecular weight/size of those species retained in a percentage larger than 90% by the membrane, hence it is not a sharply defined value.

The use of dialysis has been reported in the measurement of nanomaterial dissolution to isolate ionic species (references in Table 2). Because dialysis methods take long times to achieve equilibrium, ultrafiltration is preferred to speed up the separation process. In UF, species below the MWCO are forced to cross the membrane through the use of a centrifugal force. The use of UF has been extensively applied to study the dissolution of nanomaterials (Misra et al. 2012 and references in Table 2), and membranes with MWCO in the range of 1–100 kDa (from ca. 1 nm) are available. Depending on composition and surface functionality, ENMs and the corresponding dissolved species can interact with membrane surfaces, affecting their recoveries (Ladner et al. 2012).

Ion exchange. This technique is applicable to the separation of free metal ions and derivatized NPs and involves the equilibration of a known mass of resin with a given volume of sample, both in batch or column modes (Li et al. 2012).

Another technique involving ion exchange resins is diffusion gradients in thin films (Zhang and Davidson 1995). DGT is performed by using commercially available passive samplers that consist of a three-layer assembly: a membrane filter (0.45 μm pore size) in contact with the sample solution, a hydrogel layer and a second thin hydrogel layer containing the ion exchange resin, usually Chelex-100. The species isolated by DGT are considered as the labile fraction of the dissolved element.

Extraction. ENMs can be extracted from solid and liquid samples by using water or organic solvents preserving some of the properties of the ENMs. Recently, cloud point extraction has been proposed for the separation of nanoparticles, preserving the size and morphology of the nanoparticles in the sample and providing selectivity in the separation of NPs from dissolved species (Chao et al. 2011). CPE involves the addition of a non-ionic surfactant (e.g., Triton X114) at concentrations over the critical micellar

Table 2: Summary of reported analytical methods for the analysis of ENMs in seawater.

Nanoparticle	Coating	Size (nm)	Seawater	Analytical information	Sample treatment	Analytical technique	Reference
Ag	poly(allylamine)	20	natural	total element		ICP-MS	Doiron et al. 2012
Ag	-	< 100	natural	total element		ICP-MS	Turner et al. 2012
Ag	-	< 100	natural	total element dissolution	 filtration (0.02 μm)	AAS	Gomes et al. 2013
Ag	-	65	artificial	dissolution	UF (1 kDa)	GF-AAS	Fabrega et al. 2011
Ag	-	20 40 100	artificial	dissolution	dialysis	ICP-MS	Burchardt et al. 2012
Ag	citrate PVP	14 15	natural	dissolution	dialysis (1 kDa) UF (1 kDa)	ICP-OES	Angel et al. 2013
Ag	citrate PVP	30	reconstituted	dissolution	filtration (0.2 μm)	AAS	Wang et al. 2014
Ag	lactate	40–45	natural	labile element	DGT	GF-AAS	Buffet et al. 2013b
Ag	maltose	40	natural	labile element	DGT	GF-AAS	Buffet et al. 2014a
Ag	-	20 40 100	artificial	size aggregation	-	DCS	Burchardt et al. 2012
Ag	citrate	20	artificial	aggregation	-	DLS UV-Vis AFM	Chinnapongse et al. 2011
Au	citrate	24	artificial	total element	-	ICP-MS	García-Negrete et al. 2013
Au	PEG PEG/DNA	13	natural	aggregation	-	UV-Vis	Heo et al. 2013
CdS	-	5–6	natural	dissolution	dialysis	ICP-MS	Buffet et al. 2014b
				labile element	DGT	GF-AAS	
CeO_2	-	67 x 8	natural	total element	acid digestion	ICP-OES	Montes et al. 2012
CuO	-	30	natural	labile element	DGT	GF-AAS	Buffet et al. 2011 Buffet et al. 2013a
CuO	-	< 50	natural	total element	-	GF-AAS	Gomes et al. 2012
CuO		< 50	natural	total element dissolution	filtration 0.02 μm	GF-AAS	Gomes et al. 2011
CuO	-	30–50	artificial	dissolution	filtration (0.45 μm) + UF (10 kDa)	ICP-OES	Park et al. 2014

Table 2 Cont....

Table 2 Cont...

Nanoparticle	Coating	Size (nm)	Seawater	Analytical information	Sample treatment	Analytical technique	Reference
Fe	polyacrylamide	50	natural	dissolution aggregation	dialysis (12 kDa)	ICP-MS NTA DLS	Kadar et al. 2013
Fe oxide	- gum arabic dextran carboxymethyl dextran	-	natural	dissolution	UF (3 kDa)	ICP-MS	Kadar et al. 2014
SiO_2	-	5–15	natural	aggregation	-	DLS	Miglietta et al. 2011
TiO_2	-	32	natural	total element	acid digestion	ICP-OES	Galloway et al. 2010
TiO_2	-	7 86 28	natural artificial	aggregation	-	DLS SEM	Doyle et al. 2014
TiO_2	-	21	reconstituted	aggregation	-	DLS	Wang et al. 2014
ZnO	-	24	natural	total element	acid digestion	ICP-OES	Montes et al. 2012
ZnO	-	24	natural	dissolution	UF (3 kDa)	ICP-OES	Fairbairn et al. 2011
ZnO	-	100 200	natural	dissolution	filtration (0.02 μm)	ICP-MS	Manzo et al. 2013
ZnO	- C12–15 alkyl benzoate, polyhydroxystearic acid, and triethoxycaprylylsilane	21	artificial	dissolution formation of secondary phases	UF (1 kDa) centrifugation	- TEM XRD XPS	Gelabert et al. 2014
ZnO	-	32	natural	dissolution	UF (10 kDa)	ICP-OES	Trevisan et al. 2014
ZnO	-	20	artificial	dissolution	filtration (0.1 μm)	ICP-OES	Wong et al. 2010
ZnO	-	20–30	artificial	dissolution	filtration (0.45 μm) + UF (10 kDa)	ICP-OES	Park et al. 2014
ZnO	-	-	natural	aggregation	-	DLS	Buffet et al. 2012
ZnO	-	24	natural	aggregation	-	DLS UV-Vis	Fairbairn et al. 2011
ZnO	-	< 100	natural	aggregation	-	DLS	Miglietta et al. 2011

concentration, the incorporation of the NPs in the micellar aggregates and the separation of the surfactant phase from the aqueous one by mild heating (ca. 40°C).

Analytical methods to study the fate of ENMs in seawater

The fate of different ENMs in natural and artificial seawaters has been studied to know their behaviour in these saline media, focussing almost exclusively on their dissolution and aggregation. A number of bioaccumulation studies on marine organisms also consider the characterization and quantification of the

ENMs in the test media to correlate this information with the effects observed in the organisms. Table 2 summarizes the analytical methods used in these studies.

Up till date, the behaviour in seawater of metallic (Ag, Au, Fe) and metal oxides (CeO_2, CuO, Fe oxide, SiO_2, TiO_2, ZnO) nanoparticles, as well as quantum dots (CdS) have been studied. These studies involve the use of non-coated as well as nanoparticles coated by different substances, which influence their dissolution and aggregation behaviours.

Most of the methods are oriented to determine the total concentration of the element present in the nanoparticle. Seawater samples are analyzed directly by an atomic spectrometry technique (GF-AAS, ICP-OES, or ICP-MS) depending on the concentration expected. The dissolution of the nanoparticles in the test media has been studied by performing a separation step of the dissolved element fraction by using filtration, ultrafiltration, or dialysis with membranes of different MWCOs. These cut-offs have ranged from 0.02 μm down to 1 kDa (equivalent to ca. 1 nm). In some studies, filtration membranes of 0.1 and 0.2 μm have been used (Wong et al. 2010; Wang et al. 2014) although the usefulness of the results is questionable. In any case, the filtrates, ultrafiltrates, or dialysates were analyzed by an atomic spectrometry technique to obtain the element content, which was assumed to be the dissolved fraction, consisting of the ionic element and their complexes below the MWCO of the membrane selected. As an alternative to these membrane-based separation techniques, DGT has been used in combination with atomic spectrometry to quantify the labile fraction of the element present in the media, which accounts for the ionic as well as the weak complexed element.

It is worth pointing out that most of these studies do not report information related to the validation of the methods used. In the case of ultrafiltration, low recoveries of dissolved species have been reported due to adsorption onto the membranes (Mitrano et al. 2014). With respect to the quantification step, most of the methods analyze the seawater samples directly and just in two studies, one involving CeO_2 (Montes et al. 2012) and the other TiO_2 (Galloway et al. 2010), samples were digested prior to their analysis by ICP-OES. In these cases, the digestion of the samples is recommended since these refractory oxides can behave in ICP differently respect to the corresponding ionic standards used for quantification, which may lead to erroneous results (Fabricius et al. 2014).

In addition to dissolution, aggregation of nanoparticles is the next most significant process expected in seawater. Aggregation has been studied mainly by DLS, which also provides information about the size of the aggregates. In some studies, more specific sizing techniques, like DCS (Burchardt et al. 2012) have been used. Measurements of UV-Vis absorption vs. time have also been used to follow aggregation processes based on the variation of the spectrum produced by aggregation. UV-Vis absorption is useful for nanoparticles showing surface plasmon resonance, like Ag (Chinnapongse et al. 2011) and Au nanoparticles (Heo et al. 2013), but also with others like ZnO (Fairbairn et al. 2011).

In the study of Gelabert et al. (2014) about the fate of ZnO nanoparticles in artificial seawater, apart from the information provided about the dissolution of the NPs by using UF, they studied the formation of secondary phases by using TEM, XRD and XPS. They were able to confirm the formation of wurtzite (ZnS) along with Zn carbonate phases such as hydrozincite ($Zn_5(OH)_6(CO_3)_2$).

Analytical methods for ENMs in bioaccumulation tests

Bioaccumulation tests typically involve the exposure of the test organisms until a steady-state concentration of the test substance between the external media and the tissues of the individuals is reached. However, many parameters needed to model the bioconcentration of ENMs in organisms remain unknown, such as the assumption of equilibrium conditions to knowing the concentration of the active species or the pathways of uptake and elimination of the ENMs (Handy et al. 2012). In addition to all these problems, the ability to measure ENMs in tissues and the tested environmental media is limited. For this reason, most of the bioaccumulation tests with inorganic ENMs are limited to the determination of the total element contents, quantifying the bioaccumulation in relation to the corresponding unexposed control tests. These analyses involve the organisms under study as well as the test media.

Table 1 summarizes different types of marine organisms studied. Microorganisms include biofilm bacteria and microalgae, whereas macro-organisms extend to arthropods, annelids, molluscs, and macroalgae. Up till date, bioaccumulation studies have been performed with metallic (Ag, Au, Fe) and metal oxide (AgO, CeO_2, CuO, TiO_2, ZnO) nanoparticles, as well as quantum dots (CdS, CdSe). These studies involve the use of non-coated as well as coated nanoparticles, and in some of them soluble salts of the metals have been used to compare their effects.

As has been pointed out, most of the analyses in bioaccumulation test with inorganic ENMs involve the determination of the total element content of whole organisms, or the edible tissue in the case of bivalves, by using an atomic spectrometry technique after the digestion of the sample. Samples are digested by using different concentrated acids (HNO_3, H_2SO_4) or mixtures (HNO_3+H_2O_2, aqua regia), using conventional heating systems at low temperatures (< 100°C) for long periods of time or pressurized microwave-assisted systems.

In the case of mussels and oysters, the distribution of the metal in different organs and fluids (gill, digestive gland mantle tissues, hemolymph) has been reported (Tedesco et al. 2010a; Gomes et al. 2011; Gomes et al. 2013; Trevisan et al. 2014). When the determination of the total element in the edible tissue is combined with the analysis of the faeces, information about the uptake, accumulation, and elimination of ENMs was obtained.

In a different way, Bielmyer-Fraser et al. (2014) studied the cellular metal distribution in microalgae by analyzing cell components previously separated by differential centrifugation. They observed that metal concentrations were the highest in the algal cell wall when cells were exposed to metal oxide nanoparticles, whereas algae exposed to dissolved metals had higher proportions of metal in the organelle and endoplasmic reticulum fractions.

Future trends and recommendations for the analysis of ENMs in marine samples

Marine ecotoxicology is involved in the development of realistic ENMs exposure models to predict the impacts of ENMs in marine ecosystems (Corsi et al. 2014). Currently, the strategy followed is based on developing laboratory experiments to study the fate and transport of ENMs, together with the evaluation of their toxicity on organisms at risk of exposure in marine habitats. As a next step, results from these experiments can be checked under more realistic environmental conditions in mesocosms experiments and, finally, all the outputs can be integrated in the corresponding models. Apart from this approach, ENMs should be monitored in marine environments to support the exposure estimates from these models. In any case, the approach must be supported by reliable analytical information about the ENMs vis-á-vis the different fate, toxicity, and mesocosms studies, where the transformations, aging, and weathering of ENMs should also be considered.

While available analytical techniques and methods for the detection, characterization, and quantification of ENMs are widely known and described, as we have shown along this chapter, data available on their application in marine samples is limited. Detection of inorganic ENMs in marine samples is generally reduced to the use of atomic spectrometry techniques alone, to determine total element contents, or in combination with fractionation techniques, to differentiate dissolved and particulate forms of the element present in the ENM. Even electron microscopy techniques, like TEM or SEM, are scarcely used, apart from the characterization of the pristine ENMs. Moreover, these analyses are confined to samples from laboratory studies or ecotoxicological assays and have not been applied to field samples yet.

At the present time, marine ecotoxicology sustains a certain backwardness in relation to the application of analytical techniques and methods for the analysis of ENMs, which are currently been used in other fields. Whereas the combination of imaging techniques (typically, TEM) and elemental analysis (ICP-MS, mainly) can provide valuable information for detection/characterization/quantification of inorganic ENMs at relatively high concentrations, when working at realistic environmental concentrations, sensitive and selective approaches must be adopted to get something more than total element contents. In this sense, the use of the ICP-MS in single particle mode allows the detection of nanoparticles, as such, at concentrations at part per trillion levels, being able to provide information about size, as well as about nanoparticle number concentration. On the other hand, when ICP-MS is used as element specific detector in separation techniques like FFF or HDC, size distributions and quantitative information with respect to size can be obtained at the part per billion level, both for nanoparticles and macromolecule complexes of the corresponding elements. NTA, in spite of being a non element-specific technique, is being used to get information about number concentration and size as a more reliable alternative to DLS. Additionally, the extremely low concentration levels expected in marine environments (seawater, sediments, organism) will required the use of preconcentration methods, and those based on solid phase and cloud point extraction will be of great value.

It is expected that in the near future, all these emerging methodologies will be incorporated into the analytical schemes of any ecotoxicological study to get a deeper insight into the behaviour of inorganic ENMs in the marine environment.

Acknowledgements

This work was supported by the Spanish Ministry of Economy and Competitiveness and the European Regional Development Fund, project CTQ2015-68094-C2-1-R (MINECO/FEDER).

Keywords: Inorganic engineered nanomaterials, metal nanoparticles, metal oxide nanoparticles, quantum dots, fate and transport, toxicity, bioacummulation, analytical methods, detection, characterization, quantification

References

Angel, B.M., G.E. Batley, C.V. Jarolimek and N.J. Rogers. 2013. The impact of size on the fate and toxicity of nanoparticulate silver in aquatic systems. Chemosphere 93: 359–365.

Baalousha, M., W. How, E. Valsami-Jones and J.R. Lead. 2014. Overview of environmental nanoscience. pp. 88–89. *In*: J.R. Lead and E. Valsami-Jones [eds.]. Nanoscience and the Environment. Elsevier, Amsterdam.

Baker, T.J., C.R. Tyler and T.S. Galloway. 2014. Impacts of metal and metal oxide nanoparticles on marine organisms. Environ. Pollut. 186: 257–271.

Batley, G., J. Kirby and M. McLaughlin. 2012. Fate and risks of nanomaterials in aquatic and terrestrial environments. Accounts Chem. 46: 854–862.

Boyd, R.D., S.K. Pichaimuthu and A. Cuenat. 2011. New approach to inter-technique comparisons for nanoparticle size measurements using atomic force microscopy, nanoparticle tracking analysis and dynamic light scattering. Colloids Surf. A Physicochem. Eng. Asp. 387: 35–42.

Bielmyer-Fraser, G.K., T.A. Jarvis, H.S. Lenihan and R.J. Miller. 2014. Cellular partitioning of nanoparticulate versus dissolved metals in marine phytoplankton. Environ. Sci. Technol. 48: 13443–13450.

Bradford, A., R.D. Handy, J.W. Readman, A. Atfield and M. Mühling. 2009. Impact of silver nanoparticle contamination on the genetic diversity of natural bacterial assemblages in estuarine sediments. Environ. Sci. Technol. 43: 4530–4536.

Buffet, P.E., O.F. Tankoua, J.F. Pan, D. Berhanu, C. Herrenknecht, L. Poirier, C. Amiard-Triquet, J.C. Amiard, J.B. Bérard, C. Risso, M. Guibbolini, M. Roméo, P. Reip, E. Valsami-Jones and C. Mouneyrac. 2011. Behavioural and biochemical responses of two marine invertebrates *Scrobicularia plana* and *Hediste diversicolor* to copper oxide nanoparticles. Chemosphere 84: 166–174.

Buffet, P.E., C. Amiard-Triquet, A. Dybowska, C. Risso-de Faverney, M. Guibbolini, E. Valsami-Jones and C. Mouneyrac. 2012. Fate of isotopically labeled zinc oxide nanoparticles in sediment and effects on two endobenthic species, the clam *Scrobicularia plana* and the ragworm *Hediste diversicolor*. Ecotoxicol. Environ. Saf. 84: 191–198.

Buffet, P.E., M. Richard, F. Caupos, A. Vergnoux, H. Perrein-Ettajani, A. Luna-Acosta, F. Akcha, J. Amiard, C. Amiard-Triquet, M. Guibbolini, C. Risso-De Faverney, H. Thomas-Guyon, P. Reip, A. Dybowska, D. Berhanu, E. Valsami-Jones and C. Mouneyrac. 2013a. A mesocosm study of fate and effects of CuO nanoparticles on endobenthic species (*Scrobicularia plana*, *Hediste diversicolor*). Environ. Sci. Technol. 47: 1620–1628.

Buffet, P.E., J.F. Pan, L. Poirier, C. Amiard-Triquet, J.C. Amiard, P. Gaudin, C. Risso-de Faverney, M. Guibbolini, D. Gilliland, E. Valsami-Jones and C. Mouneyrac. 2013b. Biochemical and behavioural responses of the endobenthic bivalve *Scrobicularia plana* to silver nanoparticles in seawater and microalgal food. Ecotoxicol. Environ. Saf. 89: 117–124.

Buffet, P.E., A. Zalouk-Vergnoux, A. Châtel, B. Berthet, I. Métais, H. Perrein-Ettajani, L. Poirier, A. Luna-Acosta, H. Thomas-Guyon, C. Risso-de Faverney, M. Guibbolini, D. Gilliland, E. Valsami-Jones and C. Mouneyrac. 2014a. A marine mesocosm study on the environmental fate of silver nanoparticles and toxicity effects on two endobenthic species: the ragworm *Hediste diversicolor* and the bivalve mollusc *Scrobicularia plana*. Sci. Total Environ. 470-471: 1151–1159.

Buffet, P.E., L. Poirier, A. Zalouk-Vergnoux, C. Lopes, J.C. Amiard, P. Gaudin, C. Risso-de Faverney, M. Guibbolini, D. Gilliland, H. Perrein-Ettajani, E. Valsami-Jones and C. Mouneyrac. 2014b. Biochemical and behavioural responses of the marine polychaete *Hediste diversicolor* to cadmium sulfide quantum dots (CdS QDs): waterborne and dietary exposure. Chemosphere 100: 63–70.

Burchardt, A.D., R.N. Carvalho, A. Valente, P. Nativo, D. Gilliland, C.P. Garcìa, R. Passarella, V. Pedroni, F. Rossi and T. Lettieri. 2012. Effects of silver nanoparticles in diatom *Thalassiosira pseudonana* and cyanobacterium *Synechococcus* sp. Environ. Sci. Technol. 46: 11336–11344.

Casals, E., S. Vazquez-Campos, N. Bastús and V. Puntes. 2008. Distribution and potential toxicity of engineered inorganic nanoparticles and carbon nanostructures in biological systems. Trends Anal. Chem. 27: 672–683.

Castro-Bugallo, A., A. González-Fernández, C. Guisande and A. Barreiro. 2014. Comparative responses to metal oxide nanoparticles in marine phytoplankton. Arch. Environ. Contam. Toxicol. 67: 483–493.

Chao, J., J. Liu, S. Yu, Y. Feng, Z. Tan, R. Liu and Y. Yin. 2011. Speciation analysis of silver nanoparticles and silver ions in antibacterial products and environmental waters via cloud point extraction-based separation. Anal. Chem. 83: 6875–6882.

Chinnapongse, S.L., R.I. MacCuspie and V.A. Hackley. 2011. Persistence of singly dispersed silver nanoparticles in natural freshwaters, synthetic seawater, and simulated estuarine waters. Sci. Total Environ. 409: 2443–2450.

Cong, Y., G.T. Banta, H. Selck, D. Berhanu, E. Valsami-Jones and V.E. Forbes. 2011. Toxic effects and bioaccumulation of nano-, micron- and ionic-Ag in the polychaete, *Nereis diversicolor*. Aquat. Toxicol. 105: 403–411.

Corsi, I., G.N. Cherr, H.S. Lenihan, J. Labille, M. Hassellov, L. Canesi, F. Dondero, G. Frenzilli, D. Hristozov, V. Puntes, C. Della Torre, A. Pinsino, G. Libralato, A. Marcomini, E. Sabbioni and V. Matranga. 2014. Common strategies and technologies for the ecosafety assessment and design of nanomaterials entering the marine environment. ACS Nano 8: 9694–709.

Dai, L., K. Syberg and G. Banta. 2013. Effects, Uptake, and Depuration Kinetics of Silver Oxide and Copper Oxide Nanoparticles in a Marine Deposit Feeder, *Macoma balthica*. ACS Sustain. 1: 760–767.

Danovaro, R., L. Bongiorni, C. Corinaldesi, D. Giovannelli, E. Damiani, P. Astolfi, L. Greci and A. Pusceddu. 2008. Sunscreens cause coral bleaching by promoting viral infections. Environ. Health Perspect. 116: 441–447.

Doiron, K., E. Pelletier and K. Lemarchand. 2012. Impact of polymer-coated silver nanoparticles on marine microbial communities: a microcosm study. Aquat. Toxicol. 124-125: 22–27.

Doyle, J.J., V. Palumbo, B.D. Huey and J.E. Ward. 2014. Behavior of Titanium Dioxide Nanoparticles in Three Aqueous Media Samples: Agglomeration and Implications for Benthic Deposition. Water, Air, Soil Pollut. 225: 2106.

Fabrega, J., R. Zhang, J.C. Renshaw, W.T. Liu and J.R. Lead. 2011. Impact of silver nanoparticles on natural marine biofilm bacteria. Chemosphere 85: 961–966.

Fabricius, A.L., L. Duester, B. Meermann and T.A. Ternes. 2014. ICP-MS-based characterization of inorganic nanoparticles-sample preparation and off-line fractionation strategies. Anal. Bioanal. Chem. 406: 467–479.

Fairbairn, E.A., A.A. Keller, L. Mädler, D. Zhou, S. Pokhrel and G.N. Cherr. 2011. Metal oxide nanomaterials in seawater: Linking physicochemical characteristics with biological response in sea urchin development. J. Hazard. Mater. 192: 1565–1571.

Franklin, N.M., N.J. Rogers, S.C. Apte, G.E. Batley, G.E. Gadd and P.S. Casey. Comparative toxicity of nanoparticulate ZnO, bulk ZnO, and $ZnCl_2$ to a freshwater microalga (*Pseudokirchnerilla subcapitata*): the importance of particle solubility. Environ. Sci. Technol. 4: 1–27.

Galloway, T., C. Lewis, I. Dolciotti, B.D. Johnston, J. Moger and F. Regoli. 2010. Sublethal toxicity of nano-titanium dioxide and carbon nanotubes in a sediment dwelling marine polychaete. Environ. Pollut. 158: 1748–1755.

García-Alonso, J., N. Rodriguez-Sanchez, S.K. Misra, E. Valsami-Jones, M.N. Croteau, S.N. Luoma and P.S. Rainbow. 2014. Toxicity and accumulation of silver nanoparticles during development of the marine polychaete *Platynereis dumerilii*. Sci. Total Environ. 476-477: 688–695.

García-Negrete, C.A., J. Blasco, M. Volland, T.C. Rojas, M. Hampel, A. Lapresta-Fernández, M.C. Jiménez de Haro, M. Soto and A. Fernández. 2013. Behaviour of Au-citrate nanoparticles in seawater and accumulation in bivalves at environmentally relevant concentrations. Environ. Pollut. 174: 134–141.

Gelabert, A., Y. Sivry, R. Ferrari, A. Akrout, L. Cordier, S. Nowak, N. Menguy and M.F. Benedetti. 2014. Uncoated and coated ZnO nanoparticle life cycle in synthetic seawater. Environ. Toxicol. Chem. 33: 341–349.

Gomes, T., J.P. Pinheiro, I. Cancio, C.G. Pereira, C. Cardoso and M.J. Bebianno. 2011. Effects of copper nanoparticles exposure in the mussel *Mytilus galloprovincialis*. Environ. Sci. Technol. 45: 9356–9362.

Gomes, T., C.G. Pereira, C. Cardoso, J.P. Pinheiro, I. Cancio and M.J. Bebianno. 2012. Accumulation and toxicity of copper oxide nanoparticles in the digestive gland of *Mytilus galloprovincialis*. Aquat. Toxicol. 118-119: 72–79.

Gomes, T., O. Araújo, R. Pereira, A.C. Almeida, A. Cravo and M.J. Bebianno. 2013. Genotoxicity of copper oxide and silver nanoparticles in the mussel *Mytilus galloprovincialis*. Mar. Environ. Res. 84: 51–59.

Gottschalk, F., C. Ort, R.W. Scholz and B. Nowack. 2011. Engineered nanomaterials in rivers—Exposure scenarios for Switzerland at high spatial and temporal resolution. Environ. Pollut.

Gray, E.P., J.G. Coleman, A.J. Bednar, A.J. Kennedy, J.F. Ranville and C.P. Higgins. 2013. Extraction and analysis of silver and gold nanoparticles from biological tissues using single particle inductively coupled plasma mass spectrometry. Environ. Sci. Technol. 47: 14315–14323.

Handy, R.D., F. von der Kammer, J.R. Lead, M. Hassellöv, R. Owen and M. Crane. 2008. The ecotoxicology and chemistry of manufactured nanoparticles. Ecotoxicology 17: 287–314.

Handy, R.D., G. Cornelis, T. Fernandes, O. Tsyusko, A. Decho, T. Sabo-Attwood, C. Metcalfe, J.A. Steevens, S.J. Klaine, A.A. Koelmans and N. Horne. 2012. Ecotoxicity test methods for engineered nanomaterials: Practical experiences and recommendations from the bench. Environ. Toxicol. Chem. 31: 15–31.

Hanna, S.K., R.J. Miller, D. Zhou, A.A. Keller and h.s. Lenihan. 2013. Accumulation and toxicity of metal oxide nanoparticles in a soft-sediment estuarine amphipod. Aquat. Toxicol. 142-143: 441–446.

Heo, J.H., K.I. Kim, M.H. Lee and J.H. Lee. 2013. Stability of a Gold Nanoparticle-DNA System in Seawater. J. Nanosci. Nanotechnol. 13: 7254–7258.

Huynh, K.A. and K.L. Chen. 2011. Aggregation kinetics of citrate and polyvinylpyrrolidone coated silver nanoparticles in monovalent and divalent electrolyte solutions. Environ. Sci. Tec. 45: 5564–5571.

JRC-IRMM. 2011. Zinc Oxide NM-110, NM-111, NM-112, NM-113 Characterisation and Test Item Preparation NM-series of Representative Manufactured Nanomaterials, Luxembourg.

Kadar, E., P. Rooks, C. Lakey and D.A. White. 2012. The effect of engineered iron nanoparticles on growth and metabolic status of marine microalgae cultures. Sci. Total Environ. 439: 8–17.

Kadar, E., O. Dyson, R.D. Handy and S.N. Al-Subiai. 2013. Are reproduction impairments of free spawning marine invertebrates exposed to zero-valent nano-iron associated with dissolution of nanoparticles? Nanotoxicology 7: 135–143.

Kadar, E., I.L. Batalha, A. Fisher and A.C.A. Roque. 2014. The interaction of polymer-coated magnetic nanoparticles with seawater. Sci. Total Environ. 487: 771–777.

Kaegi, R., A. Ulrich, B. Sinnet, R. Vonbank, A. Wichser, S. Zuleeg, H. Simmler, S. Brunner, H. Vonmont, M. Burkhardt and M. Boller. 2008. Synthetic TiO_2 nanoparticle emission from exterior facades into the aquatic environment. Environ. Pollut. 156: 233–239.

Keller, A.A., H. Wang, D. Zhou, H.S. Lenihan, G. Cherr, B.J. Cardinale, R. Miller and Z. Ji. 2010. Stability and aggregation of metal oxide nanoparticles in natural aqueous matrices. Environ. Sci. Technol. 44: 1962–1967.

Klaine, S.J., P.J.J. Alvarez, G.E. Batley, T.F. Fernandes, R.D. Handy, D.Y. Lyon, S. Mahendra, M.J. McLaughlin and J.R. Lead. 2008. Nanomaterials in the environment: Behaviour, fate, bioavailability and effects. Environ. Toxicol. Chem. 27: 1825–1851.

Labille, J. and J. Brant. 2010. Stability of Nanoparticles in Water. Nanomedicine 5: 985–998.

Laborda, F., E. Bolea and J. Jiménez-Lamana. 2014. Single particle inductively coupled plasma mass spectrometry: A powerful tool for nanoanalysis. Anal. Chem. 86: 2270–2278.

Ladner, D.A., M. Steele, A. Weir, K. Hristovski and P. Westerhoff. 2012. Functionalized nanoparticle interactions with polymeric membranes. J. Hazard Mater 211-212: 288–95.

Li, L., K. Leopold and M. Schuster. 2012. Effective and selective extraction of noble metal nanoparticles from environmental water through a noncovalent reversible reaction on an ionic exchange resin. Chem. Commun. 48: 9165–9167.

Loeschner, K., M.S.J. Brabrand, J.J. Sloth and E.H. Larsen. 2014. Use of alkaline or enzymatic sample pretreatment prior to characterization of gold nanoparticles in animal tissue by single-particle ICPMS. Anal. Bioanal. Chem. 406: 3845–3851.

Manzo, S., M.L. Miglietta, G. Rametta, S. Buono and G. Di Francia. 2013. Toxic effects of ZnO nanoparticles towards marine algae Dunaliella tertiolecta. Sci. Total Environ. 445-446: 371–376.

Matranga, V. and I. Corsi. 2012. Toxic effects of engineered nanoparticles in the marine environment: model organisms and molecular approaches. Mar. Environ. Res. 76: 32–40.

McCarthy, M.P., D.L. Carroll and A.H. Ringwood. 2013. Tissue specific responses of oysters, *Crassostrea virginica*, to silver nanoparticles. Aquat. Toxicol. 138-139: 123–128.

Miao, A.J., K.A. Schwehr, C. Xu, S.-J. Zhang, Z. Luo, A. Quigg and P.H. Santschi. 2009. The algal toxicity of silver engineered nanoparticles and detoxification by exopolymeric substances. Environ. Pollut. 157: 3034–3041.

Miao, A.-J., X.-Y. Zhang, Z. Luo, C.-S. Chen, W.-C. Chin, P.H. Santschi and A. Quigg. 2010. Zinc oxide-engineered nanoparticles: dissolution and toxicity to marine phytoplankton. Environ. Toxicol. Chem. 29: 2814–2822.

Miglietta, M.L., G. Rametta, G. Francia, S. Di, Manzo, A. Rocco, R. Carotenuto, F.D.L. Picione and S. Buono. 2011. Characterization of nanoparticles in seawater for toxicity assessment towards aquatic organisms. pp. 425–429. *In*: G. Neri, N. Donato, A. D'Amico and C. Di Natale [eds.]. Sensors and Microsystems, Lecture Notes in Electrical Engineering. Springer Netherlands, Dordrecht.

Miller, R.J., H.S. Lenihan, E.B. Muller, N. Tseng, S.K. Hanna and A.A. Keller. 2010. Impacts of metal oxide nanoparticles on marine phytoplankton. Environ. Sci. Technol. 44: 7329–7334.

Miller, R.J., S. Bennett, A.A. Keller, S. Pease and H.S. Lenihan. 2012. TiO_2 nanoparticles are phototoxic to marine phytoplankton. PLoS One 7: e30321.

Misra, S.K., A. Dybowska, D. Berhanu, S.N. Luoma and E. Valsami-Jones. 2012. The complexity of nanoparticle dissolution and its importance in nanotoxicological studies. Sci. Total Environ. 438: 225–232.

Mitrano, D.M., E. Rimmele, A. Wichser, R. Erni, M. Height and B. Nowack. 2014. Presence of nanoparticles in wash water from conventional silver and nano-silver textiles. ACS Nano 8: 7208–7219.

Montaño, M.D., G.V. Lowry, F. von der Kammer, J. Blue and J.F. Ranville. 2014. Current status and future direction for examining engineered nanoparticles in natural systems. Environ. Chem. 11: 351–356.

Montes, M.O., S.K. Hanna, H.S. Lenihan and A.A. Keller. 2012. Uptake, accumulation, and biotransformation of metal oxide nanoparticles by a marine suspension-feeder. J. Hazard. Mater. 225-226: 139–145.

Park, J., S. Kim, J. Yoo, J.S. Lee, J.W. Park and J. Jung. 2014. Effect of salinity on acute copper and zinc toxicity to Tigriopus japonicus: The difference between metal ions and nanoparticles. Mar. Pollut. Bull. 85: 526–531.

Peng, X., S. Palma, N.S. Fisher and S.S. Wong. 2011. Effect of morphology of ZnO nanostructures on their toxicity to marine algae. Aquat. Toxicol. 102: 186–196.

Peralta-Videa, J.R., L. Zhao, M.L. Lopez-Moreno, G. de la Rosa, J. Hong and J.L. Gardea-Torresdey. 2011. Nanomaterials and the environment: a review for the biennium 2008–2010. J. Hazard. Mater. 186: 1–15.

Piccinno, F., F. Gottschalk, S. Seeger and B. Nowack. 2012. Industrial production quantities and uses of ten engineered nanomaterials in Europe and the world. J. Nanopart. Res. 14: 1109.

Praetorius, A., N. Tufenkji, K.U. Goss, M. Scheringer, F. von der Kammer and M. Elimelech. 2014. The road to nowhere: equilibrium partition coefficients for nanoparticles. Environ. Sci. Nano. 1: 317–323.
Stolpe, B. and M. Hassellov. 2007. Changes in size distribution of fresh water nanoscale colloidal matter and associated elements on mixing with seawater. Geochim. Cosmochim. Acta 71: 3292–3301.
Tedesco, S., H. Doyle, J. Blasco, G. Redmond and D. Sheehan. 2010a. Oxidative stress and toxicity of gold nanoparticles in *Mytilus edulis*. Aquat. Toxicol. 100: 178–186.
Tedesco, S., H. Doyle, J. Blasco, G. Redmond and D. Sheehan. 2010b. Exposure of the blue mussel, *Mytilus edulis*, to gold nanoparticles and the pro-oxidant menadione. Comp. Biochem. Physiol.—C Toxicol. Pharmacol. 151: 167–174.
Tiede, K., M. Hassellöv, E. Breitbarth, Q. Chaudhry and A.B. Boxall. 2009. Considerations for environmental fate and ecotoxicity testing to support environmental risk assessments for engineered nanoparticles. J. Chromatogr. A 1216: 503–509.
Trevisan, R., G. Delapedra, D.F. Mello, M. Arl, E.C. Schmidt, F. Meder, M. Monopoli, E. Cargnin-Ferreira, Z.L. Bouzon, A.S. Fisher, D. Sheehan and A.L. Dafre. 2014. Gills are an initial target of zinc oxide nanoparticles in oysters *Crassostrea gigas*, leading to mitochondrial disruption and oxidative stress. Aquat. Toxicol. 153: 27–38.
Tsao, T.M., Y.M. Chen and M.K. Wang. 2011. Origin, separation and identification of environmental nanoparticles: a review. J. Environ. Monit. 13: 1156–1163.
Turner, A., D. Brice and M.T. Brown. 2012. Interactions of silver nanoparticles with the marine macroalga, *Ulva lactuca*. Ecotoxicology 21: 148–154.
Wahie, S., J.J. Lloyd and P.M. Farr. 2007. Sunscreen ingredients and labelling: a survey of products available in the UK. Clin. Exp. Dermatol. 32: 359–364.
Wang, H., R.M. Burgess, M.G. Cantwell, L.M. Portis, M.M. Perron, F. Wu and K.T. Ho. 2014. Stability and aggregation of silver and titanium dioxide nanoparticles in seawater: Role of salinity and dissolved organic carbon. Environ. Toxicol. Chem. 33: 1023–1029.
Wells, M.L. 1998. Marine Colloids: A Neglected Dimension. Nature 391: 530–531.
Wong, S.W.Y., P.T.Y. Leung, A.B. Djurisić and K.M.Y. Leung. 2010. Toxicities of nano zinc oxide to five marine organisms: Influences of aggregate size and ion solubility. Anal. Bioanal. Chem. 396: 609–618.
Wong, S.W.Y. and K.M.Y. Leung. 2014. Temperature-dependent toxicities of nano zinc oxide to marine diatom, amphipod and fish in relation to its aggregation size and ion dissolution. Nanotoxicology 8: 24–35.
Xu, M., J. Li, N. Hanagata, H. Su, H. Chena and D. Fujita. 2013. Challenge to assess the toxic contribution of metal cation released from nanomaterials for nanotoxicology—the case of ZnO nanoparticles. Nanoscale 5: 4763–4769.
Zhang, H. and W. Davidson. 1995. Performance-characteristics of diffusion gradients in thin-films for the *in-situ* measurement of trace-metals in aqueous-solution. Anal. Chem. 67: 3391–3400.
Zhu, X., J. Zhou and Z. Cai. 2011. The toxicity and oxidative stress of TiO_2 nanoparticles in marine abalone (*Haliotis diversicolor supertexta*). Mar. Pollut. Bull. 63: 334–338.
Zhu, X. and Z. Cai. 2012. Behavior and effect of manufactured nanomaterials in the marine environment. Integr. Environ. Assess. Manag. 8: 566–567.

13

Pharmaceuticals in the Marine Environment

Analytical Techniques and Applications

Sara Rodríguez-Mozaz,[1,a,*] *Diana Álvarez-Muñoz*[1,b] and *Damià Barceló*[1,2]

ABSTRACT

Pharmaceutically Active Compounds (PhACs) are continuously discharged into the aquatic marine environment through sewage effluents, rivers, and aquaculture facilities located in coastal areas. Therefore they can be found in different compartments (water, sediments, and biota) in marine environment. Although still not regulated, PhACs are considered pollutants of emerging concern and therefore they need to be monitored in both freshwater and marine ecosystems. In this chapter, current analytical methodologies and studies about the presence of pharmaceuticals in the marine environment are thoroughly reviewed. The presence of several therapeutic families was evaluated, antibiotics, psychiatric drugs, and analgesics/anti-inflammatories being the groups most frequently detected in all environmental compartments. Levels found in the marine environment were in general lower than those found in the freshwater system but occasionally reached values up to 2983 ng/L of acetaminophen, 458 ng/g of ofloxacin, and 490 ng/g of salicylic acid in water, sediment, and biota respectively. Other aspects related to the presence of PhACs in marine environment, such as pollutant distribution in the environment, geographical variations, and environmental risk assessment, are also considered in this chapter.

Introduction

Pharmaceutically Active Compounds (PhACs) play an important role in assuring both humans and animals health. They include a wide variety of therapeutic families such as antibiotics, psychiatric drugs,

[1] Catalan Institute for Water Research (ICRA), Parc Científic i Tecnològic de la Universitat de Girona, C/Emili Grahit 101, Edifici H_2O, E-17003 Girona, Spain.
[a] Email: srodriguez@icra.es
[b] Email: dalvarez@icra.es
[2] Water and Soil Quality Research Group, Department of Environmental Chemistry, IDAEA-CSIC, Jordi Girona 18-26, 08034 Barcelona, Spain.
Email: dbarcelo@icra.es
* Corresponding author

analgesics/anti-inflammatories, tranquilizers, hormones, β-blockers, diuretics, etc. The main sources of PhACs contamination are sewage effluents, waste disposal, aquaculture, animal husbandry, and horticulture (Gaw et al. 2014). They provide a continuous input of these compounds into the environment and although PhACs have high transformation/removal rates, they are compensated by their introduction into the aquatic systems, therefore, they are considered as "pseudo-persistent" contaminants. They are frequently detected in aquatic monitoring on a global scale (Kuster 2014). More than 600 PhACs have been shown to be present in the environment (Kuster 2014). Most of them are usually found in the water due to their main hydrophilic character. They have been detected in wastewater, groundwater, surface and marine waters at a nanogram to microgram/L range (Collado et al. 2014; de Jongh et al. 2012; Godfrey et al. 2007; Gros et al. 2013; Loos et al. 2013a; Nodler et al. 2014). Once introduced into surface water, pharmaceuticals may undergo biodegradation, hydrolysis or photodegradation, as well as adsorption to suspended particulate matter, and finally sedimentation (Xu 2009). Sediments are natural repositories of many contaminants present in the water column and in the case of PhACs they have been identified as a major sink for antibiotics (Kim and Carlson 2007).

The presence of PhACs in the environment may result in a chronic exposure of aquatic organisms to these compounds and/or their transformation products, especially in the case of filter feeding organisms due to polar nature of PhACs which makes them directly bioavailable. Besides, PhACs are designed to target specific metabolic and molecular pathways and can act at very low concentrations which raise the concern about their potential to cause adverse effects on non-target organisms. Over the past two decades, investigations on how exposure to PhACs may affect organisms have revealed that estrogens exposure contributed to the feminization of male fish in effluent-dominated rivers (Sumpter 1995). More recently it has been observed that antidepressants in surface water alter animal behaviors that are known to have ecological and evolutionary consequences (Brodin 2013). Besides, another issue of growing concern for both the scientific and general community is the induction of antibiotic resistance in aquatic organisms from aquaculture and the potential risk to human health through the ingestion of this seafood (Sapkota 2008).

Pharmaceuticals are thus pollutants of emerging concern although they have not been regulated in the marine environment yet. However, recently, the EU commission included diclofenac, 17β-estradiol (E2), 17α-ethynylestradiol (EE2) and macrolide antibiotics (erythromycin, clarithromycin and azithromycin) in a "watch list" of emerging aquatic pollutants (EU 2015/495); which could one day be placed on the "priority list"of compounds from Water Framework directive (WFD, 2000/60/EC) and further regulated in directive 2008/105/EC. The last directive sets environmental quality standards (EQS) for these substances in surface waters in inland and coastal waters, defined as being within a distance of one nautical mile on the seaward side (Loos et al. 2013b). Chemical contaminants, including pharmaceuticals, can enter marine environment both directly and indirectly through a range of diffuse and point sources and therefore, there is a clear need to assess the state of marine environment, particularly near highly populated areas. Thus with respect to this, the aim of the EU Marine Strategy Framework Directive (MSFD) is to align a sustainable use of the marine resources with the maintenance of ecosystem health and functioning by 2020 (MSFD, 2008/56/EC), and establishes that concentrations of contaminants in marine waters "should be at levels not giving rise to pollution effects". Therefore, the monitoring and control of priority emerging pollutant concentrations is necessary to ensure the protection of coastal ecosystems. As a result, an increasing number of studies have appeared in recent years to assess the presence of pharmaceuticals in the marine environment, particularly in coastal areas (Arpin-Pont et al. 2014).

The objective of this chapter is to review the current analytical methodologies applied for the determination of pharmaceuticals in water, sediments, and biota, as well as to summarize the existing studies and occurrence data of pharmaceuticals in the marine environment.

Pharmaceuticals in Sea Water

Interest in the marine environment and its status is more recent in comparison to natural freshwater ecosystems. Pharmaceuticals have been investigated since early 1990s both in wastewater treatment plants (WWTPs) as well as in natural environments usually impacted by WWTPs effluents and other sources (mainly derived from veterinary uses). However, coastal areas are considered the ultimate sink for sewage and other by-products of human activities through direct discharges of WWTP located near the sea or through the outfall of polluted rivers. Also fish farming activities contribute to the pollution of these areas. The presence of pharmaceuticals and other pollutants in coastal waters, and particularly estuaries, harbors, lagoons, and enclosed or semi-enclosed areas have thus attracted the attention of many researches and an increasing number of monitoring studies have been published in the last years. These publications will be reviewed in next sections.

Analytical methodologies

An overview of the analytical methods employed for the determination of PhACs in marine water in the last 15 years is given in Table 1. Thirty one methodologies are listed in the table, along with details about the therapeutic groups analyzed, number of PhACs included, extraction and detection method used as well as quality control parameters such as limit of detection and quantification (LOD and LOQ), and recoveries.

All the analytical methods reported in the table are multi-residue methods; namely compounds belonging to different therapeutic or chemical families. Antibiotics and analgesics/anti-inflammatory groups are the pharmaceuticals most frequently considered in these multi-residue methodologies (24 and 25 methods reported respectively). Moreover PhACs can also be analyzed together with other contaminants of emerging concern such as endocrine disruptors, polar pesticides, and personal care products (USEPA 1970) (Klosterhaus et al. 2013; Nödler et al. 2010).

Solid phase extraction followed by liquid chromatography (LC) in combination with mass spectrometry (MS) is the methodology of choice in the majority of analytical methods. However, there are three manuscripts that focus on passive samplers as a simultaneous sampling and extraction methodology (Martinez Bueno et al. 2009; Munaron et al. 2012; Tertuliani et al. 2008) and one presenting an innovative extraction methodology SPE, the so called bag-SPE (Magner et al. 2010). Analytical performance of the methods applied to sea water has been evaluated by some authors (Borecka et al. 2015; Gros et al. 2012; Wu et al. 2010) since both, extraction efficiency and ionization efficiency in ESI might be impacted by suspended matter and particularly by salinity. In order to avoid inaccurate quantification (by overestimation or underestimation) application of isotope-labeled standards seems to be the most powerful tool to compensate matrix effects. Wu et al. (2010) also recommends to increase the volume of washing solution to completely remove the salts retained in the SPE cartridge after loading the sample.

Occurrence of PhACs in sea water

A total of forty two field studies published since 2003 till 2015 are listed in Table 2, which also provides information about the country where the study was conducted, number of sampling sites, analytical method applied as well as the occurrence concentrations in the marine water investigated. Studies have been performed all over the world China being, with 7 manuscripts, the most studied marine environment, followed by USA, Spain, France, and UK with 5, 4, 3 and 2 studies respectively. In other countries such as Poland, Ireland, Germany, Netherlands, Norway, Sweden, Belgium Greece, Portugal, Italy, Turkey, Israel, Taiwan, Singapore, and Canada, between 1 and 2 field studies were performed and thus generated further information about the levels of pharmaceuticals in coastal areas mainly.

One hundred and fourteen pharmaceuticals were found at least once within all the studies considered, and among them antibiotics are the therapeutic group more documented with 32 compounds detected in at least one study, followed by psychiatric drugs (19 compounds), and analgesics and anti-inflammatory drugs (16). On the other hand, 10 pharmaceuticals (acetaminophen; ibuprofen; diclofenac; erythromycin; clarithromycin; sulfamethoxazole; trimethoprim; carbamazepine; gemfibrozil and atenolol) were quantified at least in 10 out of the 42 studies considered here.

In most of the studies reviewed in this work, special attention has been paid to the impact of the contamination sources namely rivers, WWTP discharging treated effluents in the marine environment, or aquaculture activities located in coastal areas where several type of chemicals (including antibiotics) are administrated. Spatial distribution and attenuation of this contamination sources have been evaluated by several studies, where aspects such as dilution and hydrodynamics were also assessed (Bayen et al. 2013;

Gulkowska et al. 2007; Jiang et al. 2014; McEneff et al. 2014; Zhang et al. 2013a; Zhang et al. 2013b). In addition, in some other studies, a vertical profile of contaminant concentration was measured, although no significant difference in the occurrence of compounds was noticed (Borecka et al. 2015). Temporal and seasonal variation of pharmaceuticals' concentrations was followed by some authors (Lolic et al. 2015; Vidal-Dorsch et al. 2012). Higher concentrations of pharmaceuticals was observed in beach water in tourist areas in August and September probably due to population increase in summertime (Lolic et al. 2015).

Pharmaceuticals in Marine Sediments

Although PhACs are hydrophilic molecules when they are discharged into water bodies, they come into contact with particulate matter in suspension and interactions can occur, especially with those PhACs with lower water solubility. Particulate suspended matter is deposited in areas of low flow velocity and it is incorporated into the sediments. Sorption to sediments is a complex process influenced by simultaneously hydrophobic interactions (Van de Waals forces), ionic interactions (electrostatic interactions and complexation), hydrogen bonding, etc. (Von Oepen 1991). PhACs molecules contain dissociable functional groups whose equilibrium depends on the seawater pH, and besides, salinity has been pointed out as another factor that may control the sorption of PhACs onto marine sediment (Wang et al. 2010). This complex scenario contributes to a relatively low number of articles published regarding the occurrence of PhACs in marine sediments in comparison with its occurrence in seawater.

Analytical methodologies

An overview of the analytical methods employed for the determination of PhACs in marine sediments in the last eight years is given in Table 3. Details about the therapeutic groups analyzed, number of PhACs included, method used (extraction, clean up, and detection), and quality control parameters such as LOD, LOQ, and recoveries are provided. Ten studies have been found in the literature, all of them multi-residue analytical methods. Compounds belonging to different therapeutic families have been included except in some cases where the methods were exclusively focused on the detection of antibiotics (Na et al. 2013; Shi et al. 2014; Zhou 2011; Zhou et al. 2012b). Solid phase extraction (SPE) on Oasis HLB has been used as clean up step except in two publications where purification was not performed (Beretta 2014; Shi et al. 2014). The detection method utilized by all the authors has been liquid chromatography (LC) in combination with mass spectrometry (MS). The LOD and LOQ ranged in the low ng/g levels, however, the recoveries ranged in a very wide interval from 20 to 262%.

Occurrence of PhACs in marine sediments

Table 4 presents the studies published since 2011 regarding the occurrence of PhACs in marine sediments. Practical information about these publications like the country where the study was conducted, number of sampling sites, different environmental matrices analyzed, and the analytical methods applied are shown. A total of 12 studies have been published, in most of them the analysis was carried out in marine sediments collected from coastal areas of China, although coastal areas of the United States have been also monitored on four occasions. Brazilian, New Zealander, and Spanish coastal sediments have been also analyzed but just in one location percountry (Todos os Santos Bay, Auckland, and the lagoon of Mar Menor, respectively). Other environmental matrices like water, biota, and suspended solids have been also analyzed in some of these works giving an integrated vision of the PhACs contamination in a determined area (Chen et al. 2015; Klosterhaus et al. 2013; Lara-Martin et al. 2014; Liang et al. 2013a; Maruya 2012; Moreno-Gonzalez et al. 2015; Na et al. 2013).

Antibiotics are the therapeutic group more documented with 37 compounds positively identified mainly in the Chinese marine environment. The other groups of PhACs measured in sediments samples contain a considerably lower number of representing compounds, i.e., the second group on the ranking is analgesics/anti-inflammatories with only six positive findings. Surprisingly, only three psychiatric drugs have been identified in sediments whereas the occurrence of this group is well documented both in

Table 1: Analytical methodologies for the determination of pharmaceuticals in sea water.

Therapeutic group	Total n° of PhACs	Method Clean up	Detection	LOD/LOQ (ng/L)	Rec. (%)	Reference	year
Analgesics/Anti-inflammatory; lipid regulators	5	SPE + derivatization	GC-MS	n.m./ 0,002–0,10	42–71	Weigel	2001
Analgesics/Anti-inflammatories; Antibiotics; Psychiatric drugs; Lipid regulators; B-Blocking agents; Citotoxic; Antifungal	14	SPE	LC-MS/MS	4–20/ n.m.	4–120	Thomas	2004
Analgesics/Anti-inflammatories; Antibiotics; Psychiatric drugs; Lipid regulators; B-Blocking agents; Antihelmintics; Diuretic; Antihypertensives; Antihistamine; Calcium channel blockers; Antidiabetic; To treat asthma; Synthetic glucocorticoid; Antiespasmodic; Sedation and muscle relaxation	119 *	SPE	LC-MS/MS	0,1–177/1–200	5–200	USEPA 1694	2007
Antibiotics	9	SPE	LC-MS/MS	n.m./0,4–5	64–90	Xu	2007
Antibiotics	9	SPE	LC-MS/MS	2–13/n.m.	93–116	Gulkowska	2007
Analgesics/Anti-inflammatories; Antibiotics; Psychiatric drugs; Lipid regulators; B-Blocking agents; Citotoxics	12	SPE	LC-MS/MS	0,03–0,96/0,1–3,2	6–79	Nebot	2007
Analgesics/Anti-inflammatories; Antibiotics; Psychiatric drugs; B-Blocking agents; citotoxic;	11	SPE	LC-MS/MS	0,001–0,288/0,004–0,963	9–103	Zhang	2007
Analgesics/Anti-inflammatories; Psychiatric drugs; To treat asthma;	17	SPE	GC-MS	1,2–2,6/n.m.	54–120	Togola	2008
Analgesics/Anti-inflammatories; Lipid regulators	12	SPE	LC-MS/MS	3–46/n.m.	40–125	Comeau	2008
Analgesics/Anti-inflammatories; Antibiotics; Psychiatric drugs; Antihelmintics; Antihypertensives; Antihistamine, Anticoagulants	18	Passive Sampler (POCIS)	LC-MS/MS	n.m.	n.m.	Tertuliani	2008
Antibiotics; Antihistamine; anesthesic	13	Passive Sampler (POCIS)	LC-MS/MS	10–1500/n.m.	n.m.	Martinez-Bueno	2009
Analgesics/Anti-inflammatories; Antibiotics; Psychiatric drugs; Lipid regulators; B-Blocking agents; Diuretic; Antihypertensives; Antihistamine; Citotoxic; Antidiabetic; barbiturates; Synthetic-agonists	73	SPE	LC-MS/MS	0,01–6/0,02–20	31–149	Gros	2009
Antibiotics	16	SPE	LC-MS/MS	0,2–6/n.m.	59–98	Minh	2009

Analgesics/Anti-inflammatories; Antibiotics; Psychiatric drugs; Lipid regulators; B-Blocking agents; Diuretic; Antihypertensives; Antihistamine; To treat asthma; to treat reumathism; antiplatelet agent; proton pump inhibitor; hormones	27	SPE	LC-MS/MS	0,8–29,4	25–204	Rodriguez-Navas	2010
Analgesics/Anti-inflammatories; Antibiotics; Psychiatric drugs; Lipid regulators; B-Blocking agents; Antihistamine; iodinated contrast media, gastric acid regulator	47 *	SPE	LC-MS/MS	n.m./1,4–21	c.a. 65%	Nödler	2010
Analgesics/Anti-inflammatories; Psychiatric drugs; Lipid regulators; B-Blocking agents	10	Bag-SPE	LC-Q-TOF	1–13 /3–43	11–65	Magner	2010
Analgesics/Anti-inflammatories; Antibiotics; Psychiatric drugs; Lipid regulators; B-Blocking agents	13	SPE	LC-MS/MS	n.m./1–50	97–108	Wille	2011a
Analgesics/Anti-inflammatories; Lipid regulators	4	SPE	LC-MS/MS	0,4–1/n.m.	86–110	Wu	2010
Analgesics/Anti-inflammatories; Antibiotics; Psychiatric drugs; Antiespasmodic; B-Blocking agents; Citotoxic	8	SPE	LC-MS/MS	0,001–0.14 /0,004–0,5	51–103	Yang	2011
Analgesics/Anti-inflammatories; Antibiotics; Psychiatric drugs; Lipid regulators; B-Blocking agents; Antihelmintics; Diuretic; Antihypertensives; Antihistamine; Calcium channel blockers; Citotoxic; Antidiabetic; To treat asthma; Synthetic glucocorticoid; Antiespasmodic; Sedation and muscle relaxation; X-ray contrast agents; Prostatic hyperplasia; antiplatelet agent	81	SPE	LC-MS/MS	0 ,01–7,2/0,04–20	24–166	Gros	2012
Antibiotics	13	SPE	LC-MS/MS	n.m./0,24–4,4	68–85	Zhang	2012
Antibiotics	50	SPE	LC-MS/MS	0,2–1,8/0,5–5,9	32–292	Zhou	2012
Analgesics/Anti-inflammatories; Antibiotics; Psychiatric drugs; Lipid regulators; B-Blocking agents; Antihypertensives; X-ray contrast agents	24	SPE	LC-MS/MS & GC-MS	n.m./0,3–500	80–120	Vidal-Dorsch	2012
Analgesics/Anti-inflammatories; Psychiatric drugs; Lipid regulators; to treat asthma	21	Passive Sampler (POCIS)	LC-MS/MS	n.m.	n.m.	Munaron	2012
Analgesics/Anti-inflammatories; Antibiotics; Psychiatric drugs; Lipid regulators; B-Blocking agents	15	SPE	LC-MS/MS	n.m./0,007–0,23	45–93	Loos	2013b
Analgesics/Anti-inflammatories; Antibiotics; Psychiatric drugs; Lipid regulators; B-Blocking agents; Antihistamine; To treat asthma; anticohagulant	31	SPE	LC-MS/MS	0,09–2/n.m.	70–130	Bayen	2013

Table 1 contd....

Table 1 contd...

Therapeutic group	Total n° of PhACs	Method Clean up	Detection	LOD/LOQ (ng/L)	Rec. (%)	Reference	year
Analgesics/Anti-inflammatories; Antibiotics; Psychiatric drugs; Lipid regulators; B-Blocking agents; Antihelmintics; Diuretic; Antihypertensives; Antihistamine; Calcium channel blockers; to treat asthma	104 *	SPE	LC-MS/MS	0,5–14 /n.m.	18–156	Klosterhaus	2013
Analgesics/Anti-inflammatories; Antibiotics; Psychiatric drugs; Lipid regulators	5	SPE	LC-MS/MS	n.m./3–38	56–110	McEneff	2013
Analgesics/Anti-inflammatories; Antibiotics; Psychiatric drugs; Lipid regulators; B-Blocking agents; Proton Pump Inhibitor	18	SPE	LC-MS/MS	1–10/n.m.	70–107	Jiang	2014
Analgesics/Anti-inflammatories; Antibiotics	13	SPE	LC-MS/MS	0,2–16,7/0,5–50	61–93	Borecka	2015
Analgesics/Anti-inflammatories	11	SPE	LC-MS/MS	0,02–3,9/0,06–24,8	9,5–101	Paiga	2015

* pharmaceuticals and other contaminants
SPE: solid phase extraction
n.m. = not mentioned

seawater and marine organisms. The most recurring compounds identified in sediments were antibiotics like tetraclycine and oxytetracycline with four positive identifications, followed by fluoroquinolones like norfloxacin, ofloxacin, enrofloxacin, and ciprofloxacin with up to three identifications in different publications. In fact, sediments have been identified as a major sink for antibiotics by other authors (Kim and Carlson 2007; Yang 2010), and also more efforts have been done for the analysis of this therapeutic group in sediments. The concentrations of the PhACs detected in sediments are much lower than the ones measured in seawater ranging in the low nanograms per grams levels. The maximum levels corresponded to this kind of compounds as well, concretely up to 458 ng/g of ofloxacin were measured in the Yangtze estuary (Shi et al. 2014), 184 ng/g of chlortetracycline, and 176 ng/g of oxytetracycline in the Dan-Shui (Chen et al. 2015).

Pharmaceuticals in Marine Biota

A crucial step in the impact assessment of pharmaceuticals compounds in aquatic organisms is the establishment and validation of analytical methods for their extraction and determination in biological samples. Analytical techniques to detect PhACs at trace quantities in biota have been developed in the last years and mainly applied to fresh water species. Freshwater organisms have attracted more attention than marine organisms due to the highest environmental levels of PhACs found in continental aquatic systems. The discharge of waste water treatment plants' effluents into rivers or lakes provokes the worst exposure scenario to wild life and consequently higher bioaccumulation in freshwater organisms was expected. However, PhACs have been also identified and quantified in coastal areas (as previously reported) and the study of their bioaccumulation in marine organisms has become an important issue, especially in the case of species highly consumed by the population, and in particular those provided by aquaculture where antibiotics are used for the treatment of animals' diseases. The presence of PhACs in seafood may potentially act as a risk to the consumer either through direct effect of allergy and toxicity, or indirectly through potential microbial resistance (Cabello 2006). In order to protect public health and on the basis of the scientific assessment of the safety of pharmaceuticals, The European Community set in 2009 maximum residue limits (MRLs) for a variety of these chemicals in foodstuffs of animal origin including all food producing species (2010). However, this list only covers anti-infectious and anti-parasitic agents in the case of seafood. In order to provide more information about the presence of PhACs in marine organisms that would allow completing and updating this list, suitable analytical techniques are required. The most significant pharmaceutical families, according to their potential effects, need to be analyzed for establishing seafood quality control and simultaneous monitoring of contaminated areas through bioindicator species.

Analytical methodologies

There are some reviews focusing on the methodology applied for the determination of pharmaceuticals in biota samples. Two of them are dedicated to the specific analysis of antibiotics in fish (Canada-Canada et al. 2009; Samanidou and Evaggelopoulou 2007) and one to PhACs in general (Huerta et al. 2012b). The widespread use of antibiotics in the aquaculture industry together with the requirements of food safety legislation has attracted more efforts related to the analysis of this therapeutic family. An overview of the analytical methods employed for the determination of PhACs in marine organisms in the last eight years is given in Table 5. Details about the therapeutic groups analyzed, number of PhACs included, organism type and species, extraction, clean up and detection, together with quality control parameters such as limit of detection, limit of quantification, and recoveries of the methods are provided. As previously mentioned, among a total of 27 studies gathered the majority of them (14) are dedicated to the analysis of antibiotics whereas only three are exclusively focused on psychiatric drugs. The rest are multi-residue analytical methods covering different therapeutic groups like analgesics, anti-inflammatories, hormones, β-blockers, etc. The most used organism was fish, followed by mollusk, and crustacean. Concretely, *Sparus aurata*, commonly named sea bream, and *Mytilus* spp., mussel species, have been the most used organisms for method development. Marine organisms are complex matrices and therefore sample preparation becomes a critical issue. They are rich in biological components like lipids and proteins which could interfere with the

Table 2: Pharmaceuticals concentration (ng/L) in water. Expressed as mean (maximun). Continued

Reference	Weigel et al. 2002	Thomas et al. 2003	Weigel et al. 2004	Weigel et al. 2004	Xu et al. 2007	Gulkowska et al. 2007	Nebot et al. 2007	Comeau et al. 2008	Martinez Bueno et al. 2009	Minh et al. 2009	Magner et al. 2010	Wu et al. 2010	Wille et al. 2011a	Nödler et al. 2010	Zou et al. 2011	Yang et al. 2011	Gros et al. 2012	Munaron et al. 2012	Rodriguez-Navas et al. 2012	Vidal-Dorsch et al. 2012	Fang et al. 2012
Country	Netherlands, ,K, Norway, Denmark and Germany	UK	Norway	France	China	China	UK	Canada	France	China	Sweden	Singapore	Belgium	Germany	China	China	Spain	France	Spain	USA	Taiwan
Sampling Sites	15	5	13	13	5	8	1	11	12	20	12	6	13	1	21	6	1	13	8	10	14
Method	Weigel et al. 2002	Thomas et al. 2003	Weigel et al. 2002	Togola et al. 2008	Xu et al. 2007	Gukowska et al. 2007	Nebot et al. 2007	Comeau et al. 2008	Martinez Bueno et al. 2009	Singh et al. 2010	Magner et al. 2010	Wu et al. 2010	Wille et al. 2011a	Nödler et al. 2010	Xu et al. 2007	Yang et al. 2011	Gros et al. 2012	Munaron et al. 2012	Rodriguez-Navas et al. 2012	Vidal-Dorsch et al. 2012	Zhang and Zu 2007
1	2	3	4	5	6	7	8	9	10	11	12	13	14	15	16	17	18	19	20	21	22
Analgesics/ Anti-inflammatories																					
Acetaminophen																	23			BQL (11,0)	
Salicylic Acid								44 (140)					155 (855)				2				
Ketoprofen																		x			3,2 (6,6)
Naproxen								18,1 (130)				22,3 (30)					6			0,7 (26,0)	

Ibuprofen CUTHERE	48 (930)	0,13 (0,7)			22,6 (230)			69,5 (121)				16			BQL (30,0)	10,9 (57,1)
Ibuprofen carboxylic acid		2,1 (5,3)														
2-Hydroxylated ibuprofen		0,5 (1,5)														
Indomethacin											479 (979)	3				
Diclofenac	< 8 (195)				0,06 (6)			20 (38)			314 (843)	4	x	BQL	BQL (0,6)	8,5 (53,6)
Phenazone												2				
Oxycodone																
Nimesulide																
Meclofenamic Acid											250 (679)					
Mefenamic Acid	< 20 (196)															
Dextropropoxyphene	< 8 (80)															
Codeine												2				
Antibiotics																
Erythromycin			3,4 (5,2)			0,02 (0,03)				83 (150)						
Erythromycin-H2O				91,2 (486)			221 (1900)									
Azithromycin																
Clarithromycin									14			17				
Salinomycin																
Roxithromycin			5,6 (30,6)				6,9 (47)			24 (630)						

Table 2 contd....

Table 2 contd...

1	2	3	4	5	6	7	8	9	10	11	12	13	14	15	16	17	18	19	20	21	22
Tetracycline						18,2 (122)				28,8 (313)					7 (30)						
Oxytetracycline										3 (44)					93 (270)						
Chlortetracycline																					
Metacycline																					
Norfloxacin					10,8 (28,1)	1,25 (8)				9,1 (27)					460 (6800)						
Ofloxacin					7,6 (16,4)					95,9 (634)					390 (5100))		2				
Enoxacin																					
Enrofloxacin																					
Ciprofloxacin															110 (390)						
Ampicilllin																					
Sulfamethoxazole										5,2 (47,5)			3,9 (96)	7	97 (140)	370 (765)	9			0,5 (3,4)	
Sulfacetamide																					
Sulfadiazine															21 (41)						
Sulfamethizole																					
Sulfathiazole																					
Sulfamerazine																					
Sulfamonomethoxime																					
Sulfameter																					

Sulfamethazine				0,8 (8,6)				17 (130)					
Sulfadimethoxine													
Sulfaquinoxaline													
Sulfapyridine													
Trimethoprim	5 (569)	6 (21,8)	0,03			1,95 (29)		54 (120)		1			0,7 (2,1)
Metronidazole			13,4							4			
Cefalexin		41,4 (182)		72,5 (493)						nd			
Amoxicilin				14,6 (76)									
Psychiatric drugs													
Carbamazepine		19,1 (26,3)				27,7 (321)	BQL (4,9)		238 (675)	8	x	BQL	BQL (0,9)
10,11-epoxycarbazaepine										21			
Acridone													
Citalopram										4			
Venlafaxine										52			
Oxazepam					8,5 (12,5)								
Trazodone										1			
Fluoxetine											x		
Paroxetine													
Diazepam											x	BQL	

Table 2 contd....

Table 2 contd...

1	2	3	4	5	6	7	8	9	10	11	12	13	14	15	16	17	18	19	20	21	22
Lorazepam																					
Haloperidol																					
Amitriptyline																					
Cotinine																					
Risperidone																				nd (1,4)	
Primidone																					
Bultabital																					
Pentobarbital																					
Lipid Regulators																					
Bezafibrate													3,5 (18)				2				
Gemfibrozil								1,6 (53)				4,2 (9)					23	x		0,9 (13)	
Pravastatin																					
Fluvastatin																					
Atorvastatin																	1			BQL (0,4)	
Fenofibrate																					
Clofibric Acid	1,6 (18,6)	< 20 (180)																			10,5 (55,1)
B-Blocking agents																					
Atenolol													10,8 (293)				6		38	0,4 (11,0)	
Sotalol																	2				

Propanolol	13 (56)		3,13 (24)		41,5 (142)		
Metoprolol		37,5 (210)					BQL (8)
Nadolol						0,3	
Antihelmintics							
Albendazole							
Levamisole						1	
Antifungal							
Clotrimazole	7 (930)						
Diuretic							
Hydrochlorothiazide						9	
Furosemide							BQL (47)
Torasemide						3	
Antihypertensives							
Losartan						4	
Irbesartan						21	
Valsartan						29	
Enalapril							
Lysinopril							
X-ray Contrast Agent							
Iohexol							
Iomeprol				40			

Table 2 contd....

Table 2 cont...

1	2	3	4	5	6	7	8	9	10	11	12	13	14	15	16	17	18	19	20	21	22
Iopamidol																					
Iopromide														25							
Antihistamine																					
Loratadine														4							
Ranitidine																	0,8				
Cimetidine																			BQL (15)		
Diphenhydramine																					
Cetrizine														4							
Calcium channel blockers																					
Diltiazem																	140		BQL (11)		
Verapamil																			BQL (3)		
Tranquilizer																					
Azaperone																					
Azaperol																					
Citotoxic																					
Tamoxifen		< 8 (71)														152 (224)					
Antidiabetic																					
Glipizide																					
Glibenclamide																					

Anti-coagulant					
Warfarin					
To treat asthma					
Albuterol					
xamol			1	x	
Synthetic glucocorticoid					
Prednisolone					BQL (2)
Antiespasmodic					
Mebeverine	12,8 (50,3)	96,7 (154)			
Sedation and musclere-laxation					
Xylazine					
Prostatic hyperplasia					
Tamsulosin					
Antiplatelet Agent					
Clopidogrel					
Proton Pump Inhibitor					
Omeprazol					11,4 (18)

BQL = below quantification limit

Table 2 cont....

Table 2 contd...

Reference	**Zhang et al. 2012**	**Zhang et al. 2013a**	**Zhang et al. 2013b**	**Klosterhaus et al. 2013**	**Bayen et al. 2013**	**Loos et al. 2013**	**Jiang et al. 2014**	**Nödler et al. 2014**	**Nödler et al. 2014**	**Nödler et al. 2014**	**Nödler et al. 2014**	**Nödler et al. 2014**	**Nödler et al. 2014**	**Nödler et al. 2014**	**Nödler et al. 2014**	**McEneff et al. 2014.**	**Alvarez et al. 2014**	**Lara-Martin et al. 2014**	**Chen et al. 2015**	**Moreno-Gonzalez et al. 2015**	**Borecka et al. 2015**
Country	**China**	**China**	**China**	**USA**	**Singapore**	**Italy**	**Taiwan**	**Germany**	**Greece & Turkey**	**Italy**	**Italy**	**USA**	**USA**	**Israel**	**Spain**	**Ireland**	**USA**	**Long Island**	**China**	**Spain**	**Poland**
Sampling Sites	27	62	56	104*	8	1	53	30	73	7	5	20	10	6	2	3	10	13	39	18	17
Method	**Zhang et al. 2012**	**Zhang et al. 2012**	**Zhang et al. 2012**	**Klosterhaus et al. 2013**	**Bayen et al. 2013**	**Loos et al. 2013**	**Jiang et al. 2014**	**Nödler et al. 2010**	**Nödler et al. 2010**	**Nödler et al. 2010**	**Nödler et al. 2010**	**Nödler et al. 2010**	**Nödler et al. 2010**	**Nödler et al. 2010**	**Nödler et al. 2010**	**McEneff 2013**	**Tertuliani 2008**	**Gros et al. 2009**	**Zhou et al. 2012**	**Gros et al. 2012**	**Borecka et al. 2015**
1	**2**	**3**	**4**	**5**	**6**	**7**	**8**	**9**	**10)**	**11**	**12**	**13**	**14**	**15**	**16**	**17**	**18**	**19**	**20**	**21**	**22**
Analgesics/ Anti-inflammatories																					
Acetaminophen							8,4 (16,7)	48 (48)	39 (2983)		99 (375)	11 (85)		12 (12)				13,4			
Salicylic Acid																					
Ketoprofen							8,7 (23,3)														32,1 (135)
Naproxen				BQL (8,2)	0,6 (1,7)						8,5 (8,5)							50			

Ibuprofen				BQL (37,9)	3,7 (9,1)	0,5 (1,2)	2,6 (12,1)	109 (109)	35 (30)		37 (70)	11 (12)		7,5 (7,5)				38			BQL (34,9)
Ibuprofen carboxylic acid																					
2-Hydroxylated ibuprofen																					
Indomethacin																		0,5			
Diclofenac					3,05 (11,6)			9,2 (9,2)	4,6 (9,7)					6,1 (6,1)		279 (550)		0,1			BQL (92,6)
Phenazone								4,4 (5,9)	2,0 (2,0)											BQL	
Oxycodone																				BQL (6,8)	
Nimesulide																					
Meclofenamic Acid																					
Mefenamic Acid																246 (610)		4			
Dextropropoxyphene																					
Codeine							22,1 (63,6)													BQL (1,8)	
Antibiotics																					
Erythromycin											5,8 (5,8)	28 (217)	28 (86)							7,6 (78,4)	
Erythromycin-H2O	2,6 (8,5)	0,69 (6,7)	0,63 (25,2)	2,4 (12,1)			2,8 (88,7)												18,5 (183)		
Azithromycin	0,14 (1,2)	0,06 (0,39)	0,31 (2,5)															25		67,5 (164)	
Clarithromycin	0,19 (0,82)	0,07 (0,51)	0,30 (2,6)					14 (14)	16 (16)		8,5 (8,7)		86 (130)					14	BQL (0,15)	4,6 (9,6)	
Salinomycin																			5,5 (13)		
Roxithromycin	0,38 (1,5)	0,09 (0,26)	0,87 (6,9)					16 (16)					97 (141)								

Table 2 contd....

Table 2 contd...

1	2	3	4	5	6	7	8	9	10)	11	12	13	14	15	16	17	18	19	20	21	22
Tetracycline																			78,51 (2305)		
Oxytetracycline																			418 (15163)		
Chlortetracycline																			BQL		
Metacycline																			1,3 (2,4)		
Norfloxacin	40 (103)																		0,3 (5,04)		
Ofloxacin	0,2 (6,5)																		0,8 (13,7)		
Enoxacin	62 (209)																				
Enrofloxacin	1,8 (7,6)																		1,9 (56,7)		BQL (18)
Ciprofloxacin	31 (166)																	2	5,3 (187)		
Ampicilllin																			6,9 (88,7)		
Sulfamethoxazole	19 (82)	1 (8,3)	5,5 (50,4)	6,3 (66,7)		0,5 (1,0)		21 (42)	3,8 (11)	4,1 (5,5)	6,7 (7,2)	16 (61)	6,4 (6,4)					11,2	1,3 (6,2)	9,8 (94)	15,6 (70,1)
Sulfacetamide		0,003 (0,12)																			
Sulfadiazine	0,02 (0,43)	0,01 (0,36)	0,01 (0,24)																0,7 (2,7)		
Sulfamethizole				BQL (15,6)																	
Sulfathiazole		0,02 (0,17)																			BQL (6,3)
Sulfamerazine																					BQL (17,9)
Sulfamonomethoxime																			0,1 (1,9)		

Sulfameter																			0,1 (2,05)		
Sulfamethazine	0,1 (1,5)																	0,2	0,2 (2,7)		
Sulfadimethoxine		0,01 (0,16)	0,03 (0,34)			BQL (0,09)															0,25 (1)
Sulfaquinoxaline																			0,1 (1,3)		
Sulfapyridine																					11,7 (167)
Trimethoprim	53 (330)	1,4 (16,6)	3,05 (14,1)	BQL (4,1)												431 (870)	0.3 (2)	0,7	4,2 (36,9)	0,8 (1,5)	3,3 (14,2)
Metronidazole																					
Cefalexin																			2,0 (9,2)		
Amoxicilin																					
Psychiatric drugs																					
Carbamazepine					2,7 (10,9)	0,23 (0,36)	0,2 (3,8)	22 (157)	2,9 (22)	3,1 (3,1)	4,9 (6,5)	6,2 (13)		8,8 (8,8)		551 (1410)	2.6 (21)	10			
10,11-epoxycarbazaepine																					
Acridone																				BQL	
Citalopram													19 (27)	4,3 (4,3)							
Venlafaxine																					
Oxazepam																					
Trazodone																					
Fluoxetine													66 (90)					0,7			
Paroxetine																		0,5			

Table 2 contd....

Table 2 contd...

1	2	3	4	5	6	7	8	9	10)	11	12	13	14	15	16	17	18	19	20	21	22
Diazepam																					
Nordiazepam				BQL (0,5)																	
Lorazepam																				15,3 (41,8)	
Haloperidol													39 (56)								
Amitriptyline				BQL (0,6)																	
Cotinine									5,1 (71)		7,9 (8,8)						2.7 (6.3)				
Risperidone																					
Primidone												6,4 (10)									
Bultabital																		3			
Pentobarbital																		6			
Lipid Regulators																					
Bezafibrate						0,10 (0,14)			3,5 (3,5)		7,8 (7,8)			3,7 (3,8)				29			
Gemfibrozil	0,75 (3,7)	0,75 (3,7)	0,75 (3,7)	27 (38,2)	2,63 (19,8)	0,1 (0,18)		14 (43)	6,2 (6,2)			14 (43)	6,2 (6,2)			168 (640)		1,1		0,25 (3,9)	
Pravastatin																		0,7			
Fluvastatin																				BQL	
Atorvastatin																		3,5			
Fenofibrate																		2			
Clofibric Acid						0,1 (0,18)												0,2			

B-Blocking agents									
Atenolol		0,06 (0,17)	13 (13)	23 (194)	16 (22)	15 (57)		13	
Sotalol			37 (65)	20 (67)	11 (12)	10 (12)		0,7	BQL (0,8)
Propanolol								0,5	BQL (0,5)
Metoprolol	1,7 (26,2)	0,685	10 (158)	5,6 (6)		18 (32)	6,4 (6,7)	5,2	0,15 (0,73)
Nadolol								0,9	BQL
Antihelmintics									
Albendazole									BQL
Levamisole									
Antifungal									
Clotrimazole									
Diuretic									
Hydrochlorothiazide								10	BQL
Furosemide									
Torasemide									BQL
Antihypertensives									
Losartan									BQL (104)
Irbesartan									0,42 (16,8)
Valsartan	45,2 (92,1)								1,6 (38)
Enalapril								0,6	
Lysinopril								0,9	

Table 2 contd....

Table 2 contd...

1	2	3	4	5	6	7	8	9	10)	11	12	13	14	15	16	17	18	19	20	21	22
X-ray Contrast Agent																					
Iohexol								58 (861)	76 (76)			68 (162)		24 (26)							
Iomeprol								98 (1159)	83 (145)	29 (29)	19 (20)										
Iopamidol								52 (1027)	29 (145)	44 (61)		140 (783)	32 (100)								
Iopromide								45 (109)	109 (199)		30 (71)										
Antihistamine																					
Loratadine								4,1 (4,1)	4,3 (4,3)				33 (57)					0,5			
Ranitidine																					
Cimetidine																					
Diphenhydramine					0,68 (4,6)																
Cetrizine								6,2 (13)			2,7 (2,7)	4,1 (6,6)									
Calcium channel blockers																					
Diltiazem					0,6 (1,7)															BQL	
Verapamil																					
Tranquilizer																					
Azaperone																				BQL	
Azaperol																				BQL (0,8)	
Citotoxic																					

Tamoxifen		23 (90)	0,3	
Antidiabetic				
Glipizide			0,2	
Glibenclamide			0,2	
Anti-coagulant				
Warfarin				BQL
To treat asthma				
Albuterol	BQL (0,5)			
Salbutamol				
Synthetic glucocorticoid				
Prednisolone				
Antiespasmodic				
Mebeverine				
Sedation and musclere-laxation				
Xylazine				12,42 (13,8)
Prostatic hyperplasia				
Tamsulosin				BQL
Antiplatelet Agent				
Clopidogrel				BQL
Proton Pump Inhibitor				
Omeprazol				BQL

BQL= below quantification limit

Table 3: Analytical methodologies for the determination of pharmaceuticals in sediments.

Therapeutic group	Total n° of PhACs	Method: Extraction	Clean up	Detection	LOD/LOQ (ng/g)	Rec. (%)	Reference	year
Analgesics/Anti-inflammatories; Antibiotics; Psychiatric drugs; Lipid regulators; B-Blocking agents; Antihelmintics; Diuretic; Antihypertensives; Antihistamine; Calcium channel blockers; Citotoxic; Antidiabetic; To treat asthma; Synthetic glucocorticoid; Antiespasmodic; Sedation and muscle relaxation	119 *	Solvent extraction	SPE	LC-MS/MS	0,2–270/ 1–1000	20–262	USEPA 1694	2007
Analgesics/Anti-inflammatories; Antibiotics; Psychiatric drugs; Lipid regulators; B-Blocking agents; Antihelmintics; Diuretic; Antihypertensives; Antihistamine; Calcium channel blockers; Citotoxic; Antidiabetic; To treat asthma; Synthetic glucocorticoid; Antiespasmodic; Sedation and muscle relaxation	43	PLE	SPE	LC-MS/MS	0.01–3,2/ 0.02–10.7	35–206	Jelic	2009
Psychiatric drugs; Lipid regulators	3	Solvent extraction	SPE	LC-MS/MS	0,1–0,5./n.m.	86	Kwon	2009
Analgesics/Anti-inflammatories; Antibiotics; Psychiatric drugs; Citotoxic; Antiespasmodic	9	MAE	SPE	LC-MS/MS	0.2–64/ 1–214	43–88	Hibberd	2009
Antibiotics	17	Solvent extraction	SPE	LC-MS/MS	0,1–0,7/ 0,2-2,2	49–198	Zhou	2011
Antibiotics	48	Solvent extraction	SPE	LC-MS/MS	0,2–2/0,8–6,7	12–343	Zhou	2012
Analgesics/Anti-inflammatories; Antibiotics; Psychiatric drugs; Lipid regulators; B-Blocking agents; Antihelmintics; Diuretic; Antihypertensives; Antihistamine; Calcium channel blockers; to treat asthma;	104 *	Solvent extraction	SPE	LC-MS/MS	0,13–135 /n.m.	21–230	Klosterhaus	2013
Antibiotics	20	Solvent extraction	SPE	LC-MS/MS	0,7–3,2/n.m.	61–98	Na	2013
Analgesics/Anti-inflammatories; Antibiotics; Psychiatric drugs; B-Blocking agents	6	Solvent extraction	no	LC-MS/MS	0.1/n.m.	87–94	Beretta	2014
Antibiotics	20	Solvent extraction	no	LC-MS/MS	0,05–1,2/0,1–1,7	64–126	Shi	2014

MAE: microwave-assisted extraction
PLE: Pressurize liquid extraction
SPE: Solid phase extraction
* pharmaceuticals and other compounds
n.m. = not mentioned

Table 4: Pharmaceutical concentrations (ng/g) in sediments. Expressed as mean (maximun).

Reference	**Yang et al. 2011***	**Maruya et al. 2012**	**Klosterhaus et al. 2013**	**Na et al. 2013**	**Liang et al. 2013**	**Long et al. 2013**	**Stewart et al. 2014***	**Lara-Martin et al. 2014**	**Beretta et al. 2014**	**Shi et al. 2014**	**Chen et al. 2015**	**Moreno-Gonzalez et al. 2015**
Country	China	USA	USA	China	China	USA	New Zealand	USA	Brazil	China	China	Spain
Sampling Sites	6	4	104	20	14	40	13	13	17			18
Method	Hibberd et al. 2009	Kwon et al. 2009	Klosterhaus et al. 2014	Na et al. 2013	Zhou et al. 2011	USEPA 1694	Jelic et al. 2009	Jelic et al. 2009	Beretta et al. 2014	Shi et al.2014	Zhou et al. 2012	Jelic et al. 2009
1	2	3	4	5	6	7	8	9	10	11	12	13
Analgesics/Anti-inflammatories												
Ibuprofen						BQL			11,6 (18,8)			
Indomethacin	x											BQL (19,2)
Salicylic Acid												3,8 (14,9)
Diclofenac	x						x	< 1	0,67 (1,06)			
Mefenamic Acid								< 1				
Propoxyphene						BQL (1,7)						
Antibiotics												
Erythromycin									0,24 (2,29)	4,0 (51,5)		BQL (2,37)
Erythromycin-H2O			BQL (3,4)		5,4 (14,0)						5,1 (65,3)	
Azithromycin						1,3 (1,6)						

Table 4 contd...

Table 4 contd...

1	2	3	4	5	6	7	8	9	10	11	12	13
Antibiotics												
Clarithromycin							x				< 0.07 (0,3)	BQL (3)
Roxithromycin					9,8 (13,5)		x			0,7 (3,6)		
Tetracycline				1,7 (1,7)	2,3 (7,1)					0,2 (6,8)	< 10,6 (73,0)	
Oxytetracycline						BQL				1,7 (14)	8,1 (176)	
Doxycycline				1,3 (1,5)								
Doxycyclinehyclate										BQL (18,6)		
Chlortetracycline										0,8 (12)	11,6 (184)	
Methacycline											< 3.05	
4-Epitetracycline						BQL						
Anhydrochlorotetracycline						BQL						
Norfloxacin					7,6 (20,5)					4,3 (69,3)	1,01 (22,2)	
Ofloxacin					3,6 (13,7)					7,3 (458,2)	1,0 (19,4)	
Enrofloxacin					1,1 (1,4)					0,3 (4,8)	0,6 (2,7)	
Ciprofloxacin										3,4 (42,9)	0,2 (4,2)	
Pefloxacine											< 1.5 (1,9)	
Narasin											0,6 (7,7)	
Novobiocin											< 0.35 (0,6)	

Salinomycin									2,5 (10,0)
Sulfamethoxazole	x	BQL (0,7)						0,3 (1,1)	
Sulfapyridine								0,7 (9,1)	
Sulfacetamide			0,13 (1,4)						
Sulfadiazine			0,5 (1,68)					0,1 (0,5)	0,1 (1,7)
Sulfamethazine			0,3 (1,8)	1,2 (3,24)		x		0,9 (4,8)	
Sulfamethiazole			0,4 (1,3)						
Sulfamerazine			1,7 (3,7)					BQL (0,4)	
Sulfathiazole			0,1 (1,9)					BQL	
Sulfamonomethoxime			1,6 (7;0)						
Sulfameter			8,82 (56,6)						
Sulfamethoxypyridazine			1,5 (7,7)						
Sulfaquinoxaline								0,3 (0,9)	
Trimethoprim		BQL (18,2)				x			
Thiamphenicol								BQL	
Florophenicol			0,4 (1,2)					BQL	
Chloramphenicol			1,0 (2,3)					BQL	
Psychiatric drugs									
Carbamazepine	x					x	0,41 (0,62)		
Diazepam						x	0,39 (0,71)		
Amitriptyline					BQL				
Lipid regulators									

Table 4 contd...

Table 4 contd...

1	2	3	4	5	6	7	8	9	10	11	12	13
Bezafibrate							x					BQL (0,37)
Gemfibrozil							x	< 1				
Pravastatin							x					BQL (2,48)
B-Blocking agents												
Atenolol							x		6,1 (9,84)			
Propanolol	x							< 1				
Metoprolol							x	24				
Nadolol							x					BQL (0,09)
Antihelmintics												
Thiabendazole			BQL (9,1)									
Diuretic												
Hydrochlorothiazide							x	< 1				BQL (1,81)
Furosemide							x	10				BQL (8,06)
Triamterene			0,3 (10,8)			0,82 (3,22)						
Antihypertensives												
Losartan												BQL (6,47)
Irbesartan												BQL (1,35)

Valsartan					BQL (1,72)
Amlodipine					BQL
Antihistamine					
Famotidine			x		BQL (0,33)
Diphenhydramine		1,68 (4,8)			
Calcium channel blockers					
Verapamil		0,1 (0,3)			
Noverapamil		BQL			
Citotoxic					
Tamoxifen	x			7	
Antispasmodic					
Mebeverine	x				
Sedation and muscle relaxation					
Xylazine					12,8 (50,3)

BQL = below quantification method
*Studies where the numerical values are not reported. The positive detection of a compound is marked with an "x".

Table 5: Analytical methodologies for the determination of pharmaceuticals in marine organisms.

			Method						
Therapeutic group	**Total n° of PhACs**	**Organism type *(species)***	**Extraction**	**Clean up**	**Detection**	**LOD/LOQ (ng/g)**	**Rec. (%)**	**Reference**	**year**
Psychiatric drugs	3	**Fish** (*Ameiurus nebulosus, Dorosoma cepedianum, Morone americana*)	PLE	SPE	LC-MS/MS	0.02–0.07/ 0.07–0.24	85–99	Chu	2007
Analgesics/anti-inflammatories; Psychiatric drugs; Lipid regulators	6	**Mollusk** (*Mytilus galloprovincialis, Ruditapes philippinarum, Scrobicularia plana, Spondylus calcifer*)	Agitation, centrifugation, and microwave-assisted micellar extraction	SPE	HPLC-UV	30–220/100–730	85–119	Cueva-Mestanza	2008
Antibiotics	7	**Fish** (*Sparus aurata*)	Solvent extraction vortex	SPE	LC-MS/MS	2–2.7/6–8	95–132	Samanidou	2008
Antibiotics	7	**Fish** (*Sparus* spp.)	PLE	-	LC-MS	18–51/n.m*.	66–90	Berrada	2008
Psychiatric drugs; Lipid regulators; Hormones	5	**Fish** (*Ictalurus* spp., *Pleuronichthys verticalis*)	Solvent extraction	SPE	LC-MS/MS	n.m./4.2–12.3	72–103	Kwon	2009
Antibiotics	4	**Fish** (*Sparus aurata*)	Vortex extraction	SPE	ELISA	0.6–1/n.m.	65–95	Cháfer-Pericás	2010a
Antibiotics	3	**Fish** (*Sparus aurata*)	Vortex extraction	SPE	TR-FIA	0.08–16/n.m.	74–95	Cháfer-Pericás	2010b
Antibiotics	6	**Fish** (*Sparus aurata*)	Vortex extraction	-	HPLC-MS/MS	1.2–16/4–52	88–110	Cháfer-Pericás	2010c
Antibiotics	22	**Fish** (*Dicentrarchus labrax, Sparus spp.*)	Solvent extraction vortex and ultrasonication	-	LC-MS/MS	5.65–25.8/17.1–78.1	35–100	Dasenaki	2010
Anti-inflammatories; Antibiotics, Psychiatric drugs; Lipid regulators; β-blockers; Hormones	20	**Fish** (*Dicentrarchus labrax, Solea* spp., *Salmo* spp.)	Precipitation/ centrifugation/ filtration	SPE	GC-MS	0.0005–0.0027/n.m.	92–101	Azzouz	2011
Antibiotics	6	**Fish** (*Sparus aurata*)	Vortex extraction	-	HPLC-MS/ ELISA/TR-FIA	0.08–109/ n.m.	78–108	Cháfer-Pericás	2011
Antibiotics	15	**Mollusk** (*Mytilus* spp.) and **Fish** (*Merluccius merluccius, Solea solea, Engrulis encrasicolus*)	Enzymatic microwave assisted extraction	Centrifugation	HPLC-DAD/FL	70–250/230–1020	34–91	Fernandez-Torres	2011

Analgesics/anti-inflammatories; Antibiotics; Psychiatric drugs lipid regulators; β-blocker	11	**Mollusk** (*Mytilus edulis, Crassostrea gigas*) and **Crustacean** (*Crangon crangon*)	PLE	SPE	U-HPLC-MS	n.m./0.1–10	90–106	Wille	2011b
Antibiotics and Anti-parasitic agents	14	**Crustacean** (shrimp)	QuEChERS	QuEChERS	LC-TOF/MS	0.6–7.1/n.m.	33–118	Villar -Pulido	2011
Antibiotics	22	**Mollusk** (*Crassostrea talienwhanensis, Chlamys farreri, Amussium, Scapharca subcrenata, Meretrix merhjgntrix linneaeus, Mactra veneriformis, Mactra chinesis, Mya arenaria, Neverita didyma, Rapana venosa, Mytilus edulis*)	PLE	SPE	LC-MS/MS	0.05–0.6/0.2–1	n.m.	Li	2012
Antibiotics	3	**Fish** (*Trachinotus ovatus, Lutianus argentimaculatus, Saprus Latus, Lutianus erythroptrus, Sparus macrocephalus, Gymnocranius griseus, Siganus fuscescens, Sciaenops ocellantus*)	Solvent extraction /homogenize/ centrifuge	Deffating step	HPLC-FL	0.2–0.3/0.65–1.0	82–112	He	2012
Antibiotics	32	**Fish** (*Sparus aurata*)	QuEChERS	-	HPLC-MS/MS	3.0–15.0/10–50	69–125	Lopes	2012
Psychiatric drugs	8	**Mollusk** (*Mytilus galloprovincialis*)	QuEChERS	Centrifugation	HRMS	0.1–0.3/0.2–1	67–110	Martinez-Bueno	2013
Analgesics/anti-inflammatories; Antibiotics; Psychiatric drugs; Lipid regulators	5	**Mollusk** (*Mytilus* spp.)	PLE	SPE	LC-MS/MS	n.m./4–29	83–104	McEneff	2013
Analgesics/anti-inflammatories; Antibiotics; Psychiatric drugs; Lipid regulators; β-blockers; To treat asthma; stimulant; Anti-hypertension; Antilipemic; Anti-histamine; Diuretic	104	**Mollusk** (*Geukensia demissa*)	Solvent extraction	SPE	LC-MS/MS	0.3–60/n.m.	20–179	Klosterhaus	2013
Antibiotics	20	**Mollusk** (*Crassostrea Gigas, Patinopecten yessoensis, Chlamys farreri*)	Solvent extraction/ shake/centrifuge	SPE (for sulfonamides and tetracyclines)	HPLC-MS/MS	0.62–4.6/n.m.	59–93	Na	2013

Table 5 contd....

Table 5 contd...

Therapeutic group	Total nº of PhACs	Organism type (*species*)	Method: Extraction	Method: Clean up	Method: Detection	Method: LOD/LOQ (ng/g)	Method: Rec. (%)	Reference	year
Analgesics; Antibiotics; Psychiatric drugs; β-blockers; Anti-hypertension; stimulant; Antilipemic; Anti-histamine; anti-coagulant	13	**Mollusk** (*Mytilus* spp.)	Solvent extraction	SPE	LC-MS/MS	0.03–7.8/n.m.	65–75	Dodder	2014
Psychiatric drugs	6	**Mollusk** (*Mytilus galloprovincialis*)	QuEChERS	Centrifugation	HRMS	0.1–0.3/0.5–0.7	55–88	Martinez-Bueno	2014
Antibiotics	32	**Fish** (*Cyprinus carpio, Oreochromis niloticus*) **Crustacean** (*Metapenaeus ensis*)	Solvent extraction / homogenization/ filtration	-	LC-MS/MS	0.062–4.6/ n.m.	70–120	Federova	2014
Antibiotics	47	**Fish** (Shrimp, tilapia, rainbow trout, atlantic salmon, swai)	Sonication	SPE	LC-MS/MS	0.2–25.5/n.m.	15–137	Done	2014
Analgesics; Antibiotics; Antihelmintics; Coccidiostats; β-agonists; Steroids; Thyresotats	143	**Fish** (*Sparus aurata, Dicentrarchus labrax*)	Sonication/ centrifuge	Deffating step	UHPL-QTOF-MS	n.m.	n.m.	Dasenaki	2015
Analgesics/anti-inflammatories; Antibiotics; Psychiatric drugs; Tranquilizers; Calcium channel blockers; Diuretic Prostatic hyperplasia	23	**Mollusks** (*Crassostrea gigas, Mytilus galloprovincialis, Chamalea gallina*)	PLE	SPE	UPLC-MS/MS	0.01–1.1/0.02–2.66	40–115	Alvarez-Muñoz	2015a

*n.m. = not mentioned

analysis and with the low concentrations at which PhACs are usually present. The target analytes must be isolated and concentrated from the sample matrix and an appropriate extraction and clean-up methodology needs to be employed. Different extraction techniques have been used on marine organisms (Table 5), among them organic solvent extraction combined with either shaking, vortexing, or ultrasonication being the most recurrent one, probably because it is easy to perform (Chafer-Pericas 2011; Chafer-Pericas 2010b; Chafer-Pericas 2010c; Chafer-Pericas 2010a; Dasenaki et al. 2015; Dasenaki and Thomaidis 2010; Dodder 2014; Done and Halden 2015; Federova 2014; Klosterhaus 2013; Kwon et al. 2009; Na et al. 2013; Samanidou and Evaggelopoulou 2007). Pressurized liquid extraction (PLE) was also utilized on several occasions (Alvarez-Muñoz 2015a; Berrada et al. 2008; Chuand Metcalfe 2007; McEneff 2013; Wille et al. 2011b; Zhou et al. 2012b); QuEChERs (quick, easy, cheap, effective, rugged, and safe) (Lopes R.P. 2012; Martinez Bueno et al. 2013; Martinez Bueno et al. 2014; Villar-Pulido et al. 2011), and microwave-assisted solvent extraction (MASE), either with the addition of a surfactant concentration (Cueva-Mestanza et al. 2008) (micellar extraction) or with enzymatic extraction (Fernandez-Torres 2011) were also used. Solid phase extraction (SPE) has been the selected clean-up technique by excellence. Most of the studies were performed on Oasis HLB cartridges, although Strata-X and Oasis MCX were also used by some authors (Chuand Metcalfe 2007; McEneff 2013; Wille et al. 2011b). Alternatively, in a couple of articles a deffating step was carried out with the addition of hexane after centrifugation and discarded of the pellet (Dasenaki et al. 2015; He 2012). Sometimes a deeper clean up step is necessary and gel permeation chromatography (GPC) has been used by other authors in freshwater fish (Huerta et al. 2012a; Tanoue et al. 2014). Regarding the detection of PhACs in marine organisms most of the literature found (Table 5) is based on liquid chromatography (LC) in combination with mass spectrometry (MS) (Berrada et al. 2008; Chafer-Pericas 2010c; Chuand Metcalfe 2007; Dasenaki and Thomaidis 2010; Dodder 2014; Done and Halden 2015; Federova 2014; Klosterhaus 2013; Kwon et al. 2009; Li 2012; Lopes R.P. 2012; McEneff 2013; Na et al. 2013; Samanidou and Evaggelopoulou 2007; Wille et al. 2011b) although other kinds of detectors like fluorescence (FL), ultraviolet (UV), or diode array detector (DAD) have been also utilized (Cueva-Mestanza et al. 2008; Fernandez-Torres 2011; He 2012). These applications are, however, limited to compounds with the presence of chromophores or fluorescent groups in their molecular structure. Gas chromatography (GC) coupled to mass spectrometry was used only on one ocassion by Azzouz et al. (2011). Some other authors have used LC coupled to high resolution mass spectrometry (HRMS) as an alternative to screening of contaminants that offers high signal specificity and mass accuracy in full scan acquisition mode (Dasenaki et al. 2015; Martinez Bueno 2013; Martinez Bueno 2014; Villar-Pulido et al. 2011). Other techniques have been also applied like enzyme-linked immunosorbent assay (ELISA) and time-resolved fluoroimmunoassay (TR-FIA) (Chafer-Pericas 2010b; Chafer-Pericas 2010a).

Regarding the quality control parameters of the methods, LOD and LOQ ranged in the low ng/g levels in most of the cases except when the detection was done with UV or DAD/FL where they were considerably higher ranging from 30 to 250 ng/g LOD, and from 100 to 1020 ng/g LOQ (Cueva-Mestanza et al. 2008; Fernandez-Torres 2011). The recoveries range is very wide from 15 to 179%. Most of the method's recoveries ranged between 65 and 100% but the threshold decreases especially in multi-residue methods (Alvarez-Muñoz 2015a; Klosterhaus 2013) due to the different physicochemical properties of the therapeutic groups included (e.g., polarity, solubility, and stability). In these kind of methods it is difficult to obtain acceptable recoveries for all target compounds and it requires a compromise in the selection of the experimental conditions.

Occurrence of PhACs in marine organisms

Data regarding the occurrence of PhACs in marine organisms have become available at the same time that the analytical methods were developed and published. The majority of the publications reported in Table 5 have a final section where the method was applied for the analysis of "real samples"; wild or farm organism, directly sampled from the marine environment, or organisms bought from local supermarkets. Besides that, monitoring studies have been performed in different locations of Europe, United States, China, and Canada, providing data on PhACs concentrations in marine organisms and accounting for a total of 24 papers published since 2007. Table 6A shows general information about these publications like the organism type

Table 6A: Pharmaceuticals in biota. Study Details

Reference	Organism type	Species	Wild/farm	Origin	Method
Pojana et al. 2007	m	*Mytilus galloprovincialis*	w	Mediterranean and Black Sea	Ultrasonication+SPE+HPLC-MS
Chu et al. 2007	f	*Ameiurus nebulosus, Dorosoma cepedianum, Morone americana*	w	Hamilton Harbour Canada	Chu et al. 2007
Berradad et al. 2008	f	*Sparus* spp.	w	Supermarket Spain	Berrada et al. 2008
Kwon et al. 2009	f	*Pleuronichthys verticalis*	w	California Pacific Ocean	Kwon et al. 2009
Chafer-Pericas et al. 2010a	f	*Sparus aurata*	w	Supermarket Spain	Chafer-Pericas et al. 2010a
Chafer-Pericas et al. 2010b	f	*Sparus aurata*	w	Supermarket Spain	Chafer-Pericas et al. 2010b
Chafer-Pericas et al. 2010c	f	*Sparus aurata*	w	Supermarket Spain	Chafer-Pericas et al. 2010c
Dasenaki et al. 2010	f	*Dicentrarchus labrax, Sparus* spp.	f	Greece	Dasenaki et al. 2010
Chafer-Pericas et al. 2011	f	*Sparus aurata*	w	Supermarket Spain	Chafer-Pericas et al. 2011
Azzouz et al. 2011	f	*Dicentrarchus labrax, Solea* spp., *Salmo* spp.	f	Supermarket (Spain)	Azzouz et al. 2011
Wille et al. 2011b	m, c	*Mytilus edulis, Crassostrea gigas, Crangon crangon*	w	Atlantic, Northeast	Wille et al. 2011b
He et al. 2012	f	*Trachinotus ovatus, Lutianus argentimaculatus, Saprus Latus, Lutianus erythroptrus, Sparus macrocephalus, Gymnocranius griseus, Siganus fuscescens, Sciaenops ocellantus*	f	China	He et al. 2012
Li et al. 2012	m	*Crassostrea talienwhanensis, Chlamys farreri, Amussium, Scapharca subcrenata, Meretrix merhjgntrix linneaeus, Mactra veneriformis, Mactra chinesis, Mya arenaria, Neverita didyma, Rapana venosa, Mytilus edulis*	w	China Pacific Ocean	Li et al. 2012
Maruya et al. 2012	f	*Pleuronichthys verticalis*	w	California Pacific Ocean	Kwon et al. 2009
Klosterhaus et al. 2013	m	*Geukensia demissa*	w	Pacific, Eastern Central	Klosterhaus et al. 2013

Na et al. 2013	m	*Crassostrea Gigas, Patinopecten yessoensis, Chlamys farreri*	w	Bohai Sea China	Na et al. 2013
Martínez Bueno et al. 2013	m	*Mytilus galloprovincialis*	w	Mediterranean Southerneastern France	Martínez Bueno et al. 2013
Mc Eneff et al. 2014	m	*Mytilus edulis* and *galloprovincialis*	w	Atlantic, Northeast	Mc Eneff et al. 2013
Federova et al. 2014	m,c,f	n.s.	w,f	Czech market	Federova et al. 2014
Martínez Bueno et al. 2014	m	*Mytilus galloprovincialis*	w	Mediterranean Southerneastern France	Martínez Bueno et al. 2014
Dodder et al. 2014	m	*Mytilus* spp.	w	California Pacific Ocean	Solvent extraction/SPE/LC-MS/MS
Done et al. 2014	f	*Shrimp, tilapia, rainbow trout, atlantic salmon, swai*	n.s.	Supermarket United States	Done et al. 2014
Alvarez-Muñoz et al. 2015a	m	*Crassostrea gigas, Mytilus galloprovincialis, Chamalea gallina*	w	Spanish Mediterranean	Alvarez-Muñoz 2015
Alvarez Muñoz et al. 2015b	a,m,f	*Saccharina latissima, Laminaria digitata, Mytilus galloprovincialis, Mytilus* spp.*, Chamalea gallina, Crassostrea gigas, Liza aurata, Platichthys flesus*	w	Fureholmen Solund (Norway), Po Delta (Italy), Tagus Estuary (Portugal), Ebro Delta (Spain), Scheldt Estuary (Netherland)	Alvarez Muñoz et al. 2015[a] Huerta et al. 2013

m = mollusk
f = fish
c = crustacean
a = algae
w = wild
f = farm
n.s. = not specified

Table 6B: Pharmaceutical concentrations (ng/g) in organisms. Expressed as mean or minimum maximum interval.

Reference	Pojana et al. 2007	Chu et al. 2007	Berradad et al. 2008	Kwon et al. 2009	Chafer-Pericas et al. 2010a	Chafer-Pericas et al. 2010b	Chafer-Pericas et al. 2010c	Dasenaki et al. 2010	Chafer-Pericas et al. 2011	Azzouz et al. 2011	Wille et al. 2011b	He et al. 2012	Li et al. 2012	Maruya et al. 2012	Klosterhaus et al. 2013	Na et al. 2013	Martínez Bueno et al. 2013	Mc Eneff et al. 2014	Federova et al. 2014	Martinez Bueno et al. 2014	Dodder et al. 2014	Done et al. 2014	Alvarez-Muñoz et al. 2015a	Alvarez Muñoz et al. 2015b
1	2	3	4	5	6	7	8	9	10	11	12	13	14	15	16	17	18	19	20	21	22	23	24	25
Analgesics/Anti-inflammatories																								
Acetaminophen											n.d.–115													
Salicylic Acid											n.d–490													
Phenazone																								n.d.–2.90
Mefenamic Acid																		n.d.–23						
Phenazone																								n.d.–2.90
Antibiotics																								
Erythromycin			58–87										n.d.–31.3		n.d.–0.1				10		0.14			
Azithromycin																							1.2–3.0	1.28–13.30
Tetracycline									n.d.–13.1							1.73								
Oxytetracycline						n.d.–92	BQL-60		n.d.–178							1.25						2.7–8.6		
7-chlortetracycline monohydrochloride									n.d.–23.2															
Doxycycline																1.2								
Ofloxacin											n.d.–65		n.d.–242								1.2			

Ciprofloxacin		n.d.– 13.19	n.d.– 208				7.3– 13	
Enoxacin							3.9– 10	
Ampicillin							4.9– 10	
Flumequine							2.9– 25	
Sulfamethoxazole			n.d.– 20.1	n.d.	0.54			
Sulfacetamide					0.4			
Sulfamethoxypyridazine					1.64			
Sulfadiazine			n.n.– 2.72		2.07		3.0– 20.0	
Sulfasalazine							3.10	
Sulfamethizole				n.d.– 0.2				
Sulfathiazole					0.2			
Sulfamerazine			n.d.– 5.98		3.3			
Sulfamethiazole					0.9			
Sulfamonomethoxine			n.d.– 15.4		1.19			
Sulfachloropyridazine					0.22			
Sulfameter					8.8			
Sulfamethazine					1			24
Siulfisoxazole	45		n.d.– 71.6					
Trimethoprim				n.d.		< 4–9.22	13–15	

Table 6B contd....

Table 6B contd....

1	2	3	4	5	6	7	8	9	10	11	12	13	14	15	16	17	18	19	20	21	22	23	24	25
Dimetridazole																								n.d.–7.73
Ronidazole																							n.d.–1.8	n.d.–1.84
Chloramphenicol										n.d.						1.8								
Flofernicol										0.6–3.4						0.14								
Amithriptyline																					0			
Enrofloxacin												n.d.–30.62									1.3			
Norfloxacin												1.95–254	n.d.–370						12.0–14.0					
Lomefloxacin													n.d.–141								29			
Ormethoprim																						1		
Total macrolides													1–36.2									5.2		
Total quinolones													0.7–1575											
Total sulfonamides					n.d.–2.8								1–76.7									0.3		
Psychiatric drugs																								
Carbamazepine											n.d.–11				1.3–5.3		n.d.–3.5	n.d.–6					n.d.–2.1	n.d.–2.15
2-Hydroxycarbamazepine																							n.d.–1.3	n.d.–1.33
10,11-epoxycarbazaepine																							n.d.–1.3	n.d.–1.28
Sertraline															0.1–1.4						1.4			
Citalopram																							n.d.–1.9	n.d.–20.59

Venlafaxine							BQL–2.7		2.1–2.7	2.13–36.09
O-desmethylvenlafaxine							n.d.–3.7		n.d.–1.4	BQL–4.86
N-desmethylvenlafaxine							n.d.–3.0			
N,O-didesmethylvenlafaxine							2.5–3.5			
N,N-didesmethylvenlafaxine							3.8			
Fluoxetine	n.d.–1.02									
Norfluoxetine	n.d.–1.08									
Paroxetine	n.d.–0.58									
Diazepam		23–110			23–110	n.d.				
Alprazolam									n.d.–0.8	n.d.–0.86
Amitriptyline						n.d.–0.2				
Diltiazem						n.d.–0.1				
Amphetamine						n.d.–4.2		2.3		
Cocaine						n.d.–0.3		0.28		
Methylprednisolone								18		
B-Blocking agents										
Atenolol						n.d.–0.3		0.45		
Propanolol			n.d.	n.d.–63						
Tranquilizer										
Azaperone									n.d.–1.6	n.d.–1.60

Table 6B contd....

Table 6B contd....

1	2	3	4	5	6	7	8	9	10	11	12	13	14	15	16	17	18	19	20	21	22	23	24	25
Hormones																								
Estriol (E3)	n.d.																							
17 β-estradiol (E2)	n.d.									0.81–1.7														
Estrone (E1)	n.d.									0.52–1.3														
17α-ethinylestradiol (EE2)	< 3–38									n.d.														
Diuretic																								
Triamterene															n.d.–0.6									
Ranitidine															n.d.–0.3									
Calcium channel blockers																								
Diltiazem																							n.d.–1.5	n.d.–1.47
Dehydronifedipine															0.2–0.7									
Risperidone															n.d.–0.1									
diphenhydramine															n.d.–0.3						0.87			
Steroids																								
Digoxigenin															n.d.–9.7									
Prostatic hyperplasia																								
Tamsulosin																								n.d.–3.03

n.d. = non detected
BQL = below quantification limit
dry weight (dw)
wet weight (ww)

and species used, if they were wild or farmed, their origin, and the analytical methods employed for their analysis. Some of the publications present their own analytical methods (Dodder 2014; Pojana 2007), others have implemented methods initially developed in freshwater organisms (Álvarez-Muñoz 2015b). Most of the studies have been carried out using one type of organism although some of them have analyzed up to three types from different tropic levels (Álvarez-Muñoz 2015b; Federova 2014). Marine fishes have been the most researched organism (14 studies) followed by mollusk (12), crustacean (2), and macroalgae (1). In agreement with above mentioned *Sparus aurata* and *Mytilus* spp. have been the most analyzed species. Wild organisms have been targeted by the majority of the authors despite the fact that anti-infectious and anti-parasitic agents are widely used in aquaculture.

Table 6b shows the ranges of PhACs concentrations reported in those marine organisms. They are expressed in dry or wet weight (dw or ww), like an interval of concentrations (minimum-maximum), mean or single value. Antibiotics are the main represented pharmaceutical class with 38 detected molecules, followed by psychiatric drugs with 20. Lower down in number of positive identified compounds come other therapeutic groups like analgesics/anti-inflammatories, calcium channel blockers, β-blockers, diuretics, etc. The most ubiquitous pharmaceuticals have been carbamazepine (detected up to six times) (Alvarez-Muñoz 2015a; Álvarez-Muñoz 2015b; Klosterhaus et al.; Martinez Bueno 2013; McEneff 2014; Wille et al. 2011b), and oxytetracycline (detected up to five times) (Chafer-Pericas 2011; Chafer-Pericas 2010b; Chafer-Pericas 2010c; Done and Halden 2015; Na et al. 2013). The rest of the identified compounds have been detected in marine species between 1 and 3 times. The concentrations reported usually ranged in the low nanogram per gram levels with few exceptions. The highest concentration found for a single compound corresponded to salicylic acid reaching up to 490 ng/g dw in *Mytilus edulis* transplanted to cages and deployed in the Belgian coast (Wille et al. 2011b). Li et al. (Li et al. 2012) found up to 370 ng/g dw of norfloxacin, 242 ng/g dw of ofloxacin, 208 ng/g dw of ciprofloxacin, and 147 ng/g dw of enrofloxacine in wild mollusk collected from the Bohai Sea, China. These fluoroquinolones are antibiotics widely used. Norfloxacin is used in both human and veterinary medicines whereas ofloxacine is used only by humans. Enrofloxacin has also a single use but in veterinary medicine. It is highly susceptible to photodegradation in the environment and can be transformed into ciprofloxacin in target organisms (Liang et al. 2013b).

Current and Future Research Trends

Analytical Challenges

Multi-residue methods are able to provide a comprehensive view about contamination by micropollutants. However, in order to have an accurate and broad determination of all plausible contaminants present there is a need of a battery of methodologies, as well as chemical standards for all target compounds. Screening methodologies for the rapid identification of chemicals that do not require the use of standards are being explored as an alternative methodology able to identify and confirm non-target analytes. Liquid chromatography coupled to High Resolution Mass Spectrometry (HRMS) is the technique of choice for such applications. HRMS allows researchers to perform the so-called "suspect screening", where public and home-made libraries of a broad set of compounds suspected to be present in the sample can be used. This type of methodology also allows focusing on metabolites and transformation products of the pharmaceutical compounds, both generated in the organisms after drug administration and also as a consequence of biotic and abiotic transformations that take place during WWTPs and in natural water systems. Some authors have proposed workflows for target, suspect, and non-target screening of emerging pollutants and their transformation products (Bletsou et al. 2015; Krauss et al. 2010; Zedda and Zwiener 2012) although such methodologies have been scarcely applied to marine environment yet (Magner et al. 2010; Wille et al. 2011a).

Robust and well-established methods based on chromatography coupled to mass spectrometry are time consuming, expensive, and require specialized personnel and instrumentation. Additionally they usually require extensive sample preparation and clean-up. Alternative methodologies, able to monitor contaminants of environmental relevance as quickly and as cheaply as possible, and even the possibility of allowing on-site field monitoring have also been developed in the last years. In this respect, sensors and biosensors have demonstrated a great potential in recent years and thus arise as proposed analytical tools for effective monitoring in monitoring programs (Rodriguez Mozaz et al. 2006). The main advantages offered by biosensors over conventional analytical techniques are the possibility of portability, of miniaturization and working on-site, and the ability to measure pollutants in complex matrices with minimal sample preparation. Although many of the systems developed cannot compete with conventional analytical methods in terms of accuracy and reproducibility, they can be used by regulatory authorities and by industry to provide enough information for routine testing and screening of samples. Examples of sensors and biosensors applied to detection of micropollutants in marine monitoring have recently been reviewed by (Justino et al. 2015) although only two biosensors have been identified which can specifically detect the contamination by pharmaceuticals diclofenac and estradiol in seawater (Arvand et al. 2012; Ou et al. 2009). To this aim, an example of current research initiatives is the EU project Sea-on-a-chip, which aims to develop and implement an early warning system based on immunosensors operated by remote control (FP7 Ocean 2013-614168). The sensors are like little "floating laboratories" that can monitor up to eight contaminants related to aquaculture activities such as antibiotics and pesticides.

Integrated studies

Most of the literature available about contamination of aquatic environment by pharmaceuticals has focused on sea water (42 studies, as reported in Table 2); whereas studies addressing sediment contamination are rather scarce (12 studies as reported in Table 4). Literature about contamination of marine aquatic biota by pharmaceuticals is also abundant (24 studies as reported in Table 6a and 6b) probably due to human health implications derived from the consumption of seafood. Not only occurrence but also distribution and partition of contaminants in different environmental compartments, water, sediments, particulate matter, and also uptake by biota has been addressed by some of the studies reviewed.

There is a number of articles devoted to studying water-sediment distribution (Chen et al. 2015; Liang et al. 2013a; Yang et al. 2011). In some studies, concentrations of antibiotics were lower in sediments than in water (Chen et al. 2015; Lara-Martin et al. 2014; Yang et al. 2011), whereas in other studies sediments accumulated high levels of compounds such as antibiotics, demonstrating that sediments can act as a reservoir of these compounds and as a potential source of them to the aquatic environments with changes in environmental conditions (Chen et al. 2015; Liang et al. 2013a). A couple of studies deal with the distribution of pharmaceuticals not only between sediments and water but also in suspended particulate matter (SPM) and colloids (Lara-Martin et al. 2014; Yang et al. 2011). Some pharmaceuticals were detected in suspended solids at concentrations ranging from low ng/g to μg/g, and aspects such as particulate organic carbon (POC) and physic-chemical properties of pollutants were pointed out as driving factors of their sorption. On the other hand, sorption capacity of pharmaceuticals in the colloidal phase was 3–4 orders of magnitude higher than in SPM (Yang et al. 2011).

With regards aquatic organisms, integrated studies taking into account water and sediment concentrations of pharmaceuticals (Maruya et al. 2012; Na et al. 2013). Klosterhaus et al. (2013) and McEneff et al. (2014) have attempted to establish bioaccumulation factors (BAF) for several contaminants and suggested that some antibiotics are bioaccumulative (BAF > 5000 L/Kg) (Na et al. 2013). Lipid partitioning processes do not sufficiently explain the bioaccumulation observed and the uptake of ionizable compounds needs to be considered as other mechanisms of uptake (Klosterhaus et al. 2013).

Environmental Risk Assessment

Environmental Risk Assessment (ERA) of pharmaceuticals has been mostly applied to freshwater ecosystems since WWTPs (one of the main sources of pharmaceuticals in the environment) are mainly discharging their effluents in surface water. As reviewed in this work, many studies have reported the dilution of pollutant concentration from rivers or WWTPs to the sea. However pharmaceuticals occur widely in coastal areas and can still pose an ecological risk to aquatic organisms and therefore several studies have assessed the environmental risk of these contaminants (Bayen et al. 2013; Chen et al. 2015;

Jiang et al. 2014; Lolic et al. 2015; McEneff et al. 2014; Zhang et al. 2013a; Zhang et al. 2013b; Zhang et al. 2012). Most of these studies calculated the Hazard or Risk Quotients (HQ) of pharmaceuticals to evaluate their potential to exert biological effects (Hernando et al. 2006) (TGD 93/67/EEC). Several antibiotics and one anti-inflammatory exhibited high risk (HQ > 1) to relevant sensitive aquatic organisms in some of these studies: enrofloxacine, ciprofloxacin, and sulfamethoxazole in Zhang et al. (2012), norfloxacine, erythromycin and oxytetracycline in Chen et al. (2015), sulfamethoxazol, erythromycin, and chlaritromycin in Zhang et al. (2013a), and the anti-inflammatory diclofenac in Lolic et al. (2015).

Antibiotics

As it has been highlighted in previous sections antibiotics have been thoroughly studied in the marine environment in comparison to other therapeutic groups and there is already a group of specific studies about the occurrence of antibiotics in water, sediments, and biota (Chen et al. 2015; Gulkowska et al. 2007; Liang et al. 2013a; Minh et al. 2009; Na et al. 2013; Shi et al. 2014; Xu et al. 2007; Zhang et al. 2013a; Zhang et al. 2013b; Zhang et al. 2012; Zheng et al. 2012; Zhou et al. 2012a). Fish farming activities are one of the major sources of antibiotic contamination in the marine environment. Unintentional distribution of antibiotic drugs from an aquaculture system into the surrounding environment is unavoidable when the drugs are administered through medicated feed because it dissolves in the water or spreads on particles transported in the water (Sorum 2008). Antibiotics pose a significant risk to environmental and human health, as indicated in previous section, where several antibiotics where highlighted as contaminants of high risk in ERA studies. In addition, the overuse and misuse of antibiotics has led to the emergence of antibiotic-resistant bacteria, compromising the effectiveness of antimicrobial therapy (Marti et al. 2014). Marine bacteria exposed to antibiotics inside or outside the fish farming environment can acquire antimicrobial resistance and can cause changes in the coastal environments (Labella et al. 2013). Drug resistant bacteria in the environment can be a major source of antibiotic resistance genes, which can be transferred to pathogens of fish and other animals, including humans (Sorum 2008). Emergence and spread of antibiotic resistance bacteria has been indeed classified by the World Health Organization (WHO) as one of the three biggest threats to public health in the 21st century. As a result of this public concern, studies about the prevalence and resistance transfer mechanisms in marine environment have appeared in the last five years (Germond and Kim 2015; Labella et al. 2013; Leonard et al. 2015; Na et al. 2014; Zhang et al. 2011), and an increase in this type of studies is foreseen. Monitoring antimicrobial drug usage and resistance will help to identify trends, assess environmental risk, and to establish a linkage of antimicrobial usage to antimicrobial resistance; and as a basis for interventions among other purposes (Sorum 2008).

Conclusions

Analytical methodologies applied for the determination of PhACs in different marine environmental compartments (water, sediments, and biota) have been reviewed in this chapter. As expected, most of the literature available about the contamination of marine environment by PhACs focus on sea water, in particular from coastal areas, considered the most polluted areas in terms of PhACs inputs (through direct discharges of WWTP and outfall of polluted rivers). Up to 42 studies address sea water pollution whereas studies addressing sediment and biota contamination are less abundant (12 and 24 studies respectively). Different organisms were considered concerning biota samples: fish, mollusk, crustacean, and also algae.

Levels found in the marine environment were in general lower than those found in freshwater system with values ranging between low ng/L till occasionally low µg/L in the case of water, and from low ng/g till near five hundred ng/g in sediments and biota. Antibiotics, followed by psychiatric drugs, and analgesics/anti-inflammatories were the therapeutic groups most frequently detected in all environmental compartments. Antibiotics were also identified as the compounds posing a higher potential risk for the environment based on the studies reported in the chapter. In addition, the presence of PhACs compounds in marine aquatic organism can have some human health implications derived from the consumption of seafood.

Acknowledgements

The research leading to this work has received funding from the European Union Seventh Framework Programme (FP7/2007-2013) under the ECsafeSEAFOOD project (grant agreement n°. 311820) and SEA-on-a-CHIP project (grant agreement n°. 614168). This work was partly supported by the Generalitat de Catalunya (Consolidated Research Group: Catalan Institute for Water Research 2014SGR291).

Keywords: Pharmaceuticals, coastal marine environment, environmental compartments, seawater, Sediment, Biota

References

Alvarez-Muñoz, D., B. Huerta, M. Fernandez-Tejedor, S. Rodríguez-Mozaz and D. Barceló. 2015a. Multi-residue method for the analysis of pharmaceuticals and some of their metabolites in bivalves. Talanta 136: 174–182.

Álvarez-Muñoz, D., S. Rodríguez-Mozaz, A.L. Maulvault, A. Tediosic, M. Fernández-Tejedor, F. Van den Heuvele, M. Kottermanf, A. Marques and D. Barceló. 2015b. Occurrence of pharmaceuticals and endocrine disrupting compounds in macroalgaes, bivalves, and fish from coastal areas in Europe. Environ. Res. 143: 56–64.

Arpin-Pont, L., M. Bueno, E. Gomez and H. Fenet. 2014. Occurrence of PPCPs in the marine environment: a review. Environ. Sci. Poll. Res. 1–14.

Arvand, M., T.M. Gholizadeh and M.A. Zanjanchi. 2012. MWCNTs/Cu (OH) 2 nanoparticles/IL nanocomposite modified glassy carbon electrode as a voltammetric sensor for determination of the non-steroidal anti-inflammatory drug diclofenac. Mater. Sci. Eng. C 32: 1682.

Azzouz, A., B. Souhail and E. Ballesteros. 2011. Determination of residual pharmaceuticals in edible animal tissues by continuous solid-phase extraction and gas chromatography-mass spectrometry. Talanta 84(3): 820–8.

Bayen, S., H. Zhang, M.M. Desai, S.K. Ooi and B.C. Kelly. 2013. Occurrence and distribution of pharmaceutically active and endocrine disrupting compounds in Singapore's marine environment: Influence of hydrodynamics and physical-chemical properties. Environ. Poll. 182(0): 1–8.

Beretta, M., B. Britto, T.M. Tavares, S.M.T. Da Silva and A.L. Pletsch. 2014. Occurrence of pharmaceuticals and personal care products (PPCPs) in marine sediments in the Todos os Santos Bay and the north coast of Salvador, Bahia, Brazil. J. Soil Sedim. 14: 1278–1286.

Berrada, H., F. Borrull, G. Font and R.M. Marce. 2008. Determination of macrolide antibiotics in meat and fish using pressurized liquid extraction and liquid chromatography-mass spectrometry. J. Chromatogr. A 1208(1-2): 83–9.

Bletsou, A.A., J. Jeon, J. Hollender, E. Archontaki and N.S. Thomaidis. 2015. Targeted and non-targeted liquid chromatography-mass spectrometric workflows for identification of transformation products of emerging pollutants in the aquatic environment. Trends Anal. Chem. 66(0): 32–44.

Borecka, M., G. Siedlewicz, L.P. Halinski, K. Sikora, K. Pazdro, P. Stepnowski and A. Bialk-Bielinska. 2015. Contamination of the southern Baltic Sea waters by the residues of selected pharmaceuticals: Method development and field studies. Mar. Poll. Bull. 94: 62–71.

Brodin, T., J. Fick, M. Johnsson and J. Klaminder. 2013. Dilute concentrations of a psychiatric drug alter behaviour of fish from natural populations. Science 339: 814–815.

Cabello, F.C. 2006. Heavy use of prophylactic antibiotics in aquaculture: a growing problem for human and animal health and for the environment. Environ. Microbiol. 8(7): 1137–44.

Canada-Canada, F., A. Munoz de la Pena and A. Espinosa-Mansilla. 2009. Analysis of antibiotics in fish samples. Anal. Bioanal. Chem. 395(4): 987–1008.

Cháfer-Pericás, C., A. Maquieira, R. Puchades, J. Miralles and A. Moreno. 2010a. Fast screening immunoassay of sulfonamides in commercial fish samples. Anal. Bioanal. Chem. 396(2): 911–21.

Chafer-Pericas, C., A. Maquieira, R. Puchades, J. Miralles, A. Moreno, N. Pastor-Navarro and F. Espinosa. 2010b. Immunohemical determination of oxythetracycline in fish: comparison between enzymatic and time-resolved fluorometric assays. Anal. Chim. Acta 662: 177–185.

Chafer-Pericas, C.M., A. Puchades, Compañy, B- R. Miralles, J. Moreno, A. 2010c. Multiresidue determination of antibiotics in aquaculture fish samples by HLPC-MS/MS. Aquac. Res. 41: 217–255.

Chafer-Pericas, C., A. Maqueira, R. Puchades, J. Miralles and A. Moreno. 2011. Multiresidue determination of antibiotics in feed and fish samples for food safety evaluation. Comparison of immunoassay vs. LC-MS-MS. Food Control 22: 993–999.

Chen, H., S. Liu, X.-R. Xu, G.-J. Zhou, S.-S. Liu, W.-Z. Yue, K.-F. Sun and G.-G. Ying. 2015. Antibiotics in the coastal environment of the Hailing Bay region, South China Sea: Spatial distribution, source analysis and ecological risks. Mar. Poll. Bull. (0).

Chu, S. and C.D. Metcalfe. 2007. Analysis of paroxetine, fluoxetine and norfluoxetine in fish tissues using pressurized liquid extraction, mixed mode solid phase extraction cleanup and liquid chromatography-tandem mass spectrometry. J. Chromatogr. A 1163(1-2): 112–8.

Collado, N., S. Rodriguez-Mozaz, M. Gros, A. Rubirola, D. Barcelo, J. Comas, I. Rodriguez-Roda and G. Buttiglieri. 2014. Pharmaceuticals occurrence in a WWTP with significant industrial contribution and its input into the river system. Environ. Pollut. 185: 202–12.
Cueva-Mestanza, R., M.E. Torres-Padron, Z. Sosa-Ferrera and J.J. Santana-Rodriguez. 2008. Microwave-assisted micellar extraction coupled with solid-phase extraction for preconcentration of pharmaceuticals in molluscs prior to determination by HPLC. Biomed. Chromatogr. 22(10): 1115–22.
Dasenaki, M.E. and N.S. Thomaidis. 2010. Multi-residue determination of seventeen sulfonamides and five tetracyclines in fish tissue using a multi-stage LC-ESI-MS/MS approach based on advanced mass spectrometric techniques. Anal. Chim. Acta 672(1-2): 93–102.
Dasenaki, M.E., A.A. Bletsou, G.A. Koulis and N.S. Thomaidis. 2015. Qualitative multiresidue screening method for 143 veterinary drugs and pharmaceuticals in milk and fish tissue using liquid chromatography quadrupole-time-of-flight mass spectrometry. J. Agric. Food. Chem. 63(18): 4493–508.
de Jongh, C.M., P.J.F. Kooij, P. de Voogt and T.L. ter Laak. 2012. Screening and human health risk assessment of pharmaceuticals and their transformation products in Dutch surface waters and drinking water. Sci. Total Environ. 427-428: 70–77.
Dodder, N.G., K.A. Maruya, P. Lee Ferguson, R. Grace, S. Klosterhaus, M.J. La Guardia, G.G. Lauenstein and J. Ramirez. 2014. Occurrence of contaminants of emerging concern in mussels (*Mytilus* spp.) along the California coast and the influence of land use, storm water discharge, and treated wastewater effluent. Mar. Pollut. Bull. 81(2): 340–6.
Done, H.Y. and R.U. Halden. 2015. Reconnaissance of 47 antibiotics and associated microbial risks in seafood sold in the United States. J. Hazard. Mater. 282: 10–7.
EC. 2008. Directive 2008/105/EC of the European Parliament and of the Council on Environmental Quality Standards in the Field of Water Policy, Amending and Subsequently Repealing Council Directives 82/176/EEC, 83/513/EEC, 84/156/EEC, 84/491/EEC, 86/280/EEC and Amending Directive 2000/60/EC of the European Parliament and of the Council, OJ L 348/84 of 16 December 2008.
EC. 2010. Commission regulation (EU) No. 37/2010 on pharmacologically active substances and their classification regarding maximum residue limits in foodstuff of animal origin. Commission regulation (EU) No. 37/2010.
Federova, G., V. Nebesky, T. Randak and R. Grabic. 2014. Simultaneous determination of 32 antibiotics in aquaculture products using LC-MS/MS. Chem. Pap. 68: 29–36.
Fernandez-Torres, R., M.A. Bello Lopez, M. Olias Consentino and M. Callejon Mochon. 2011. Simultaneous determination of selected veterinary antibiotics and their main metabolites in fish and mussel samples by high-performance liquid chromatography with diode array-fluorescence (HPLC-DAD-FLD) detection. Anal. Lett. 44(14): 2357–2372.
Ferrara, F., N. Ademollo, M. Delise, F. Fabietti and E. Funari. 2008. Alkylphenols and their ethoxylates in seafood from the Tyrrhenian Sea. Chemosphere 72(9): 1279–85.
Gaw, S., K.V. Thomas and T.H. Hutchinson. 2014. Sources, impacts and trends of pharmaceuticals in the marine and coastal environment. Philos. Trans. R Soc. Lond. B Biol. Sci. 369(1656).
Germond, A. and S.-J. Kim. 2015. Genetic diversity of oxytetracycline-resistant bacteria and tet(M) genes in two major coastal areas of South Korea. J.G.A.R.
Godfrey, E., W.W. Woessner and M.J. Benotti. 2007. Pharmaceuticals in on-site sewage effluent and ground water, Western Montana. Ground Water 45(3): 263–71.
Gros, M., S. Rodriguez-Mozaz and D. Barcelo. 2012. Fast and comprehensive multi-residue analysis of a broad range of human and veterinary pharmaceuticals and some of their metabolites in surface and treated waters by ultra-high-performance liquid chromatography coupled to quadrupole-linear ion trap tandem mass spectrometry. J. Chromatogr. A 1248(0): 104–121.
Gros, M., S. Rodriguez-Mozaz and D. Barcelo. 2013. Rapid analysis of multiclass antibiotic residues and some of their metabolites in hospital, urban wastewater and river water by ultra-high-performance liquid chromatography coupled to quadrupole-linear ion trap tandem mass spectrometry. J. Chromatogr. A 1292: 173–88.
Gulkowska, A., Y. He, M.K. So, L.W.Y. Yeung, H.W. Leung, J.P. Giesy, P.K.S. Lam, M. Martin and B.J. Richardson. 2007. The occurrence of selected antibiotics in Hong Kong coastal waters. Mar. Poll. Bull. 54(8): 1287–1293.
He, X., Z. Wang, X. Nie, Y. Yang, D. Pan, A.O.W. Leung, Z. Cheng, Y. Yang, K. Li and K. Chen. 2012. Residues of fluoroquinolonas in marine aquaculture environment of the Pearl River Delta, South China. Environ. Geochem. Health 34: 323–335.
Hernando, M.D., M. Mezcua, A.R. Fernandez-Alba and D. Barceló. 2006. Environmental risk assessment of pharmaceutical residues in wastewater effluents, surface waters and sediments. Talanta 69(2): 334–342.
Huerta, B., A. Jakimska, M. Gros, S. Rodriguez-Mozaz and D. Barcelo. 2012a. Analysis of multi-class pharmaceuticals in fish tissues by ultra-high-performance liquid chromatography tandem mass spectrometry. J. Chromatogr. A 1288: 63–72.
Huerta, B., S. Rodriguez-Mozaz and D. Barcelo. 2012b. Pharmaceuticals in biota in the aquatic environment: analytical methods and environmental implications. Anal. Bioanal. Chem. 404(9): 2611–24.
Jiang, J.-J., C.-L. Lee and M.-D. Fang. 2014. Emerging organic contaminants in coastal waters: Anthropogenic impact, environmental release and ecological risk. Mar. Poll. Bull. 85(2): 391–399.
Justino, C.I.L., A.C. Freitas, A.C. Duarte and T.A.P.R. Santos. 2015. Sensors and biosensors for monitoring marine contaminants. Trends Environ. Anal. Chem. (0).
Kim, S.C. and K. Carlson. 2007. Temporal and spatial trends in the occurrence of human and veterinary antibiotics in aqueous and river sediment matrices. Environ. Sci. Technol. 41(1): 50–7.

Klosterhaus, S.L., R. Grace, M.C. Hamilton and D. Yee. 2013. Method validation and reconnaissance of pharmaceuticals, personal care products, and alkylphenols in surface waters, sediments, and mussels in an urban estuary. Environ. Int. 54(0): 92–99.

Krauss, M., H. Singer and J. Hollender. 2010. LC-high resolution MS in environmental analysis: from target screening to the identification of unknowns. Anal. Bioanal. Chem. 397: 943–951.

Kuster, A. and N. Adler. 2014. Pharmaceuticals in the environment: scientific evidence of risks and its regulation. Philos. Trans. R Soc. Lond. B Biol. Sci. 369(1656).

Kwon, J.W., K.L. Armbrust, D. Vidal-Dorsch and S.M. Bay. 2009. Determination of 17alpha-ethynylestradiol, carbamazepine, diazepam, simvastatin, and oxybenzone in fish livers. J. AOAC Int. 92(1): 359–69.

Labella, A., M. Gennari, V. Ghidini, I. Trento, A. Manfrin, J.J. Borrego and M.M. Lleo. 2013. High incidence of antibiotic multi-resistant bacteria in coastal areas dedicated to fish farming. Mar. Poll. Bull. 70: 197–203.

Lara-Martin, P.A., E. Gonzalez-Mazo, M. Petrovic, D. Barcelo and B.J. Brownawell. 2014. Occurrence, distribution and partitioning of nonionic surfactants and pharmaceuticals in the urbanized Long Island Sound Estuary (NY). Mar. Poll. Bull. 85(2): 710–719.

Leonard, A.F.C., L. Zhang, A.J. Balfour, R. Garside and W.H. Gaze. 2015. Human recreational exposure to antibiotic resistant bacteria in coastal bathing waters. Environ. Int. 82(0): 92–100.

Li, W., Y. Shi, L. Gao, J. Liu and Y. Cai. 2012. Investigation of antibiotics in mollusks from coastal waters in the Bohai Sea of China. Environ. Pollut. 162: 56–62.

Liang, X., B. Chen, X. Nie, Z. Shi, X. Huang and X. Li. 2013a. The distribution and partitioning of common antibiotics in water and sediment of the Pearl River Estuary, South China. Chemosphere 92(11): 1410–1416.

Liang, X., B. Chen, X. Nie, Z. Shi, X. Huang and X. Li. 2013b. The distribution and partitioning of common antibiotics in water and sediment of the Pearl River Estuary, South China. Chemosphere 92(11): 1410–6.

Liu, S., H. Chen, X.R. Xu, S.S. Liu, K.F. Sun, J.L. Zhao and G.G. Ying. 2015. Steroids in marine aquaculture farms surrounding Hailing Island, South China: Occurrence, bioconcentration, and human dietary exposure. Sci. Total Environ. 502: 400–7.

Lolic, A., P. Paiga, L.H.M.L.M. Santos, S. Ramos, M. Correia and C. Delerue-Matos. 2015. Assessment of non-steroidal anti-inflammatory and analgesic pharmaceuticals in seawaters of North of Portugal: Occurrence and environmental risk. Sci. Total Environ. 508(0): 240–250.

Loos, R., R. Carvalho, D.C. Antonio, S. Comero, G. Locoro, S. Tavazzi, B. Paracchin, M. Ghiani, T. Lettieri, L. Blaha, B. Jarosova, S. Voorspoels, K. Servaes, P. Haglund, J. Fick, R.H. Lindberg, D. Schwesig and B.M. Gawlik. 2013a. EU-wide monitoring survey on emerging polar organic contaminants in wastewater treatment plant effluents. Water Res. 47(17): 6475–87.

Loos, R., S. Tavazzi, B. Paracchini, E. Canuti and C. Weissteiner. 2013b. Analysis of polar organic contaminants in surface water of the northern Adriatic Sea by solid-phase extraction followed by ultrahigh-pressure liquid chromatography-QTRAP® MS using a hybrid triple-quadrupole linear ion trap instrument. Anal. Bioanal. Chem. 405: 5875–5885.

Lopes, R.P., R.C. Reyes, R. Romero-González, J.L. Vidal and A.G. Frenich. 2012. Multireside determination of veterinary drugs in aquaculture fish samples by ultra high perfomance liquid chromatography coupled to tandem mass spectrometry. J. Chromatogr. B 895-896: 39–47.

Magner, J., M. Filipovic and T. Alsberg. 2010. Application of a novel solid-phase-extraction sampler and ultra-performance liquid chromatography quadrupole-time-of-flight mass spectrometry for determination of pharmaceutical residues in surface sea water. Chemosphere 80(11): 1255–1260.

Marti, E., Variatza, E., J.L, B. 2014. The role of aquatic ecosystems as reservoirs of antibiotic resistance. Trends Microbiol. 22: 36–41.

Martinez Bueno, M.J., M.D. Hernando, A. Agüera and A.R. Fernandez-Alba. 2009. Application of passive sampling devices for screening of micro-pollutants in marine aquaculture using LC-MS/MS. Talanta 77(4): 1518–1527.

Martinez Bueno, M.J., C. Boillot, H. Fenet, S. Chiron, C. Casellas and E. Gomez. 2013. Fast and easy extraction combined with high resolution-mass spectrometry for residue analysis of two anticonvulsants and their transformation products in marine mussels. J. Chromatogr. A 1305: 27–34.

Martinez Bueno, M.J., C. Boillot, D. Munaron, H. Fenet, C. Casellas and E. Gomez. 2014. Occurrence of venlafaxine residues and its metabolites in marine mussels at trace levels: development of analytical method and a monitoring program. Anal. Bioanal. Chem. 406(2): 601–10.

Maruya, K.A., D.E. Vidal-Dorsch, S.M. Bay, J.W. Kwon, K. Xia and K.L. Armbrust. 2012. Organic contaminants of emerging concern in sediments and flatfish collected near outfalls discharging treated wastewater effluent to the Southern California Bight. Environ. Toxicol. Chem. 31(12): 2683–2688.

McEneff, G., L. Barron, B. Kelleher, B. Paull and B. Quinn. 2013. The determination of pharmaceutical residues in cooked and uncooked marine bivalves using pressurised liquid extraction, solid-phase extraction and liquid chromatography-tandem mass spectrometry. Anal. Bioanal. Chem. 405(29): 9509–21.

McEneff, G., L. Barron, B. Kelleher, B. Paull and B. Quinn. 2014. A year-long study of the spatial occurrence and relative distribution of pharmaceutical residues in sewage effluent, receiving marine waters and marine bivalves. Sci. Total Environ. 476-477: 317–326.

Minh, T.B., H.W. Leung, I.H. Loi, W.H. Chan, M.K. So, J.Q. Mao, D. Choi, J.C.W. Lam, G. Zheng, M. Martin, J.H.W. Lee, P.K.S. Lam and B.J. Richardson. 2009. Antibiotics in the Hong Kong metropolitan area: Ubiquitous distribution and fate in Victoria Harbour. Mar. Poll. Bull. 58(7): 1052–1062.

Moreno-Gonzalez, R., S. Rodriguez-Mozaz, M. Gros, D. Barcelo and V.M. Leon. 2015. Seasonal distribution of pharmaceuticals in marine water and sediment from a mediterranean coastal lagoon (SE Spain). Environ. Res. 138(0): 326–344.
Munaron, D., N. Tapie, H. Budzinski, B. Andral and J.-L. Gonzalez. 2012. Pharmaceuticals, alkylphenols and pesticides in Mediterranean coastal waters: Results from a pilot survey using passive samplers. Estuar. Coast. Shelf S. 114(0): 82–92.
Na, G., X. Fang, Y. Cai, L. Ge, H. Zong, X. Yuan, Z. Yao and Z. Zhang. 2013. Occurrence, distribution, and bioaccumulation of antibiotics in coastal environment of Dalian, China. Mar. Pollut. Bull. 69(1-2): 233–7.
Na, G., W. Zhang, S. Zhou, H. Gao, Z. Lu, X. Wu, R. Li, L. Qiu, Y. Cai and Z. Yao. 2014. Sulfonamide antibiotics in the Northern Yellow Sea are related to resistant bacteria: Implications for antibiotic resistance genes. Mar. Poll. Bull. 84: 70–75.
Nödler, K., T. Licha, K. Bester and M. Sauter. 2010. Development of a multi-residue analytical method, based on liquid chromatography tandem mass spectrometry, for the simultaneous determination of 46 micro-contaminants in aqueous samples. J. Chromatogr. A 1217(42): 6511–6521.
Nodler, K., D. Voutsa and T. Licha. 2014. Polar organic micropollutants in the coastal environment of different marine systems. Mar. Pollut. Bull. 85(1): 50–9.
Ou, H., Z. Luo, H. Jiang, H. Zhou, X. Wang and C. Song. 2009. Indirect inhibitive immunoassay for estradiol using surface plasmon resonance coupled to online in-tube SPME. Song. Anal. Lett. 42: 2758.
Podlipna, D. and M. Cichna-Markl. 2007. Determination of bisphenol A in canned fish by sol-gel immunoaffinity chromatography, HPLC and fluorescence detection. Eur. Food Res. Technol. 224: 629–634.
Pojana, G., A. Gomiero, N. Jonkers and A. Marcomini. 2007. Natural and synthetic endocrine disrupting compounds (EDCs) in water, sediment and biota of a coastal lagoon. Environ. Int. 33(7): 929–36.
Rodriguez Mozaz, S., M. Lopez de Alda and D. Barcelo. 2006. Biosensors as useful tools for environmental analysis and monitoring. Anal. Bioanal. Chem. 386: 1025–1041.
Samanidou, V.F. and E.N. Evaggelopoulou. 2007. Analytical strategies to determine antibiotic residues in fish. J. Sep. Sci. 30(16): 2549–69.
Sapkota, A., M. Kucharski, J. Burke, S. McKenzie, P. Walker and R. Lawrence. 2008. Aquaculture practices and potential human health risks: current knowledge and future priorities. Environ. Int. 34(8): 1215–26.
Shi, H., Y. Yang, M. Liu, C. Yan, H. Yue and J. Zhou. 2014. Occurrence and distribution of antibiotics in the surface sediments of the Yangtze Estuary and nearby coastal areas. Mar. Pollut. Bull. 83(1): 317–23.
Sorum, H. 2008. Antibiotic resistance associated with veterinary drug use in fish farms. pp. 157–182. *In*: Ø. Lie [ed.]. Improving Farmed Fish Quality and Safety. Woodhead Publishing Series in Food Science, Technology and Nutrition: ISBN: 978-1-84569-299-5.
Sumpter, J.P. 1995. Feminized responses in fish to environmental estrogens. Toxicol. Lett. 82-83: 737–42.
Tanoue, R., K. Nomiyama, H. Nakamura, T. Hayashi, J.W. Kim, T. Isobe, R. Shinohara and S. Tanabe. 2014. Simultaneous determination of polar pharmaceuticals and personal care products in biological organs and tissues. J. Chromatogr. A 1355: 193–205.
Tertuliani, J.S., D.A. Alvarez, E.T. Furlong, M.T. Meyer, S.D. Zaugg and G.F. Koltun. 2008. Occurrence of organic wastewater compounds in the Tinkers Creek watershed and two other tributaries to the Cuyahoga River, Northeast Ohio. U.S. Geological Survey Scientific Investigations Report 5173: 60.
US Environmental Protection Agency. 2007. Method 1694: Pharmaceuticals and personal care products in water, soil, sediment, and biosolids by HPLC/MS/MS. EPA 821/R-08/002. Washington, DC.
Vidal-Dorsch, D.E., S.M. Bay, K. Maruya, S.A. Snyder, R.A., Trenholm and B.J. Vanderford. 2012. Contaminants of emerging concertn in municipal wastewater effluents and marine receiving water. Environ. Toxicol. Chem. 31: 2674–2682.
Villar-Pulido, M., B. Gilbert-Lopez, J.F. Garcia-Reyes, N.R. Martos and A. Molina-Diaz. 2011. Multiclass detection and quantitation of antibiotics and veterinary drugs in shrimps by fast liquid chromatography time-of-flight mass spectrometry. Talanta 85(3): 1419–27.
Von Oepen, B., W. Kördel and W. Klein. 1991. Sorption of nonpolar and polar compounds to soils: processes, measurements and experience with the applicability of the modified OECD gudeline 106. Chemosphere 22: 285–304.
Wang, J., J. Hu and S. Zhang. 2010. Studies on the sorption of tetracycline onto clays and marine sediment from seawater. J. Colloid Interface Sci. 349(2): 578–82.
Wille, K., M. Claessens, K. RappÃ©, E. Monteyne, C.R. Janssen, H.F. De Brabander and L. Vanhaecke. 2011a. Rapid quantification of pharmaceuticals and pesticides in passive samplers using ultra high performance liquid chromatography coupled to high resolution mass spectrometry. J. Chromatogr. A 1218(51): 9162–9173.
Wille, K., J.A.L. Kiebooms, M. Claessens, K. Rappé, J. Vanden Bussche, H. Noppe, M. Van Praet, E. De Wulf, P. Van Caerter, R.C. Janssen, H.F. De Brabander and L. Vanhaecke. 2011b. Development of analytical strategies using U-HPLC-MS/MS and LC-ToF-MS for the quantification of micropollutants in marine organisms. Anal. Bioanal. Chem. 400: 1459–1472.
Wu, J., X. Qian, Z. Yang and L. Zhang. 2010. Study on the matrix effect in the determination of selected pharmaceutical residues in seawater by solid-phase extraction and ultra-high-performance liquid chromatography electrospray ionization low-energy collision-induced dissociation tandem mass spectrometry. J. Chromatogr. A 1217(9): 1471–1475.
Xu, W.-h., G. Zhang, S.-c., Zou, X.-d. Li and Y.-c. Liu. 2007. Determination of selected antibiotics in the Victoria Harbour and the Pearl River, South China using high-performance liquid chromatography-electrospray ionization tandem mass spectrometry. Environ. Poll. 145(3): 672–679.
Xu, W.H., G. Zhang, O.W.H. Wai, S.C. Zou and X.D. Li. 2009. Transport and asorption of antibiotics by marine sediments in a dynamic environment. Soil Sediment. 9: 364–373.

Yang, J.F., G.C. Ying, J.L. Zhao, R. Tao, H.C. Su and S.L. Lui. 2010. Simultaneous determination of four classes of antibiotics in sediments of the Pearl River using RRLC-MS/MS. Sci. Total. Environ. 408: 3424–3432.

Yang, Y., J. Fu, H. Peng, L. Hou, M. Liu and J.L. Zhou. 2011. Occurrence and phase distribution of selected pharmaceuticals in the Yangtze Estuary and its coastal zone. J. Hazard. Mater. 190(1-3): 588–596.

Zedda, M. and C. Zwiener. 2012. Is nontarget screening of emerging contaminants by LC-HRMS successful? A plea for compound libraries and computer tools. Anal. Bioanal. Chem. 403: 2493–2502.

Zhang, R., G. Zhang, Q. Zheng, J. Tang, Y. Chen, W. Xu, Y. Zou and X. Chen. 2012. Occurrence and risks of antibiotics in the Laizhou Bay, China: Impacts of river discharge. Ecotox. Environ. Safe. 80(0): 208–215.

Zhang, R., J. Tang, J. Li, Z. Cheng, C. Chaemfa, D. Liu, Q. Zheng, M. Song, C. Luo and G. Zhang. 2013a. Occurrence and risks of antibiotics in the coastal aquatic environment of the Yellow Sea, North China. Sci.Total Environ. 450â€“451(0): 197–204.

Zhang, R., J. Tang, J. Li, Q. Zheng, D. Liu, Y. Chen, Y. Zou, X. Chen, C. Luo and G. Zhang. 2013b. Antibiotics in the offshore waters of the Bohai Sea and the Yellow Sea in China: Occurrence, distribution and ecological risks. Environ. Poll. 174(0): 71–77.

Zhang, Y.B., Y. Li and X.L. Sun. 2011. Antibiotic resistance of bacteria isolated from shrimp hatcheries and cultural ponds on Donghai Island, China. Mar. Poll. Bull. 62(11): 2299–2307.

Zheng Q, R. Zhang, Y. Wang, X. Pan, J. Tang and G. Zhang. 2012. Occurrence and distribution of antibiotics in the Beibu Gulf, China: Impacts of river discharge and aquaculture activities. Mar. Environ. Res. 78(0): 26–33.

Zhou, L.J., Ying G.G., Zhao, J.L., Yang, J.F., Wang, L., Yang, B., L, S. 2011. Trends in the occurrence of human and veterinary antibiotics in the sediments of the yellow river, hai river and liao river in northen china. Environ. Poll. 159: 1877–1885.

Zhou, L.-J., G.-G. Ying, S. Liu, J.-L. Zhao, F. Chen, R.-Q. Zhang, F.-Q. Peng and Q.-Q. Zhang. 2012a. Simultaneous determination of human and veterinary antibiotics in various environmental matrices by rapid resolution liquid chromatography-electrospray ionization tandem mass spectrometry. J. Chromatogr. A 1244(0): 123–138.

Zhou, L.J., G.G. Ying, S. Liu, J.L. Zhao, F. Chen, R.Q. Zhang, F.Q. Peng and Q.Q. Zhang. 2012b. Simultaneous determination of human and veterinary antibiotics in various environmental matrices by rapid resolution liquid chromatography-electrospray ionization tandem mass spectrometry. J. Chromatogr. A 1244: 123–38.

14

Biological Effects of Pharmaceuticals in Marine Environment

Julián Blasco,[1,a,*] *Miriam Hampel,*[2] *Olivia Campana*[3] and *Ignacio Moreno-Garrido*[1,b]

ABSTRACT

Pharmaceuticals are present in water (river, estuaries, coastal waters) and sediment. The available information about their levels in the coastal ecosystems is scarce. In this chapter, a review is carried out about the ecotoxicity of pharmaceuticals focused on marine environment, in organisms from different trophic levels—from bacteria to fish. The biomarker approach and new omic technologies are considered as tools for environmental risk assessment of these compounds. In fact, the –omic technologies is a promising tool for evaluation of global molecular and cellular responses of an organisms to pharmaceuticals exposure. Nevertheless, their employment in marine organisms are scarce because of limited genomic resources.

Introduction

Although the contamination by pharmaceuticals in water (rivers, estuaries, and coastal ecosystems) has not been considered one of the main environmental problems at the global scale yet, the inputs of these compounds and their effect of non-target species need to be considered. The occurrence of pharmaceuticals in surface water is wide and variable depending on the compound and the environmental compartment. Their concentrations are in the range between ng L^{-1} to μg L^{-1} (Blasco and DelValls 2008). The entry of pharmaceuticals to the environment is mainly by regular domestic use, because after their consumption they are excreted and partially metabolized and released into the aquatic environment via wastewater environment. The Committee for Property Medicinal Products (CPMP) has summarized the scenarios

[1] Department of Ecology and Coastal Management, Instituto Ciencias Marinas de Andalucía (CSIC), Campus Río San Pedro, 11510 Puerto Real (Cádiz), Spain.

[a] Email: julian.blasco@csic.es

[b] Email: ignacio.moreno@icman.csic.es

[2] Dpt. Physical Chemistry, Universidad de Cádiz, Andalusian Center for Marine Science and Technology, CACYTMAR, Edificio Institutos de Investigación, Campus Universitario Puerto Real, 11510 Puerto Real (Cádiz), Spain. Email: miriam.hampel@uca.es

[3] Environment Department, University of York, York, YO10 5DD, UK. Email: olivia.campana@york.ac.uk

* Corresponding author

for inputs of pharmaceuticals in the environment. Nevertheless, the knowledge about the occurrence and levels in coastal ecosystems and sediments are scarce (Fang et al. 2012; Vazquez-Roig et al. 2012; Pintado-Herrera et al. 2013). Due to its pseudo-persistent character, they represent a growing concern about the negative effects on the wildlife.

Recently, as result of the Mistra Pharma Research project (2008–2015) whose objectives are to identify human pharmaceuticals which can impact aquatic ecosystems, several areas have been pointed out as relevant for research: (a) the antibiotic resistance, (b) identification of high risk substances, (c) to improve environmental classification systems, and (d) developing efficient technologies for depuration. Currently, three pharmaceuticals (17α-ethinylestradiol (EE2), 17 β-estradiol (E2) and Diclofenac) have been proposed as priority substances for the Water Framework Directive (WFD). In USA, the Environmental Protection Agency (EPA) has identified the pharmaceuticals as hazardous substances recently (USEPA 2009, 2013a, 2013b, 2013c). A wide-ranging consensus has been reached about the risk that occurrence of pharmaceuticals can exert on wildlife and even human health. Besides the mentioned research areas, the effect of pharmaceuticals mixtures and the joint action with other legacy and emergent pollutants should be taken in account to assess the environmental risk.

In summary, although the knowledge about the fate, behavior, and effects associated with the occurrence of pharmaceuticals in fresh, ground, seawater, soil, and sediments have improved in the last years, a lack of information remains in the scientific arena. In this chapter, a review of the available information about the ecotoxicity of pharmaceuticals focused on marine environment at different levels of hierarchical organization will be presented: First, the effect of pharmaceuticals at different trophic levels from bacteria to fish in water; second the occurrence and effect in sediment; third the use of biomarkers approach; and finally, the use of–omics techniques as a new and relevant approach to improve in-depth knowledge about the mechanisms affected by this type of pollution.

Toxicity of Pharmaceuticals in Marine and Estuarine Biota

In estuarine and coastal waters of developed countries and emergent economies, levels of pharmaceuticals substances seldom surpass the threshold of the ng L^{-1} concentrations. Nevertheless, those low concentrations do not mean that all pharmaceuticals will be innocuous (Carlsson et al. 2006; Fent et al. 2006; Cooper et al. 2008; Fang et al. 2012; Jiang et al. 2014). O'Brien and Keough (2014) recently stated, in a meta-analysis of experimental marine studies, that only 22% of the studies about toxicity on marine invertebrates used ecologically relevant concentrations of pollutants. In any case, as emergent pollutants, knowledge about the effects of pharmaceuticals on aquatic biota is scarce, and mechanisms of action as well as effect on specific physiological processes in non-target organisms are not well understood. Thus, high concentrations are often used in order to get information about such action mechanisms, and "negative" results (it means, effects caused by concentrations quite higher than expected environmental concentrations) published. Nevertheless, for some pharmaceuticals, which can act as endocrine disrupters (for instance), effective concentrations could be very low and this implies an important social concern (Waring and Harris 2005). Ferrari et al. (2004) summarized a good part of the existing data on toxicity results on aquatic organisms from different levels (from bacteria to fish) for a variety of pharmaceuticals (carbamezapine, diclofenac, ofloxacin, propanolol, and sulphamethoxazole), and stated an (expected) difference in sensitivity for chronic and acute toxicity tests, the first ones being more sensitive. Most of the tests found in the literature are acute, and very rarely chronic (Santos et al. 2010). Due to the high number of potential drugs and the many different organisms which could be non-specific targets, it would be necessary to predict which of them will enter the environment in non-negligible quantities (normally more than 10 ng L^{-1}, with some exceptions such as endocrine disruptors), and then to select the potentially hazardous to be assayed (Christen et al. 2010; Caldwell et al. 2014). But even though other pollutants seem to have more weight than pharmaceuticals in toxicity, the effect of these substances, over all, when present in cocktails, should not be underestimated (Damásio et al. 2011; Udovyk and Gilek 2014; Vasquez et al. 2014; Green et al. 2015). On the other hand, it seems that the scientific community tends to investigate substances which already have been identified in previous studies (Daughton 2014). Thus, toxicity of several emergent contaminantes, such as a wide list of marine-untested pharmaceuticals, should be checked.

In this section, a survey on the toxic effect of the most conspicuous pharmaceuticals on marine organisms of different structural complexity levels will be performed.

Bacteria

Most of data on marine bacteria toxicity of pharmaceuticals came from the use of the luminescent bacteria *Vibrio fischeri*, widely employed in the standardized test Microtox. In many cases, this bacterial species did not show high sensitivity (EC50 values higher than 100 mg L^{-1}) to some human pharmaceuticals such as ß-blockers metoprolol, propanolol, or nadolol (Maszkowska et al. 2014), as expected, due to the mechanisms of action of those drugs. Nunes et al. (2014) also calculated high EC50 values (92.2 mg L^{-1}) for paracetamol on these luminescent bacteria. Isidori et al. (2006) did not find toxicity of furosemide on this species at 200 mg L^{-1}. Schmidt et al. (2011) calculated EC50 values for gemfibrozil (57 mg L^{-1}) and diclofenac (28 mg L^{-1}) in the same organism. Nevertheless, Carballeira et al. (2012) recommend the use of this test (as a representant of marine bacteria) when measuring toxicity of effluents to coasts, combined with algal bioassays and other complementary toxicity tests involving animals (as minimum).

Studies on other marine bacteria are not so frequent. Peele et al. (1981) studied the effect of pharmaceutical wastes in a marine dump site, finding that there was an evident drifting in the natural marine bacterial population near those points. Toxicity of antibiotics on marine bacteria could be a topic of concern in the future, as those organisms seem to be able to acquire resistance to drugs (Bound et al. 2006; Liu and Wong 2013).

Marine and coastal bacterial degradation of some pharmaceutical compounds have been studied, and this process seems to be slower than suspected for some compounds (Benotti and Brownawell 2009), but other pharmaceuticlas could be metabolized or degraded very quickly, by both physical and biological reactions (Robinson and Hellou 2009).

There is a lack of studies on the effect of pharmaceuticals on cyanophyte marine species, and this taxonomic group could be sensitive to these pollutants. Populations of the freshwater cyanophyte *Anabaena* spp. were exposed to wastewater and demonstrated to be very sensitive to it, in contrast with cladocerans or *Vibrio fischerii* (Rosal et al. 2010). No specific toxicity bioassays on marine cyanophytes and pharmaceuticals have been found by the authors in the literature till the data.

Biofilms and periphyton

Biofilms formed in marine submerged surfaces is a key compartment of freshwater and coastal environments (Proia et al. 2013). It is formed by an assemblage of bacteria, fungus, and microalgae, which can also be colonized in later stages of development by other more complex organisms. Some bioassays have been designed in order to check toxicity of pollutants in this ecological compartment. One of these bioassays is described in Porsbring et al. (2007), and named as SWIFT test. Basically, it consists of disposable glass discs (1.5 cm^2 surface) mounted on polyetilene racks able to hold many of those discs in a submerged location, where natural periphyton communities establish after some time (around a week). After colonization, the discs are transferred to the lab and there exposed to the selected substances to be checked, in 300 mL containers shacked and exposed to continuous light, setting the temperature to that found in the sampling site. Medium is replaced every 24 hours for four days. The test medium is filtered seawater with nutrients added to reach 0.7 μM and 8 mM of phosphate and nitrate, respectively. After incubation, different parameters can be measured in the periphytic community, such as pigment profiles or taxonomic diversity. Using this technique, Porsbring et al. (2009) checked the effect of clotrimazole on the marine periphytic community and found that environmentally relevant concentrations of this fungicide (500 pMol L^{-1}, it means 172 ng L^{-1}) of clotrimazole increased the concentration of sterols in the periphytic community (the sterols metabolism is important in diatoms as crustaceans and nemathodes do not synthetize them *de novo*) and at the same concentration there is a decrease in chlorophyll contents. In any case, drifting of the microalgal periphytic populations can occur at smaller concentrations than those provoking pigment content decrease (Porsbring et al. 2007). Similar approach has been used by Johansson et al. (2014b), checking the effect of ciprofloxacin and sulfamethoxazole on periphyton attached to the same technique (SWIFT periphyton test system). For ciprofloxacin, total carbon source utilization EC10 and EC50 were 15 and 163 μg L^{-1}, respectively, and for sulfamethoxazole, EC10 and EC50 were 14 and 272 μg L^{-1}. The same authors (Johansson et al. 2014a) also checked the effect of triclosan in periphytic experimental communities in two different seasons (spring and summer), and they found that spring periphytic community was near eight fold more sensitive than the summer community (EC50 values of 11.4 and 86.7 μg L^{-1} triclosan for spring and summer, respectively). In these experiments the authors stated that algal periphytic communities were more sensitive than bacterial communities, as effect on attached algae were stated at 3.1 μg L^{-1} triclosan in spring and 9.5 μg L^{-1} triclosan in summer. A summary of toxicity values for different pharmaceuticals on bacteria and biofilms can be found in Table 1.

Algae

In general, algae are not quite sensitive to most pharmaceuticals, as target tissues or biochemical pathways able to be affected by human-employed pharmaceuticals are not present. For instance, EEC (2007) does not even include bioassays for endocrine disrupters in algae. Fibrates seem to have also low incidence in the physiology of microalgae (Isidori et al. 2007). Nevertheless, some type of pharmaceuticals, such as anti-bacterials, should be tested on macro and microalgae as chloroplasts could suffer the effect of those substances. Carballeira et al. (2012) assume that in a minimal set of bioassays developed in order to assess toxicity of effluents in marine media at least a microalgal species should be present, due to the importance of this biological compartment in marine trophic nets. Those authors recommend the use of microalgal species such as *Phaeodactylum tricornutum* or *Isochrysis galbana*.

Claessens et al. (2013) established the effect thresholds for many pharmaceutical compounds occurring in coastal Belgian waters on the marine diatom *Phaeodactylum tricornutum*, species more sensitive to toxicants than others, according to these authors, but in any case those authors stated that there is no risk for acute effects on this microalgal species for most of the pharmaceuticals tested (as EC50 values for salicylic acid, paracetamol, carbamezapine, atenolol, benzafibrate and trimethoprim were in the concentration range of mg L^{-1}), but in some locations such as harbors, levels of other pharmaceuticals (propanolol, for instance, with and EC50 of around 0.3 mg L^{-1} and EC10 values of around 90 μg L^{-1}) could have chronic effects on photosynthetic microbiota. Special attention should be taken with respect to effect of pharmaceutical mixtures (Claessens et al. 2013). Aguirre-Martínez et al. (2015) exposed populations of *Isochrysis galbana* to ibuprofen, carbamezapine, and novobiocin, finding EC50 concentrations to be 22.6 mg L^{-1}, 0.03 mg L^{-1}, and 14.8 mg L^{-1}, respectively (far away from environmentally realistic concentrations). Saçan and Balcıoğlu (2006) exposed populations of the chlorophyte *Dunaliella tertiolecta* to effluents from aluminium and pharmaceutical industries (producing analgesics and anti-inflammatory drugs) finding that low concentrations (it means, high dilutions) of the effluents in cultures could inhibit or increase cellular densities, but high concentrations of the effluents clearly inhibit cellular growth. This highlighted that the water effluent treatments were inefficient, as even being in the permitted range reported by the Water Pollution Control Act of the country where the research was performed, there were a deleterious effect on photosynthetic microbiota. Schmidt et al. (2011) found low toxicity of gemfibrozil and diclofenac to the marine diatom *Skeletonema costatum* (IC50 values of 56 and 5 mg L^{-1}, respectively, for both pharmaceuticals). Petersen et al. (2014) exposed *Skeletonema pseudocostatum* to combinations of pharmaceuticals and other potential pollutants, finding additive effects for all tested mixtures. Among the used pharmaceuticals, fluoxetine, triclosan and propanolol showed EC50 values of near 18 μg·L^{-1}, 27.5 μg L^{-1} and 237 μg L^{-1} respectively on this species. Nunes et al. (2005) measured toxicity of diazepam, clofibrate, and clofibric acid to *Tetraselmis chuii*, finding IC50 values of 16.5 mg L^{-1}, 39.7 mg L^{-1} and 316.2 mg L^{-1} respectively, all values far away from the environmentally relevant concentrations. Ferreira et al. (2007) stated IC50 values of 11.2 mg L^{-1} and 6.1 mg L^{-1} for oxytetracycline and florphenicol, respectively, on the same species (*T. chuii*). Seoane et al. (2014) also calculated EC50 values for chloramphenicol, florphenicol, and oxytetracycline of 11.16, 9.03, and 17.25 mg L^{-1}, respectively, on *Tetraselmis suecica*.

Marine sediments act as sink for many xenobiotics, and thus the idea of testing pharmaceuticals susceptible to be sediment accumulated on microbiota present in this biotope should be encouraged.

Table 1: Toxicity of pharmaceuticals on marine bacteria and biofilms (assemblage between algae and bacteria).

Organism	Pharmaceutical	Endpoint	Concentration	Reference
Vibrio fischerii (Bacteria)	Metoprolol	EC50 (Microtox)	> 100 $mg \cdot L^{-1}$	Maszkowska et al. 2014
Vibrio fischerii (Bacteria)	Propanolol	EC50 (Microtox)	> 100 $mg \cdot L^{-1}$	Maszkowska et al. 2014
Vibrio fischerii (Bacteria)	Nadolol	EC50 (Microtox)	> 100 $mg \cdot L^{-1}$	Maszkowska et al. 2014
Vibrio fischerii (Bacteria)	Paracetamol	EC50 (Microtox)	92.2 $mg \cdot L^{-1}$	Nunes et al. 2014
Vibrio fischerii (Bacteria)	Furosemide	EC50 (Microtox)	92.2 $mg \cdot L^{-1}$	Isidori et al. 2006
Vibrio fischerii (Bacteria)	Gemfibrozil	EC50 (Microtox)	57 $mg \cdot L^{-1}$	Schmidt et al. 2011
Vibrio fischerii (Bacteria)	Diclofenac	EC50 (Microtox)	28 $mg \cdot L^{-1}$	Schmidt et al. 2011
Biofilms (Bacteria+Algae)	Clotrimazole	SWIFT test, pigment content	127 $ng \cdot L^{-1}$	Porsbring et al. 2009
Biofilms (Bacteria+Algae)	Triclosan	SWIFT test, pigment content, spring/ summer	3.1 $\mu g \cdot L^{-1}$/9.5 $\mu g \cdot L^{-1}$	Johansson et al. 2014a
Biofilms (Bacteria+Algae)	Ciprofloxacin	SWIFT test, C utilization, EC10/EC50	15/163 $\mu g \cdot L^{-1}$	Johansson et al. 2014b
Biofilms (Bacteria+Algae)	Sulfamethoxazole	SWIFT test, C utilization, EC10/EC50	14/272 $\mu g \cdot L^{-1}$	Johansson et al. 2014b

Hagenbuch and Pinkney (2012) exposed two marine benthic diatoms (*Cylindrotheca closterium*, -formerly *Nitzschia closterium*- and *Navicula ramosissima*) to three antibiotics: tylosin, lincomycin, and ciprofloxacin, and to their mixtures. Single IC50 values for tylosin, lincomycin, and ciprofloxacin were 0.27, 14.16, and 55.43 mg L^{-1} respectively for *C. closterium* and 0.99, 11.08, and 72.12 mg L^{-1} respectively for *N. ramosissima*, all of these values quite higher than environmentally relevant concentrations. The authors stated that mixtures had a synergistic effect on *C. closterium* but additive for *N. ramosissima*, thus sustaining that monochemical bioassays were not useful in order to predict interactions in mixtures, and this assertion seems to be valid not only for microalgae (Alsop and Wood 2013).

Cytotoxic techniques (such as the Comet assay) related to pharmaceuticals have recently been applied to microalgal populations from freshwater species (Prado et al. 2015). Those techniques are still waiting to be applied in marine microalgae for those potential pollutants.

Few studies on toxicity of pharmaceuticals have been developed on macroalgae. Oskarsson et al. (2012) assayed the effect of three pharmaceuticals (propanolol, diclofenac, and ibuprofen) on the brown algae *Fucus vesiculosus*. Concentrations used were too high (up to 1000 µg L^{-1}), but even in this case, propanolol demonstrated to be more toxic to the algae than the other two drugs, which did not showed adverse effects. Nevertheless, measuring chlorophyll fluorescence, those authors stated that effect of propanolol was not only dose-dependent, but also time-dependent, which is very important in a species able to live for some years, such as taxons from the genus Fucus.

As explained, toxicity of many pharmaceuticals in marine algae or even bacteria is not expected, but they should be included in bioassays for pharmaceuticals-type in order to ensure safety for those important components of the biota. In the case of the invertebrates, on the other hand, effects began to be expected for some pharmaceuticals (Depledge and Billinghurst 1999; Oetken et al. 2004). A summary of toxicity values for different pharmaceuticals on algae can be found in Table 2.

Rotifers

Due to their characteristics, rotifers (Phylum Platyzoa) are an excellent material for toxicity bioassays as primary consumers (grazers): they are a key component of freshwater, estuarine, and coastal ecosystems, easy to culture (presenting a predominant parthenogenetic reproduction), with short life cycle, and sensitive to a high number of toxic substances (Dahms et al. 2011). Three decades ago, Nogrady and Keshmiriam (1986), detected effects on oviposition in the freshwater rotifer *Philodinia acutiformis* caused by the presence of cholinergic drugs. More recently, DellaGreca et al. (2004), found low sensitivity EC50 values in the range of in mg L^{-1} for prednisolone and dexametasone on estuarine *Brachionus calyciflorus*, but they stated that photoderivates of the drugs could be more toxic than the original drugs. This was also stated tamoxifen (DallaGreca et al. 2007). Rhee et al. (2013) studied the effect of different pharmaceuticals (acetaminophen, atenolol, carbamezapine, oxytetracycline, sulfamethoxazole, and trimethoprim) on acetylcholynesterase activity (AchE) in the rotifer *Brachionus koreanus*. Effects were found only for acetaminophen, carbamezapine, and trimethoprim at high concentrations. Oxytetracycline and sulfamethoxazole showed a slight decrease at the highest concentrations (1000 µg L^{-1}). Freshwater rotifers do not seem to be more sensitive to pharmaceuticals than the marine or estuarine species (Parrella et al. 2014). Swimming velocities of rotifers can be also used as a parameter suitable to be measured in toxicity tests (Chen et al. 2014; Chen and Guo 2015). As primary consumers, rotifers could be used as a suitable model for food mediated drugh transference, but literature on this topic is extremely scarace.

Polychaete

Due to their characteristics, marine polychaetes are important organisms to be taken into account in sediment toxicity tests. Species such as *Hediste diversicolor* (formerly *Nereis diversicolor*) or species from the genus *Capitella* have been used in sediment toxicity bioassays (Solé et al. 2009; Kalman et al. 2010), but few studies have been carried out on pharmaceuticals. Méndez et al. (2013) exposed populations of the detritivorous marine polychaete *Capitella teleta* to spiked sediments containing fluoxetine. Those authors did not find effects on egestion rate or body weight, but they stated that fluoxetine favoured the

Table 2: Toxicity of pharmaceuticals on marine algae.

Organism	Pharmaceutical	Endpoint	Concentration	Reference
P. tricornutum	Salycilic acid	Growth, EC10/EC50	96.7/255.5 $mg \cdot L^{-1}$	Claessens et al. 2013
P. tricornutum	Paracetamol	Growth, EC10/EC50	93.4/265.8 $mg \cdot L^{-1}$	Claessens et al. 2013
P. tricornutum	Carbamezapine	Growth, EC10/EC50	42.2/62.5 $mg \cdot L^{-1}$	Claessens et al. 2013
P. tricornutum	Atenolol	Growth, EC10/EC50	6.9/311.9 $mg \cdot L^{-1}$	Claessens et al. 2013
P. tricornutum	Propanolol	Growth, EC10/EC50	0.09/0.228 $mg \cdot L^{-1}$	Claessens et al. 2013
P. tricornutum	Bezafibrate	Growth, EC10/EC50	> Water solubility	Claessens et al. 2013
P. tricornutum	Timethroprim	Growth, EC10/EC50	2.4/5.1 $mg \cdot L^{-1}$	Claessens et al. 2013
Isochrysis galbana	Ibuprofen	Growth, EC50	22.6 $mg \cdot L^{-1}$	Aguirre-Martínez et al. 2015
Isochrysis galbana	Carbamezapine	Growth, EC50	0.03 $mg \cdot L^{-1}$	Aguirre-Martínez et al. 2015
Isochrysis galbana	Novobiocin	Growth, EC50	14.8 $mg \cdot L^{-1}$	Aguirre-Martínez et al. 2015
Skeletonema costatum	Gemfibrozil	Growth, EC50	56 $mg \cdot L^{-1}$	Schmidt et al. 2011
Skeletonema costatum	Diclofenac	Growth, EC50	5 $mg \cdot L^{-1}$	Schmidt et al. 2011
Skeletonema pseudocostatum	Fluoxetine	Growth, EC50	18 $\mu g \cdot L^{-1}$	Petersen et al. 2014
Skeletonema pseudocostatum	Triclosan	Growth, EC50	27.5 $\mu g \cdot L^{-1}$	Petersen et al. 2014
Skeletonema pseudocostatum	Propanolol	Growth, EC50	237 $\mu g \cdot L^{-1}$	Petersen et al. 2014
Tetraselmis chuii	Diazepam	Growth, EC50	16.5 $mg \cdot L^{-1}$	Nunes et al. 2005
Tetraselmis chuii	Clofibrate	Growth, EC50	39.7 $mg \cdot L^{-1}$	Nunes et al. 2005
Tetraselmis chuii	Clofibric acid	Growth, EC50	316.2 $mg \cdot L^{-1}$	Nunes et al. 2005
Tetraselmis chuii	Oxytetracycline	Growth, EC50	11.2 $mg \cdot L^{-1}$	Ferreira et al. 2007
Tetraselmis chuii	Florefenicol	Growth, EC50	6.1 $mg \cdot L^{-1}$	Ferreira et al. 2007
Tetraselmis suecica	Chloramphenicol	Growth, EC50	11.16 $mg \cdot L^{-1}$	Seoane et al. 2014
Tetraselmis suecica	Florphenicol	Growth, EC50	9.03 $mg \cdot L^{-1}$	Seoane et al. 2014
Tetraselmis suecica	Oxytetracycline	Growth, EC50	17.25 $mg \cdot L^{-1}$	Seoane et al. 2014
Nitzschia closterium	Tylosin	Growth, EC50	0.27 $mg \cdot L^{-1}$	Hagenbuch and Pinkney 2012
Nitzschia closterium	Lincomycin	Growth, EC50	14.16 $mg \cdot L^{-1}$	Hagenbuch and Pinkney 2012
Nitzschia closterium	Ciprofloxacin	Growth, EC50	55.43 $mg \cdot L^{-1}$	Hagenbuch and Pinkney 2012
Navicula ramosissima	Tylosin	Growth, EC50	0.99 $mg \cdot L^{-1}$	Hagenbuch and Pinkney 2012
Navicula ramosissima	Lincomycin	Growth, EC50	11.08 $mg \cdot L^{-1}$	Hagenbuch and Pinkney 2012
Navicula ramosissima	Ciprofloxacin	Growth, EC50	72.12 $mg \cdot L^{-1}$	Hagenbuch and Pinkney 2012

occurrence of males with abnormal genital spines, and this could have important reproductive implications. Maranho et al. (2014a) exposed *Hediste diversicolor* individuals to sediments experimentally spiked with five different pharmaceuticals (carbamezapine, ibuprofen, fluoxetine, 17α-ethinylestradiol, and propanolol) for two weeks, finding low sensitivity of this polychaete to most of the substances exposed, but in some cases, non-linear dose-responses were also found for certain pharmaceuticals. For instance, carbamezapine activated the GST enzyme at concentrations as low as 0.05 μg kg^{-1} and at high concentrations (50 and 100 μg kg^{-1}) but the response of the enzyme at 0.5 and 5 μg kg^{-1} was not different from controls. On the other hand, carbamezapine activated the activity of AChE at 0.5 μg kg^{-1}, but not at lower (0.05 μg kg^{-1}) or higher (5, 50 or 500 μg kg^{-1}) concentrations. The same trend could be observed for this pharmaceutical respect to parameters such as DNA damage, although in this case differences were not statistically significant.

Toxicity values for different pharmaceuticals on rotifers and polychaetes are summarized in Table 3.

Molluscs

Control of reproduction in some molluscs could be mediated by steroids (Croll and Wang 2007), and thus endocrine disruption in those organisms should be expected when exposed some pharmaceuticals (Ketata et al. 2008). There is still an open debate about whether vertebrate sex steroids can be used by molluscs as reproductive hormones or not, although it is now clear that some substances (not pharmaceuticals), such as TBT, clearly act as endocrine disruptors for this animal group (Scott 2013).

In relation with other types of pharmaceuticals, Gaume et al. (2012) stated toxicity of the antibiotics triclosan and its derivate methyl triclosan on hemocytes and cultures gill cells of the European abalone (*Haliotis tuberculata*) finding toxicity at 2.3 mg L^{-1} for triclosan and 1.2 mg L^{-1} for the methyl derivate of this drug. Minguez et al. (2014) exposed the same species to four antidepressants (amitriptyline, clomipramine, citalopram, and paroxetine), also finding LC50 levels of mg L^{-1}. In certain cases those authors also found non-linear dose-dependent response, for instance for the ROS production in haemocytes. Petridis et al. (2009), measuring genotoxicity by a comet assay, stated that xeno-estrogens resulted to be genotoxic to *Scrobicularia plana* but at high concentrations: 100 ng L^{-1} for 17ß-oestradiol and 1 μg L^{-1} for ethinylestradiol. Wessel et al. (2007) observed that 0.5 μg L^{-1} of 17α-ethinylestradiol had no a clear embryotoxic effect on *Crassostrea gigas*, but ten times less concentrations (50 ng L^{-1}) of the same substance increased the concentration of vitelogenin in males of *Saccostrea glomerata*, and provoke the occurrence of intersex individuals. Concentrations of 1 μg L^{-1} delayed the spermatogenesis in males of this species (Andrew et al. 2008). In an experiment focused to detect changes in immunological parameters of *Crassostrea gigas*, Luna-Acosta et al. (2012) exposed populations of this oyster to a mix of herbicides (diuron and isoproturon) with ibuprofen. They stated that toxicity of the diuron was very high even at 1 μg L^{-1}, but no additive effect of ibuprofen to a mixture of the two herbicides could be assessed. Matozzo et al. (2012) also studied the effect of ibuprofen on the immune parameters of the clam *Ruditapes philippinarum*. After seven days exposure, these authors found no differences in haemocyte diameter or volume even at 1000 μg L^{-1}, but a decrease of the density of haemocytes per mL of hemolimph and an increase of lactate dehydrogenase activity only at the highest concentration (1000 μg L^{-1}). An increase of haemocyte proliferation was also observed at 500 μg L^{-1} of ibuprofen. On the same organism, Munari et al. (2012) studied the effects of seawater acidification on the activity of diclofenac (at low concentrations: 0.05 and 0.5 μg L^{-1}) on immune parameters of this clams. In those experiments, effect of acidification (lowered up to pH 7.4) on biomarkers was too high to establish clear effects due to the anti-inflammatory drug. *Ruditapes philippinarum* was also used by Trombini et al. (2012) as subject on bioassays studying the effects of low concentrations (15 μg L^{-1}) of carbamezapine, diclofenac, and ibuprofen, stating that there was a time-dependent response of a battery of biomarkers related with oxidative stress when the organisms were exposed even to low concentrations of those drugs. Maranho et al. (2015b) also alert about the seasonality on the oxidative stress biomarkers response of this clam (*R. philippinarum*) to different pollutants, varying the response from some type of pollutants to other in function of summer or winter sampled individuals. Sellami et al. (2015) checked the effect of permethrin in mixture with other pollutants on shell components of the clam *Venerupis decussata*, finding that after exposure to this drug, there was a transition phase in the shell from aragonite to calcite, indicating thus a loss of the shell structure.

Table 3: Toxicity values for different pharmaceuticals on rotifers and polychaetes.

Organism	Pharmaceutical	Endpoint	Concentration	Reference
Brachionus calyciflorus (Rotifera)	Tamoxifen	EC50	0.97 $mg \cdot L^{-1}$	DallaGreca et al. 2004
Brachionus koreanus (Rotifera)	Acetaminophen	Effect on AchE	100 $\mu g \cdot L^{-1}$	Rhee et al. 2013
Brachionus koreanus (Rotifera)	Atenolol	Effect on AchE	> 1000 $\mu g \cdot L^{-1}$	Rhee et al. 2013
Brachionus koreanus (Rotifera)	Carbamezapine	Effect on AchE	100 $\mu g \cdot L^{-1}$	Rhee et al. 2013
Brachionus koreanus (Rotifera)	Oxytetracyline	Effect on AchE	1000 $\mu g \cdot L^{-1}$	Rhee et al. 2013
Brachionus koreanus (Rotifera)	Sulfamethoxazole	Effect on AchE	1000 $\mu g \cdot L^{-1}$	Rhee et al. 2013
Brachionus koreanus (Rotifera)	Timethoprim	Effect on AchE	100 $\mu g \cdot L^{-1}$	Rhee et al. 2013
Capitella teleta (Polychaeta)	Fluoxetine	Effect on feeding activity	> 3.3 $\mu g \cdot g^{-1}$ (sediment)	Méndez et al. 2013
Hediste diversicolor (Polychaeta)	Carbamezapine	Significant mortality, 14 d	50 $ng \cdot g^{-1}$ (sediment)	Maranho et al. 2014a
Hediste diversicolor (Polychaeta)	Carbamezapine	Effect on AchE	0.5 $ng \cdot g^{-1}$ (sediment)	Maranho et al. 2014a
Hediste diversicolor (Polychaeta)	Ibuprofen	Significant mortality, 14 d	> 500 $ng \cdot g^{-1}$ (sediment)	Maranho et al. 2014a
Hediste diversicolor (Polychaeta)	Fluoxetine	Significant mortality, 14 d	1 $ng \cdot g^{-1}$ (sediment)	Maranho et al. 2014a
Hediste diversicolor (Polychaeta)	17α-ethynylestradiol	Significant mortality, 14 d	10 $ng \cdot g^{-1}$ (sediment)	Maranho et al. 2014a

Mussels have been widely used for monitoring contamination, and they demonstrated an excellent material for toxicity testing (Klosterhaus et al. 2013; Álvarez-Muñoz et al. 2015; Franzellitti et al. 2014; Binelli et al. 2015; Kasiotis et al. 2015). This bivalve, *Mytilus galloprovincialis*, has been used in laboratory studies (Gomez et al. 2012) as well as in *in situ* experiments (de los Ríos et al. 2012). The last type of tests reduces repeatability of bioassays, but greatly increases environmental relevance of the results.

Cortez et al. (2012) exposed the mussel *Perna perna* to different concentrations of triclosan. They stated that while fertilization assays and embryo-larval development assays rend low sensitivity (IC50 values of 0.490 and 0.135 mg L^{-1}, respectively), the bioassays based in the neutral red retention time (NRRT) as a measurement of the physiological stress (test based in the premise that only lysosomes of healthy cells take and retain this dye), found differences respect to the control at levels of ng L^{-1}, it means, realistic, environmentally relevant concentrations. The freshwater mussel *Lampsilis siliquoidea* has been also used in toxicity testing moxifloxacin (Gilroy et al. 2014; Leonard et al. 2014). Halem et al. (2014) demonstrated endocrine disruption of the Atlantic ribbed mussel *Geukensia demissa* exposed to urban wastewaters, and found a correlation between low oxygen and endocrine disruption, hypothesizing that anaerobiosis facilitate the allosteric inhibition of the cytochrome P450 aromatase, which aids the conversion of testosterone in estradiol.

Among all genera of mussels, *Mytilus* has been one of the most used, and literature on effect and accumulation of pharmaceuticals on these organisms can be widely found (Widdows et al. 2002), in laboratory as well in field experiments (McEneff et al. 2014). Zenker et al. (2014) pointed out the importance of studies about bioaccumulation and biomagnifications potential of pharmaceuticals in aquatic biota, and provided a list of drugs susceptible to be accumulated by aquatic organisms. In general, studies on biomagnification of those pollutants are still scarce, but some data recommend to not obviate the possibility of occurrence for this processes (Du et al. 2015).

Canty et al. (2009) found an inverse relation between clearance rate (capacity of filtering food from water) of individuals of the species *Mytilus edulis* and genetic damage produced by different concentrations of the anti-cancer cyclophosphamide. Schmidt et al. (2011) injected gemfibrozil and diclofenac into the posterior adductor muscle of individuals of the species *Mytilus* and measured the response taking biomarkers as endpoints (lipid peroxidation, DNA damage, activity of antioxidant enzymes), stating that response of the organisms was evident at lower doses than those causing toxicity for other organisms (marine bacteria *Vibrio fischeri*, marine microalgae *Skeletonema costatum* and marine copepod *Tisbe battagliai*), but exposition route is clearly different and results should not be compared. Gonzalez-Rey and Bebianno (2011) exposed populations of *Mytilus galloprovincialis* to realistic concentrations of ibuprofen (250 ng L^{-1}) for two weeks, renewing the media (and re-establishing ibuprofen concentrations) each 48 h. These authors found clear differences on most of toxicity biomarkers studied after seven days incubation, but the response was not so clear after 15 days except for catalase activity and glutation reductase activity. The same authors (Gonzalez-Rey and Bebianno 2014) studied the condition index (ratio between soft tissue weight and total weight of each individual), antioxidant enzymes, and lipid peroxidation of *M. galloprovincialis* populations exposed to diclofenac in the laboratory at near environmentally relevant concentrations (250 ng L^{-1}) and stated that CAT activity and lipid peroxidation were good biomarkers in digestive glands for this pharmaceutical, while SOD and GR activities were useful as biomarkers in gills. Besides, Franzellitti et al. (2013) found an increase of serotonin levels and some specific messenger RNAs in the same organism after seven days exposure to very low, environmental concentrations (0.3 ng L^{-1}) of fluoxetine and propanolol.

Seasonality is a factor to be considered in molluscs toxicity bioassay (Moreno-González et al. 2014). Ciocan et al. (2010) exposed populations of *Mytilus edulis* to different endocrine disrupters (17ß-estradiol, ethinyl estradiol, and estradiol benzoate) at environmentally relevant concentrations (5 to 200 ng L^{-1}) and they found that the effect on the vielogenin concentration and production of messenger RNA *ER2* greatly depends on the maturity stage of the individuals: exposed mature mussels did not show any apparent difference with controls, while individuals at early stages of gametogenesis showed a significant increase in those two parameters. Schmidt et al. (2013) also found great variations in the biomarkers expression in *Mytilus* spp. as a function of seasonality, alerting about possible mistakes in toxicity bioassays due to this aspect.

Gonzalez-Rey et al. (2014) exposed populations of *Mytilus galloprovincialis* to mixtures of pharmaceuticals (diclofenac, ibuprofen, and fluoxetine) with a stressor toxic metal (copper). Mixtures (with or without Cu) demonstrated to have a xenostrogenic effect on *M. galloprovincialis* (which was not observed in single-pharmaceutical exposition bioassays).

Toxicity bioassays on embryos of *Mytilus galloprovincialis* have been also adapted to pharmaceuticals (Fabbri et al. 2014), but after exposition to different concentrations of ibuprofen, diclofenac, or bezafibrate, exposition concentrations of 1000 μg L^{-1} did not reach 50% inhibition. In this experiment, only for diclofenac, data of lower effective concentration (LOEC) was in the range of environmentally realistic values (0.01 μg L^{-1}), but authors reported a non-monotonic response, as higher diclofenac concentrations did not provoke responses different from controls. Ericson et al. (2010) exposed populations of *Mytilus edulis* to different concentrations of diclofenac, ibuprofen, and propanolol, as persistent model pharmaceuticals (Robinson and Hellou 2009). Those authors found lower sizes and lower byssus strength in organisms exposed to the highest concentrations of the selected pharmaceuticals and stated the capacity of *M. edulis* to accumulate these substances at concentrations higher than the surrounding media. Gomez et al. (2012) also stated the high capacity of *Mytilus galloprovincialis* to accumulate diazepam and tetrazepam. Kookana et al. (2013) also found high accumulation capacities for pharmaceuticals on this organism exposed to triclosan and methyl-triclosan (in the laboratory) or field exposed in cages receiving metropolitan waste waters. This is important, as accumulation of xenobiotics is a topic of concern for marine organisms (Brackers de Hugo et al. 2013). Martinez-Bueno et al. (2013) stated the accumulation capacity of this mussel for carbamezapine and oxcarbamezapine, demonstrated the suitability of this species as biomonitors.

Behavior bioassays have been recently developed for other molluscs. Fong and Molnar (2013) described an increase of foot detachment in different marine snails from both the Pacific and Atlantic coasts of North America when exposed to μg L^{-1} levels of anti-depressants, finding higher sensitivities in trochids and turbinids than in the muricids species. Fong et al. (2105) also stated that antidepressants such as venlafaxine or fluoxetine: first one increased crawling and second one slowed it, both processes precluding foot detachment of those organisms from submerged surfaces.

Higher evolved invertebrates could be most exposed to human pharmaceuticals due to the complexicity of their nervous systems. Di Poi et al. (2013) studied the perinatal exposure to fluoxetine on the memory processing of the cuttlefish *Sepia officinalis*, finding that exposition to this anti-depressant presented a non-unimodal dose-dependent response in the behavior of the cuttlefish: very low concentrations (1 ng L^{-1}) altered the efficiency of learning of young cuttlefish, but exposition to higher concentrations (100 ng L^{-1}) did not provoke different behavior from the controls. Authors stated that very low antidepressant concentrations in marine waters could lead to cephalopods to inappropriate predatory behaviors. Some aspects of this work have been vividly discussed in the recent literature (Franzellitti et al. 2013; Di Poi and Bellanger 2014; Sumpter and Margiotta-Casaluci 2014), but the work has an indubitable value as the first approach to effects of pharmaceuticals on memory of a taxonomic group probably holding the most advanced brain in invertebrates.

A summary of toxicity values for different pharmaceuticals on molluscs can be found in Table 4.

Crustaceans

In freshwater environments, the cladocerans *Daphnia* spp. and *Ceriodaphnia* spp. have been widely used in toxicity tests, sometimes focused on pharmaceuticals (Kar and Roy 2010) and results of those bioassays have been extrapolated to other organisms. In marine environments, cladocerans were not used to be considered as key organisms and few efforts have been done in order to find a marine equivalent to daphnia, with the exception of *Moina salina* (Anderson-Carnahan 1994). In marine environments most of the bioassays have been developed on copepods, anostraceans, and amphipods, and some of them on decapods.

Moore and Stevenson (1991) observed higher intersex occurrence in the marine copepod *Paraphiascella hyperborea* in populations living in contaminated coastal areas. *Tisbe battagliai* does not seem to be very sensitive to gemfibrozil or diclofenac, as LC50 values of 128 and 16 mg L^{-1}, respectively for both pharmaceuticals, have been calculated (Schmidt et al. 2011). Andersen et al. (1999) exposed copepod populations (*Acartia tonsa*) to potential endocrine disruptors finding estrogenic effects at

Table 4: Toxicity of pharmaceuticals on marine molluscs.

Organism	Pharmaceutical	Endpoint	Concentration	Reference
Haliotis tuberculata	Triclosan	Toxicity on haemocytes	2.3 $mg \cdot L^{-1}$	Gaume et al. 2012
Haliotis tuberculata	Methyl triclosan	Toxicity on haemocytes	1.2 $mg \cdot L^{-1}$	Gaume et al. 2012
Haliotis tuberculata	Amytriptyline	Toxicity on haemocytes, LD50	45.24 $mg \cdot L^{-1}$	Minguez et al. 2014
Haliotis tuberculata	Clomipramine	Toxicity on haemocytes, LD50	4.76 $mg \cdot L^{-1}$	Minguez et al. 2014
Haliotis tuberculata	Citalopram	Toxicity on haemocytes, LD50	32.94 $mg \cdot L^{-1}$	Minguez et al. 2014
Haliotis tuberculata	Paroxetine	Toxicity on haemocytes, LD50	4.16 $mg \cdot L^{-1}$	Minguez et al. 2014
Scrobicularia plana	17ß-oestradiol	Comet assay	100 $ng \cdot L^{-1}$	Petridis et al. 2009
Scrobicularia plana	ethynylestradiol	Comet assay	1 $\mu g \cdot L^{-1}$	Petridis et al. 2009
Crassostrea gigas	17α-ethilylestradiol	Embryotoxicity	> 0.5 $\mu g \cdot L^{-1}$	Wessel et al. 2007
Saccostrea glomerata	17α-ethilylestradiol	Increase of vitelogenin in males	50 $ng \cdot L^{-1}$	Andrew et al. 2008
Ruditapes philippinarum	Ibuprofen	Increase of haemocyte proliferation	500 $mg \cdot L^{-1}$	Matozzo et al. 2012
Ruditapes philippinarum	Carbamezapine	Immune parameters	15 $\mu g \cdot L^{-1}$	Trombini et al. 2012
Ruditapes philippinarum	Diclofenac	Immune parameters	15 $\mu g \cdot L^{-1}$	Trombini et al. 2012
Ruditapes philippinarum	Ibuprofen	Immune parameters	15 $\mu g \cdot L^{-1}$	Trombini et al. 2012
Venerupis decussate	Permethrin	Shell chemical structure	100 $\mu g \cdot L^{-1}$	Sellami et al. 2015
Perna Perna	Tricolsan	Fertilization assay	0.490 $mg \cdot L^{-1}$	Cortez et al. 2012
Perna Perna		Embryotoxicity	0.135 $mg \cdot L^{-1}$	Cortez et al. 2012
Perna Perna		Neutral red retention time (NRRT) 24 h	120 $ng \cdot L^{-1}$	Cortez et al. 2012
Mytilus galloprovincialis	Diclofenac	Antioxidant enzymes	250 $ng \cdot L^{-1}$	Gonzalez-Rey and Bebianno 2011
Mytilus galloprovincialis	Fluoxetine	Specific mRNA	0.3 $ng \cdot L^{-1}$	Fanzellitti et al. 2013
Mytilus galloprovincialis	Propanolol	Specific mRNA	0.3 $ng \cdot L^{-1}$	Fanzellitti et al. 2013
Mytilus galloprovincialis	Ibuprofen	Embryotoxicity, LC50	> 1$g \cdot L^{-1}$	Fabbri et al. 2014
Mytilus galloprovincialis	Diclofenac	Embryotoxicity, LC50	> 1$g \cdot L^{-1}$	Fabbri et al. 2014
Mytilus galloprovincialis	Bezafibrate	Embryotoxicity, LC50	> 1$g \cdot L^{-1}$	Fabbri et al. 2014
Mytilus edulis	17ß-estradiol	Vitelogenin levels, mature mussels	> 200 $ng \cdot L^{-1}$	Ciocan et al. 2010
Mytilus edulis	17ß-estradiol	Vitelogenin levels, immature mussels	5 $ng \cdot L^{-1}$	Ciocan et al. 2010
Mytilus edulis	Diclofenac	Less strength in byssus	10 $mg \cdot L^{-1}$	Ericson et al. 2010
Mytilus edulis	Ibuprofen	Less strength in byssus	10 $mg \cdot L^{-1}$	Ericson et al. 2010
Mytilus edulis	Propanolol	Less strength in byssus	10 $mg \cdot L^{-1}$	Ericson et al. 2010
Sepia officinalis	Fluoxetine	Ability of learning	1 $ng \cdot L^{-1}$	Di Poi et al. 2013

23 μg L^{-1} of 17ß-estradiol and 20 μg L^{-1} of bisphenol, but they did not find effects for 2,3, dichlorophenol. Hutchinson et al. 1999, assayed a battery of natural and synthetic steroids on the copepod *Tisbe battalgliai* and found low effects (not observed effect concentrations (NOECs) of 100 μg L^{-1}) for 17ß-estradiol and, 17α-ethinylestradiol, using 20-hydroxyecdisone as a positive control. This hormone in often used as reference for its endocrine disruption effect on crustaceans (Rodríguez et al. 2007).

Behavioral bioassays (respect to swimming characteristics) have been also performed on copepods related to certain pollutants or effluents (Michalec et al. 2013), but no specific references have been found by the authors related to those organisms and pharmaceuticals.

Nunes et al. (2005) measured toxicity of diazepam, clofibrate, and clofibric acid to anostracean *Artemia parthenogenetica*, finding LC50 values of 12.2 mg L^{-1}, 36.6 mg L^{-1}, and 87.22 mg L^{-1} respectively. Measurement of the cholinesterase activity seems to be more sensitive for this species and the same pharmaceuticals (Nunes et al. 2006). Ferreira et al. (2007), testing mortality of the same species (*A. parthenogenetica*) stated LD50 values of more than 800 mg L^{-1} for oxytetracycline, and they did not find toxicity for florfenicol on this crustacean.

With respect to amphipods, Lee and Arnold (1983) stated that more than 1% of waste water affects survival of the marine species *Amphitoe valida*, although no analysis of the components of the wastewater was performed. Vanderbergh et al. (2003) found that 17α-ethinylestradiol had a clear effect on the second generation of the freshwater species *Hyalella azteca*, provoking histological aberrations, hermaphroditism, and other effects at concentrations as low as 0.1 μg L^{-1}. Ford et al. (2006) found an increase of intersex occurrence in *Echinogammarus marinus* exposed to effluents with industrial potential pollutants. Guler and Ford (2010) exposed populations of the same species to selected concentrations of fluoxetine, carbamezapine, and diclofenac, finding that exposed individuals (although without a clear response-dose distribution) spent more time exposed to light than control individuals. The same effect is also caused by acanthocephalan parasites. On the other hand, intersex individuals were also more susceptible to suffer parasitic infections (Ford et al. 2006), probably risking in this other way the non-intersex populations. The enhancement of parasitic infections due to potential weakness caused by pollutants, including some pharmaceuticals, has been reviewed by Morley (2009). Bossus et al. (2014), also researching on *Echinogammarus marinus*, exposed populations of this amphipod to selected concentrations of fluoxetine and sertraline, finding higher differences respect to controls in swimming velocity after 1 hour exposed to 0.01 μg·L^{-1} sertraline or after 1 day exposition to 0.1 μg·L^{-1} fluoxetine plus 0.01 μg·L^{-1} sertraline, but again there were not linear dose-response results. Neuparth et al. (2014) exposed populations of the amphipod *Gammarus locustus* to simvastatin, finding a significant decrease in survival at 8 μg L^{-1}, a decrease in the average of ovigerous females at 1.6 μg L^{-1}, and a dramatic decrease of number of newborns per female at 320 ng L^{-1}, a concentration below the 630 ng L^{-1} predicted in the environment. This drug, a hypocholesterolaemic, seems to deeply affect the cholesterol metabolism of this crustacean.

Cirripeds have been also used in order to tests the effect of pharmaceuticals. Five anti-histaminics (loratadine, antazoline, clemastine, azelastine, and triprolidine) have demonstrated the ability to alter the capacity of colonizing submerged surfaces of the cirriped *Amphybalanus Amphitrite*, as stated by Jin et al. (2014) after checking thirty-five potential anti-fouling pharmaceuticals. The authors pointed out that the process of attaching larvae to surfaces has a close relationship with neurotransmitters.

In relation to decapods, González-Ortegón et al. (2013) exposed the marine shrimp *Palaemon serratus* to low concentrations of diclofenac, clofibric acid, and clotrimazole, checking the effect of those pharmaceuticals on growth, development, and body mass of larval stages at different salinities and temperatures. The two first pharmaceuticals did not show effect at the assayed concentrations, but clotrimazole demonstrated to be toxic at the higher concentration used (2.78 μg L^{-1}) at seawater salinity (32 PSU), but also at the lower concentration (0.14 μg L^{-1}) when larvae were reared at less salinity (20 PSU). Thus, authors stated that co-stressing factors could enhance toxicity of certain pharmaceuticals. The same authors, still working on *P. serratus*, did not find toxicity of diclofenac or clofibric acid in the larval development of this shrimp, but stated that clotrimazole, in addition to food limitation, had a negative effect on larval development even at low concentrations (0.07 μg L^{-1}) (González-Ortegón et al. 2015).

Carcinus maenas have been used in ecotoxicology for a long time (Brian 2005; Ricciardi et al. 2010; Rodrigues and Pardal 2014). This crustacean showed impairment of osmoregulatory capacity when exposed to realistic concentrations of diclofenac (Eades and Waring 2010). Chronic effects of ibuprofen, carbamezapine, and novobiocin were also stated for this crab by Aguirre-Martínez et al. (2013a,b), who measured a battery of oxidative stress biomarkers. In those experiments, neutral red retention time assay (NRRT) also demonstrated to be a very a sensitive tool in order to measure toxicity of ibuprofen, finding differences with respect to the control at low (5 μg L^{-1}) concentrations. For carbamezapine and novobiocin, other biomarkers such as EROD, GST, or GPX seem to be more sensitive than NRRT, revealing differences respect to the controls at drug concentrations of 10 μg L^{-1}.

The role of pharmaceutical pollution in intersex occurrence in crustaceans is not still clear (Stentiford 2012), as many factors can cause this phenomenon (parasitism, among others). Development and study of genomic biomarkers could help in this field (Ford 2012). A recent revision on the biological effects of antidepressants in crustaceans and molluscs can be found in Fong and Ford (2014).

A summary of toxicity values for different pharmaceuticals on crustaceans can be found in Table 5.

Echinoderms

Echinoderms are more related to chordates than any other invertebrate groups (except Ascidians, as they actually *are* chordates), and thus this invertebrate group should be more sensitive to drugs designed to act on humans than others (Lavado et al. 2006). Carballeira et al. (2012) recommend the use of sea urchin larval development toxicity tests (on species such as *Paracentrotus lividus* or *Arbacia lixula*) as one of the minimum set of bioassays necessary to establish toxicity in coastal effluents.

Aguirre-Martínez et al. (2015), working on embryo-larval development of *Paracentrotus lividus*, found EC50 values for ibuprofen, carbamezapin, and novobiocin of 0.01 μg L^{-1}, 0.01 μg L^{-1}, and 1 mg L^{-1}, respectively. Ribeiro et al. (2015), by their part, in the same type of bioassay on the same organism, found abnormalities in embryos at concentrations of 12.5, 5, 0.64, and 2.0 μg L^{-1} for diclofenac, propanolol, sertraline, and simvastatin, respectively. Lavado et al. (2006) found that individuals of the crinoid *Antedon mediterranea* exposed to 1 μg L^{-1} of 17α-ethinylestradiol for two weeks increased their levels of testosterone, and they also observed an increasing trend (although non-statistically demonstrated) of estradiol levels when the organisms were exposed to concentrations as low as 10 ng L^{-1}. But, as in many other cases, there was not a clear dose-response relationship, finding non-monomodal response curves: 10 ng L^{-1} of 17α-ethinylestradiol provoked a more acute response than 1000 ng L^{-1}.

Righting time (time employed in turn right and individual disposed upside-down) is a classic sub-lethal assay for echinoderms. Individuals of the species *Asterias rubens* increased their righting time when exposed to 18 mg L^{-1} of cyclophosphamide, and it has been stated that there is a relation between this parameter and genotoxic effects of this substance (Canty et al. 2009).

Ascidians

Tunicates are invertebrate belonging to the Phylum Chordata, a group closely related with vertebrates and thus more susceptible to be affected by pharmaceuticals. In spite of this, few references can be found in the literature about pharmaceutical potential toxicity for those organisms. Akers (1969) stated that low concentrations (10 μg L^{-1}) of epinephrine and norepinefrine increase the contraction rate of the tunicate *Ciona intestinalis*, also stating that α-blockers and ß-blockers interfere with the effect of those two neurotransmitter hormones. Matozzo et al. (2014) checked the effect of ibuprofen on immune parameters of the ascidian *Botryllus schlosseri*, finding a low sensitivity of this organism to this pharmaceutical (effects were found just at concentrations as high as 1000 μg L^{-1}).

Fish

As vertebrates, fish could be susceptible to be affected by pharmaceuticals (Trudeau et al. 2005). For this group, steroids are important hormones in sex differentiation control (Chang et al. 1999). Tangtian et al. (2012) found liver vitelogenin induction in the marine medaka *Oryzias melastigma* after 35 days exposed to 0.01 mg L^{-1} of benzotriazole, stating that this drug could act as a potential estrogen. Estuarine

Table 5: Toxicity of pharmaceuticals on marine crustaceans.

Organism	Pharmaceutical	Endpoint	Concentration	Reference
Tisbe battagliai	Gemfibrozil	LC50	128 mg·L^{-1}	Schmidt et al. 2011
Tisbe battagliai	Diclofenac	LC50	16 mg·L^{-1}	Schmidt et al. 2011
Tisbe battagliai	17ß-estradiol	NOEC	100 µg·L^{-1}	Hutchinson et al. 1999
Tisbe battagliai	17α-ethinylestradiol	NOEC	100 µg·L^{-1}	Hutchinson et al. 1999
Acartia tonsa	17ß-estradiol	Endocrine disruption	23 µg·L^{-1}	Andersen et al. 1999
Acartia tonsa	bisphenol	Endocrine disruption	20 µg·L^{-1}	Andersen et al. 1999
Artemia parthenogenetica	Diazepam	LC50	12.2 mg·L^{-1}	Nunes et al. 2005
Artemia parthenogenetica	Clofibrate	LC50	36.6 mg·L^{-1}	Nunes et al. 2005
Artemia parthenogenetica	Clofibric acid	LC50	87.22 mg·L^{-1}	Nunes et al. 2005
Artemia parthenogenetica	Oxytetracycline	LC50	> 800 mg·L^{-1}	Ferreira et al. 2007
Artemia parthenogenetica	Florfenicol	LC50	> 1400 mg·L^{-1}	Ferreira et al. 2007
Hyalella azteca	17α-ethinylestradiol	Histological aberrations	0.1 µg·L^{-1}	Vanderbergh et al. 2003
Echinogammarus marinus	Fluoxetine	Swimming velocity, 1 hour exposition	0.01 µg·L^{-1}	Bossus et al. 2014
Echinogammarus marinus	Sertraline	Swimming velocity, 1 day exposition	0.1 µg·L^{-1}	Bossus et al. 2014
Gammarus locustus	Simvastatin	Affect survival	8 µg·L^{-1}	Neuparth et al. 2014
Gammarus locustus	Simvastatin	Percentage of ovigerous females	1.6 µg·L^{-1}	Neuparth et al. 2014
Gammarus locustus	Simvastatin	Decrease of newborns per female	320 ng·L^{-1}	Neuparth et al. 2014
Amphybalanus Amphitrite	Loratadine	Larval settlement EC50%	< 1.6 µg·L^{-1}	Jin et al. 2014
Amphybalanus Amphitrite	Antazoline	Larval settlement EC50%	< 1.6 µg·L^{-1}	Jin et al. 2014
Amphybalanus Amphitrite	Clemastine	Larval settlement EC50%	< 1.6 µg·L^{-1}	Jin et al. 2014
Amphybalanus Amphitrite	Azelastine	Larval settlement EC50%	< 1.6 µg·L^{-1}	Jin et al. 2014
Amphybalanus Amphitrite	Triprolidine	Larval settlement EC50%	< 1.6 µg·L^{-1}	Jin et al. 2014
Palaemon serratus	Diclofenac	Larval development	> 1600 µg·L^{-1}	González-Ortegón et al. 2013
Palaemon serratus	Clofibric acid	Larval development	> 1000 µg·L^{-1}	González-Ortegón et al. 2013
Palaemon serratus	Clotrimazole	Larval development	2.78 µg·L^{-1}	González-Ortegón et al. 2013
Carcinus maenas	Diclofenac	Osmoregulation	10 ng·L^{-1}	Eades and Waring 2010
Carcinus maenas	Ibuprofen	Neutral Red Retention Assay	5 µg·L^{-1}	Aguirre-Martínez et al. 2013a
Carcinus maenas	Carbamezapine	Oxidative stress biomarkers	10 µg·L^{-1}	Aguirre-Martínez et al. 2013b
Carcinus maenas	Novobiocin	Oxidative stress biomarkers	10 µg·L^{-1}	Aguirre-Martínez et al. 2013 b

sand goby (*Pomatoschistus minutus*) was induced to express vitelogenin (egg protein in male individuals and thus an impairment in reproduction by very low concentrations (11 ng L^{-1}) of 17α-ethinylestradiol (Humble et al. 2013)).

With respect to other pharmaceuticals, Nunes et al. (2005) measured toxicity of diazepam, clofibrate, and clofibric acid to *Gambusia holbrooki*, a fish able to colonize estuarine and hypersaline environments, finding LC50 values of 12.7 mg L^{-1}, 7.7 mg L^{-1}, and 526.5 mg L^{-1} respectively. There are clear signs about the species-dependent sensitivity of fish to pharmaceuticals. Ellesat et al. (2011) compared the sensitivity of hepatocytes from three marine fishes: plaice (*Pleuronectes platessa*), long rough dab (*Hippoglossoides platessoides*) and Atlantic cod (*Gadus morhua*) to statins, finding that hepatocytes from cods were more resistant to this drug than hepatocytes from the other two flat fish.

Behavioural bioassays are becoming more and more important in recent years, as their responses are more sensitive than classical LD50 bioassays, and authors can infer ecological consequences from those bioassays. Pollutants which could be considered as not dangerous in classical (based on death) bioassays could be actually harmful for the populations if some behaviors are altered (as positive/negative phototaxis, aggression, courtship or other reproductive behaviors) (Brooks et al. 2003; Gaworecki and Klaine 2008; Brooks 2014). Behavior studies in freshwater are more developed than in marine environments. Colman et al. (2009) stated that concentrations as low as 50 ng L^{-1} of 17α-ethinylestradiol inhibited the aggressive behavior of male freshwater zebrafish (*Danio rerio*) and even lower concentrations (0.5 ng L^{-1}) diminished the courtship behavior of males of the same species, but Brodin et al. (2013) observed that the freshwater fish species *Perca fluviatilis* became more aggressive when exposed to oxacepam. Antidepressants such as fluoxetine and fluoxamine seem to be inhibitors of key enzymatic activities in the synthesis of androgens in male fish (Fernandes et al. 2011). Brandão et al. (2013) studied the oxidative stress and scototaxis (dark-light preference, Maximino et al. 2010) of freshwater fish *Lepomis gibbosus* exposed to carbamezapine, finding no effects in biomarkers or behavior at the concentrations used.

A summary of toxicity from different pharmaceuticals on echinoderms, ascidians, and fish can be found in Table 6.

Big fish, marine birds, and marine mammals

Due to the nature of the animals involved, few studies have been performed on pharmaceutical toxicity on birds, big fish, and marine mammals, but those animals, due to their longer lives, are able to suffer chronic effects of pharmaceuticals.

A study on occurrence of pharmaceuticals in bull sharks (*Carcharinus leucas*) inhabiting the mouths of rivers in Florida (this is one of the few species able to occupy estuarine environments) was performed by Gelsleichter and Szabo (2013). Blood was sampled from captured individuals and which were then returned alive to the media after blood sampling. Few samples contained measurable concentrations of pharmaceuticals, but in some years of sampling, eight different pharmaceuticals were found in the plasma of at least one individual in certain locations, although detection rates varied with the year of capture.

Contaminants of diverse types have been found in eggs of marine birds in the Wadden Sea (Laane et al. 2013), but in general there is a lack of data about occurrence and effects of pharmaceuticals in this type of organisms, although some few data can be found on other pollutants (Morales et al. 2012).

Occurrence of triclosan (which can suffer photoderivation to chlorophenols) has been reported on plasma in Atlantic bottlenose dolphins (*Tursiops truncatus*) (Fair et al. 2009). Toxicity experiments have been also performed on fibroblasts and blood cells of the same species (Frenzilli et al. 2014). Perhaps this would be a good approach to investigate effects of certain pharmaceuticals (with evident limitations) on such a difficult material as marine mammals. Other pollutants (no pharmaceuticals), such as organochlorines, have been measured in the liver of minke whales (*Balaenoptera acutorostrata*) captured by the Japanese Whale Research Program (Niimi et al. 2014). Data on pollutants such as mercury can be also found on whales, as in some countries this marine mammals are still hunted and consumed by humans, and mercury biomagnifications is human-concerning. No data about pharmaceuticals in whales have been found by these authors (perhaps it would be possible to establish mechanisms for sampling and measuring potential

Table 6: Toxicity of pharmaceuticals on marine echinoderms, ascidians, and fishes.

Organism	**Pharmaceutical**	**Endpoint**	**Concentration**	**Reference**
Paracentrotus lividus (Echinoderma)	Ibuprofen	EC50	0.01 $\mu g \cdot L^{-1}$	Aguirre-Martínez et al. 2015
Paracentrotus lividus (Echinoderma)	Carbamezapin	EC50	0.01 $\mu g \cdot L^{-1}$	Aguirre-Martínez et al. 2015
Paracentrotus lividus (Echinoderma)	Novobiocin	EC50	1 $mg \cdot L^{-1}$	Aguirre-Martínez et al. 2015
Paracentrotus lividus (Echinoderma)	Diclofenac	Embryos abnormalities	12.5 $\mu g \cdot L^{-1}$	Ribeiro et al. 2015
Paracentrotus lividus (Echinoderma)	Propanolol	Embryos abnormalities	5 $\mu g \cdot L^{-1}$	Ribeiro et al. 2015
Paracentrotus lividus (Echinoderma)	Sertraline	Embryos abnormalities	0.64 $\mu g \cdot L^{-1}$	Ribeiro et al. 2015
Paracentrotus lividus (Echinoderma)	Simvastatin	Embryos abnormalities	2 $\mu g \cdot L^{-1}$	Ribeiro et al. 2015
Antedon mediterranea (Echinoderma)	17α-ethinylestradiol	Increased levels of testosterone	1 $\mu g \cdot L^{-1}$	Lavado et al. 2006
Antedon mediterranea (Echinoderma)	17α-ethinylestradiol	Slight increase in levels of estradiol	10 $ng \cdot L^{-1}$	Lavado et al. 2006
Asterias rubens (Echinoderma)	Cyclophosphamide	Righting time assay	18 $mg \cdot L^{-1}$	Canty et al. 2009
Ciona intestinalis (Tunicata)	Epinefrine	Contraction rate	10 $\mu g \cdot L^{-1}$	Akers 1969
Ciona intestinalis (Tunicata)	Norepinefrine	Contraction rate	10 $\mu g \cdot L^{-1}$	Akers 1969
Botryllus schlosseri (Tunicata)	Ibuprofen	Immune parameters	1000 $\mu g \cdot L^{-1}$	Matozzo et al. 2014
Oryzias melastigma (Fish)	Benzotriazole	Vitelogenin induction	0.01 $mg \cdot L^{-1}$	Tangtian et al. 2012
Pomatoschistus minutus (Fish)	17α-ethinylestradiol	Vitelogenin induction	11 $ng \cdot L^{-1}$	Humble et al. 2013
Gambusia holbrooki (Fish)	Diazepam	LD50	12.7 $mg \cdot L^{-1}$	Nunes et al. 2005
Gambusia holbrooki (Fish)	Clofibrate	LD50	7.7 $mg \cdot L^{-1}$	Nunes et al. 2005
Gambusia holbrooki (Fish)	Clofibric acid	LD50	526.5 $mg \cdot L^{-1}$	Nunes et al. 2005

pharmaceuticals only in stranded individuals). Effect of plasticizers on thyroid hormones in marine mammals is just beginning to be understood (Mathieu-Denoncourt et al. 2015).

Sediment Toxicity

An understanding of the partitioning of pharmaceuticals between water and sediment is critical for the prediction of their fate in the environment and consequently, of their bioavailability to marine organisms. Unfortunately, current knowledge on the partitioning of pharmaceuticals is very limited and just few studies (Klosterhaus et al. 2013; Lai et al. 2000; Maranho et al. 2014a, 2015a) investigated the response of the organisms to sediment-associated pharmaceuticals in marine environments.

Most pharmaceuticals are hydrophilic and biodegradable and have short half-lives. However, some of them with particularly high octanol-water (log K_{ow}) partition coefficient may show high affinity to particulate matter, as sediment, sludge, or soil. K_{ow} partition coefficient is frequently used to indicate the tendency of the compounds to partition organic matter; the greater the coefficient value is, the greater its tendency to associate to particulate matter.

For examples, carbamazepine (CBZ) with a log K_{ow} of 2.48 it is not supposed to adsorbed to organic matter resulting in less than 10% of CBZ being removed in standard wastewater treatment processes (Zhang et al. 2008). Klosterhaus et al. (2013) investigated the partitioning of CBZ among surface waters, sediment, and in the tissues of the marine mussel *Geukensia demissa* in San Francisco Bay. CBZ was detected in surface waters (concentration range: 5.2–44.2 ng L^{-1}) and in the bivalve tissue (concentration range: 1.3–5.3 ng g^{-1}) but, despite its resistance to degradation in the environment - with a half-life on the order of 100 days in surface water -, CBZ was not detected in sediment demonstrating its low affinity to particulate matter.

Other compounds as gemfibrozil (Khan and Ongerth 2002), diclofenac (Avdeef et al. 2000), ibuprofen (Khan and Ongerth 2002) and many estrogens such as estrone, 17β-estradiol, and 17α-ethinylestradiol (Ying et al. 2002) have log K_{ow} higher than 3.0 and are expected to be significantly adsorbed onto sediment. Nevertheless, persistence and biodegradability are other important factors that affect the kinetics of sorption of these compounds to sediment. For example, Buser et al. (1998) observed that in lake water fortified with diclofenac and exposed to natural sunlight, this compound was subjected to rapid photo-degradation with a first-order kinetics and half-life of less than 1 hour, showing a negligible adsorption onto sediment particles. However, at higher salinity levels, removal of pharmaceuticals from aqueous phase by particulate matter might occur. Lai et al. (2000) demonstrated that even a slight increase of salinity levels (from 0.04 to 3.5) in sediments with low-moderate total organic carbon content (2.0 percentage) increases the sorption of estrogens to sediment particles. They suggested that increased removal of estrogens from water was due to aggregation and flocculation in the higher ionic strength medium. Clearly, this is a typical condition of estuarine and marine areas, where it is more likely that estrogens sorbed to suspended or dissolved organic material are deposited in sediments.

To date, information about the adverse effects of pharmaceuticals bound to sediments in marine environments is almost nonexisting.

In a recent study Mendez et al. (2013) investigated the effect of fluoxetine on the feeding activity (through egestion rates) and growth (body weight) of adult individuals of the detritivorous marine polychaete *Capitella teleta* to spiked sediments containing 0.001 to 3.3 μg g^{-1} of fluoxetine. After 18 days of exposure they did not observed any significant effect on egestion rates, growth rates, or size-specific egestion rates but, at the highest concentration, male individuals developed abnormal genital spines. This effect could have important reproductive consequences in natural environments because it could possibly be causing a fertilization impediment.

Two of more recent and comprehensive studies were conducted by Maranho et al. (2014b, 2015a). They evaluated the biological response of the polychaete *Hediste diversicolor* and the amphipod *Ampelisca brevicornis* to different pharmaceuticals bound to marine sediments applying a battery of biomarkers to assess bioavailability, oxidative stress, genotoxicity, and neurotoxicity. Bioavailability was assessed analyzing the activities of the biotransformation enzymes of the phase I (ethoxyresorufin O-deethylase—EROD and dibenzylfluorescein—DBF) and phase II (glutathione S-transferase—GST) of the metabolism, which are activated to transform the parent compound in an easier extractable one. The impact of the oxidative stress was evaluated by the lipid peroxidation (LPO) and glutathione peroxidase (GPx) and reductase (GR) activities. Finally, neurotoxicity and genetic damage were determined analyzing the acetylcholinesterase activity (AChE) and DNA strand breaks, respectively.

Carbamazepine-spiked sediments did not elicit any significant induction on the enzymes of the phase I and II of the metabolism in *A. brevicornis* and in *H. diversicolor* only GST showed a significant increase at environmental concentrations (0.05 ng g^{-1}) and at highest exposures (50, 500 ng g^{-1}). The antioxidant system was activated by CBZ both in amphipods and polychaetes and a concentration of 0.5 ng g^{-1} resulted as neurotoxic for both organisms. On the other hand, this compound did not cause any DNA damage in both species.

In polychaetes exposed to ibuprofen (IBP)-spiked sediments phase I and II enzymes were not induced; conversely, environmental concentrations of ibuprofen (0.05, 0.5 ng g^{-1}) were metabolized by phase I and phase II metabolism in the amphipod—showing a significant increase of DBF and the antioxidant enzyme activity of GPx, which resulted in an increase of the lipid peroxidation activity. In *H. diversicolor* the oxidative stress did not seem to be reduced by the antioxidant system but IBP exposure concentrations did lead to an increase of the LPO and genotoxic damage and showed a neurotoxic damage at the highest exposure (500 ng g^{-1}). In *A. brevicornis* IBP activated the antioxidant system only at the lowest exposure (0.05 ng g^{-1}) without any sign of genotoxic or neurotoxic damage.

In polychaetes exposed to fluoxetine (FX)-spiked sediment, FX was detoxified by the phase I (EROD) enzymatic activity. Among antioxidant enzymes only GR activity showed a significant activation after exposure to 1 ng g^{-1} but LPO activity was significantly higher than the control only at the highest exposure (100 ng g^{-1}). In *A. brevicornis* FX caused a significant decrease of phase I and II metabolism activities but it did not elicit any clear antioxidant response in the amphipod. FX sediment exposure determined the same response in both species inducing a significant decrease of DNA strand breaks at the whole concentration range (0.01 to 100 ng g^{-1}) and leading to a neurotoxic damage at the highest concentrations (10, 100 ng g^{-1}).

The 17α-ethinylestradiol (EE2) bound to sediments clearly induced different responses in the species exposed. In the amphipods, the detoxification metabolism and the antioxidant system were significantly affected by a decrease of the EROD and GST activities and GPx and LPO activities, respectively. The exposure to EE2 also resulted in a significant decrease of DNA strand breaks at the lowest environmental concentrations (0.01, 0.1 ng g^{-1}) but did not produce any neurotoxic response. In *H. diversicolor*, EE2 was detoxified by the phase I of the metabolism (EROD activity) but phase II was not activated. Among antioxidant enzymes only GR activity was induced at 0.1 ng g^{-1} leading to a significant increase of LPO activity only at the lowest exposure (0.01 ng g^{-1}). As for the amphipods, genotoxic damage was significantly reduced in exposed polychaetes compared to the control but they did show increased neurotoxic response.

Propranolol (PRO) bioaccumulated by *H. diversicolor* was significantly degradated by the phase I of the metabolism (EROD); on the other hand, DBF and GST activities were significantly inhibited against almost all concentration range (0.05 to 500 ng g^{-1}). PRO did not induce any antioxidant response, except an increase of the LPO at environmental concentrations (0.05 to 5 ng g^{-1}), nor caused any DNA damage or was neurotoxic to polychaetes. Likewise, PRO was significantly detoxified by phase I metabolism (EROD and DBF activities) in *A. brevicornis* but did not induce any response in the phase II metabolism. A clear antioxidant response was observed in amphipods exposed at all concentrations (GR and LPO activities). As for the polychaetes, PRO did not cause any DNA damage in *A. brevicornis* and did not lead to a clear neurotoxic response.

Maranho et al. (2015a) also evaluated the biological response to caffeine (CAF)-spiked sediments in the amphipod *A. brevicornis* applying the same battery of biomarkers. In this case, CAF exposure didn't produce a clear induction of phase I and II system nor of the antioxidant system but resulted in membrane impairment (LPO) in amphipods. Any DNA damage was observed in organisms exposed but CAF was neurotoxic at the lowest environmental concentration (0.15 ng g^{-1}).

Based on these laboratory studies, pharmaceuticals products associated to sediments at concentrations currently found in the marine environment are sufficient to cause a wide variety of effects to marine organisms.

Biomarkers

The term biomarker is defined as changes in biological responses—ranging from molecular through cellular, and from physiological to behavioural changes—that can be related to exposure to, or toxic effects of, environmental chemicals (van der Oost et al. 2005). Biomarker responses have been employed in biomonitoring programs (Solé et al. 2009) and as tools for initial screening as early warning of toxic chemical effects on organisms (Cajaraville et al. 2000; Martín-Díaz et al. 2004). And their use in environmental risk assessment of pharmaceuticals should be considered (Blasco and DelValls 2008).

The pharmaceutical effects are well known in mammalians and analogous metabolic pathway effects can be reported on aquatic organisms which present similar mode of actions (MoA) (Boxall et al. 2012). Lysosome membrane stability (LMS) is a general indicator de cellular well-being in crabs (*Carcinus maenas*) and the exposure for 28 days at ecological relevant concentrations for four pharmaceuticals (caffeine—15 μg L^{-1}, carbamezapine—1 μg L^{-1}, ibuprofen—5 μg L^{-1}, and novobiocin—0.1 μg L^{-1}) reduced neutral retention time (NRR) after 28 days. In fact, LMS has been considered as a sensitive tool for assessing exposure to concentrations of selected drugs as screening biomarker (Aguirre-Martínez et al. 2013b). Several biomarkers (ethoxyresorufin-O-deethylse (EROD), dibenzylfluorescein dealkylase (DBF), glutathione-S-transferase (GST), glutathione peroxidase (GPX), lipid peroxidation (LPO) and DNA damage) in a toxicity test study with caffeine and ibuprofen in the range 0.1–50 μg L^{-1} in gills, hepatopancreas, and muscles were analysed; the exposure of these products activated biotransformation enzymes, altered the oxidation state of the cells, provoked oxidative stress and was associated with DNA damage (Aguirre-Martínez et al. 2013a). Chemotherapeutic agents (methotrexate-MTX and tamoxifen-TMF) have been reported in environmental compartments; a risk assessment for both compounds was carried out according to the EMEA guidelines (EMEA 2006) and two-tier approach proposed by Viarengo (2007) for Tier III. The exposure of *Ruditapes philippinarum* to both chemicals for 14 days in a range 0.1–50 μg L^{-1} showed significant effects for clams in digestive gland with induction of EROD, GST, GPX, GR, and LPO levels (Aguirre-Martínez 2014). Ibuprofen (IBF) is a pharmaceutical widely distributed which alters the antioxidant system in bivalves (Martín-Díaz et al. 2009) and induces structural damage in membrane integrity (González-Rey and Bebianno 2011, 2012). Carbamezapine (CBZ) on Mediterranean mussels affected haemocyte membrane lysosome stability, increased neutral lipid and lipofucsin accumulation in digestive gland and lipid peroxidation; in agreement with human MoA resulted in a significant reduction in cAMP levels and lowered the mRNA expression of genes encoding three different MXR-related transporters (Franzellitti et al. 2013). In the clam, *Ruditapes philippinarum,* the exposure to CBZ for 28 days in the range 0–9 μg L^{-1} showed different responses depending on the threshold concentration, thus concentrations higher than 3.0 μg L^{-1} provoked a toxic effect, affecting biotransformation and antioxidant enzymes, although more significant effects were observed at 9.0 μg L^{-1} (Almeida et al. 2015), Diclofenac is another non-steroidal anti-inflammatory drug (NSAID) that is reported in surface waters (Blasco and DelValls 2008) and it has been included in the list of priority substances under the Water Framework Directive (EU 2012a, 2012b). Its effect is related to increasing in reactive oxygen species production (ROS) (González-Rey and Bebianno 2014). The study about the effect of carbamezapine (CBZ) on the marine algal species *Dunaliella tertiolecta* at concentrations (1–200 mg L^{-1}) showed an increase of lipid peroxidation products (Tsiaka et al. 2013). The IBU exposure to the oyster *Crassostrea gigas* showed a significant upregulation of *CYP2AU1, CYP356, CYP3071A1, GST-ω-like, GST-π-like, COX- like*, and *FABP-like*, which is indicative of increased transcription of certain genes and modifications in the antioxidant and auxiliary enzymes and to cause damages in the exposed organisms. The lipid lowering agent gemfibrozil and NSAID, diclofenac were injected to mussel *Mytilus* spp. induced stress biomarkers (GST and metallothionein) and biomarkers of damage (lipid peroxidation and DNA damage) and alkyl labile phosphate assay indicating the potential oxidative stress. LPO were induced by DCF after 96 h indicating tissue damage (Schmidt et al. 2011). *Hediste diversicolor* was exposed to sediment spiked with CBZ, IBF, and FX, and sublethal responses were examined and authors recommended the use of a battery of biomarkers, using enzyme assay in association with other biomarkes as LMS and lipofucsin accumulation (Maranho et al. 2014a). The interactive effect of environmental variables (temperature—16 and 24°C) and occurrence of antibiotic oxytetracycine (OTC) on cellular and molecular parameters in the mussel *Mytilus galloprovincialis* was assessed (Banni et al. 2015). The results showed a clear interaction between both stressors, increasing LMS and MDA accumulation and reducing hsp27 gene expression and suggest an increase in the risk due to temperature in contaminated waters.

Although realistic scenarios involve mixture toxicity of pharmaceuticals, the information available for mixture of these compounds is limited. González-Rey et al. (2014) exposed mussels to a mixture of three compounds (ibuprofen, diclofenac, and fluoxetine) using a multiobiomarker approach reporting that antioxidant enzyme activities and LPO levels were tissue and time-specific dependent. In *Carassius auratus*, the exposure to caffeine and sulfamethoxazlole produced similar effects for pharmaceuticals individually and as a mixture: EROD, SOD, and GST activities were increased. The joint effect of caffeine/sulfamethoxazole was additive with regard to AChE and GST activity variation and was antagonistic with regard to EROD and SOD induction (Li et al. 2012). The freshwater mussel, *Dreissena polyphorma* was exposed to three NSAIDs (ibuprofen, diclofenac, and paracetamol) and the multibiomarker approach was employed to identify the cytogenotoxic effects and the imbalance of the oxidative stress. The individual treatment induced significant celular stress. Moreover, the mixture provoked a significant enhacement of DNA fragmentation which preluded genetic damage (Parolini and Binelli 2011). Vasquez et al. (2014) pointed out some concerning issues about the design of mixture toxicity, e.g., taking in account the MoA, thus sinergystic or antagonystic effect can be expected depending on similarity or dissamilirity of MoA, occurrence of other pollutants in the environment; key metabolic pathways and biomarkers should be considered. In fact, the new –omics technologies have a great potential for characterization of new biomarkers. Although downstream of WWTP can represent a realistic scenario for mixture of pharmaceuticals compounds, changes in biomarkers can not be related to pharmaceutical compounds in a clear way, because it is difficult to asssociate these changes only to pharmaceuticals, because a cocktail of contaminants are present in surface water (Jasinska et al. 2015) and there is a lack of specific biomarkers for these compounds.

Omic Techniques for Assessing Pharmaceutical Effects

Biological responses to toxicants can range from the subcellular to the population scales and extend beyond single species to multiple trophic levels. In the environmental risk assessment, the goal of many current research initiatives is to bridge the gap between molecular initiating events (such as receptor binding) and resulting adverse outcomes at higher and more complex levels of biological organization, particularly those that determine individual survival and reproduction, population growth and abundance, and aquatic community structure and function (Brander et al. 2015). Moreover, molecular initiating events are the first response to the challenge and the ability to measure them provides an early warning tool for the detection of environmental hazard.

The term "omic techniques" indicates "a totality of some sort". In biology, it is used for very large-scale data collection and analysis, and can be divided in three main categories: genomics, proteomics, and metabolomics/metabonomics (Gehlenborg et al. 2010). Omic profiling is the measurement of the activity or expression of thousands of genes (transcriptomics), proteins (proteomics), or metabolites (metabolomics/metabonomics) at the same time to create a global picture of cellular function. By measuring these expression levels, omics allows researchers to study the detailed mode of action of compounds (Dulin et al. 2013).

Apart from many other external and internal stimuli that produce alteration in gene, protein and/or metabolite expression, omic technologies offer the possibility to assess the effects of potential toxic compounds (including pharmaceuticals) on the overall cell functioning. One major promise of these techniques is that they will further increase our knowledge about toxic mechanisms of actions on the basis of which the hazard and potential risk of a compound can be assessed. As the effects on gene expression alterations precede clinical effects, mechanisms of action can be detected after short exposure times of animals and thus allow the prediction of expected higher level effects resulting in a reduction of both test animals and suffering (Currie 2012; Hartung et al. 2012). Thus, toxicogenomics integrates the expression analysis of thousands of genes, proteins, or metabolites with classical methods in toxicology, as all physiological and pathological changes of the exposed organism are related to different types of toxicity at the molecular level.

Transcriptomics

Generally, transcriptomics is the most frequently used omics technique in environmental toxicology, the predominant platform being cDNA microarrays or oligo arrays (Hartung and McBride 2011). Transcription levels of single genes have been measured frequently by quantitative reverse transcription PCR. This technique however is based on informed guesswork of expected effects and does not provide information on the interaction of a huge number of genes at the time nor is it useful for the detection of unexpected and unknown processes and pathways. A DNA microarray is a collection of microscopic DNA spots attached to a solid surface. Microarray techniques require that the targets, mRNA molecules, have to be transcribed to copy DNA (cDNA) that bind to the probes spotted on the chip. An important drawback of microarray techniques though is that obtained results are limited by the features of interest (number and nature) spotted onto the array, and the possibility of "loosing" interesting responses in unexpected features. This is now being overcome by new promising techniques to measure gene expression such as next generation and RNA sequencing tools (NGS and RNAseq). These techniques produce thousands or millions of sequences at once. RNAseq is probably to substitute microarray based transcriptomics in the future when this technique will become more and more affordable. Although RNAseq is still a technology under active development, it offers several key advantages over existing technologies. Most important is the fact that, unlike hybridization based approaches such as cDNA microarrays, RNAseq is not limited to detecting transcripts that correspond to existing genome sequences. This makes RNAseq particularly attractive for non-model organisms with genomic sequences that are yet to be determined (Wang et al. 2009).

Proteomics

Proteomics is a bundle of techniques that attempt to separate, quantify, and identify many hundreds of proteins in a sample simultaneously. Proteomics deals with cell and tissue-wide protein expression. Thus, while genomics is mainly concerned with gene expression, proteomics analyses the protein products of the genes. Two-dimensional polyacrylamide gel electrophoresis (2D-PAGE) and mass spectrometry (MS) has been used since the late 1960s as a powerful tool for protein separation and identification, respectively, and to compare protein expression between controls and diseased or chemically-exposed biological samples (Martyniuk et al. 2012). The most significant innovation for proteomics was coupling mass spectrometry with 2D gels as a way to identify differentially expressed proteins (Lilley et al. 2001; Beranova-Giorgianni 2003; Herbert et al. 2001). The introduction of Differential in Gel Electrophoresis (DIGE) methods has increased the standard for this proteomics approach, allowing co-separation of a control set of proteins with proteins isolated from a treatment or disease (Tonge et al. 2001). The separation in the first dimension allows resolution of proteins that differ by a single positive or negative charge making this method ideal for distinguishing protein isoforms and proteins that are post-translationally modified, in addition to quantifying protein expression changes. More recently, gel independent techniques are being employed based on mass spectrometry with differently labeled protein fragments. In this context, protein quantification through incorporation of stable isotopes has become a central technology in modern proteomics research. Quantification by iTRAQ (**i**sobaric **t**ags for **r**elative and **a**bsolute **q**uantification) is one of several techniques in toxicology to monitor relative changes in protein abundance across perturbed biological systems (Ross et al. 2004). Application of iTRAQ method enables comparing of up to four different samples in one mass spectrometry-based experiment and avoids the much less sensitive in gel separation of proteins.

Metabolomics

The metabolites, or small molecules, within a cell, tissue, organ, biological fluid, or entire organism constitute the metabolome (Miller 2007). Metabolomics aims to measure the global, dynamic metabolic response of living systems to biological stimuli by identifying and quantifying all the small molecules in a biological system (cell, tissue, organ, and organism). Most commonly used metabolomics techniques are liquid chromatography–mass spectrometry (LC-MS), gas chromatography–mass spectrometry (GC-MS), and nuclear magnetic resonance (NMR) (Nicholson and Lindon 2008). Though metabolomics is viewed as complementary to other omic techniques it may actually provide a solution to the many shortcomings that are encountered with the other omic methods (Griffin and Bollard 2004; Bilello 2005; van Ravenzwaay et al. 2007). Measurements of changes in gene expression and protein production are subjected to a variety of homeostatic controls and feedback mechanisms; these changes are amplified at the level of metabolome. This may result in metabolomics being a more sensitive indicator of the external stress than other omic technologies (Nicholson et al. 1999; Ankley et al. 2006; van Ravenzwaay et al. 2007). Although methods have been developed to detect changes in genomic, transcriptomic, and proteomic profiles, the basic information required to make meaningful interpretations based on these data are sometimes not readily available (Ankley et al. 2006). Alternatively, the structure and function of most metabolites is fairly well characterized and may be lower in number than genes and proteins (van Ravenzwaay et al. 2007).

Thus, omic data offers a great potential for shedding more light onto the mechanisms triggered by the exposure to environmental pollutants (among many others), however, often the huge amount of data generated is not fully exploited. Recent advances in 'omic' technologies have created unprecedented opportunities for biological research, but current software and database resources are extremely fragmented (Henry et al. 2014) and there is an urgent need of organizing the bioinformatics resources (Cannata et al. 2005). Moreover and despite all enthusiasm on the new possibilities, it is not the use of a particular technique per se but it still needs a sound study design and interpretation. As put by Miracle et al. (2003): "If a well thought-out approach is neglected during experimental design and data interpretation (of molecular technologies), then we are simply left with standard toxicology in Technicolor". In addition, tool details and access often change following the original publication (Dellavalle et al. 2003), rendering it more and more challenging for research groups to stay current (for example van der Ven et al. 2005, 2006). Moreover, design and performance of these studies vary significantly in exposure times and regimes, doses applied, and development stages employed of a relatively high number of different model organisms. Thus, this data is difficult to interpret and unify for making joint conclusions. Therefore, initiatives should be encouraged that allow the wide variety of data to be eventually integrated into a single model of biological function that would act as a "simulation space." An "omic" simulation would contain for each gene an estimate of each critical parameter, as well as rules for interactions at each level to provide networks of interactions and allow estimations of biological functions like an in silico biological laboratory (Evans 2000).

Marine Genomic Resources and Applications to Effect Studies of Pharmaceuticals

Most of the currently available (eco-)toxicogenomic data regarding the molecular effects of exposure to pharmaceuticals are limited to species used in regulatory testing or freshwater species (Handy et al. 2008a; Handy et al. 2008b; Blaise et al. 2008; Federici et al. 2007; Warheit et al. 2007; Lovern and Klaper 2006). The lack of previous genetic information on most marine fish and invertebrate species frequently employed in toxicity testing and risk assessment has been a major drawback for a more general application of the different omic technologies currently available. At present, advances in genome sequencing of marine non model species are being made, and partly or complete genomes are publicly available for the following marine organisms frequently employed in classical toxicity testing: the Japanese puffer, *Fugu rubripes* (http://www.fugu-sg.org/); the Killifish or Mummichog, *Fundulus heteroclitus* (http://www.ccs.miami.edu/cgi-bin/Fundulus/Fundulus_home.cgi); the Pacific oyster, *Crassostrea gigas* (http://www.oysterdb.com/FrontHomeAction.do?method=home), the Mediterranean mussel, *Mytilus galloprovincialis* (http://mussel.cribi.unipd.it); the Japanese clam, *Ruditapes philippinarum* (http://compgen.bio.unipd.it/ruphibase), the European flounder, *Platichthys flesus*, the Sea bream, *Sparus aurata* (http://www.nutrigroup-iats.org/seabreamdb), the Sea bass, *Dicentrarchus labrax*, among others, including some anadromous species such as Atlantic salmon, *Salmo salar*, and the Stickleback, *Gasterosteus aculeatus*.

However and despite the availability of genomic resources in marine species, no studies have been published so far on the molecular effects of exposure to pharmaceuticals in this habitat. Hampel et al.

(in preparation) have presented preliminary results in the brain of the Gilthead seabream, *Sparus aurata*, exposed to environmentally relevant concentrations of three representative human pharmaceuticals: the antipyretic Acetaminophen (APAP), the antidepressant Carbamazepine (CBZ), and the β-blocker Atenolol (AT). Markedly more genes were activated or silenced during CBZ exposure with 612 features, reaffirming the brain as main target organ of this drug. APAP activated or silenced 411, whereas exposure to AT only altered 7 features. Enrichment analysis of these features showed that after APAP treatment, the most significantly enriched Biological Process (BP) GO terms were related with transcription and RNA metabolic processes, epithelium development and morphogenesis, and several further morphogenesis and development related terms. Predominant Cellular Compartment terms (CC) were organelle and mitochondrial membrane related and enriched Molecular Function terms (MF) were represented by a high number of (ribo-)nucleotide related terms. After CBZ treatment, enriched BP GO terms were related to glycosylation and glycoproteins, as well as RNA metabolic processes, DNA dependent transcription, morphogenesis and development related BPs. Enriched CCs and MFs were related with organelle and mitochondrial membranes and with development, steroid dehydrogenase activity, transcription and monovalent inorganic cation transmembrane transporter activity, respectively. The low number of differently expressed genes after AT treatment did not allow any enrichment analysis. KEGG identified one enriched pathway for each, APAP and CBZ treatment: the lysosome pathway and the androgen and estrogen metabolism pathway, respectively.

During CBZ exposure, a 3,105-fold induction ($p < 0.05$) of the FKBP5 binding protein was observed. The FK506 binding protein 5 (FKBP5) plays a central role in the regulation of the HPA axis (Binder et al. 2004) and it was previously identified to be one of the key genes associated with changes in the body's normal response to antidepressant treatment in humans (Horstmann and Binder 2009). Although in freshwater, Park et al. (2012) observed a significant down-regulation of FK506 in zebrafish, *Danio rerio*, after exposure to the antidepressants fluoxetine and sertraline. Given the simultaneous observation of induction of FKBP5 after different treatments with antidepressants and its importance in antidepressant mediated processes and its high sensitivity towards low level exposure, FKBP5 could be a candidate as biomarker of contamination by antidepressant pharmaceutical compounds. Another study on the effects on protein expression in *Mytilus galloprovincialis* exposed to the lipid regulator gemfibrozil and the non-steroidal anti-inflammatory drug diclofenac revealed differential protein expression signatures (PES) in the digestive gland (Schmidt et al. 2014). Twelve spots were significantly increased or decreased by gemfibrozil and/or diclofenac, seven of which were successfully identified by liquid chromatography-tandem mass spectrometry (LC-MS/MS). These proteins were involved in energy metabolism, oxidative stress response, protein folding, and immune responses.

This striking lack of literature and studies addressing the molecular aspects of contaminant challenge in marine organisms compared to freshwater systems underlines the need to pay more attention to this sensitive habitat. Although levels of these contaminants are often low due to dilution processes in open sea, coastal waters and estuaries receive important amounts of treated and/or non-treated waste water discharges that could locally affect the organisms living within. Additionally, issues and shortcomings that still need to be resolved or improved for efficient incorporation of genomics in drug development and environmental toxicology research include data analysis, data interpretation tools, and accessible data repositories, with the effective use of this information by project teams remaining often a challenge.

Gaps and Future

Currently, in literature many studies are available that investigated the hazard posed by pharmaceuticals at environmental concentrations. However, the great majority of this research concerns the occurrence of pharmaceuticals in surface waters or sewage treatment plant effluents and is limited to freshwater environments and/or species. Further research is urgently needed about the occurrence, the characterization of physico-chemical factors that affect the partitioning of pharmaceuticals, and their ecotoxicological effects on marine environments. Information about the effect of mixture of pharmaceuticals, and other pollutants and pharmaceuticals are lacking in the scientific literature; in order to improve the knowledge about the risk of chemical pollutants in realistic scenarios these mixtures should be considered. Although omic technologies are promising techniques that allow researchers to gain information about mechanisms of pharmaceutical toxic effects, the lack of previous genetic information on most marine fish and invertebrate species has been a major drawback for a more general application of the different omic technologies currently available. Increasing to include scientific efforts in genomic resources for non-target species are necessary to implement this approach in a general environmental risk assessment scheme.

Acknowledgements

This work has been supported by the project CTM2012-38720-C03-03 of Spanish Ministry of Economy and Competitiveness and EU Feder funds.

Keywords: Pharmaceuticals, seawater, sediment, toxicity, aquatic species, biomarkers, omics technologies

References

Aguirre-Martínez, G. 2014. Diseño de un herramienta integrada para la evaluación y gestión de la calidad ambiental de sistemas afectados por vertidos de productos farmacéuticos. Tesis Doctoral. Universidad de Cádiz, p. 512.

Aguirre-Martínez, G.V., T.A. Del Valls and M.L. Martín-Díaz. 2013a. Identification of biomarkers responsive to chronic exposure to pharmaceuticals in target tissues of *Carcinus maenas*. Mar. Environ. Res. 87-88: 1–11.

Aguirre-Martínez, G.V., T.A. Del Valls and M.L. Martín-Díaz. 2013b. Early responses measured in the brachyuran crab *Carcinus maenas* exposed to carbamezapine and novobiocin: application of a 2-yier approach. Ecotox. Environ. Safe. 97: 47–58.

Aguirre-Martínez, G.V., M.A. Owuor, C. Carrido-Pérez, M.J. Salamanca, T.A. DelValls and M.L. Martín-Díaz. 2015. Are standard tests sensitive enough to evaluate effects of human pharmaceuticals in aquatic biota? Facing changes in research approaches when performing risk assessment of drugs. Chemosphere 120: 75–85.

Akers, T.K. 1969. Effects of epinephrine, norepinifrine and some blocking agents on tunicates smooth muscle. Comp. Biochem. Physiol. 29: 813–819.

Almeida, Â., R. Freitas, V. Calisto, V.I. Esteves, R.J. Schneider, A.M.V.M. Soares and E. Figueira. 2015. "Chronic toxicity of the antiepileptic carbamazepine on the clam Ruditapes philippinarum." Comp. Biochem. Physiol. C Toxicol. Pharmacol. 172-173: 26–3.

Alsop, D. and C.M. Wood. 2013. Metal and pharmaceutical mixtures: is ion loss the mechanism underlying acute toxicity and widespread additive toxicity in zebrafish? Aquat. Tox. 140-141: 257–267.

Álvarez-Muñoz, D., B. Huerta, M. fernandez-Tejedor, A. Rodríguez-Mozaz and D. Barceló. 2015. Multi-residue method for the analysis of pharmaceuticals and some of their metabolites in bivalves. Talanta (in press).

Andersen, H.R., B. Halling-Sørensen and K.O. Kusk. 1999. A parameter for detecting estrogenic exposure in the copepod *Acartia tonsa*. Ecotox. Environ. Safe. 44: 56–61.

Anderson-Carnahan, L. 1994. Development of methods for culturing and conducting aquatic toxicity tests with the australian cladoceran *Moina australensis*. CSIRO. Division of Water resources. Water Resources Series, nº 13.

Andrew, M.N., R.H. Dunstan, W.A. O'Connor, L. Van Zwieten, B. Nixon and G.R. MacFarlane. 2008. Effects of 4-nonylphenol and 17alpha-ethinylestradiol exposure in the Sydney rock oyster, *Saccostrea glomerata*: Vitellogenin induction and gonadal development. Aquat. Tox. 88: 39–47.

Ankley, G.T., G.P. Daston, S.J. Degitz, N.D. Denslow, R.A. Hoke, S.W. Kennedy, A.L. Miracle, E.J. Perkins, J. Snape, D.E. Tillitt, C.R. Tyler and D. Versteeg. 2006. Toxicogenomics in regulatory ecotoxicology. Environ. Sci. Technol. 40: 4055–4065.

Avdeef, A., C. Berger and C. Brownell. 2000. pH-Metric Solubility. 2: Correlation between the acid-base titration and the saturation shake-flask solubility-pH methods. Pharm. Res. 17: 85–89.

Banni, M., S. Sforzini, S. Franzellitti, C. Oliveri, A. Viarengo and E. Fabbri. 2015. Molecular and cellular effects induced in Mytilus galloprovincialis treated with oxytetracycline at different temperatures. PLoS ONE 10: 6.

Benotti, M.J. and B.J. Brownawell. 2009. Microbial degradation of pharmaceuticals in estuarine and coastal seawater. Environ. Pollut. 157: 994–1002.

Beranova-Giorgianni, S. 2003. Proteome analysis by twodimensional gel electrophoresis and mass spectrometry: strengths and limitations. Trac-Trend Anal. Chem. 5: 273–281.

Bilello, J.A. 2005. The agony and ecstasy of "OMIC" technologies in drug development. Curr. Mol. Med. 5: 39–52.

Binder, E.B., D. Salyakina, P. Lichtner, G.M. Wochnik, M. Ising, B. Putz, S. Papiol, S. Seaman, S. Lucae, M.A. Kohli, T. Nickel, H.E. Kunzel, B. Fuchs, M. Majer, A. Pfennig, N. Kern, J. Brunner, S. Modell, T. Baghai, T. Deiml, P. Zill, B. Bondy, R. Rupprecht, T. Messer, O. Kohnlein, H. Dabitz, T. Bruckl, N. Muller, H. Pfister, R. Lieb, J.C. Mueller, E. Lohmussaar, T.M. Strom, T. Bettecken, T. Meitinger, M. Uhr, T. Rein, F. Holsboer and B. Muller-Myhsok. 2004. Polymorphisms in FKBP5 are associated with increased recurrence of depressive episodes and rapid response to antidepressant treatment. Nat. Gen. 36: 1319–1325.

Binelli, A., C. Della Torre, S. Magni and M. Parolini. 2015. Does Zebra mussel (*Dreissena polymorpha*) represent the freshwater counterpart of *Mytilus* in ecotoxicological studies? A critical review. Environ. Pollut. 196: 386–403.

Blaise, C., F. Gagné, J.F. Férard and P. Eullaffroy. 2008. Ecotoxicity of selected nano-materials to aquatic organisms. Environ. Toxicol. 23: 591–598.

Blasco, J. and T.A. DelValls. 2008. Impact of emergent contaminants in the environment: Environmental risk assessment. pp. 169–188. *In*: D. Barceló and M. Petovic [eds.]. Emerging Contaminants from Industrial and Municipal Waste. The Handbook of Environmental Chemistry 5 S1, Springer-Verlag, Berlin.

Bossus, M.C., Y.Z. Guler, S.J. Short, E.R. Morrison and A.T. Ford. 2014. Behavioural and transcriptional changes in the amphipod *Echinogammarus marinus* exposed to two antidepressants fluoxetine and sertraline. Aquat. Tox. 151: 46–56.

Bound, J.P., K. Kitsou and N. Voulvoulis. 2006. Household disposal of pharmaceuticals and perception of risk to the environment. Environ. Tox. Pharm. 21: 301–307.

Boxall, A.B., M.A. Rudd, B.W. Brooks, D.J. Caldwell, K. Choi, S. Hickmann, E. Innes, K. Ostapyk, J.P. Staveley, T. Verslycke, G.T. Ankley, K.F. Beazley, S.E. Belanger, J.P. Berninger, P. Carriquiriborde, A. Coors, P.C. DeLeo, S.D. Dyer, J.F. Ericson, F. Gagné, J.P. Giesy, T. Gouin, L. Hallstrom, M.V. Karlsson, D.G. Larsson, J.M. Lazorchak, F. Mastrocco, A. McLaughlin, M.E. McMaster, R.D. Meyerhoff, R. Moore, J.L. Parrott, J.R. Snape, R. Murray-Smith, M.R. Servos, P.K. Sibley, J.O. Straub, N.D. Szabo, E. Topp, G.R. Tetreault, V.L. Trudeau and G. Van Der Kraak. 2012. Pharmaceuticals and personal care products in the environment: What Are the Big Questions? Environ. Health Perspect. 120: 1221–1229.

Brackers de Hugo, A., B. Sylvie, D. Alain, G. Jérôme and P. Yves. 2013. Ecotoxicological risk assessment linked to the discharge by hospitals of bio-accumulative pharmaceuticals into aquatic media: the case of mitotane. Chemosphere 93: 2365–2372.

Brandão, F.P., S. Rodrigues, B.B. Castro, F. Gonçalves, S.C. Antunes and B. Nunes. 2013. Short-term effects of neuroactive pharmaceutical drugs on a fish species: biochemical and behavioral effects. Aquat.- Tox.- 144-145: 218–229.

Brander, S., S. Hecht and K. Kuivila. 2015. ET&C Perspectives. Environ. Toxicol. Chem. 34: 459–466.

Brian, J.V. 2005. Inter-population variability in the reproductive morphology of the shore crab (*Carcinus maenas*): evidence of endocrine disruption in a marine crustacean? Mar. Pollut. Bull. 50: 410–416.

Brodin, T., J. Fick, M. Johnsson and J. Klaminder. 2013. Dilute concentrations of a psychiatric drug alter behavior of fish from natural populations. Science 339: 814–815.

Brooks, B.W., C.M. Foran, S.M. Richards, J. Weston, P.K. Turner, J.K. Stanley, K.R. Solomon, M. Slattery and T.W. La point. 2003. Aquatic ecotoxicology of fluoxetine. Tox. Letters 142: 169–183.

Brooks, B.W. 2014. Fish on Prozac (and Zoloft); Ten years later. Aquat. Tox. 151: 61–67.

Buser, H.-R., T. Poiger and M.D. Müller. 1998. Occurrence and fate of the pharmaceutical drug diclofenac in surface waters: rapid photodegradation in a lake. Environ. Sci. Technol. 32: 3449–3456.

Cajaraville, M.P., M.J. Bebianno, J. Blasco, C. Porte, C. Sarasquete and A. Viarengo. 2000. The use of biomarkers to assess the impact of pollution in coastal environments of the Iberian Peninsula: A practical approach. Sci. Total Environ. 247: 295–311.

Caldwell, D.J., F. Mastrocco, L. Margiotta-Casaluci and B.W. Brooks. 2014. An integrated approach for prioritizing pharmaceuticals found in the environment for risk assessment, monitoring and advanced research. Chemosphere 115: 4–12.

Cannata, N., E. Merelli and R.B. Altman. 2005. Time to organize the bioinformatics resourceome. PLoS Comput. Biol. 1: 76–81.

Canty, M.N., T.H. Hutchinson, R.J. Brown, M.B. Jones and A.N. Jha. 2009. Linking genotoxic responses with cytotoxic and behavioural or physical consequences: differential sensitivity of echinoderms (*Asterias rubens*) and marine mollusks (*Mytilus edulis*). Aquat. Tox. 94: 68–76.

Carballeira, C., M.R. De Orte, I.G. Viana and A. Carballeira. 2012. Implementation of a minimal set of biological tests to assess the ecotoxic effects of effluents from land-based marine fish farms. Ecotox. Environ. Safe. 78: 148–161.

Carlsson, C., A.K. Johansson, G. Alvan, K. Bergman and T. Kühler. 2006. Are pharmaceuticals potent environmental pollutants? Part I: Environmental risk assessment of selected active pharmaceutical ingredients. Sci. Total Environ. 364: 67–87.

Chang, C.-G., C.-H. Hung, M.-C. Chiang and S.C. Lan. 1999. The concentrations of plasma sex steroids and gonadal aromatase during controlled sex differentiation in grey mullet, *Mugil cephalus*. Aquaculture 177: 37–45.

Chen, J., Z. Wang, G. Li and R. Guo. 2014. The swimming speed alteration of two freshwater rotifers *Brachionus calyciflorus* and *Asplanchna brightwelli* under dimethoate stress. Chemosphere 95: 256–260.

Chen, J. and R. Guo. 2015. The process-dependent impacts of dimethoate on the feeding behavior of rotifer. Chemosphere 119: 318–325.

Christen, V., S. Hickmann, B. Rechenberg and K. Fent. 2010. Highly active human pharmaceuticals in aquatic systems: A concept for their identification based on their mode of action. Aquat. Toxicol. 96: 167–181.

Ciocan, C.M., E. Cubero-León, A.M. Puinean, E.M. Hill, C. Minier, M. Osada, K. Fenlon and J.M. Rotchell. 2010. Effects of estrogen exposure in mussels, *Mytilus edulis*, at different stages of gametogenesis. Environ. Pollut. 158: 2977–2984.

Claessens, M., L. Vanhaecke, K. Wille and C.R. Janssen. 2013. Emerging contaminants in Belgian marine waters: single toxicants and mixture risks of pharmaceuticals. Mar. Pollut. Bull. 71: 41–50.

Colman, J.R., D. Baldwin, L.L. Johnson and N.L. Scholz. 2009. Effects of the synthetic estrogen, 17alpha-ethinylestradiol, on aggression and courtship behavior in male zebrafish (*Danio rerio*). Aquat. Toxicol. 91: 346–354.

Cooper, E.R., T.C. Siewicki and K. Phillips. 2008. Preliminary risk assessment database and risk ranking of pharmaceuticals in the environment. Sci. Total Environ. 398: 26–23.

Cortez, F.S., C.D.S. Pereira, A.R. Santos, A. Cesar, R.B. Choueri, G.A. Martini and M.B. Bohrer-Morel. 2012. Biological effects of environmentally relevant concentrations of the pharmaceutical Triclosan in the marine mussel *Perna perna* (Linnaeus, 1758). Environ. Pollut. 168: 145–150.

CPMP (Committee for Propieraty Medicinal Products). 2003. Note for Guidance on Environmental Risk Assessment of Medicinal Products for Human Use. CPMP/SWP/4447/00. EMEA. London.
Croll, R.P. and C. Wang. 2007. Review. Possible roles of sex steroids in the control of reproduction in bivalve mollusks. Aquaculture 272: 76–86.
Currie, R.A. 2012. Toxicogenomics: the challenges and opportunities to identify biomarkers, signatures and thresholds to support mode-of-action. Mutat. Res. 746: 97–103.
Dahms, H.-U., A. Hagiwara and J.-S. Lee. 2011. Review—Ecotoxicology, ecophysiology and mechanistics studies with rotifers. Aquat. Tox. 101: 1–12.
Damásio, J., D. Barceló, R. Brix, C. Postigo, M. Gross, M. Petrovic, S. Sabater, H. Guasch, M. Lopez de Alda and C. Barata. 2011. Are phramaceuticals more harmful than other pollutants to aquatic invertebrate species: a hypothesis tested using multi-biomarker and multi-species responses in field collected and transplanted organisms. Chemosphere 85: 1548–1554.
Daughton, C.G. 2014. The Matthew Effect and widely prescribed pharmaceuticals lacking environmental monitoring: case study of an exposure-assessment vulnerability. Sci. Total Environ. 466-467: 315–325.
de los Ríos, A., J.A. Juanes, M. Ortiz-Zarragoitia, M. López de Alda, D. Barceló and M.P. Cajaraville. 2012. Assessment of the effects of a marine urban outfall discharge on caged mussels using chemical and biomarker analysis. Mar. Pollut. Bull. 64: 563–573.
DellaGreca, M.M., A. Fiorentino, M. Isidori, M. Lavorgna, L. Previtera, M. Rubino and F. Temussi. 2004. Toxicity of prednisolone, dexamethasolone and their photochemical derivates on aquatic organisms. Chemosphere 54: 629–637.
DellaGreca, M.M., M.R. Iesce, M. Isidori, A. Nardelli, L. Previtera and M. Rubino. 2007. Phototransformation products of tamoxifen by sunlight in water. Toxicity of the drug and its derivates on aquatic organisms. Chemosphere 67: 1933–1939.
Dellavalle, R.P., E.J. Hester, L.F. Heilig, A.L. Drake, J.W. Kuntzman, M. Graber and L.M. Schilling. 2003. Information science. Going, going, gone: lost Internet references. Science 302: 787–788.
Depledge, M.H. and Z. Billinghurst. 1999. Ecological significance of endocrine disruption in marine invertebrates. Mar. Pollut. Bull. 39: 32–38.
Di Poi, C., A.S. Darmaillacq, L. Dickel, M. Bouluard and C. Bellanger. 2013. Effects on perinatal exposure to waterborne fluoxetine on memory processing in the cuttlefish *Sepia officinalis*. Aquat. Tox. 132-133: 84–91.
Di Poi, C. and C. Bellanger. 2014. Response to commentary on are some invertebrates exquisitely sensitive to the human pharmaceutical fluoxetine? Aquat. Toxicol. 146: 261–263.
Du, B., S.P. Haddad, W.C. Scott, C.K. Chambliss and B.W. Brooks. 2015. Pharmaceutical bioaccumulation by periphyton and snails in an effluent-dependent stream during an extreme drought. Chemosphere 119: 927–934.
Dulin, D., J. Lipfert, M.C. Moolman and N.H. Dekker. 2012. Studying genomic processes at the single-molecule level: introducing the tools and applications. Nat. Rev. Genet. 14: 9–22.
Eades, C. and C.P. Waring. 2010. The effects of diclofenac on the physiology of the green shore crab *Carcinus maenas*. Mar. Environ. Res. 69: 546–548.
EEC. 2007. Commision staff working document on the implementation of the community Strategy for Endocrine Disrupters a range of substances suspected of interfering with the hormone system of humans and wildlife. EEC, Brussels, Belgium.
Ellesat, K.S., M. Yazdani, T.F. Holth and K. Hylland. 2011. Species-dependent sensitivity to contaminants: an approach using primary hepatocyte cultures with three marine fish species. Mar. Environ. Res. 72: 216–224.
EMEA (European Medicine Agency). 2006. Guideline on the environmental risk assessment. Assessment of the medicional products for Human Use. Doc. Ref. EMEA/CHMP/SWP/4447/00. Committee for Medicinal Products for Human Use. London. UK.
Ericson, H., G. Thorsén and L. Kumblad. 2010. Physiological effects of diclofenac, ibuprofen and propranolol on Baltic Sea blue mussels. Aquat. Toxicol. 99: 223–231.
European Commission. 2012a. Proposal for a Directive of the European Parliament and of the Council Amending Directives 2000/60/EC and 2008/105/EC as regards priority substances I in the field of water policy. Brussels: European Environment Agency, 35 p.
European Commission. 2012b. Report from the Commission to the European Parliament and the Council of the outcome of the review annex X to Directive 2000/60/EC of the European Parliament and of the Council of the priority substances in the field of the water policy. Brussels: European Environment Agency, 6 p.
Evans, G.A. 2000. Designer science and the "omic" revolution. Nature Biotechnology 18: 127.
Fabbri, R., M. Montagna, M.T. Balbi, E. Raffo, F. Palumbo and L. Canesi. 2014. Adaptation of the bivalve embryotoxicity assay for the high throughput screening of emerging contaminants in *Mytilus galloprovincialis*. Mar. Environ. Res. 99: 1–8.
Fair, P.A., H.-B. Lee, J. Adams, C. Darling, G. Pacepavicius, M. Alaee, G.D. Bossart, N. Henry and D. Muir. 2009. Occurrence of triclosan in plasma of wild Atlantic bottlenose dolphins (*Tursiops truncatus*) and in their environment. Environ. Pollut. 157: 2248–2254.
Fang, T.-S., F.-H Nan, T.-S. Chin and H.-M. Feng. 2012. The occurrence and distribution of pharmaceutical compounds in the effluents of a major sewage treatment plant in Northern Taiwan and the receiving coastal waters. Mar. Pollut. Bull. 64: 1435–1444.
Federici, G., B.J. Shaw and R.D. Handy. 2007. Toxicity of titanium dioxide nanoparticles to rainbow trout (*Oncorhynchus mykiss*): Gill injury, oxidative stress, and other physiological effects. Aquat. Toxicol. 84: 415–430.
Fent, K., A.A. Weston and D. Caminada. 2006. Ecotoxicology of human pharmaceuticals. Review. Aquat. Tox. 76: 122–159.

Fernandes, D., S. Schnell and C. Porte. 2011. Can pharmaceuticals interfere with the synthesis of active androgens in male fish? An *in vitro* study. Mar. Pollut. Bull. 62: 2250–2253.

Ferrari, B., R. Mons, B. Vollat, B. Fraysse, N. Paxéus, R. Lo Giudice, A. Pollio and J. Garric. 2004. Environmental risk assessment of six human pharmaceuticals: are the current environmental risk assessment procedures sufficient for the protection of the aquatic environment? Environ. Tox. Chem. 23: 1344–1354.

Ferreira, C.S.G., B.A. Nunes, J.M.M. Henriques-Almeida and L. Guilhermino. 2007. Acute toxicity of oxytetracycline and florfenicol to the microalgae *Tetraselmis chuii* and to the crustacean *Artemia parthenogenetica*. Ecotox. Environ. Safe. 67: 452–458.

Fong, P.P. and N. Molnar. 2013. Antidepressants cause foot detachment from substrate in five species of marine snail. Mar. Environ. Res. 84: 24–30.

Fong, P.P. and A.T. Ford. 2014. The biological effects of antidepressants on the molluscs and crustaceans: a review. Aquat. Toxicol. 151: 4–13.

Fong, P.P., T.B. Bury, A.D. Dworkin-Brodsky, C.M. Jasion and R.C. Kell. 2015. The antidepressants venlafaxine (Effexor) and fluoxetine (Prozac) produce different effects on locomotion in two species of marine snail, the oyster drill (*Urosalpinx cinerea*) and the starsnail (*Lithopoma americanum*). Mar. Environ. Res. 103: 89–94.

Ford, A.T., T.F. Fernandes, C.D. Robinson, I.M. Davies and P.A. Read. 2006. Can industrial pollution cause intersexuality in the amphipod *Echinogammarus marinus*? Mar. Pollut. Bull. 53: 100–106.

Ford, A.T. 2012. Intersexuality in Crustacea: an environmental issue? Aquat. Toxicol. 108: 125–129.

Franzelliti, S., S. Buratti, P. Valbonesi and E. Fabbri. 2013. he mode of action (MOA) approach reveals interactive effects of environmental pharmaceuticals on *Mytilus galloprovincialis*. Aquat. Toxicol. 140-141: 249–256.

Franzellitti, S., S. Buratti, M. Capolupo, B. Du, S.P. Haddad, C.K. Chambliss, B.W. Brooks and E. Fabbri. 2014. An exploratory investigation of various modes of action and potential adverse outcomes of fluoxetine in marine mussels. Aquat. Toxicol. 151: 14–26.

Frenzilli, G., M. Bernardeschi, P. Guidi, V. Scarcelli, P. Lucchesi, L. Marsili, M.C. Fossi, A. Brunelli, G. Pojana, A. marcomini and M. Nigro. 2014. Effects of *in vitro* exposure to titanium dioxide on DNA integrity of bottlenose dolphin (*Tursiops truncatus*) fibroblasts and leukocytes. Mar. Environ. Res. 100: 68–73.

Gaume, B., N. Bourgougnon, S. Auzoux-Bordenave, B. Roig, B. Le Bot and G. Bedoux. 2012. *In vitro* effects of triclosan and methyl-triclosan on the marine gastropod *Haliotis tuberculata*. Comp. Biochem. Physiol. C 156: 87–94.

Gaworecki, K.M. and S.J. Klaine. 2008. Bevavioral and biochemical responses of hybrid striped bass during and after fluoextine exposure. Aquat. Tox. 88: 207–213.

Gehlenborg, N., S. O'Donoghue, N.S. Baliga, A. Goesmann, M.A. Hibbs, H. Kitano, O. Kohlbacher, H. Neuweger, R. Schneider, D. Tenenbaum and A.C. Gavin. 2010. Visualization of omics data for systems biology. Nat. Methods 7: 56–68.

Gelsleichter, J. and N.J. Szabo. 2013. Uptake of human pharmaceuticals in bull sharks (*Carcharhinus leucas*) inhabiting a wastewater-impacted river. Sci. Total Environ. 456-457: 196–201.

Gilroy, E.A.M., J.S. Klink, S.D. Campbell, R. McInnis, P.L. Gillis and S.R. de Solla. 2014. Toxicity and bioconcentration of the pharmaceuticals moxifloxacin, rosuvastatin, and drospirenone to the unionid mussel *Lampsilis siliquoidea*. Sci. Total Environ. 487: 537–544.

Gomez, E., M. Bachelot, C. Boillot, D. Munaron, S. Chiron, C. Casellas and H. Fenet. 2012. Bioconcentration of two pharmaceuticals (benzodiazepines) and two personal care products (UV filters) in marine mussels (*Mytilus galloprovincialis*) under controlled laboratory conditions. Environ. Sci. Pollut. Res. Int. 19: 2561–2569.

González-Ortegón, E., J. Blasco, L. Le Vay and L. Giménez. 2013. A multiple stressor approach to study the toxicity and sub-lethal effects of pharmaceutical compounds on the larval development of a marine invertebrate. J. Hazard. Mater. 263 Pt. 1: 233–238.

González-Ortegón, E., L. Giménez, J. Blasco and L. Le Vay. 2015. Effects of food limitation and pharmaceutical compounds on the larval development and morphology of *Palaemon serratus*. Sci. Total. Environ. 503-504: 171–178.

Gonzalez-Rey, M. and M.J. Bebianno. 2011. Non-steroidal anti-inflammatory drug (NSAID) ibuprofen distresses antioxidant defense system in mussel *Mytilus galloprovincialis* gills. Aquat. Toxicol. 105: 264–269.

Gonzalez-Rey, M. and M.J. Bebianno. 2012. Does non steroidl anti-inflammatory (NSAID) ibuprofen induces antioxidant stress and endocrine disruption in mussel, *Mytilus galloprovincialis*. Environ. Toxicol. Pharmacol. 33: 361–371.

Gonzalez-Rey, M. and M.J. Bebianno. 2014. Effects of non-steroidal anti-inflammatory drug (NSAID) diclofenac exposure in mussel *Mytilus galloprovincialis*. Aquat. Toxicol. 148: 221–230.

Gonzalez-Rey, M., J.J. Matos, C.E. Piazza, A.C.D. Bainy and M.J. Bebianno. 2014. Effects of active pharmaceutical ingredients mixtures in mussel *Mytilus galloprovincialis*. Aquat. Toxicol. 153: 12–26.

Green, C., J. Brina, R. Kanda, M. Scholze, R. Williams and S. Jobling. 2015. Environmental concentrations of anti-androgenic pharmaceuticals do not impact sexual disruption in fish alone or in combination with steroid oestrogens. Aquat. Toxicol. 160C: 117–127.

Griffin, J.L. and M.E. Bollard. 2004. Metabonomics: its potential as a tool in toxicology for safety assessment and data integration. Curr. Drug Metab. 5: 389–398.

Guler, Y. and A.T. Ford. 2010. Anti-depressants make amphipods see the light. Aquat. Tox. 99: 397–404.

Hagenbuch, I.M. and J.L. Pinckney. 2012. Toxic effect of the combined antibiotics ciprofloxacin, lincomycin, and tylosin on two species of marine diatoms. Wat. Res. 46: 5028–5036.

Halem, Z.M., D.J. Ross and R.L. Cox. 2014. Evidence for intraspecific endocrine disruption of *Geukensia demissa* (Atlantic ribbed mussel) in an urban watershed. Comp. Biochem. Physiol. A Mol. Integr. Physiol. 175: 1–6.

Hampel, M., M. Milan, J. Blasco, S. Ferraresso and L. Bargelloni. Transcriptome analysis of the brain of the sea bream (*Sparus aurata*) after exposure to environmental concentrations of human pharmaceuticals. In preparation.

Handy, R.D., T.B. Henry, T.M. Scown, B.D. Johnstone and C.R. Tyler. 2008a. Manufactured nanoparticles: their uptake and effects on fish—a mechanistic analysis. Ecotoxicology 17: 396–409.

Handy, R.D., F. Von der Kammer, J.R. Lead, M. Hassellöv, R. Owen and M. Crane. 2008b. The ecotoxicology and chemistry of manufactured nanoparticles. Ecotoxicology 17: 287–314.

Hartung, T. and M. McBride. 2011. Food for thought... on mapping the human toxome. ALTEX 28: 83–93.

Hartung, T., F. van Vliet, J. Jaworska, L. Bonilla, N. Skinner and R. Thomas. 2012. Systems toxicology. ALTEX 29: 119–28.

Henry, V.J., A.E. Bandrowski, A.S. Pepin, B.J. Gonzalez and A. Desfeux. 2014. OMICtools: an informative directory for multi-omic data analysis. Database: article ID bau069; doi:10.1093/database/bau069.

Herbert, B.R., J.L. Harry, N.H. Packer, A.A. Gooley, S.K. Pederson and K.L. Williams. 2001. What place for polyacrylamide in proteomics? Trends Biotechnol. 19: S3–9.

Horstmann, S. and E.B. Binder. 2009. Pharmacogenomics of antidepressant drugs. Pharmacol. Ther. 124: 57–73.

Humble, J.L., E. Hands, M. Saaristo, K. Lindström, K.K. Lehtonen, O. Diaz de Cerio, I. Cancio, G. Wilson and J.A. Craft. Characterisation of genes transcriptionally upregulated in the liver of sand goby (*Pomatoschistus minutus*) by 17alpha-ethinyloestradiol: identification of distinct vitellogenin and zona radiata protein transcripts. Chemosphere 90: 2722–2729.

Hutchinson, T.H., N.A. Pounds, M. Hampel and T.D. Williams. 1999. Impact of natural and synthetic steroids on the survival, development and reproduction of marine copepods (*Tisbe battagliai*). Sci. Tot. Environ. 233: 167–179.

Isidori, M., A. Nardelli, A. Parrella, L. Pascarella and L. Previtera. 2006. A multispecies study to assess the toxic and genotoxic effect of pharmaceuticals: Furosemide and its photoproduct. Chemosphere 63: 785–793.

Isidori, M., A. Nardelli, L. Pascarella, M. Rubino and A. Parrella. 2007. Toxic and genotoxic impact of fibrates and their photoproducts on non-target organisms. Environ. Int. 33: 635–641.

Jasinska, E.J., G.G. Goss, P.L. Gillis, G.J. Van Der Kraak, J. Matsumoto, A.A. de Souza Machado, M. Giacomin, T.W. Moon, A. Massarsky, F. Gagné, M.R. Servos, J. Wilson, T. Sultana and C.D. Metcalfe. 2015. Assessment of biomarkers for contaminants of emerging concern on aquatic organisms downstream of a municipal wastewater discharge. Sci. Total Environ. 530-531: 140–153.

Jiang, J.J., C.L. Lee and M.D. Fang. 2014. Emerging organic contaminants in coastal waters: anthropogenic impact, environmental release and ecological risk. Mar. Pollut. Bull. 85: 391–399.

Jin, C., J. Qiu, L. Miao, K. Feng and X. Zhou. 2014. Antifouling activities of anti-histamine compounds against the barnacle *Amphibalanus* (=*Balanus*) *amphitrite*. J. Exp. Mar. Biol. Ecol. 452: 47–53.

Johansson, C.H., L. Janmar and T. Backhaus. 2014a. Triclosan causes toxic effects to algae in marine biofilms, but does not inhibit the metabolic activity of marine biofilm bacteria. Mar. Pollut. Bull. 84: 208–212.

Johansson, C.H., L. Janmar and T. Backhaus. 2014b. Toxicity of ciprofloxacin and sulfamethoxazole to marine periphytic algae and bacteria. Aquat. Toxicol. 156: 248–258.

Kalman, J., B.D. Smith, I. Riba, J. Blasco and P.S. Rainbow. 2010. Biodynamic modelling of the accumulation of Ag, Cd and Zn by the deposit-feeding polychaete *Nereis diversicolor*: Inter-population variability and a generalised predictive model. Mar. Environ. Res. 69: 363–373.

Kar, S. and K. Roy. 2010. First report on interspecies quantitative correlation of ecotoxicity of pharmaceuticals. Chemosphere 81: 738–747.

Kasiotis, K.M., C. Emmanouil, P. Anastasiadou, A. Papadi-Psyllou, A. Papadopoulos, O. Okay and K. Machera. 2015. Organic pollution and its effects in the marine mussel *Mytilus galloprovincialis* in Eastern Mediterranean coasts. Chemosphere 119 Suppl: S145–152.

Ketata, I., X. Denier, A. Hamza-Chaffai and C. Minier. 2008. Endocrine-related reproductive effects in molluscs. Comp. Biochem. Physiol. C Toxicol. Pharmacol. 147: 261–270.

Khan, S.J. and J.E. Ongerth. 2002. Estimation of pharmaceutical residues in primary and secondary sewage sludge based on quantities of use and fugacity modelling. Water. Sci. Technol. 46: 105–13.

Klosterhaus, S.L., R. Grace, M.C. Hamilton and D. Yee. 2013. Method validation and reconnaissance of pharmaceuticals, personal care products, and alkylphenols in surface waters, sediments and mussels in an urban estuary. Environ. Int. 54: 92–99.

Kookana, R.S., A. Shareef, M.B. Fernandes, S. Hoare, S. Gayland and A. Kumar. 2013. Bioconcentration of triclosan and methyl-triclosan in marine mussels (*Mytilus galloprovincialis*) under laboratory conditions and in metropolitan waters of Gulf St. Vincent, South Australia. Mar. Pollut. Bull. 74: 66–72.

Laane, R.W.P.M., A.D. Vethaak, J. Gandrass, K. Vorkamp, A. Köhler, M.M. Larsen and J. Strand. 2013. Chemical contaminants in the Wadden Sea: Sources, transport, fate and effects. J. Sea Res. 82: 10–53.

Lai, K.M., K.L. Johnson, M.D. Scrimshaw and J.N. Lester. 2000. Binding of Waterborne Steroid Estrogens to Solid Phases in River and Estuarine Systems. Environ. Sci. Technol. 34: 3890–3894.

Lavado, R., A. Barbaglio, M.D.C. Carnevali and C. Porte. 2006. Steroids levels in crinoid echinoderms are altered by exposure to model endocrine disruptors. Steroids 7 l: 489–497.

Lee, W.Y. and C.R. Arnold. 1983. Chronic toxicity of ocean-dumped pharmaceutical wastes to the marine amphipod *Ampithoe valida*. Mar. Pollut. Bull. 14:150–153.

Leonard, J.A., W.G. Cope, M.C. Barnhart and R.B. Bringolf. 2014. Metabolomic, behavioral, and reproductive effects of the synthetic estrogen 17 alpha-ethinylestradiol on the unionid mussel *Lampsilis fasciola*. Aquat. Toxicol. 150: 103–116.

Li, Z., G. Lu, X. Yang and C. Wang. 2012. Single and combined effects of selected pharmaceuticals at sublethal concentrations on multiple biomarkers in *Carassius auratus*. Ecotoxicology 21: 353–361.

Lilley, K.S., A. Razzaq and P. Dupree. 2001. Two-dimensional gel electrophoresis: recent advances in sample preparation, detection and quantitation. Curr. Opinion Chem. Biol. 6: 46–50.

Liu, J.-L. and M.-H. Wong. 2013. Pharmaceuticals and personal care products (PPCPs): a review on environmental contamination in China. Environ. Int. 59: 208–224.

Lovern, S.B. and R. Klaper. 2006. Daphnia magna mortality when exposed to titanium dioxide and fullerene nanoparticles. Environ. Toxicol. Chem. 25: 1132–1137.

Luna-Acosta, A., T. Renault, H. Thomas-Guyon, N. Faury, D. Saulnier, H. Budzinski, K. Le Menach, P. Pardon, I. Fruitier-Arnaudin and P. Bustamante. 2012. Detection of early effects of a single herbicide (diuron) and a mix of herbicides and pharmaceuticals (diuron, isoproturon, ibuprofen) on immunological parameters of Pacific oyster (*Crassostrea gigas*) spat. Chemosphere 87: 1335–1340.

Maranho, L., C. André, T.A. Del Valls, F. Gagné and L. Martín-Díaz. 2014a. Toxicological evaluation of sediment samples spiked with human pharmaceutical products: Energy status and neuroendocrine effects in marine polychaetes Hediste diversicolor. Ecotox. Environ. Safe. 118: 27–36.

Maranho, L.A., R.M. Baena-Nogueras, P.A. Lara Martín, T.A. DelValls and M.L. Martín-Díaz. 2014b. Bioavailability, oxidative stress, neurotoxicity and genotoxicity of pharmaceuticals bound to marine sediments. The use of the polychaete *Hediste diversicolor* as bioindicator species. Environ. Res. 134: 353–365.

Maranho, L.A., R.M. Baena-Nogueras, P.A. Lara-Martin, T.A. DelValls and M.L. Martin-Diaz. 2015a. A candidate short-term toxicity test using *Ampelisca brevicornis* to assess sublethal responses to pharmaceuticals bound to marine sediments. Arch. Environ. Contam. Toxicol. 68: 237–58.

Maranho, L.A., T.A. delValls and M.L. Martín-Díaz. 2015b. Assessing potential risks of wastewater discharges to benthic biota: An integrated approach to biomarker responses in clams (*Ruditapes philippinarum*) exposed under controlled conditions. Mar. Pollut. Bull. (in press).

Martín-Díaz, L., E. Gagné and C. Blaise. 2009. The use of biochemical responses to assess ecotoxicological effects of pharmaceuticals and care products (PPPCPs) after the injection in the mussel. *Elliptio complanata*. Env. Toxicol. Pharmacol. 28: 237–242.

Martín-Díaz, M.L., J. Blasco, D. Sales and T.A. DelValls. 2004. Biomarkers as tools to assess sediment quality. Laboratory and field surveys. TrAC—Trends in Analytical Chemistry 23: 807–818.

Martinez Bueno, M.J., C. Boillot, H. Fenet, S. Chiron, C. Casellas and E. Gómez. 2013. Fast and easy extraction combined with high resolution-mass spectrometry for residue analysis of two anticonvulsants and their transformation products in marine mussels. J. Chromatogr. A 1305: 27–34.

Martyniuk, C.J., A. Alvarez and N.D. Denslow. 2012. DIGE and iTRAQ as biomarker discovery tools in aquatic toxicology. Ecotox. Environ. Safe. 76: 3–10.

Maszkowska, J., S. Stolte, J. Kumirska, P. Łukaszewicz, K. Mioduszewska, A. Puckowski, M. Caban, M. Wagil, P. Stepnowski and A. Białk-Bielińska. 2014. Beta-blockers in the environment: part II. Ecotoxicity study. Sci. Total Environ. 493: 1122–1126.

Mathieu-Denoncourt, J., S.J. Wallace, S.R. de Solla and V.S. Langlois. 2014. Plasticizer endocrine disruption: Highlighting developmental and reproductive effects in mammals and non-mammalian aquatic species. Gen. Comp. Endocrinol. (in press).

Matozzo, V., S. Rova and M.G. Marin. 2012. The nonsteroidal anti-inflammatory drug, ibuprofen, affects the immune parameters in the clam *Ruditapes philippinarum*. Mar. Environ. Res. 79: 116–121.

Matozzo, V., S. Rova and M.G. Marin. 2014. *In vitro* effects of the nonsteroidal anti-inflammatory drug, ibuprofen, on the immune parameters of the colonial ascidian *Botryllus schlosseri*. Toxicol. *In Vitro* 28: 778–783.

Maximino, C., T. Marques de Brito, C.A.G. de Mattos Dias, A. Gouoveia, Jr. and S. Morato. 2010. Scototaxis as anxiety-like behavior in fish. Nature Protocols 5: 221–228.

McEneff, G., L. Barron, B. Kelleher, B. Paull and B. Quinn. 2014. A year-long study of the spatial occurrence and relative distribution of pharmaceutical residues in sewage effluent, receiving marine waters and marine bivalves. Sci. Total Environ. 476-477: 317–326.

Méndez, N., S. Lacorte and C. Barata. 2013. ffects of the pharmaceutical fluoxetine in spiked-sediments on feeding activity and growth of the polychaete *Capitella teleta*. Mar. Environ. Res. 89: 76–82.

Michalec, F.G., M. Holzner, D. Menu, J.-S. Hwang and S. Souissi. 2013. Behavioral responses of the estuarine calanoid copepod *Eurytemora affinis* to sub-lethal concentrations of waterborne pollutants. Aquat. Toxicol. 138-139: 129–138.

Miller, M.G. 2007. Environmental metabolomics: a SWOT analysis (strengths, weaknesses, opportunities, and threats). J. Proteome Res. 6: 540–545.

Minguez, L., M.P. Halm-Lemeille, K. Costil, R. Bureau, J.M. Lebel and A. Serpentini. 2014. Assessment of cytotoxic and immunomodulatory properties of four antidepressants on primary cultures of abalone hemocytes (*Haliotis tuberculata*). Aquat. Toxicol. 153: 3–11.

Miracle, A.L., G.P. Toth and D.L. Lattier. 2003. The path from molecular indicators of exposure to describing dynamic biological systems in an aquatic organism: microarrays and fathead minnow. Ecotoxicology 12: 457–462.

Mistra Pharma. 2012. Pharmaceuticals in a Healthy Environment, Mistra Pharma Reasearch Project 2008–2011 (Ingvar Brandt, Magnus Breitholtz, Jes la Cour Jansen, Karin Liljelund, Joakim Larsson, Christina Rudén and Mats Tysklind, eds.). Mistra Pharma (Stockolm) 70 p.

Moore, C.J. and J.M. Stevenson. 1991. The occurrence of intersexuality in Harpacticoid copepods and its relationship with pollution. Mar. Pollut. Bull. 22: 72–74.

Morales, L., M.G. Martrat, J. Olmos, J. Parera, J. Vicente, A. Bertolero, M. Ábalos, S. Lacorte, F.J. Santos and E. Abad. 2012. Persistent organic pollutants in gull eggs of two species (*Larus michahellis* and *Larus audouinii*) from the Ebro delta Natural Park. Chemosphere 88: 1306–1316.

Moreno-González, R., S. Rodríguez-Mozaz, M. Gros, E. Pérez-Cánovas, D. Barceló and V.M. León. 2014. Input of pharmaceuticals through coastal surface watercourses into a Mediterranean lagoon (Mar Menor, SE Spain): sources and seasonal variations. Sci Total Environ. 490: 59–72.

Morley, N.J. 2009. Review—Environmental risk and toxicology of human and veterinary waste pharmaceutical exposure to wild aquatic host-parasite relationships. Environ. Tox. Pharm. 27: 161–175.

Munari, M., V. Matozzo, G. Chemello and M.G. Marin. 2012. Combined effects of seawater acidification and diclofenac on immune parameters of the clam *Ruditapes philipinarum*. Comp. Biochem. Physiol. Part A 163: 54–59.

Neuparth, T., C. Martins, C.B. de los Santos, M.H. Costa, I. Martins, P.M. Costa and M.M. Santos. 2014. Hypocholesterolaemic pharmaceutical simvastatin disrupts reproduction and population growth of the amphipod *Gammarus locusta* at the ng/L range. Aquat. Toxicol. 155: 337–347.

Nicholson, J.K. and J.C. Lindon. 2008. Systems biology: Metabonomics. Nature 455(7216): 1054–1056.

Niimi, S., M. Imoto, T. Kunisue, M.X. Watanabe, E.-Y. Kim, N. Nakayama, G. Yasunaga, Y. Fujise, S. Tanabe and H. Iwabata. 2014. Effects of persistent organochlorine exposure on the liver transcriptome of the common minke whale (*Balaenoptera acutorostrata*) from the North Pacific. Ecotox. Environ. Safe. 108: 95–105.

Nogrady, T. and J. Keshmirian. 1986. Rotifer neuropharmacology–I. Cholinergic drug affects on ovoposition of *Philodina acuticornis* (Rotifera, Aschelminthes). Comp. Biochem. Physiol. 83C: 335–338.

Nunes, B., F. Carvalho and L. Guilhermino. 2005. Acute toxicityof widely used pharmaceuticals in aquatic species: *Gambusia holbrooki*, *Artemia parthenogenetica* and *Tetraselmis chuii*. Ecotox. Environ. Safe. 61: 413–419.

Nunes, B., F. Carvalho and L. Guilhermino. 2006. Effects of widely used pharmaceuticals and a detergent ton oxidative stress biomarkers of the crustacean *Artemia parthenogenetica*. Chemosphere 62: 581–594.

Nunes, B., S.C. Antunes, J. Santos, L. Martins and B.B. Castro. 2014. Toxic potential of paracetamol to freshwater organisms: a headache to environmental regulators? Ecotox. Environ. Safe. 107: 178–185.

O'Brien, A.L. and M.J. Keough. 2014. Ecological responses to contamination: a meta-analysis of experimental marine studies. Environ. Pollut. 195: 185–191.

Oetken, M., J. Bachmann, U. Schulte-Oehlmann and J. Oehlmann. 2004. Evidence of endocrine disruption in invertebrates. Internat. Rev. Cytol. 326: 1–44.

Oskarsson, H., A.-K.E. Wiklund, K. Lindh and L. Kumblad. 2012. Effect studies of human pharmaceuticals on *Fucus vesiculosus* and *Gammarus* spp. Mar. Environ. Res. 74: 1–8.

Park, J.W., T.P. Heah, J.S. Gouffon, T.B. Henry and G.S. Sayler. 2012. Global gene expression in larval zebrafish (Danio rerio) exposed to selective serotonin reuptake inhibitors (fluoxetine and sertraline) reveals unique expression profiles and potential biomarkers of exposure. Environ. Pollut. 167C: 163–170.

Parolini, M. and A. Binelli. 2012. Sub-lethal effects induced by a mixture of three non-steroidal anti-inflammatory drugs (NSAIDs) on the freshwater bivalve Dreissena polymorpha. Ecotoxicology 21: 379–392.

Parrella, A., M. Lavorgna, E. Criscuolo, C. Russo, V. Fiumano and M. Isidori. 2014. Acute and chronic toxicity of six anticancer drugs on rotifers and crustaceans. Chemosphere 115: 59–66.

Peele, E.R., F.L. Singleton, J.W. Deming, B. Cavari and R.R. Colwell. 1981. Effects of pharmaceuticals wastes on microbial populations in surface waters at the Puerto Rico dump site in the Atlantic Ocean. Appl. Environ. Microbiol. 41: 873–879.

Petersen, K., H.H. Heiass and K.E. Tollefsen. 2014. Combined effects of pharmaceuticals, personal care products, biocides and organic contaminants on the growth of *Skeletonema pseudocostatum*. Aquat. Toxicol. 150: 45–54.

Petridis, P., A.N. Jha and W.J. Langston. 2009. Measurements of the genotoxic potential of (xeno-)oestrogens in the bivalve mollusc *Scrobicularia plana*, using the Comet assay. Aquat. Toxicol. 94: 8–15.

Pintado-Herrera, M.G., E. González-Mazo and P.A. Lara-Martín. 2013. Environmentally friendly analysis of emerging contaminants by pressurized hot water extraction–stir bar sorptive extraction–derivatization and gas chromatography–mass spectrometry. Anal. Bioanal. Chem. 405: 401–411.

Porsbring, T., Å. Arrhenius, T. Backhaus, M. Kuylenstierna, M. Scholze and H. Blanck. 2007. The SWIFT periphyton test for high-capacity assessment of toxicant effects on microalgal community development. J. Exp. Mar. Biol. Ecol. 349: 299–312.

Porsbring, T., H. Blank, H. Tjellström and T. Backhaus. 2009. Toxicity of the pharmaceutical clotrimazole to marine microalgal communities. Aquat. Tox. 91: 203–211.

Prado, R., R. García, C. Rioboo, C. Herrero and Á. Cid. 2015. Suitability of cytotoxicity endpoints and test microalgal species to disclose the toxic effect of common aquatic pollutants. Ecotox. Environ. Safe. 114C: 117–125.

Proia, L., V. Osorio, S. Soley, M. Köck-Schulmeyer, S. Pérez, D. Barceló, A.M. Romaní and S. Sabater. 2013. Effects of pesticides and pharmaceuticals on biofilm in a highly impacted river. Environ. Pollut. 178: 220–228.

Rhee, J.-S., B.-M. Kim, C.-B. Jeong, H.G. park, K.M.Y. Leung, T.-M. Lee and J.-S. Lee. 2013. Effect of pharmaceuticals exposure on acetylcholinesterase (AchE) activity and on the expression of AchE gene in the monogonont rotifer, *Brachionus koreanus*. Comp. Biochem. Physiol. C Toxicol. Pharmacol. 158: 216–224.

Ribeiro, S., T. Torres, R. Martins and M.M. Santos. 2015. Toxicity screening of Diclofenac, Propranolol, Sertraline and Simvastatin using Danio rerio and Paracentrotus lividus embryo bioassays. Ecotox. Environ. Safe. 114C: 67–74.

Ricciardi, F., V. Matozzo, A. Binelli and M.G. Marin. 2010. Biomarker responses and contamination levels in crabs (*Carcinus aestuarii*) from the lagoon of Venice: an integrated approach in biomonitoring estuarine environments. Wat. Res. 44: 1725–1736.

Robinson, B.J. and J. Hellou. 2009. Biodegradation of endocrine disrupting compounds in harbour seawater and sediments. Sci. Total Environ. 407: 5713–5718.

Rodrigues, E.T. and M.Â. Pardal. 2014. The crab *Carcinus mae*nas as a suitable experimental model in ecotoxicology. Environ. Int. 70: 158–182.

Rodriguez, E.M., D.A. Medesani and M. Fingerman. 2007. Endocrine disruption in crustaceans due to pollutants: a review. Comp. Biochem. Physiol. A Mol. Integr. Physiol. 146: 661–671.

Rosal, R., I. Rodea-Palomares, K. Boltes, F. Fernández-Piñas, F. Leganés, S. Gonzalo and A. Petre. 2010. Ecotoxicity assessment of lipid regulators in water and biologically treated wastewater using three aquatic organisms. Environ. Sci. Pollut. Res. 17: 135–144.

Ross, P.L., Y.N. Huang, J.N. Marchese, B. Williamson, K. Parker, S. Hattan, N. Khainovski, S. Pillai, S. Dey, S. Daniels, S. Purkayastha, P. Juhasz, S. Martin, M. Bartlet-Jones, F. He, A. Jacobson and D.J. Pappin. 2004. Multiplexed protein quantitation in *Saccharomyces cerevisiae* using amine-reactive isobaric tagging reagents. Mol. Cell. Proteomics 3: 1154–1169.

Saçan, M.T. and I.A. Balcıoğlu. 2006. A case study on algal response to raw and treated effluents from an aluminium plating plant and a pharmaceutical plant, Ecotox. Environ. Safe. 64: 234–243.

Santos, L.H.M.L.M., A.N. Araújo, A. Fachini, A. Pena, C. Deleure-Matos and M.C.B.S.M. Montenegro. 2010. Review—Ecotoxicological aspects related to the presence of pharmaceuticals in the aquatic environment. J. Hazard. Mater. 175: 45–95.

Schmidt, W., K. O'Rourke, R. Hernan and B. Quinn. 2011. Effects of the pharmaceuticals gemfibrozil and diclofenac on the marine mussel (*Mytilus* spp.) and their comparison with standardized toxicity tests. Mar. Pollut. Bull. 62: 1389–1395.

Schmidt, W., E. Power and B. Quinn. 2013. Seasonal variations of biomarker responses in the marine blue mussel (*Mytilus* spp.). Mar. Pollut. Bull. 74: 50–55.

Schmidt, W., L.C. Rainville, G. McEneff, D. Sheehan and B. Quinn. 2014. A proteomic evaluation of the effects of the pharmaceuticals diclofenac and gemfibrozil on marine mussels (*Mytilus* spp.): evidence for chronic sublethal effects on stress-response proteins. Drug Test Anal. 6: 210–9.

Scott, A.P. 2013. Do mollusks use vertebrate sex steroids as reproductive hormones? II. Critical review of the evidence that steroids have biological effects. Steroids 78: 268–281.

Sellami, B., A. Kharzi, A. Mezni, H. Louati, M. Delladi, P. Aissa, E. Mahmoudi, H. Beyrem and D. Sheehan. 2015. Effect of permethrin, anthracene and mixture exposure on shell components, enzymatic activities and proteins status in the Mediterranean clam *Venerupis decussata*. Aquat. Toxicol. 158: 22–32.

Seoane, M., C. Rioboo, C. Herrero and Á. Cid. 2014. Toxicity induced by three antibiotics commonly used in aquaculture on the marine microalga *Tetraselmis suecica* (Kylin) Butch. Mar. Environ. Res. 101: 1–7.

Solé, M., J. Kopecka-Pilarczyck and J. Blasco. 2009. Pollution biomarkers in two estuarine invertebrates, *Nereis diversicolor* and *Scrobicularia plana*, from a marsh ecosystem in SW Spain. Environ. Int. 35: 523–531.

Stentiford, G.D. 2012. Histological intersex (ovotestis) in the European lobster *Homarus gammarus* and a commentary on its potential mechanistic basis. Dis. Aquat. Organ. 100: 185–190.

Sumpter, J.P. and L. Margiotta-Casaluci. 2014. Are some invertebrates exquisitely sensitive to the human pharmaceutical fluoxetine? Aquat. Toxicol. 146: 259–260.

Tangtian, H., L. Bo, L. Wenhua, P.K.S. Shin and R.S.S. Wu. 2012. Estrogenic potential of benzotriazole on marine medaka (*Oryzias melastigma*). Ecotox. Environ. Safe. 80: 327–332.

Tonge, R., J. Shaw, B. Middleton, R. Rowlinson, S. Rayner, J. Young, F. Pognan, E. Hawkins, I. Currie and M. Davison. 2001. Validation and development of fluorescence two-dimensional differential gel electrophoresis proteomics technology. Proteomics 1: 377–396.

Trombini, C., M. Hampel and J. Blasco. 2012. Effects of the exposure of the clam *Ruditapes philippinarum* to carbamezapine, diclofenac and ibuprofen. Evaluation of enzymatic and molecular endpoints. Com. Biochem. Physiol. Part A 163: S22.

Trudeau, V.L., C.D. Metcalfe, C. Mimeault and T.W. Moon. 2005. Pharmaceuticals in the environment: drugged fish? *In*: T.P. Mommsen and T.W. Moon [eds.]. Biochemistry and Molecular Biology of Fishes. Elsevier. Amsterdam 6: 475–493.

Tsiaka, P., V. Tsarpali, I. Ntaikou, M.N. Kostopoulou, G. Lyberatos and S. Dailianis. 2013. Carbamazepine mediated pro-oxidant effects on the unicellular marine algal species *Dunaliella tertiolecta* and the hemocytes of mussel *Mytilus galloprovincialis*. Ecotoxicology 22: 1208–1220.

Udovyk, O. and M. Gilek. 2014. Participation and post-normal science in practice? Reality check for hazardous chemicals management in the European marine environment. Futures 63: 15–25.

USEPA. 2009. Fact Sheet. Final Third Drinking Water Contaminant List (CCL 3). EPA 815F09001. Office of Water. U.S. Environmental Protection Agency. Available at http://www.epa.gov/safewater.

USEPA. 2013a. Pharmaceuticals and Personal Care Products (PPCPs): Research and Development. PPCPs. U.S. Environmental Protection Agency. Available at http://www.epa.gov/ppcp/basic2.html.

USEPA. 2013b. Pharmaceuticals and Personal Care Products (PPCPs) in Water. Water, Science and Technology, Surface Water Standards. U.S. Environmental Protection Agency. Available at http://www.epa.gov/scitech/swguidance/ppcp/index.cfm.

Vanderbergh, G.F., D. Adriaens, T. Verslycke and C.R. Janssen. 2003. Effects of 17α-ethinylestradiol on sexual development of the amphipod *Hyalella azteca*. Ecotox. Environ. Safe. 54: 216–222.

van der Oost, R., C. Porte and N.W. van den Brink. 2005. Biomarkers in environmental asssessment. pp. 87–152. *In*: P.J. den Besten and M. Munawar [eds.]. Ecotoxicological Testing of Marine and Freshwater Ecosystems. Taylor & Francis, Boca Ratón (FL).

van der Ven, K., M. DeWit, D. Keil, L. Moens, K. Van Leemput, B. Naudts and W.M. De Coen. 2005. Development and application of a brain-specific cDNA microarray for effect evaluation of neuro-active pharmaceuticals in zebrafish (*Danio rerio*). Comp. Biochem. Physiol. B 141: 408–417.

van der Ven, K., D. Keil, L.N. Moens, P. VanHummelen, P. Van Remortel, M. Maras and W.M. De Coen. 2006. Effects of the antidepressant mianserin in zebrafish: Molecular markers of endocrine disruption. Chemosphere 65: 1836–1845.

van Ravenzwaay, B., G.C.P. Cunha, E. Leibold, R. Looser, W. Mellert, A. Prokoudine, T. Walk and J. Wiemer. 2007. The use of metabolomics for the discovery of new biomarkers of effect. Toxicol. Lett. 172: 21–28.

Vasquez, M.I., A. Lambrianides, M. Schenider, K. Kümmerer and D. Fatta-Kassinos. 2014. Environmental side effects of pharmaceutical cocktails: what we know and what we should know. J. Hazard. Mater. 279: 169–189.

Vazquez-Roig, P., V. Andreu, C. Blasco and Y. Picó. 2012. Risk Assessment of the presence of Pharmaceuticals in Sediments, Soils and Waters of the Pego-Oliva Marshlands (Valencia, Eastern Spain). Sci. Total Environ. 440: 24–32.

Viarengo, A., D. Lowe, C. Bolognesi, E. Fabbri and A. Koehler. 2007. The use of biomarkers in biomonitoring. A 2 tier approach ssessing the level of pollutant-induced stress syndrome in sentinel organisms. Comp. Biochem. Physiol. C 146: 281–300.

Wang, Z., M. Gerstein and M. Snyder. 2009 RNA-seq: a revolutionary tool for transcriptomics. Nature Reviews Genetics 10: 57–63.

Warheit, D.B., P.J.A. Borm, C. Hennes and J. Lademann. 2007. Testing strategies to establish the safety of nanomaterials: conclusions of an ECETOC workshop. Inhal. Toxicol. 19: 631–643.

Waring, R.H. and R.M. Harris. 2005. Endocrine disrupters: A human risk? Mol. Cell. Endoc. 244: 2–9.

Wessel, N., S. Rousseau, X. Caisey, F. Quiniou and F. Akcha. 2007. Investigating the relationship between embryotoxic and genotoxic effects of benzo[a]pyrene, 17alpha-ethinylestradiol and endosulfan on *Crassostrea gigas* embryos. Aquat. Toxicol. 85: 133–142.

Widdows, J., P. Donkin, F.J. Staff, P. Matthiessen, R.J. Law, Y.T. Allen, J.E. Thain, C.R. Allchin and B.R. Jones. 2002. Measurement of stress effects (scope of growth) and contaminant levels in mussels (*Mytilus edulis*) collected from the Irish Sea. Mar. Environ. Res. 53: 327–356.

Ying, G.-G., R.S. Kookana and Y.-J. Ru. 2002. Occurrence and fate of hormone steroids in the environment. Environ. Int. 28: 545–551.

Zhang, Y., S.-U. Geißen and C. Gal. 2008. Carbamazepine and diclofenac: Removal in wastewater treatment plants and occurrence in water bodies. Chemosphere 73: 1151–1161.

Zenker, A., M.R. Cicero, F. Prestinaci, P. Bottoni and M. Carere. 2014. Bioaccumulation and biomagnification potential of pharmaceuticals with a focus to the aquatic environment. J. Environ. Manage. 133: 378–387.

Index

L

M

N

O

P

Q

R

S

T